Decisions under Uncertainty

Probabilistic Analysis for Engineering Decisions

Decision-making and risk assessment are essential aspects of every engineer's life, whether this be concerned with the probability of failure of a new product within the warranty period or the potential cost, human and financial, of the failure of a major structure such as a bridge. This book helps the reader to understand the tradeoffs between attributes such as cost and safety, and includes a wide range of worked examples based on case studies. It introduces the basic theory from first principles using a Bayesian approach to probability and covers all of the most widely used mathematical techniques likely to be encountered in real engineering projects. These include utility, extremes and risk analysis, as well as special areas of importance and interest in engineering and for understanding, such as optimization, games and entropy. Equally valuable for senior undergraduate and graduate students, practising engineers, designers and project managers.

Ian Jordaan is University Research Professor in the Faculty of Engineering and Applied Science at the Memorial University of Newfoundland, Canada, and held the NSERC-Mobil Industrial Research Chair in Ocean Engineering at the university from 1986 to 1996. He is also president of Ian Jordaan and Associates Inc. and an advisor to C-CORE, an engineering R&D company based in St. John's. Prior to joining the university in 1986, he was Vice-President, Research and Development at Det Norske Veritas (Canada) Ltd., and before this he was a full professor in the Department of Civil Engineering at the University of Calgary, Canada. He is extensively involved in consulting activities with industry and government, and has pioneered methods of risk analysis for engineering in harsh environments. Recently he served on an expert panel of the Royal Society of Canada, studying science issues relative to the oil and gas developments offshore British Columbia. He is the recipient of a number of awards, including the Horst Leipholz Medal and the P. L. Pratley Award, both of the Canadian Society for Civil Engineering.

Decisions under Uncertainty

Probabilistic Analysis for Engineering Decisions

Ian Jordaan

Faculty of Engineering and Applied Science
Memorial University of Newfoundland
St John's, Newfoundland, Canada

CAMBRIDGE
UNIVERSITY PRESS

CAMBRIDGE UNIVERSITY PRESS
Cambridge, New York, Melbourne, Madrid, Cape Town,
Singapore, São Paulo, Delhi, Tokyo, Mexico City

Cambridge University Press
The Edinburgh Building, Cambridge CB2 8RU, UK

Published in the United States of America by Cambridge University Press, New York

www.cambridge.org
Information on this title: www.cambridge.org/9780521369978

First published 2005
First paperback edition 2011

A catalogue record for this publication is available from the British Library

Library of Congress Cataloguing in Publication data

Jordaan, Ian J., 1939–
Decisions under uncertainty: probabilistic analysis for engineering decisions / Ian J. Jordaan.
  p.   cm.
Includes bibliographical references and index.
ISBN 0 521 78277 5
1. Engineering–Statistical methods.   2. Decision making–Statistical methods.   3. Probabilities.   1. Title.
TA340.J68   2004
620'.007'27–dc22   2004043587

ISBN 978-0-521-78277-7 Hardback
ISBN 978-0-521-36997-8 Paperback

For Christina

Contents

9   Extremes

10   Risk, safety and reliability

Illustrations

The artistic illustrations that grace the cover and appear periodically in the text have been prepared by Grant Boland, to whom I am most grateful.

Preface

Probabilistic reasoning is a vital part of engineering design and analysis. Inevitably it is related to decision-making – that important task of the engineer. There is a body of knowledge profound and beautiful in structure that relates probability to decision-making. This connection is emphasized throughout the book as it is the main reason for engineers to study probability. The decisions to be considered are varied in nature and are not amenable to standard formulae and recipes. We must take responsibility for our decisions and not take refuge in formulae. Engineers should eschew standard methods such as hypothesis testing and think more deeply on the nature of the problem at hand. The book is aimed at conveying this line of thinking. The search for a probabilistic 'security blanket' appears as futile. The only real standard is the subjective definition of probability as a 'fair bet' tied to the person doing the analysis and to the woman or man in the street. This is our 'rule for life', our beacon. The relative weights in the fair bet are our odds on and against the event under consideration.

It is natural to change one's mind in the face of new information. In probabilistic inference this is done using Bayes' theorem. The use of Bayesian methods is presented in a rigorous manner. There are approximations to this line of thinking including the 'classical' methods of inference. It has been considered important to view these and others through a Bayesian lens. This allows one to gauge the correctness and accuracy of the approximation. A consistent Bayesian approach is then achieved. The link to decisions follows from this reasoning, in which there are two fundamental concepts. In our decision-making we use probability, a concept in which the Bayesian approach plays a special rôle, and a second concept, utility. They are related dually, as are other concepts such as force and displacement. Utility is also subjective: probabilities and utilities attach to the person, not to the event under consideration.

The book was written over many years during a busy life as a practising engineer. It is perhaps a good moment to present the perspective of an engineer using probabilistic concepts and with direct experience of engineering decision-making. Whilst a consistent approach to probability has been taken, engineering practice is often 'messy' by comparison to the mathematical solutions presented in the book. This can best be dealt with by the engineer – numerical methods, approximations and judgement must come into play. The book focusses on the principles underlying these activities. The most important is the subjective nature of the two concepts involved in decision-making. The challenge very often is to obtain consensus between engineers on design parameters, requiring a judicious addition of conservatism – but not too much! – where there are uncertainties which have not been formally analysed.

An important aspect for engineers is the link to mechanics. One can think of probability distributions as masses distributed over possible outcomes with the total mass being unity; mean values then become centres of mass; variances become moments of inertia; and other analogies appear. The Stieltjes integral is a natural way to obtain the mean values and moments; it is potentially of great usefulness to the engineer. It unifies common situations involving, for example, concentrated or distributed masses in mechanics or probability. But this has been included as an option and the usual combination of summation for discrete random quantities and Riemann integration for continuous ones has been used in most instances. Inference is treated from the Bayesian standpoint, and classical inference viewed as a special case of the Bayesian treatment is played down in favour of decision theory. This leads engineers to think about their problem in depth with improved understanding. Half hearted or apologetic use of Bayesian methods has been avoided. The derivation of the Bayesian results can be more demanding than the corresponding classical ones, but so much more is achieved in the results.

Use of this book

The book is intended as an introduction to probability for engineers. It is my sincere hope that it will be of use to practising engineers and also for teaching in engineering curricula. It is also my hope that engineering schools will in the future allow more time in their programs for modelling of uncertainty together with associated decision-making. Often probability is introduced in one course and then followed by courses such as those in 'Civil Engineering Systems'. This allows sufficient time to develop the full approach and not just pieces of 'statistics'. The present work will be of use in such situations and where it is considered beneficial for students to understand the full significance of the theory. The book might be of interest also in graduate courses.

I have often used extended examples within the text to introduce ideas, for example the collision probability of the Titanic in Chapter 2. More specific examples and illustrations are identified as 'examples' in the text. The use of urns, with associated random walks through mazes, assists in fundamental questions of assigning probabilities, including exchangeability and extremes. Raiffa's urn problem brings decision theory into the picture in a simple and effective manner. Simplified problems of decision have been composed to capture the essence of decision-making. Methods such as hypothesis testing have a dubious future in engineering as a result of their basic irrationality if applied blindly. Far better to analyse the decision at hand. But we do need to make judgements, for example, with regard to quality control or the goodness-of-fit of a particular distribution in using a data set. If wisely used, confidence intervals and hypothesis testing can assist considerably. The conjugate prior analysis in inference leads naturally to the classical result as a special case. Tables of values for standard distributions are so readily available using computer software that these have not been included.

Optimization has been introduced as a tool in linear programming for game theory and in the maximization of entropy. But it is important in engineering generally, and this introduction might become a springboard into other areas as well. Chapter 11, dealing with data, linear regression and Monte Carlo simulation is placed at the end and collects together the techniques needed for the treatment of data. This may require a knowledge of extremes or the theory of inference, so that these subjects are treated first. But the material is presented in such a way that most of it can be used at any stage of the use of the book.

Acknowledgements

Special mention should be made of two persons no longer alive. John Munro formerly Professor at Imperial College, London, guided me towards studying fundamental aspects of probability theory. Bruce Irons, of fame in finite element work, insisted that I put strong effort into writing. I have followed both pieces of advice but leave others to judge the degree of success. My former students, Maher Nessim, Marc Maes, Mark Fuglem and Bill Maddock have taught me possibly more than I have taught them. My colleagues at C-CORE and Memorial University have been most helpful in reviewing material. Leonard Lye, Glyn George and Richard McKenna have assisted considerably in reviewing sections of the book. Recent students, Paul Stuckey, John Pond, Chuanke Li and Denise Sudom have been of tremendous assistance in reviewing material and in the preparation of figures.

I am grateful to Han Ping Hong of the University of Western Ontario who reviewed the chapter on extremes; to Ken Roberts of Chevron-Texaco and Lorraine Goobie of Shell who reviewed the chapter on risks; and to Sarah Jordaan who reviewed the chapter on entropy. Paul Jowitt of Heriot-Watt University provided information on maximum-entropy distributions in hydrology. Marc Maes reviewed several chapters and provided excellent advice.

Neale Beckwith provided unstinting support and advice.

Many of the exercises have been passed on to me by others and many have been composed specially. Other writers may feel free to use any of the examples and exercises without special acknowledgement.

1 Uncertainty and decision-making

If a man will begin with certainties, he shall end in doubts; but if he will be content to begin with doubts, he shall end in certainties.

Francis Bacon, *Advancement of learning*

Life is a school of probability.

Walter Bagehot

1.1   Introduction

Uncertainty is an essential and inescapable part of life. During the course of our lives, we inevitably make a long series of decisions under uncertainty – and suffer the consequences. Whether it be a question of deciding to wear a coat in the morning, or of deciding which soil strength to use in the design of a foundation, a common factor in these decisions is uncertainty. One may view humankind as being perched rather precariously on the surface of our planet, at the mercy of an uncertain environment, storms, movements of the earth's crust, the action of the oceans and seas. Risks arise from our activities. In loss of life, the use of tobacco and automobiles pose serious problems of choice and regulation. Engineers and economists must deal with uncertainties: the future wind loading on a structure; the proportion of commuters who will use a future transportation system; noise in a transmission line; the rate of inflation next year; the number of passengers in the future on an airline; the elastic modulus of a concrete member; the fatigue life of a piece of aluminium; the cost of materials in ten years; and so on.

In order to make decisions, we weigh our feelings of uncertainty, but our decision-making involves quite naturally another concept – that of utility, a measure of the desirability, or otherwise, of the various consequences of our actions. The two fundamental concepts of probability and utility are related dually, as we shall see. They are both subjective, probability since it is a function of our information at any given time, and utility – perhaps more obviously so – since it is an expression of our preferences. The entire theory is behavioural, because it is motivated by the need to make decisions and it is therefore linked to the exigencies of practical life, engineering and decision-making.

The process of design usually requires the engineer to make some big decisions and a large number of smaller ones. This is also true of the economic decision-maker. The consequences of all decisions can have important implications with regard to cost or safety. Formal decision analysis would generally not be required for each decision, although one's thinking should be guided by its logic. Yet there are many engineering and economic problems which justify considerable effort and investigation. Often the decisions taken will involve the safety of entire communities; large amounts of money will have been invested and the decisions will, when implemented, have considerable social and environmental impact. In the planning stages, or when studying the feasibility of such large projects, decision analysis offers a coherent framework within which to analyse the problem.

Another area of activity that is assisted by an approach based on decision theory is that of the writing of design standards, rules or codes of practice. Each recommendation in the code is in itself a decision made by the code-writer (or group of writers). Many of these decisions could be put 'under the microscope' and examined to determine whether they are optimal. Decision theory can answer questions such as: does the additional accuracy obtained by method 'X' justify the expense and effort involved? Codes of practice have to be reasonably simple: the simplification itself is a decision.

The questions of risk and safety are at the centre of an engineer's activities. Offshore structures, for example, are placed in a hostile environment and are buffeted by wind, wave and even by large masses of floating ice in the arctic and sub-arctic. It is important to assess the magnitude of the highest load during the life of the structure – a problem in the probabilistic analysis of **extremes**. Engineering design is often concerned with tradeoffs, such as that between safety and cost, so that estimating extreme demands is an important activity. Servicing the incoming traffic in a computer or other network is a similar situation with a tradeoff between cost and level of service.

Extremes can also be concerned with the minimum value of a set of quantities; consider the set of strengths of the individual links in an anchor chain. The strength of the chain corresponds to the minimum of this set. The strength of a material is never known exactly. Failure of a structural member may result from accumulated damage resulting from previous cycles of loading, as in fatigue failure. Flaws existing in the material, such as cracks, can propagate and cause brittle fracture. There are many uncertainties in the analysis of fracture and fatigue – for example, the stress history which is itself the result of random load processes. Because of the dependence of fracture on temperature, random effects resulting from its variation may need to be taken into account, for example in arctic structures or vessels constructed of steel.

In materials science, we analyse potential material behaviour and this requires us to make inferences at a microscopic or atomic level. Here, the 'art of guessing' has developed in a series of stages and entropy – via its connection with information – is basic to this aspect of the art. The logic of the method is the following. Given a macroscopic measurement – such as temperature – how can one make a probabilistic judgement regarding the energy level of a particular particle of the material? Based on the assumption of equilibrium we state there is no change in average internal energy; we can then deduce probability values for all possible energy levels by maximizing entropy subject to the condition that the weighted sum of energies is constant. The commonly used distributions of energy such as Boltzmann's may be derived in this way. We shall also discuss recent applications of the entropy concept in new areas such as transportation. This approach should not be regarded as a panacea, but rather as a procedure with a logic which may or may not be appropriate or helpful in a particular situation.

1.2   The nature of a decision

We introduce the factors involved in decision-making in a relaxed and informal manner. Let us consider the problem of deciding whether or not to wear a coat in the morning: suppose that we are spending a winter in Calgary, a city near the Canadian Rockies. Warm westerly winds – the Chinook – transform bitterly cold weather ($-20\,^{\circ}\text{C}$, say) to relatively warm weather ($10\,^{\circ}\text{C}$, say!). It occasionally happens that the warm front drifts backwards and forwards so that the temperature changes quite abruptly between the two levels, cold and warm. Under such conditions – given, say, a warm morning – one is in some doubt as to whether to wear a coat or not.

One's dilemma is illustrated in Figure 1.1. The possible actions comprise an elementary action space: $\alpha_1 = $ *wear a coat*, $\alpha_2 = $ *don't wear a coat*. For each action, there are various 'states of nature' – here simplified to two states of nature for each action, defined by two symbols, θ_1 and θ_2. The symbol $\theta_1 = $ '*it turns cold during the day*', and $\theta_2 = $ '*it stays warm*

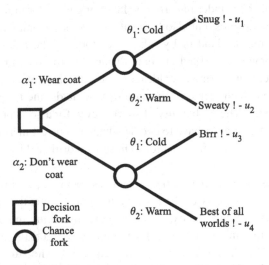

Figure 1.1   Uncertainty about the weather and its relation to wearing a coat in the morning

during the day'. Thus, there are four combinations of action and state of nature, and corresponding to each of these, there is a 'consequence', u_1, u_2, u_3 and u_4 as shown in Figure 1.1. A consequence will be termed a utility, U, a measure of value to the individual. Higher values of utility will represent consequences of greater value to the individual. The estimation of utilities will be dealt with later in Chapter 4; the concept is in no way arbitrary and requires careful analysis.

It is not difficult to reach the conclusion that the choice of decision (α_1 or α_2) depends on **two** sets of quantities only. These are the set of **probabilities** associated with the states of nature, $\{\mathbf{Pr}(\theta_1), \mathbf{Pr}(\theta_2)\}$, and the set of **utilities**, $\{u_1, u_2, u_3$ and $u_4\}$. The formal analysis of decisions will be dealt with subsequently, but it is sufficient here to note the dependence of the choice of decision on the two sets of quantities – probabilities and utilities. This pattern repeats itself in all decision problems, although the decision tree will usually be more complex than that given in Figure 1.1.

This example can also be used to illustrate a saying of de Finetti's (1974):

Prevision is not prediction.

Prevision has the same sense as 'expected value'; think of the centre of gravity of one's probabilistic weights attached to various possibilities. I recall an occasion when the temperature in Calgary was oscillating regularly and abruptly between $10\,°\mathrm{C}$ and $-20\,°\mathrm{C}$; the weather forecast was $-5\,°\mathrm{C}$. Yet nobody really considered the value of $-5\,°\mathrm{C}$ as at all likely. If one attached probabilistic 'weights' of 50% at $10°$ and $-20°$, the value $-5°$ would represent one's prevision – but certainly not one's 'prediction'.

The decision problem just discussed – to wear a coat or not – is mirrored in a number of examples from industry and engineering. Let us introduce next the oil wildcatter who literally makes his or her living by making decisions under uncertainty. In oil or gas exploration, areas of land are leased, primarily by big companies, but also by smaller scale operators. Oil wells may be drilled by the company leasing the land or by other operators, on the basis of an agreement. Exploratory wells are sometimes drilled at some distance from areas in which there are producing wells. These wells are termed 'wildcats'. The risk element is greater, because there is less information about such wells; in the case of new fields, the term 'new field wildcats' is used to denote the wells. The probability of success even for a good prospect is often estimated at only 40%. Drilling operators have to decide whether to invest in a drilling venture or not, on the basis of various pieces of information. Those who drill 'wildcat' wells are termed 'wildcatters'.

We are only touching on the various uncertainties facing oil and gas operators; for instance, the flow from a well may be large or small; it may decline substantially with time; the drilling operation itself can sometimes be very difficult and expensive, the expense being impossible to 'predict'. There may be an option to join a consortium for the purpose of a drilling, lessening risk to an individual operator. The classic work is by Grayson (1960) who studied the implementation of decision theory to this field. He analysed the actual behaviour in the field of various operators.

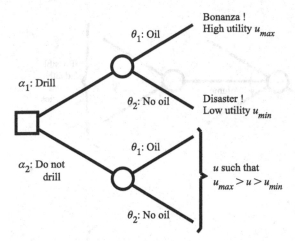

Figure 1.2   Oil wildcatter's dilemma

The wildcatter must decide whether or not to drill at a certain site before the lease expires. The situation is illustrated in the form of a decision tree in Figure 1.2. She realizes that if drilling is undertaken, substantial costs are involved in the drilling operation. On the other hand, failure to drill may result in a lost opportunity for substantial profits if there is oil present in the well. The decision must reflect an assessment of the probability of there being oil in the well. This assessment will almost invariably take into account a geologist's opinion. Thus, the action space is $A = \{\alpha_1 = drill\ for\ oil, \alpha_2 = do\ not\ drill\}$. The uncertain states of nature are $\Theta = \{\theta_1, \theta_2\}$ where $\Theta = \{\theta_1 = oil\ present$ and $\theta_2 = no\ oil\ present\}$. The choice of decision is again seen to depend on the values of U and the values $\{\mathbf{Pr}(\theta_1), \mathbf{Pr}(\theta_2)\}$.

In some decisions the states of nature differ, depending on the decision made. Consider the question of relocation of a railway that passes through a city. The trains may include cars containing chlorine, so that there is the possibility of release of poisonous gas near a human population. Figure 1.3 shows a decision tree for a problem of this kind. The consequences in terms of possible deaths and injuries are quite different for the two branches and might be quite negligible for the upper branch. Figure 1.3 illustrates a tradeoff between cost of relocation and risk to humans. This tradeoff between cost and safety is fundamental to many engineering decisions. Other tradeoffs include for example that between the service rate provided for incoming data and the queuing delay in data transmission. This is shown in Figure 1.4. Traffic in computer networks is characterized by 'bursts' in the arrival of data. The incoming traffic is aggregated and the bursts can lead to delays in transmission.

Figure 1.5 illustrates an everyday problem on a construction site. Is a batch of concrete acceptable? One might, for instance, measure air void content to ascertain whether the freeze–thaw durability is potentially good or bad. Bayes' theorem tells us how to incorporate this new information (Section 2.4). Non-destructive testing of steel structures is also rich in problems of this kind.

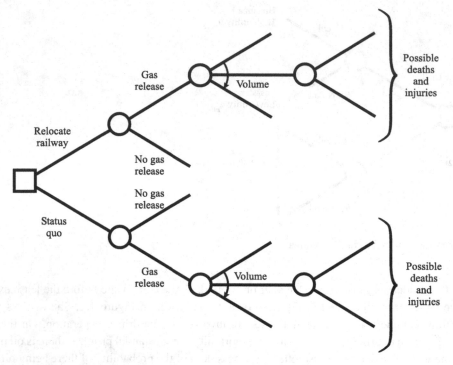

Figure 1.3   Rail relocation decision

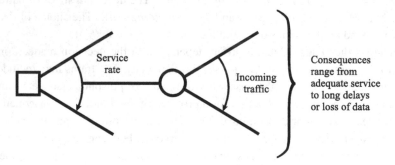

Figure 1.4   Decision on service rate for computer network

Let us return to the problems facing a code-writing committee. One of the many dilemmas facing such a committee is the question: should one recommend a simplified rule for use by engineers in practice? An example is: keep the span-to-depth ratio of a reinforced concrete slab less than a certain specified value, to ensure that the deflection under service loads is not excessive (rather than extensive analysis of the deflection, including the effects of creep, and so on). A further example is: should the 100-year wave height and period be the specified

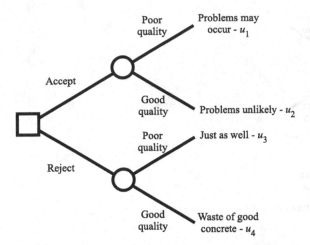

Figure 1.5   To accept or reject a batch of concrete posed as a problem of decision-making. Traditionally this is dealt with by hypothesis testing

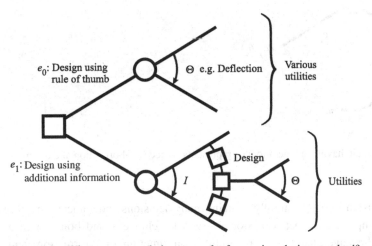

Figure 1.6   What recommendations to make for use in a design standard?

environmental indicators in the design of an offshore structure, in place of a more detailed description of the wave regime? These are numerous decisions of this kind facing the code-writer that can be analysed by decision theory. The decision tree would typically be of the kind illustrated in Figure 1.6. Rather than using the simplified rule, the committee might consider recommending the use of information 'I' which will be known at the time of design (e.g. the grade of steel or concrete, or a special test of strength if the structural material is available). This quantity could take several values and is therefore, from the point of view of the committee, a random quantity with associated probabilities.

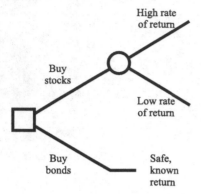

Figure 1.7   Risky and safe investment decisions

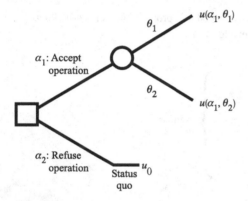

Figure 1.8   Dilemma in having an operation: θ_1 represents a successful operation and θ_2 an unsuccessful one

Economic decision-making naturally contains many decisions under uncertainty. Figure 1.7 illustrates the simple choice between stocks – a risky choice – and bonds with a known rate of return. We have stated that probability is a function of information, and consequently the better our information, the better our assessment of probabilities. The extreme of this would correspond to insider trading – betting on a sure thing, but not strictly legal or ethical!

Decisions in normal life can be considered using the present theory. Consider the following medical event.

'Patient x survives a given operation.'

Suppose that the patient (you, possibly) has an illness which causes some pain and loss of quality of life. Your doctors suggest an operation which, if successful, will relieve the pain; there is the possibility that you will not survive the operation and its aftermath. Your dilemma is illustrated in Figure 1.8. This example is perhaps a little dramatic, but the essential point is that again one's decision depends on the probabilities $\mathbf{Pr}(\theta_1)$ and $\mathbf{Pr}(\theta_2)$ (Figure 1.8) and one's utilities. One would be advised to carry out research in assessing the probabilities; the utilities must clearly be the patient's. The example points to the desirability of strong participation by patients in medical decisions affecting them.

1.3   Domain of probability: subjectivity and objectivity

The domain of probability is, in a nutshell, that which is possible but not certain. Probabilities can be thought of as 'weights' attached to what one considers to be possible. It therefore becomes essential to define clearly what one considers to be possible. The transition from 'what is possible' to 'certainty' will result from a well-defined measurement, or, more generally, a set of operations constituting the measurement. The specification or statement of the 'possible outcomes' must then be objective in the sense just described. We must be able, at least in principle, to determine its truth or falsity by means of measurement. To take an example, consider the statement:

E: *'We shall receive more than 1 cm of rain on Friday.'*

We assume that it is Monday now. We need to contrive an experiment to decide whether or not there will be more than 1 cm of rain on Friday. We might then agree that statement E will be true or false according to measurements made using a rain gauge constructed at a meteorological laboratory.

Having specified how we may decide on the truth or falsity of E we are in a position to consider probabilities. Since it is now Monday, we can only express an opinion regarding the truth or falsity of E. This 'opinion' is our probability assignment. It is subjective since we would base it on our knowledge and information at the time. There is no objective means of determining the truth or falsity of our probability assignment. This would be a meaningless exercise. If a spectacular high pressure zone develops on Thursday, we might change our assignment.

The same logic as above should be applied to any problem that is the subject of probabilistic analysis. The statement of the possible outcomes must be objective, and sufficient for the problem being analysed. It is instructive to ponder everyday statements in this light. For example, consider 'his eyes are blue', 'the concrete is understrength', 'the traffic will be too heavy', 'the fatigue life is adequate'... We should tighten up our definitions in these cases before we consider a probabilistic statement.

The approach that we have followed in defining what is possible echoes Bridgman's operational definition of a concept: the concept itself is identified with the set of operations which one performs in order to make a measurement of the concept; see Bridgman (1958). For example, length is identified with the physical operations which are used in measuring length, for instance laying a measuring rod along the object with one end coinciding with a particular mark, taking a reading at the other end, and so on. We shall suggest also that probability can be defined using the operational concept, but tied to the decision-maker's actions in making a fair bet, as described in the next chapter.

1.4 Approach and tools; rôle of judgement: indifference and preference

The theory that we are expounding deals with judgements regarding events. These judgements fall into two categories, corresponding to the two fundamental quantities introduced in Section 1.2 that are involved in decision-making: probabilities and utilities. The basic principle in dividing the problem of decision into two parts is that probability represents our judgement – regardless of our desires – as to the likelihood of the occurrence of the events under consideration, whereas utility represents our judgement of the relative desirability of the outcomes of the events. For example, we assess the likelihood (probability) of an accident in an impartial manner, taking into account all available information. The consequences of the accident may be highly undesirable, possibly involving loss of life, leading to our estimate of utility. But these factors do not affect our judgement of probability.

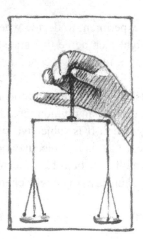

Probability will be defined formally in the next chapter; for the present introductory purposes, we introduce one of our tools, the urn containing balls of different colours. The urn could be a jar, a vase, a can or a bag – or one's hand, containing coloured stones. Our primitive ancestors would probably have used stones in their hand. It is traditional in probability theory to use urns with coloured balls, let us say red and pink, identical – apart from colour – in weight, size and texture. Then the balls are mixed and one is drawn without looking at the colour.

(a) Urn containing coloured balls

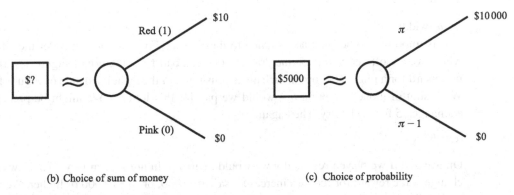

(b) Choice of sum of money   (c) Choice of probability

Figure 1.9   Tools for probability

The urn of Figure 1.9(a) contains an equal number of balls of each of two colours, designated '0' (say, pink) and '1' (say, red). We would agree that the likelihood of drawing a ball with a '0' or a '1' is the same since the numbers of '0' and '1' balls are the same. The probability of either is then 0.5. We can use the urn to conceive of any value of probability that corresponds to a rational number by placing the required number of balls into the urn. A probability of $(57/100)$ corresponds to an urn containing 100 well-shuffled balls of which 57 are red and the others pink.

To put into effect our judgement in a range of situations, it is useful to consider 'order relations'. These represent the results of our judgement, in terms of the relationship between the objects under consideration. Consider a set of objects $\{a_1, a_2, \ldots, a_n\}$; these could be sets of numbers, people, consequences of actions, or probabilities. If we have a relationship between two of them, this is a binary relation, written as $a_1 R a_2$. For example, the relation R might be that a_1 is bigger than a_2 in the case of real numbers, typified by the familiar relationship $>$. Other familiar relationships are $=$ and $<$. Another relationship R might be that a_1 is a member of the same family as a_2, in the case of persons.

We shall now introduce **negation**, indicated by $\sim$, and two order relationships: **indifference**, written as $\approx$, and **preference**, written as $\succ$. Negation represents the complement of a statement; for example, if $A = $ 'the steel strength is greater than 400 MPa' then the negation, written either

$\sim A$ or $\tilde{A}$, represents 'the steel strength is less than or equal to 400 MPa'. The tilde $\sim$ can be read as 'not', e.g. $\tilde{A}$ is not-A. The same notation is used for actions. If $\alpha = $ 'accept', then $\tilde{\alpha} = $ 'reject'.

Since we have used $\sim$ for negation, we shall use the double tilde $\approx$ for indifference. This concept is used frequently in our theory and can be read as 'twiddles'. For example, if we are indifferent between the purchase of certain stocks (Figure 1.7), termed option 'a', and the purchase of bonds, termed option 'b', then for us, we might say that we are indifferent between a and b, or

$$a \approx b,$$

or 'a twiddles b'.

It is important to become accustomed to these ideas. For example, consider the situation where we are offered \$10 upon the drawing of a red ball from the urn of Figure 1.9(a), with no reward for a pink one – option a. This is illustrated in the right hand side of Figure 1.9(b). What sum of money – option b – would we pay for this lottery? We might be prepared to exchange \$5 for the lottery. Then again

$$a \approx b.$$

On average if we played repeatedly we would neither gain nor lose money. Our views might change if the reward for red was increased, say to \$1000, or to \$10 000 or higher. We would be less likely to offer half of the reward to join when the sums are large. We would not wish to pay \$5000 for the lottery with an even chance only of gaining an extra \$5000 but with the same chance of losing the payment. In this case, option b – keep the \$5000 for sure – might be preferred to a – \$5000 payment with a reward of \$10 000 if a red ball appears but with no reward and loss of the original \$5000 if pink appears. Here we write

$$b \succ a.$$

This aversion to risk is considered in some detail in Chapter 4. We might prefer wine, option a, to beer, option b, then $a \succ b$. Preference and indifference are similar, but not the same as, $>$ and $=$ in algebra.

The question in the lottery can be posed in terms of probabilities, rather than consequences. Consider the lottery of Figure 1.9(c). For what value of probability π would one feel indifferent between the two options, that is, exchange one for the other, and *vice versa*? We might wish π to be greater than half to compensate for the possibility of losing our investment of \$5000. We can conceive any rational value of probability using our urn device.

1.5   Probability and frequency

There are two situations in which we need to analyse uncertainty. The first concerns that associated with 'repetitions of the same experiment', such as tossing a coin or repeated drawing with subsequent replacement from urns of the kind in Figure 1.9(a). We can obtain the probability of heads or tails by tossing the coin many times and taking the proportion of heads. This leads to the 'frequency' definition of probability. We might, on the other hand, be considering an event that happens only once or rarely, such as the inflation rate next year or the success of an early space shuttle. This one-off kind of event is perhaps more appropriate to the real world of our lives.

The concern with repetitive random phenomena stems from the nature of certain processes such as wind loading applied to a structure, or repeated tosses of a coin. The uncertainty is related to a particular future event of the same kind, for instance the maximum wind loading in a year, or result of the next toss of the coin. We shall view this kind of uncertainty from the same standpoint as any other kind, in particular one's uncertainty regarding an event that occurs once only, and treat the record of the past (wind speeds, heads and tails) as **information** which we should use to estimate our probabilities. We use frequencies to evaluate, not define a probability.

To sum up: probability is not a frequency – by definition. We use frequencies to estimate probabilities; this distinction is important. The approach taken here is often termed 'Bayesian', after Thomas Bayes who initiated it in the eighteenth century. One consequence of the position that probability is not a frequency is that it is not an objective 'property' of the thing under study, to be measured in some kind of lengthy 'random experiment'.[1] Consider the difficulty of constructing a random experiment to express our uncertainty about the future rate of inflation. A 'thought-experiment' or simulation can instead be used to construct a theory using frequency. We shall use frequency data just as often as the strict 'frequentist' statistician, and we shall see that 'frequentist' results can be recaptured in the Bayesian analysis. We can often replace a real repetitive experiment with an imaginary one in cases where a frequency interpretation is desired. For any of our probability assignments, one can imagine a repetitive experiment.

In everyday thinking, it is often convenient to solve problems in terms of frequency ratios. Consider the case of a rare disease. The probability of contracting the disease is 1/10 000. A person is tested by a method that is fool-proof if the person has the disease, but gives a false result 6% of the time if the person is free of the disease, that is, 6% false positive indications. We assume that we have no information on symptoms. If the person obtains a positive reading, what is the probability that he or she has the disease? It is easy to consider 10 000 persons, one of whom has the disease. Of the disease-free persons, 6%, or almost 600, will have a positive indication. Therefore, the probability that a person with a positive indication has the disease is only one in 601! But we know that the probabilities are 1/10 000 and 1/601 and that one would not necessarily find exactly one diseased person in a group of 10 000.

The situations discussed relate also to the classical dichotomy between risk and uncertainty which existed in economic theory for some time.[2] In this approach, 'risk' was considered to

---

[1] Jaynes (2003) gives an entertaining account of the difficulty in constructing such an experiment – particularly with respect to coin-tossing. He discusses the mechanics of tossing a coin and suggests that it is possible to cheat by skilful tossing.

[2] See, for example, M. Machina and M. Rothschild's article 'Risk' in Eatwell *et al.* (1990).

be involved when numerical probabilities were used; if probabilities were not assigned, the alternative possible outcomes were said to involve 'uncertainty'. Bernstein (1996) describes the development of this line of thinking. We shall not make this distinction. It seems to run counter to the normal use of the words risk and uncertainty and is perhaps losing importance in economic thinking, too. The same structure in decision-making applies whether we carry out a calculation of probability or not.

As a first step in evaluating probabilities, it is useful to judge everyday events from the point of view of indifference as outlined in the preceding section. At what value would we judge the rate of inflation next year to be equally likely to be exceeded or not? As an example, in early 1980 I judged the value to be 12% for the year; therefore my probabilities were 50% for the exceedance and non-exceedance of this value. (The change in the Consumer Price Index turned out to be 11.2%: a reminder that inflation can return!) Tweedledum and Tweedledee illustrate our attitude that a bet on heads in the next toss of a coin and a bet on the rate of inflation – as just described – are essentially equivalent. Certainly our uncertainty is the same, instances of the same phenomenon.

Tweedledum and Tweedledee

The principal benefit of the present theory is that one can view other definitions – frequentist, classical, . . . – as special cases of the one espoused here. One can, in addition, tie the problem to its *raison d'être*, the need to make a decision and to act in a rational way. The behavioural view of probability teaches one to ponder the nature of a problem, to try to idealize it, to decide whether it is appropriate to use or to massage frequency data. The theory then makes one learn to express one's opinions in as accurate a way as possible and consequently one is less inclined to take refuge in a formula or set of rules.

1.6   Terminology, notation and conventions

Most of what will be said in the following arises from conventions suggested in many cases by de Finetti. These changes assist considerably in orienting oneself and in posing problems correctly, making all the difference between a clear understanding of what one is doing and a half-formed picture, vaguely understood. As noted previously, by far the most common problem arises from the fact that much previous experience in probability theory results from dealing with 'repetitive random phenomena': series of tosses of a coin, records of wind speeds, and in general records of data. We do not exclude these problems from consideration but the repetitive part is clearly our **record of the past**. The expression 'random data' is a contradiction of terms. We should attempt to apply probability theory to events where we do not have the outcome, using the records, with judgement, to make probabilistic statements.

Examples of correctly posed problems are the following with the understanding that where measurements are to be made, the appropriate systems for this purpose are in place:

E_1: 'heads on the tenth toss of a specified coin in a series of tosses about to start'
E_2: 'heads on the tenth toss in a series of tosses made yesterday, but for which the records are misplaced'
E_3: 'the wind pressure on a specified area on April 2, 2050 at 8:00 a.m. will be in the range x_1 to x_2 Pa (specified)'
E_4: 'the wind pressure on surfaces in (location) will exceed X Pa (specified) in the next 30 years'
E_5: 'the maximum load imposed by ice on this arctic structure during its defined life will be greater than y MN (specified)'
E_6: 'the number of packets of information at a gateway in a network during a specified period will exceed 500 million'.

The events to which we apply probability theory should be distinct, clearly defined events, as above. It then becomes meaningless to talk of 'random variables'. The events are, or turn out to be, either true or false: they are not themselves varying. The adjective 'random' expresses our uncertainty about the truth or falsity of the event. The sense that one should attach to the word is 'unknown', referring to truth or falsity and, in general, 'uncertainty'. With regard to the common term 'random variable', consider the following statement:

Q_1: *'the elastic modulus of this masonry element is $X = x$ GPa'*.

Here, we permit X to take a range of values, a particular value being denoted by the lower case letter x. The quantity X is what is normally termed a 'random variable', implying that it varies; this is incorrect. X does **not** vary; it will take on a particular value, which we do not know. That is why we need probability theory. We shall therefore adopt the term '**random quantity**' as suggested by de Finetti rather than 'random variable'. As noted, 'random' is retained with the understanding that the meaning is taken as being close to that of the phrase 'whose value is uncertain'.

Bruno de Finetti used the word '**prevision**' for what would usually be described as the **mean** or **expected value** of a quantity, but with much more significance. The most vivid, and appealing, interpretation of expected value (or prevision) for an engineer is by means of centre of gravity. We can think of probability as a 'weight' distributed over the possible events, the sum of the weights being unity. An example was given in Section 1.2, in considering two possible temperatures of 10 and $-20\,°C$; if the 'weights' attached to these possibilities are $1/2$ each, the mean or prevision (**not** the value 'expected' or a 'prediction') of the temperature is $-5\,°C$. In fact, probability can be seen as a special case of prevision. De Finetti uses the symbol $\mathbf{P}(\cdot)$ for both probability and prevision. In this book, we shall use $\mathbf{Pr}(E)$ for probability of event E. We shall use $\langle X \rangle$ for the mean, or expected value, of the argument (X). The term 'expected value' is often used but only has meaning as an average in a long run experiment. We shall occasionally use 'prevision' to recognize the clarification brought by de Finetti.

1.7   Brief historical notes

A definitive statement on the history of probability would certainly be an interesting and demanding project, far beyond the intention of the present treatment. Stigler (1986) gives a scholarly analysis of the development of statistics. Bernstein (1996) gives an entertaining account of the development of the use of probability and the concepts of risk analysis applied particularly to investment. Judgements of chance are intuitive features of the behaviour of humans and indeed, animals, so that it is likely that the ideas have been around during the development of human thinking. There is mention of use of probability to guide a wise person's life by Cicero; the use of odds is embedded in the English language, for example Bacon in 1625: 'If a man watch too long, it is odds he will fall asleepe'.

Early work in mathematical probability theory dates back to the seventeenth century. The early focus of theory was towards the solution of games of chance, and the **classical** view of probability arose from this. An example is the famous correspondence between Pascal and

Fermat on games with dice, summarized in Section 9.2. In cases where one can classify the possible outcomes into sets of 'equally likely' events, probabilities can be assigned on this basis. It is now commonplace to assign probabilities for games of cards, dice, roulette, and so on, using classical methods. The extension of this to account for observations was the next task. The title of Jacob Bernoulli's work, 'Ars Conjectandi', or the art of conjecturing, summarizes much of the spirit of dealing with uncertainty. In Bayes' original paper (Bayes, 1763) one finds the definition of probability in terms of potential behaviour – as outlined in the next chapter. Laplace (1812, 1814) applied these ideas, together with Bayes' method of changing probability assignments in the light of new evidence (Bayes' theorem) to practical problems, such as in astronomy.

The 'relative frequency' view of probability was developed in the latter part of the nineteenth and first half of the twentieth century; see for instance Venn (1888) or von Mises (1957). This definition is based on a long series of repetitions of the same experiment, for instance tossing a coin repeatedly and counting the proportion of heads to obtain the probability of heads in a future trial. As noted in Section 1.5, frequency ratios are useful in certain problems; in many applications, frequencies enter into problems in a natural way. We take issue with defining probability as always being a frequency that has to be measured. Changes in theories, even as a result of well-determined experiments, can be a lengthy process. Consider the rejection of the caloric 'fluid' as a result of the discovery of the inexhaustible generation of frictional heat, as measured by Count Rumford. The natural tendency to resist change resulted in the persistence of the theory of heat as being a fluid for a considerable period of time. The frequency view of probability is a mechanistic one in which models are postulated that function repetitively in a mechanical sense. The inductive world is much wider and permits the updating of knowledge and thought using evidence in an optimal manner. It points us in the right direction for the next step in understanding.

It is interesting to note that utility theory also had its early development during the same period as mathematical probability theory; we shall give an example in Chapter 4, still pertinent today, of Daniel Bernoulli's utility of money. The parallel development of utility and probability had its flowering in the twentieth century. The resurgence of Bayesian ideas contains two main streams. They both represent subjective views but are rather different conceptions: the necessary view and the personal view. The former tries to attain the situation where two persons will arrive at exactly the same assignment of probabilities. The latter accepts the possibility that two people modelling a real situation may obtain somewhat different assignments. Idealized problems, will, at the same time, tend to lend themselves to identical solutions. The necessary view was quite prominent at Cambridge University – for instance Jeffreys (1961). Jaynes (1957, 2003 for example) adopts a necessary point of view and has done most inspiring work on the connection between entropy and probability.

Several authors including de Finetti (1937) and Savage (1954) based their approach on the idea of defining probabilities by means of **potential behaviour**, generally using 'fair' bets. De Finetti has adopted a personalist conception which seems to me to include the necessary view as a rather idealized special case. I find that the work of this writer provides the most flexible and rational viewpoint from which we can regard all probabilities. Good (1965, for example) has made significant contributions to Bayesian methods, and to the use of entropy. The significant contribution of von Neumann and Morgenstern (1947) provided the link in the structure of utility and probability theory to their close relative, games theory.

Bruno de Finetti

Several writers have focussed on the needs of statisticians, or 'users' of probability theory. The work of Lindley (1965) put the use of traditional distributions, such as 'student's' t-distribution and the use of confidence intervals, in conformity with the rest of the personalist theory. This was of great value for users of the theory who wished to apply but also to understand. Work at Harvard University in the decades after the second World War, exemplified by Raiffa (1968), Schlaifer (1959, 1961) and Raiffa and Schlaifer (1961), has made significant impact on practical applications. From an engineering standpoint, works such as Benjamin and Cornell (1970) and Cornell (1969a), which stress decision-making and are sympathetic to the Bayesian approach, were instrumental in pointing the way for many users, including the present one.

The towering figure of twentieth century probability theory is Bruno de Finetti. Lindley states in the foreword to the book by de Finetti (1974) that it is 'destined ultimately to be recognized as one of the great books of the world'. I strongly agree.

1.8 Exercise

1.1 Draw a decision tree, involving choice and chance forks, for each of

   (a) a personal decision (for example, for which term should one renew a mortgage; one, three or five years?);
   (b) an engineering problem.

2 The concept of probability

Iacta alea est.

Julius Caesar, at the crossing of the Rubicon

2.1 Introduction to probability

Probability and utility are the primary and central concepts in decision-making. We now discuss the basis for assigning probabilities. The elicitation of a probability aims at a quantitative statement of opinion about the uncertainty associated with the event under consideration. The theory is mathematical, so that ground rules have to be developed. Rather than stating a set of axioms, which are dull and condescend, we approach the subject from the standpoint of the potential behaviour of a reasonable person. Therefore, we need a person, you perhaps, who wishes to express their probabilistic opinion. We need also to consider an event about which we are uncertain. The definition of possible events must be clear and unambiguous.

It is a general rule that the more we study the circumstances surrounding the event or quantity, the better will be our probabilistic reasoning. The circumstances might be physical, as in the case where we are considering a problem involving a physical process. For example, we might be considering the extreme wave height during a year at an offshore location. Or the circumstances might be psychological, for example, when we are considering the probability of acceptance of a proposal to carry out a study in a competitive bid, or the probability of a person choosing to buy a particular product. The circumstances might contain several human elements in the case of the extreme incoming traffic in a data network. The basic principle to be decided is: how should one elicit a probability?

We shall adopt the simple procedure of using our potential behaviour in terms of a fair bet – put one's money where one's mouth is, in popular idiom. By asking how we would bet, we are asking how we would **act** and it is this potential behaviour which forces us to use available information in a sensible down-to-earth manner. This provides a link with everyday experience, with the betting behaviour of women or men in the street. It is one of the great strengths of the present theory. We shall consider small quantities of money so as to avoid having to deal with the aversion to the risk associated with the possibility of losing large quantities of money. Risk aversion was alluded to in Section 1.4 and is dealt with in detail in Chapter 4.

Let us suppose that YOU wish to express your opinion by means of a probability, and that I am acting as your advisor on probabilistic matters. My first piece of advice would be to focus your thoughts on a **fair bet**. To explain further, let us use an example. We shall assume that YOU are the engineer on a site where structural concrete is delivered. You wish to assign probabilities to the various possible strengths of a specimen taken from a batch of concrete just delivered to the site. In doing this, you should take into account all available information. The extent to which we search for new information and the depth to which we research the problem of interest will vary, depending on its importance, the time available and generally on our judgement and common sense. We have some knowledge of the concrete delivered previously, of a similar specified grade. To be precise: we cast a cylindrical specimen of the concrete, 15 cm in diameter and 30 cm long (6 by 12 inches) by convention in North America. Since its strength is dependent on time and on the ambient temperature and humidity, we specify that the strength is to be measured at 28 days after storing under specified conditions of temperature and humidity. The strength test at 28 days will be in uniaxial compression. You now have 28 days to ponder your probability assignment![1]

The first step is to decide which strengths you consider to be possible: I suggest that you take the range of strengths reasonably wide; it is generally better to assign a low probability to an unlikely event than to exclude it altogether. Let us use the interval 0–100 MPa. Let us agree to divide the interval of possible strengths into discrete subintervals, say 0–20 MPa, 20–40 MPa, and so on. (In the event of an 'exact' value at the end of a range, e.g. 40 MPa, we assign the value to the range of values below it, i.e. to 20–40 MPa in the present instance.) We shall concentrate on a particular event which we denote as E. We use in this example

E: 'The strength lies in the subinterval 40–60 MPa.'

(The arguments that follow could be applied to any of the subintervals.) If E turns out to be true, we write $E = 1$, if E is false, we write $E = 0$. The notation $1 = \text{TRUE}, 0 = \text{FALSE}$ will be used for random events, which can have only these two possible outcomes. Hence, E is a **random event**. On the other hand, concrete strength is a **random quantity**, since it can take any of a number of specific values, such as the strength 45.2 MPa, in a range, whereas an event can be either TRUE (1) or FALSE (0). The events or quantities are random since we don't know which value they will take.

We shall elicit your probability of E, which we write as $\mathbf{Pr}(E)$, in the following way. I offer you a contract for a small amount 's', say \$100, which you receive if the strength is in the specified interval, that is if E is true. If E is not true, you get nothing. The crux of the matter is as follows: you must choose the maximum amount $\$(sp) = \$(100p)$ which you would pay for the contract. We specify 'maximum' because the resulting bet must be a 'fair' one, in the accepted sense of the word, and anything less than your maximum would be in your favour or even represent a bargain for you. We assume also that you are not trying to give money

---

[1] In the reality of engineering practice, the discovery of concrete that fails the strength specification once it has hardened poses difficulties with regard to its removal!

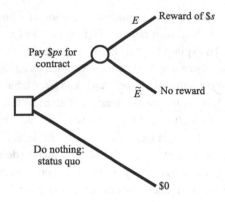

(a) Definition of probability: when one judges the upper and lower branches to be equivalent, then the value 'p' represents the probability of E. The concept works on the basis of a serious but fair bet: think of small values of money so that aversion to losing large sums does not complicate the issue. For example, consider the above with $s = \$100$

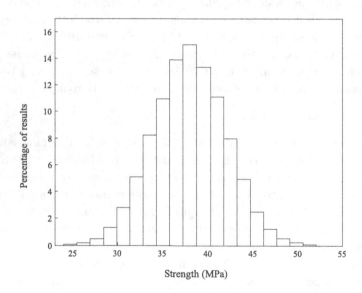

(b)   Histogram of past records of concrete cylinder tests

Figure 2.1   Evaluation of probability

away and that your maximum therefore represents a fair bet. Essentially we want the two main branches in Figure 2.1(a) to be of same value.

We noted in Section 1.4 that the symbol E with the tilde added before or above it, $\sim E$ or $\tilde{E}$, in Figure 2.1(a) denotes 'not-E', 'the complement of E', 'the negation of E', or if one wishes, 'it is not the case that E'. For example, if F = 'the concrete strength is greater than 40 MPa', then $\tilde{F}$ indicates 'the concrete strength is not greater than 40 MPa' or 'the concrete strength is less than or equal to 40 MPa'. Hence, $E + \tilde{E} = 1$, since either E or $\tilde{E}$ is true (1) and the other is false (0). The equivalence of the two branches in Figure 2.1 implies that

$$-100p + 100E \approx 0 \tag{2.1}$$

where $\approx$ indicates 'indifference'.[2] This means simply that our definition of probability implies that we are indifferent as to whether we obtain the left or right hand side of the relation sign $\approx$. Remember that E is random (uncertain) in the relationship (2.1) taking the values 1 or 0 depending on whether E is true or false respectively. The value '0' on the right hand side of the indifference relationship indicates the status quo, resulting in no loss or gain, shown in the lower branch of Figure 2.1(a).

As a vehicle for assigning your probabilities, you might wish to use an urn such as that in Figure 1.9(a). For example, if 100 balls were place in the urn with 25 being red, this would denote a probability of 0.25. It is good general advice to use any past records and histograms that you have in your possession to obtain your value of 'ps'. Figure 2.1 shows that choosing a probability is also a decision. Let us assume that we have in the past tested a large number (say, in the thousands) of similar concrete cylinders. The result is indicated in Figure 2.1(b) in terms of a frequency histogram indicating the percentage of results falling in the corresponding ranges. There is nothing to indicate that the present cylinder is distinguishable or different in any respect from those used in constructing the past records. We observe from detailed analysis of the data of Figure 2.1(b) that 34% of results fall in the range of values corresponding to event E.

The probability of E is now defined in terms of your behaviour by

$$\mathbf{Pr}(E) = p = 0.34. \tag{2.2}$$

A value of \$34 as an offer for the contract indicates that $\mathbf{Pr}(E) = p = 0.34$. This would be a reasonable offer based on the histogram of Figure 2.1(b). All the data falls into the two class intervals 20–40 MPa and 40–60 MPa; we might therefore wish to make an allowance for a small probability of results falling outside these limits, possibly by fitting a distribution. This will be deferred to later chapters. We have gone out of our way to say that probability is not a frequency, yet we use a frequency to estimate our probability. This is entirely consistent with our philosophy that the fair bet makes a correct **definition** of probability, but then to use whatever information is at hand, including frequencies, to estimate our probability.

The definition of probability has been based on small amounts of money; the reason for this is the non-linear relationship between utility and money which will become clear in Chapter 4. For the present, this aspect is not of importance; the non-linearity stems from aversion to risking large amounts of money. The concept of probability remains the same when we have to deal with large amounts of money; we then define probability in terms of small quantities of money, or in terms of utility, rather than money, a procedure which is equivalent to the above.

---

[2] Twiddles ($\approx$) was introduced in Chapter 1, and exemplified by Tweedledum and Tweedledee.

In order to measure probability, all that we need is a person (you, in this case!) who is prepared to express their opinion in the choice of $\mathbf{Pr}(E) = p$. It is important to stress that $\mathbf{Pr}(E)$ expresses your opinion regarding the likelihood of E occurring – an opinion that should be entirely divorced from what you **wish** to happen. The latter aspect is included in our measure of the utility of the outcome. The oil wildcatter hopes desperately for a well that is not dry, but the probability assignment should reflect her opinion regarding the presence of oil, disregarding aspirations or wishes. It should also be noted that a contract could be purchased for a reward of, say \$100, to be paid if $\tilde{E}$, rather than E, occurs. Given $\mathbf{Pr}(E) = 0.34$, we would expect the contract on $\tilde{E}$ to be valued at \$66; this will be discussed in some detail at the end of this section and the beginning of the next.

Odds

It is common in everyday language to use 'odds', that is, the ratio of the probability for an event, say E, to the probability of its complement $\tilde{E}$. Writing the odds as 'r', we have

$$r = \frac{\mathbf{Pr}(E)}{\mathbf{Pr}(\tilde{E})} = \frac{\text{loss if } \tilde{E}}{\text{gain if } E},\tag{2.3}$$

where the loss to gain ratio follows from the fact that the upper branch of Figure 2.1(a) can be represented by the equivalent fork shown in Figure 2.2(a). Indeed for the probabilist this should become one's 'rule for life'. For example, given odds of 5 to 2 (or $r = 2.5$) on an event H, the probability of H is 5/7. Formally, the probability of an event E is expressed as

$$\mathbf{Pr}(E) = \frac{r}{r+1}.\tag{2.4}$$

(a) Equivalence of upper branch of Figure 2.1(a) (left in the figure above) and the gain and loss shown on the right of the figure

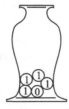

(b) Urn with $p = 4/5, \tilde{p} = 1/5$

Figure 2.2   Gain and loss in fair bet; urn equivalent

This can also be written as $r = p/(1 - p)$, or

$$r = \frac{p}{\tilde{p}},$$ (2.5)

where $\tilde{p}$ is **defined** as $(1 - p)$.

This definition of $\tilde{p}$ just given for numerical quantities is consistent with the definition for events given earlier, since $\mathbf{Pr}(E) = p = 1$ corresponds to $E = 1$ (E is certain). If $p = 0$ then $E = 0$ since E is impossible. The converse is also true; $E = 0$ implies that $p = 0$. We can write $E + \tilde{E} = 1$ and the corresponding expression $p + \tilde{p} = 1$; the basis for the latter will be discussed in more detail in Section 2.2. For example, p could be $4/5$ and then $\tilde{p} = 1/5$. It is also useful to think of the ratio $p/\tilde{p}$ as a ratio of **weights** of probability. This is illustrated with a pair of scales with $p = 4/5$, $\tilde{p} = 1/5$. Probability is a balance of judgement with 'weights' representing the odds. The weights in each scale can also be interpreted as the stakes in the fair bet. The bet on E is in the left hand scale and this is lost if $\tilde{E}$ occurs; similar reasoning applies to the bet on $\tilde{E}$, which is lost if E occurs. In this sense, one's probability can be thought of as a bet with oneself. Figure 2.2(b) shows the urn equivalent to $p = 4/5$, $\tilde{p} = 1/5$.

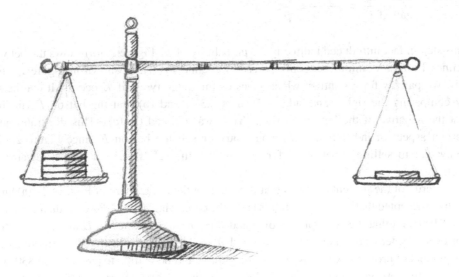

Rule for Life with example probabilities of $4/5$ and $1/5$

The notion of defining probability behaviourally is not new. The definition in terms of odds is given in the original paper by Bayes (1763):

'If a person has an expectation depending on the happening of an event, the probability of the event is to the probability of its failure as his loss if it fails to his gain if it happens'.

Implied in the definition of odds is the condition that $\mathbf{Pr}(E) = p$ and $\mathbf{Pr}(\tilde{E}) = \tilde{p}$ add up to unity. It is not always appreciated that the normalization of probabilities (i.e. that they sum to unity) follows from a very simple behavioural requirement – coherence. This means that you would not wish to make a combination of bets that you are sure to lose.

2.2   Coherence: avoiding certain loss; partitions; Boolean operations and constituents; probability of compound events; sets

It would be expected that the percentages shown in the histogram of Figure 2.1(b) should add up to unity since each is a percentage of the total number of results. If we throw a die, and judge that each face has equal probability of appearing on top, we naturally assign probabilities of 1/6 to each of the six faces. The probabilities calculated using our urn will also add up to unity. But there is a deeper behavioural reason for this result, termed 'coherence' which we now explore.

Let us consider first an event E and its complement $\tilde{E}$. These could be, for example, the events discussed in Section 2.1 or events such as $E = $ 'the die shows 1' and $\tilde{E} = $ 'the die shows 2, 3, 4, 5 or 6'. We introduced the definition of probability, $\mathbf{Pr}(E)$, as a fair bet and then went on to discuss odds on E as

$$\frac{\mathbf{Pr}(E)}{\mathbf{Pr}(\tilde{E})} = \frac{\text{loss if } \tilde{E}}{\text{gain if } E} = \frac{p}{1-p} = \frac{p}{\tilde{p}}.$$

This step in fact introduced indirectly the probability of $\tilde{E}$. Figure 2.3(a) shows the bet which defines $\mathbf{Pr}(E)$. A similar bet could be used to define $\mathbf{Pr}(\tilde{E})$, as shown in Figure 2.3(b). In this, we pay $\tilde{p}s$ for a contract which gives us the quantity 's' if $\tilde{E}$ occurs. It can be seen, by comparing the right hand side of Figures 2.3(a) and (b) that the bet on $\tilde{E}$ (in (b)) is just the negative of the bet on $\tilde{E}$ when $\mathbf{Pr}(E)$ was defined (in (a)). This illustrates an important aspect of indifference: we would **buy or sell** the bet on E since Figure 2.3(b) is equivalent to **selling** a bet on E for 'ps' with a fine of 's' (rather than a reward) if E occurs.

The crux of the present matter is that if we accept the assignment of Figures 2.3(a) and (b) we have accepted that $p + \tilde{p} = 1$. If this is not the case, what happens? Let us then assume that $s = \$100$, and that the probabilities of p and $\tilde{p}'(\neq \tilde{p})$ are assigned to E and $\tilde{E}$ respectively. For example, let us assume that $p = 0.8$ and $\tilde{p}' = 0.3$. This is illustrated in Figure 2.4. The assignment of probability means that the assignor would be willing to pay \$80 and \$30, a total of \$110, for bets on E and $\tilde{E}$ respectively, and receive \$100 in return should either happen, a sure loss of \$10! Figure 2.3(c) shows this in symbolic form.

If the assignor decides on values such that $\tilde{p}' < \tilde{p}$, for example, 0.3 and 0.5 as the probabilities of E and $\tilde{E}$ respectively, then it is again possible to find a value of 's' such that the assignor always loses. In this case where the sum of probabilities is less than unity, we would **buy** the two contracts from the assignor. This implies a negative 's', say $s = -\$100$, i.e. \$100 paid **by** the assignor if E or $\tilde{E}$ occur, with costs to us of the contracts for these two bets of \$30 and \$50 (on E and $\tilde{E}$ respectively) paid **to** the assignor. This again results in a certain loss to the assignor. This behaviour is termed **incoherence**. The implication is that if your probabilities do not normalize, it is always possible to make a combination of bets that you are sure to lose.

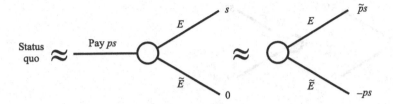

(a) Definition of probability of E

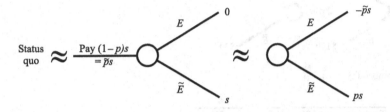

(b) Definition of probability of $\tilde{E}$

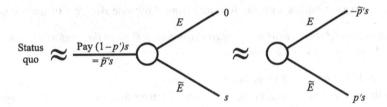

(c) Incoherent definition of probability of $\tilde{E}; \tilde{p}' \neq \tilde{p}$

Figure 2.3   Coherent (a,b), and incoherent (c) probabilities of event E and its complement $\tilde{E}$

The only way to avoid this is to ensure that

$$\mathbf{Pr}(E) + \mathbf{Pr}(\tilde{E}) = 1. \tag{2.6}$$

We can think of the assignment of probabilities as a linear operation $\mathbf{Pr}$ on $(E + \tilde{E} = 1)$.

Let us take this a little further, with YOU assigning probabilities. It is as well to think of a 'protagonist' with whom you are going to make your bets (with my advice for the present). The protagonist would probably be an experienced person such as a 'bookie' in horse-racing. Of course, a bookmaker has the objective of making money by arranging the odds in his favour. The same applies to insurance, which constitutes a bet on some future event, with the premiums

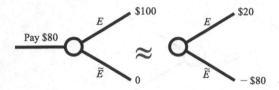

(a) Probability of E

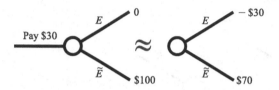

(b) Incoherent probability of $\widetilde{E}; \widetilde{p}' \neq \widetilde{p}$

Total paid: $110	Total received: $100	Net return: $-$10

(c) Totals from (a) and (b)

Figure 2.4   Example of probabilistic incoherence resulting in certain loss of $10

arranged so that the insurer makes a profit. To quote from Ambrose Bierce,[3] insurance is

'an ingenious modern game of chance in which the player is permitted to enjoy the comfortable conviction that he is beating the man who keeps the table'.

This was written in 1911 but is still valid today.

The situation that we are analysing is the question of **automatic** profit to the protagonist, apart from any 'arranging of odds' as is the case in bookmaking and insurance. Let us continue with the example of the previous section, concerning the 28-day strength of a concrete cylinder. Let the outcomes corresponding to the intervals under consideration (i.e. 0–20 MPa, 20–40 MPa, etc.) be denoted $E_1, E_2, \ldots, E_n$. Since these outcomes are **incompatible**, or **mutually exclusive**, because only one happens, and **exhaustive**, because one of them must happen, they constitute by definition a **partition**. Partitions occur frequently; other examples are the tossing of a die, where the possible results are $\{1, 2, 3, \ldots, 6\}$. An urn containing balls with n different colours would yield possible results $\{E_1, E_2, \ldots, E_n\}$ upon a draw, where E_1 corresponds to colour 1, and so on. In all of these cases,

$$E_1 + E_2 + \cdots + E_n = 1,$$

since only one of the E_i is true (1) and the others are false (0).

---

[3] *The Devil's Dictionary.*

The bookie sets stakes equal to $s_1, s_2, \ldots, s_n$ corresponding to the outcomes $E_1, E_2, \ldots, E_n$; hence, if E_i occurs, you receive s_i. You have agreed, as before, to purchase the rights to these contracts by paying the bookie the prices $p_1 s_1, p_2 s_2, \ldots, p_n s_n$ and in general $p_i s_i$. The stakes are quite arbitrary sums, positive or negative. The latter would correspond to the case where you pay the bookie the prize s_i if E_i occurs and $p s_i$ would be the amount, fair in your opinion, that he would pay you for the contract. The bookie's gain if the outcome E_h turns out to be true is

$$m_h = \sum_{i=1}^{n} p_i s_i - s_h; \ h = 1, 2, \ldots, n. \tag{2.7}$$

Let us consider the case where $s_1 = s_2 = \cdots = s_n = s$. The case where the s_i are different is treated in Exercise 2.6 at the end of the chapter, based on the elegant analysis of de Finetti (1937). Given that the s_i are all equal to s, Equations (2.7) become

$$m_h = s \left(\sum_{i=1}^{n} p_i - 1 \right), \text{ for each } h. \tag{2.8}$$

We assume initially that your probabilities do not add up to one. In this case, the bookie can regard (2.8) as a set of equations[4] in which the m_h are known. As a result, he can preassign his gain (given the p_i), and then solve for s. Thus, you will stand to lose, no matter which E_h turns out to be true. This is termed 'making a Dutch book' against you – no doubt in Holland it is an English book, or a French book. In any event, you will not find this situation acceptable. How can you stop the bookie?

If you make the assignment of probability such that $\sum_{i=1}^{n} p_i = 1$, this would avoid the possibility that the bookie could make arbitrary gains m_h. The bookie cannot invert Equation (2.8) to solve for s. Since you have normalized your probabilities by adjusting their values, then $p_i = \mathbf{Pr}(E_i)$, and we have that

$$\sum_{i=1}^{n} \mathbf{Pr}(E_i) = 1, \tag{2.9}$$

a result which is sometimes termed the **theorem of total probabilities**. We term this also the **normalization** of probabilities; your attitude, in avoiding the possibility of making a combination of bets that you are sure to lose, is termed **coherence**, which is embodied in Equation (2.9). This equation can also be thought of as the operation **Pr** applied to the equation

$$\sum_{i=1}^{n} E_i = 1.$$

We can simplify things by making the sum 's' above, in Equations (2.8), equal to unity. Then we can think of the bet $\mathbf{Pr}(E_i)$ on event E_i as a 'price' with $\mathbf{Pr} \equiv$ price, or stake paid with a reward of 'unity' if E_i occurs. A series of bets $\mathbf{Pr}(E_1) = p_1, \mathbf{Pr}(E_2) = p_2, \ldots, \mathbf{Pr}(E_i) = p_i, \ldots, \mathbf{Pr}(E_n) = p_n$, on events that constitute a partition can be interpreted as a series of prices that add up to unity. We receive the same amount back because one of the E_i must happen.

---

[4] Bookies are known for their expertise in mathematics.

If we consider two mutually exclusive events, say E_1 and E_2, we can consider the event $B = $ 'either E_1 or E_2 is true'. We can write

$B = E_1 + E_2$

since $B = 1$ if either E_1 or E_2 are true (and therefore equal to 1) or else $B = 0$ if E_1 and E_2 are both false (0). Now if we are prepared to exchange the status quo for a bet at price $\mathbf{Pr}(E_1) = p_1$ on E_1 or a bet at price $\mathbf{Pr}(E_2) = p_2$ on E_2, then as a result of this indifference we will be prepared to bet on both E_1 and E_2 separately or at the same time; consequently

$\mathbf{Pr}(B) = \mathbf{Pr}(E_1 + E_2) = \mathbf{Pr}(E_1) + \mathbf{Pr}(E_2).$

This amounts to buying one contract on B that is equivalent to two contracts, one on E_1 and one on E_2. In general

$$\mathbf{Pr}(E_i + E_{i+1} + \cdots + E_j) = \mathbf{Pr}(E_i) + \mathbf{Pr}(E_{i+1}) + \cdots + \mathbf{Pr}(E_j) \tag{2.10}$$

for any incompatible events $E_i, E_{i+1}, \ldots, E_j$.

These results all follow from coherence, and de Finetti has shown that we can extend the idea in a very simple way. Suppose that we have the following linear relationship between the events $\{E_1, E_2, \ldots, E_n\}$, which do not need to be incompatible:

$$a_1 E_1 + a_2 E_2 + \cdots + a_n E_n = a. \tag{2.11}$$

This relationship holds no matter which of the E_i turn out to be true; an example is the equation $\sum E_i = 1$ above for the special case of partitions. Another example is the following. Suppose that we have a device that fails when 5 out of 15 components fail (but we don't know which). Given failure of the device and if $E_i = $ '*the ith component fails*', then

$$\sum_{i=1}^{15} E_i = 5.$$

Or it might happen that in the analysis of an electrical circuit or structural component, we know the sum of several currents or the resultant of several forces by measuring a reaction. Then the sum of the currents or forces is known without knowledge of the individual values. Strictly speaking, these are examples of random quantities, which are a generalization of random events insofar as the random quantity may take any value in a range, and is not restricted to 0 or 1. The present theorem carries over and applies to random quantities or events. A last example might be a secret ballot: we have that the total votes cast for a proposition are m in number. If $E_i = $ '*individual i voted in favour of the proposition*', then $\sum E_i = m$. Again we don't know which E_i are true, which individuals voted in favour of the proposition.

We now show that coherence requires that we assign probabilities that are consistent with the following expression:

$$a_1 p_1 + a_2 p_2 + \cdots + a_n p_n = a, \tag{2.12}$$

where we have again used the notation $\mathbf{Pr}(E_i) = p_i$. To demonstrate (2.12), let us assume that you make the assignment

$$a_1 p_1 + a_2 p_2 + \cdots + a_n p_n = a + \delta, \tag{2.13}$$

where δ is any quantity (positive or negative). Subtracting Equation (2.11) from (2.13) we have

$$a_1(p_1 - E_1) + a_2(p_2 - E_2) + \cdots + a_n(p_n - E_n) = \delta. \tag{2.14}$$

Now, in order to make the amount δ for sure, I merely have to set stakes of $a_1, a_2, \ldots, a_n$ to be paid if $E_1, E_2, \ldots, E_n$ occur respectively. This follows because your assignment of probabilities would oblige you to pay $a_1 p_1, a_2 p_2, \ldots, a_n p_n$ for the various bets and consequently (2.14) represents my profit. If $\delta < 0$, we have to have negative stakes (and therefore paid to me), and you accept the amounts $a_i p_i$ for the contracts. To ensure coherence δ must $= 0$ and Equation (2.12) must be satisfied. De Finetti has also pointed out that if an inequality replaces the equality in Equation (2.11), the same inequality must appear in (2.12).

2.2.1 Boolean operations; constituents

We have used quite extensively the notation $1 = \text{TRUE}$, $0 = \text{FALSE}$. For a partition, in which one and only one of the events $E_1, E_2, \ldots, E_i, \ldots, E_n$ will occur, we write $\sum_{i=1}^{n} E_i = 1$. Boolean operations, the logical product and sum, denoted $\wedge$ and $\vee$ respectively, assist in studying events. The logical product

$A \wedge B$ means 'A and B',

that is, both A and B occur. Using the 0:1 notation, we can write

$A \wedge B = AB$;

the logical product and the arithmetic product coincide. For example, let

$A = $ '*the Titanic enters a certain degree-rectangle*'

and

$B = $ '*there is an iceberg in the same degree-rectangle*'.

Then $A \wedge B = AB$ is the event that both the Titanic and the iceberg are in the degree-rectangle. (We use the term degree-rectangle to denote an area bounded by degrees of latitude and longitude; it is not exactly rectangular!) If $A = 1$ and $B = 1$ then $AB = 1$ (both logically and arithmetically). Similarly if A or B or both are equal to 0, then $AB = 0$, as required. We shall use either AB or $A \wedge B$ for the logical product.

The logical sum

$A \vee B$ means 'A or B or both'.

In terms of the example above, either the Titanic or the iceberg, or both, are in the degree-rectangle. The notation $A + B$ for the logical sum is not consistent with the notation $A \vee B$, which is an event with values 0 or 1 depending on its falsity or truth. The value $A + B$ can be 0, 1, or 2 depending on whether neither, one of, or both the Titanic and the iceberg are in the degree-rectangle, respectively. The value 2 is not consistent with the Boolean notation. Therefore, $A + B$ **cannot** replace $A \lor B$.

To summarize, we shall use the notation

AB or $A \land B$,

and

$A \lor B$.

Finding partitions using constituents

We have considered situations where the events of interest may be sub-divided into a set of n events $\{E_1, E_2, \ldots, E_n\}$. These have so far constituted a partition. Three examples will suffice to illustrate that the idealization may be extended to cover other cases of practical interest.

(1) The case of repeated trials of the same phenomenon, such as n tosses of a coin, n tests of a set of 'identical' circuits with each result classified into two alternatives (e.g. $\geq x$ Volts, $< x$ Volts, with x specified), or the question of a set of n commuters with two transportation modes, with $E_1 = $ commuter 1 takes mode 1, $E_2 = $ commuter 2 takes mode 1, and so on. The sequence of compatible events is written

$\{E_1, E_2, E_3, \ldots, E_i, \ldots\}$.

A particular result or realization could include E_i or $\tilde{E}_i$, for example

$\{E_1, \tilde{E}_2, \tilde{E}_3, E_4, \ldots\}$,

or alternatively, if we wish, $\{1, 0, 0, 1, \ldots\}$.

(2) Repeated trials, as in (1), but with the trials not necessarily of the same phenomenon. Thus, E_1 could be 'the strength of the cylinder of concrete is $\geq x$ MPa', E_2 could be 'commuter Jones takes the light rail transit to work', E_3 could be 'this train contains hazardous substance y', and so on. We need not restrict our attention to compatible events; part of the series

$\{E_1, E_2, \ldots, E_n\}$

could consist of incompatible events. For instance, E_1 could be 'the strength of the cylinder is < 20 MPa', E_2 could be 'the strength of the same cylinder is ≥ 20 MPa and < 40 MPa', and E_3 could be 'the strength of the same cylinder is ≥ 40 MPa and < 60 MPa'.

(3) Taking the case just described to the extreme, the entire set

$\{E_1, E_2, \ldots, E_n\}$

could consist of incompatible events. In fact, this is a partition, where one and only one of a set of outcomes can occur. An example of this is the one we have already used at the end of the previous case: $E_1 =$ 'strength is < 20 MPa', $E_2 =$ 'the strength is ≥ 20 MPa and $<$ 40 MPa', and so on, when testing a single specimen.

The problem to be addressed here is: how do we deal with cases (1) and (2) above? They do not form partitions, since the events are not mutually exclusive. But can we form a partition? The answer is yes; we have to find the **constituents** which form a partition. Let us start with the example of two events E_1 and E_2. The constituents are

$$\{E_1 E_2, \tilde{E}_1 E_2, E_1 \tilde{E}_2, \tilde{E}_1 \tilde{E}_2\}$$

which is also a partition. (We have used the fact that for events, the logical product $E_1 E_2$ and the arithmetic product $E_1 E_2$ are equivalent.) If the events E_1 and E_2 themselves form a partition, then the constituents are $\{E_1 \tilde{E}_2, \tilde{E}_1 E_2\}$ or simply $\{E_1, E_2\}$.

There are at most 2^n constituents for the sequence of events

$$\{E_1, E_2, \ldots, E_n\}.$$

In the case where all the events are compatible, there are two possibilities $\{E_i, \tilde{E}_i\}$ for each i, and 2^n combinations of these. If any of the events are incompatible (mutually exclusive), certain constituents must consequently be omitted. For example, if $n = 3$, and if E_1 and E_2 are incompatible (only one of them can occur), the constituents are $E_1 \tilde{E}_2 E_3$, $E_1 \tilde{E}_2 \tilde{E}_3$, $\tilde{E}_1 E_2 E_3$ and $\tilde{E}_1 E_2 \tilde{E}_3$. If all the n events are incompatible, there are n constituents. The numbers of constituents, say k, is therefore such that $n \leq k \leq 2^n$. To summarize, we have shown that any set of events $\{E_1, E_2, \ldots, E_n\}$ can be expressed in terms of a partition of constituents $C_1, C_2, \ldots, C_k$, where k is in the range just noted. The probabilities of the constituents must normalize, as before.

2.2.2   Venn diagrams; probability of compound events

Events can be illustrated by means of Venn diagrams. These assist considerably in visualizing and understanding relationships using logical notation. Figure 2.5(a) shows a simple event: one can think of points within the area F corresponding to the results that are associated with event F. For example, the points could correspond to any of the concrete cylinder strengths in a range, say 20–30 MPa, which constitute event F. Other examples can be imagined for the diagrams of Figure 2.5(b) and (c). In (b), F could be within the range 20–30 and E within 25–28; in (c), F could be within 20–30 again with D in the range 25–35. We can also associate events with the 0:1 notation as in Figure 2.5(d).

We have used the notation $\sim E = \tilde{E} = 1 - E$. When we write $A \wedge B = AB$, or $\tilde{E} = 1 - E$, we are using logical notation on the left hand side, with an equivalent arithmetical expression on the right hand side, based on the 0:1 notation. We can verify using the Venn diagram of Figure 2.5(c) that $D \vee F = \sim (\tilde{D} \wedge \tilde{F})$. This can also be written as an arithmetical expression as follows:

$$D \vee F = 1 - (1 - D)(1 - F)$$
$$= D + F - DF. \tag{2.15}$$

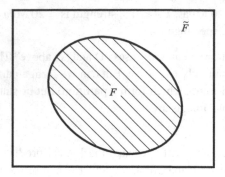

(a): Venn diagram for event F

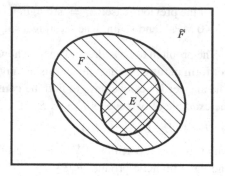

(b): Event E is contained in event F. Therefore E implies F. E and $\widetilde{F}$ are incompatible

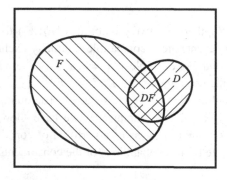

(c): Logical products (intersection) and sum (union). The total hatched area is $D \vee F$; the area with double hatching is $D \wedge F$ (or DF)

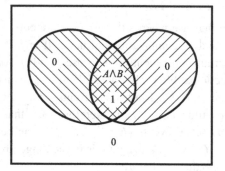

(d): Logical product $A \wedge B$ using 0:1 notation

Figure 2.5   Venn diagrams

We may interpret this as

$$D \vee F - D - F + DF = 0,$$

an equation of the type (2.11). Hence, using Equation (2.12) we deduce the well-known result

$$\mathbf{Pr}(D \vee F) = \mathbf{Pr}(D) + \mathbf{Pr}(F) - \mathbf{Pr}(DF). \tag{2.16}$$

We may visualize this in the Venn diagram of Figure 2.5(c) as suggesting that the probability of $D \vee F$ is the sum of all probabilities in D plus all those in F minus those in $D \wedge F$ since these have been counted twice (once in D and once in F).

Equation (2.15) can be generalized to

$$E_1 \vee E_2 \vee E_3 \vee \cdots \vee E_n = 1 - \Pi_{i=1}^{n}(1 - E_i). \tag{2.17}$$

This is the event that at least one of the E_i occur. For any value of n, the expression corresponding to (2.15) can be worked out by expanding the right hand side of Equation (2.17). Then, using its linear form, and the fact that probabilities must follow the same linear form, the result embodied in Equations (2.11) and (2.12), we can write an expression for the logical sum in terms of the products E_1, $E_1 E_2$, $E_1 E_2 E_3$, and so on. The results all derive from coherence.

2.2.3   Set theory

Set theory was developed for a broad class of situations – sometimes broader than the events or quantities of interest in probability theory. The sets of integers, of rational numbers, of real numbers, are all of interest in our work; sets of higher cardinality are not of relevance here. But sets can be used in probability theory: we have always a set of possible outcomes in our subject of study. We may then use set theory to illustrate our argument, but this theory is in no way fundamental to probabilistic reasoning.[5]

The set S of all outcomes of interest, the **possibilities**, is termed the **sample space** and the individual possible outcomes are the **sample points**. In tossing a die, $S = \{1, 2, 3, 4, 5, 6\}$, where 3, for instance, is a sample point. If the maximum number of vehicles that can wait in the turning bay at an intersection is 9, then the sample space corresponding to the counting of vehicles waiting to turn is $S = \{1, 2, 3, \ldots, 9\}$. These are examples of discrete sample spaces. If we are studying the depth to a certain layer of soil in a foundation problem, then we might consider S as the continuous sample space $S = \{x : 10\,\mathrm{m} < x < 100\,\mathrm{m}\}$ corresponding to the case where the soil layer can only be in the interval denoted. A further example is the set of points on the unit circle, which can be written $S = \{(x, y) : x^2 + y^2 = 1\}$. Here, the set consists of the ordered pairs (x, y) which might represent the position of an object in a problem in mechanics. These last two examples represent continuous sample spaces. The points within the Venn diagrams of Figure 2.5 are sample spaces so that the rectangles could be denoted as 'S'.

An event is defined, for the purposes of set theory, as a collection of sample points, for example $\{1, 2, 3\}$ in the turning lane problem mentioned above, or an interval such as $\{11\,\mathrm{m} < x \leq 22\,\mathrm{m}\}$ in the problem of the soil layer also mentioned above. A single sample point is termed a simple event, whereas a compound event includes two or more sample points. An event as defined in set theory is consistent with the usage in earlier sections above. There is an important difference when compared to the Boolean notation, in which, for example, $E = 1$ implies that E is true. Set theory does not automatically include this notation, but the idea can

[5] Difficulties in the axioms of set theory are exemplified by Russell's barber paradox: 'In a certain town (Seville?) every man either shaves himself or he is shaved by the (male) barber. Does the barber shave himself?' If the barber shaves himself, he is a man who shaves himself and therefore is not shaved by the barber. If the barber does not shave himself, he is a man who does not shave himself and therefore he is shaved by the barber. (This is a variation of 'All Cretans are liars' quoted by Epimenides, a Cretan himself.) The paradox is related to difficulties in specifying a set of all things which have the property of containing themselves as elements. This does not lead to difficulties in our analysis. We **insist** that the possible outcomes of the subject of study are well defined.

be used with set theory by means of an 'indicator function'. For example, if X is the set of real numbers, and x is a particular value, then the indicator function $I(x)$ could be defined such that

$$I(x) = 1 \text{ if } X \le x,$$

and

$$I(x) = 0 \text{ if } X > x.$$

Then $I(x) \equiv E$ if E is the event $X \le x$ in Boolean notation.

The relation between Boolean notation and an indicator function of sets is illustrated as follows:

Boolean notation	Indicator function on sets
$A \wedge B$ or AB	$I(A \cap B)$
$A \vee B$	$I(A \cup B)$
$A \wedge B \vee C$	$I(A \cap B \cup C)$

We shall use the Boolean notation without an indicator function in the remainder of this book. The following notation is often used.

1. $s \in S$; sample point s is a member of S.
2. $s \notin S$; s is not a member of S.
3. $A \subset B$; A is a subset of B; A is contained in B; A implies B; see Figure 2.5(b).
4. $A \cup B = \{s : s \in A \text{ or } s \in B\}$; those sample points s that are in A or B; the union of A and B; see Figure 2.5(c). The word 'or' is interpreted as meaning those sample points that are in A and B or both A and B. $A \cup B$ is sometimes written as $A + B$, but this contradicts our logical notation so that we shall not use this notation.
5. $A \cap B = \{s : s \in A \text{ and } s \in B\}$; those sample points s that are in both A and B; the intersection of A and B; see Figure 2.5(c). The intersection $A \cap B$ is sometimes written as AB, which is consistent with our logical notation.

For sets and for logical operations $\vee$ and $\wedge$, both union and intersection follow the associative, commutative and distributive rules, for example

$$A \cup B = B \cup A, (A \cap B) \cap C = A \cap (B \cap C),$$

and so on. See Exercise 2.7. The complement $\tilde{A}$ of an event A consists of all points in the sample space not included in A; see Figure 2.5(a). Thus, $\tilde{A} = \{s : s \in S \text{ and } s \notin A\}$. The difference between two sets, $A - B$, is the set containing all those elements of A which do not belong in B; again, this notation does not work with our logical notation (consider the case $A = 0, B = 1$) so that we shall not use it.

The null event $\emptyset$ contains no sample points (impossible event); two events A and B are **incompatible**, or **mutually exclusive**, or **disjoint**, if they have no points in common: $A \cap B = \emptyset$. This is of course consistent with our earlier definition of incompatibility. The null set is considered to be a member of any set S. The set S is sometimes termed the universal event or the certain event. If S contains n elements, it contains 2^n subsets including $\emptyset$ and S itself (each element is either 'in' or 'out' of a subset – two possibilities for each of the n elements).

2.3   Assessing probabilities; example: the sinking of the Titanic

We have gone to some trouble to point out that the assignment of a probability is an expression of an individual's judgement. Any set of coherent probabilities represents a possible opinion. Differences in the information available to two persons making an assessment of the probability of an event, or in the interpretation thereof, will lead quite naturally to differences of 'probabilistic opinion'. Yet there are many cases where we would agree on values of probability. For example, we might agree that the probability of heads in the 'honest' toss of an 'honest' coin is $1/2$. Or consider the urn model introduced in Section 1.4 and below under 'Necessary assignments'.

A general point should be made with regard to the question of personal judgement in assigning probabilities. In designing an engineering project, or for making planning decisions, a consensus is needed in order to set design values or to decide on parameters. The value of a 100-year design flood or maximum traffic may have to be specified in the design of a spillway or a network. As a result, it is necessary to share information and to discuss its interpretation so as to achieve the required consensus amongst engineers and decision-makers. The assignment of probabilities is assisted by techniques which are classified into necessary methods and relative frequencies. Thought and good modelling are the keys to general situations that might appear intractable initially.

2.3.1   Necessary assignments

The term 'necessary' is intended here to apply to cases where most human opinion would agree upon an assignment. (At the same time, one always has the right to disagree!) An example would be an urn containing two balls identical in size, weight, and texture; one is red and the other is pink. If one removes a ball without looking, most of us would agree that the probability of either red or pink is $1/2$. Other situations where we judge that an assignment necessarily follows are coin-tossing, throwing dice, roulette wheels, card games and games of chance in general. These can also be termed 'classical' assignments, in the sense that they were studied by the originators of the theory of probability in the sixteenth and seventeenth centuries.

Other 'necessary' concepts include the use of geometrical configurations; for example the probability that a pin dropped 'at random' in a unit square containing a circle of unit diameter lands in the circle is $\pi/4$. We interpret 'at random' as meaning that each point in the unit square is judged to be an 'equally likely' landing place for the pin. Geometric probability is used in the example of the Titanic in the sequel.

The case of 'equally likely' events is sometimes analysed using the 'principle of insufficient reason'. For example, in taking a card from a pack of 52, we can think of no reason to favour one card over another. The same argument applies to the coloured balls in urns and the other cases above. In reality distinct events (red or pink balls, heads or tails, the faces of a die, the different cards from a pack and so on) must be different, otherwise we would not be able to distinguish them. We **judge** that the factors which distinguish the events are unimportant in the assignment of probabilities. A further version of the 'necessary' method should be mentioned here; this involves the use of entropy maximization (Chapter 6).

2.3.2   Relative frequencies; exchangeability

The use of frequencies derives from 'repeated trials of the same experiment'. The experiments may consist of tossing a coin, throwing a pair of dice, or drawing balls from an urn. These experiments all involve cases where the probability might be considered to be 'well defined' by the idealization of the experiment. In some ways such an idealization can be thought of as a conceptual experiment but it can come close to reality even in practical situations (for example, the 'Titanic' problem below).

There are very different situations at two limits of knowledge (and many in between) that have the same basic idealization. The first limit corresponds to the case where the probability is thought to be 'known' or 'well defined'. For example, we would expect that in tossing a coin heads would occur about 50% of the time, that a particular face would appear $1/6$ of the time in throwing a die, or that a pink ball would occur $5/7$ of the time in drawings with replacement after a draw from an urn with five pink and two red balls. The other limit occurs when we have, again, repeated trials of the same 'experiment', but the probabilities are not well defined *a priori*. For example, we might be faced with an urn containing red and pink balls but otherwise of unknown composition. Or we might construct a transistor using a new compound of which we have little experience. We have little idea how the device will perform. Or we are designing a new space shuttle and do not know the probability of success. The difference from the 'first limit' above is that we do not now have a 'well-defined' probability, and we wish to use the results of experiments to **infer** the probabilities. The problem then is one of inference. A full treatment is deferred to Chapter 8, and is based on Bayes' theorem (Section 2.4).

We would expect that in the first limit noted, repeated trials of the same experiment would 'in the long run' give us relative frequency ratios that are close to our 'well-defined' values. We would of course become suspicious on obtaining the 100th, 1000th, . . . head in the same number of tosses of a supposedly honest coin! Such suspicions can be analysed probabilistically: see Example 2.7 (in Section 2.4.1) and Chapter 8.

We have dealt with two limiting cases of 'repetitions of the same experiment'. The fact that we obtain different results in each trial of the experiment such as different numbers in tosses of a die shows that the trials are not the same. The point is that we cannot distinguish them. What is common to each set of experiments whether we consider the probability 'well defined' or not? The simplest interpretation is that the **order of the results is not important**. In other words if we permute the set of results, for example, convert {heads, tails, heads} to {heads, heads, tails}, and judge these two sets of results to be equivalent, to have the same probability of occurrence, then the order is not important. This is the essence of the judgement of **exchangeability.** We can thus have exchangeable events, exchangeable sets of results of an experiment and we sometimes refer to exchangeable populations. Indeed, Feller (1968) defines a 'population' as an aggregate of elements without regard to their order.

2.3.3   Summarizing comments on the assignment of probability

Our fundamental definition of probability is based on our behaviour in placing a fair bet. This guides our thinking and we should then use all techniques at hand to assess the value of a probability. Probabilities are not relative frequencies but these may be used to evaluate probabilities. We therefore differ from some traditional theory in which probability is defined as a relative frequency. The following example shows how a probability is assessed in a case where there is not a long series of trials that could be conducted, except in a simulation. The method is based on geometric arguments. Our definition of probabilities pushes us naturally in the right direction.

2.3.4   Example: the Titanic

On April 15, 1912, just after 2:00 a.m., the Titanic sank with a loss of more than 1500 lives, having struck an iceberg at 11:40 p.m. the previous night. Icebergs and sea ice continue to present a hazard to shipping and other offshore activities in Eastern Canada. Figure 2.6 shows the region of interest and the position where the Titanic sank as well as the location of some Canadian offshore activities. The Hibernia, Terra Nova and White Rose oil fields are being developed and other fields are under consideration. These operations have to take into account the presence of icebergs. The 13 km bridge across the Northumberland Strait joining Prince Edward Island to New Brunswick has to resist forces due to moving sea ice. Methods have been developed, especially during the last two decades, to account for ice loads on offshore structures and vessels. These take into account both the probability of occurrence of the important interactions with ice, as well as the associated mechanics.

Since the Titanic sank, there have been detailed surveys of the area, including flights over it by the International Ice Patrol. Figure 2.6 shows that in fact the location of the sinking of the Titanic is generally far south of the usual southern limit of icebergs – although the location and number of icebergs varies considerably from year to year. As suggested above, a key aspect of the design of a vessel to resist forces from icebergs is the probability of collision. We shall calculate this value using methods that give convincing results based on geometrical arguments. In reality, information on positions of icebergs from ice patrol flights, from other

The Titanic

vessels, and from radar plays an important part in the safety of vessels. It is well known that the Titanic actually received repeated warnings of the presence of ice, including icebergs, directly ahead of the course of the ship.

The situation that we shall consider is that of a single interaction between a ship and an iceberg. For our problem, we shall consider an area bounded by two degrees of latitude and of longitude. This curved surface can be closely approximated as a rectangle, as illustrated in Figure 2.7(a). The dimensions of the rectangle are 111.12 km in the north–south direction and $111.12\cos(\theta + 1/2)°$ in the east–west direction, where θ is the latitude of the southern boundary. This will be termed a 'degree-rectangle'. The path of the vessel is from A to B. Head-on collisions will be considered, as illustrated in Figure 2.7(b). By this, it is meant that the vessel collides with the iceberg by virtue of its forward motion. There is, in fact, a small probability that a collision with the side of the vessel will take place as a result of the forward motion of the iceberg, but this is quite small compared to the probability of the situation shown in Figure 2.7(b), and will be ignored in the present analysis.

The value L_B is the longest waterline length of the iceberg, usually measured; its underwater dimension will generally be larger. Let $\bar{L}$ denote the dimension presented to the vessel, at right angles to its approach direction. For the present purpose, we can think of an average underwater

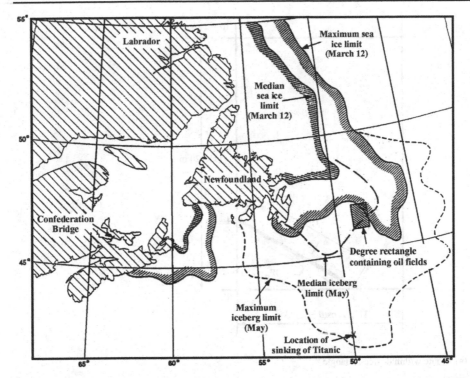

Figure 2.6 Geographical area of concern: iceberg data (1946–1971); sea ice data (1963–1973) based on Markham (1980)

dimension, or, if one wishes, a spherical berg with a diameter equal to $\bar{L}$. Then the collision will take place if the position of the iceberg is such that either edge of the berg intrudes upon the width W_s of the ship. We assume that the draft of the vessel is sufficient to contact the berg at its widest underwater dimension.

The position of the iceberg is defined for present purposes by the position of its centre of mass. If we assume that the iceberg is anywhere within the degree-rectangle (with equal probability), then the collision will take place if the centre of the iceberg is in the shaded area shown in Figure 2.7(a), i.e. within an area of $(d) \cdot (W_s + \bar{L})$, where W_s is the width (beam) of the vessel. One can think of the iceberg having been removed, and dropped back in. If the centre falls within a distance $\bar{L}/2$ of either side of the vessel, then a collision takes place.

Let $C =$ the event of a collision. Then, for a vessel traversing the distance 'd' shown in Figure 2.7(a),

$$\mathbf{Pr}(C) = \frac{(W_s + \bar{L})d}{A} \tag{2.18}$$

where $A =$ area of the degree-rectangle. For a vessel travelling from east to west, and for the degree-rectangle in which the Titanic sank, $d = 83.2$ km. Using $W_s = 28$ m (92 ft beam for

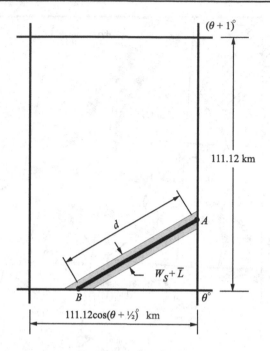

(a) Route within degree-rectangle

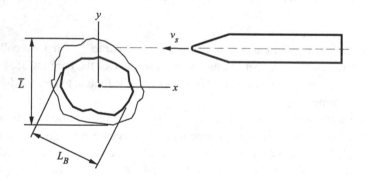

(b) Geometry of approaching collision

Figure 2.7   Vessel collision with iceberg

the Titanic), $\bar{L} = 110$ m, a reasonable average for bergs greater than $10\,000$ tonnes, then

$\mathbf{Pr}(C) = 1.24 \times 10^{-3}$.

Our calculation has involved the use of 'necessary' methods in the geometrical argument. The assumption of a single iceberg in a degree-rectangle represents a 'frequency' of occurrence. Yet the overall direction given to us by the definition of a probability as a fair bet pushes us in the right direction, that is, to use whatever tools are appropriate. A series of trials of 'Titanics' sent into the icefield to obtain a frequency can only exist in a plausible numerical or mental experiment.

The calculation has been completed for a single iceberg in the degree-rectangle. For this case, or for a few icebergs in the degree-rectangle, the probability of collision is a relatively small number, but nevertheless represents an unacceptable risk, especially given the number of people at risk (see the section on risks to human life in Chapter 10). The calculation can also be carried out to yield the probabilities per unit time. It should also be mentioned that we have not taken into account any measures taken by the vessel to avoid a collision. In fact, there were orders on board the Titanic when the iceberg was sighted to turn the wheel 'hard-a-starboard' (which would cause the ship to turn to port) and the engines were stopped and then reversed. The ship did veer to port; it might be argued that this resulted in the sinking of the vessel by increasing the number of watertight compartments breached. In any event, avoidance can be built into the model of collision.

Since the sinking of the Titanic, maps have been produced by the International Ice Patrol, showing the position of icebergs and sea ice. Figure 2.8 shows a map on the same day as the Titanic sinking (but a different year). It can be seen that icebergs drifted quite far south (compare Figure 2.6) but not as far as in the year 1912. The maps represent very useful information on ice conditions and their variation from year to year.

We have emphasized that probability is a function of information. This will be treated in a formal sense in Section 2.4, in which the methods for updating probabilities upon receipt of new information will be introduced. In closing, it is interesting to emphasize that the Titanic did indeed receive messages regarding ice and icebergs; for example, at 1:30 p.m. on the day that the Titanic struck the iceberg, the captain had seen a message from the steamer *Baltic*: 'Greek steamer *Athinai* reports passing icebergs and large quantities of field ice today in latitude 41°51′ N, longitude 49°52′ W ... Wish you and *Titanic* all success'.

This latitude was several minutes of arc north of the position where the Titanic was found on the seafloor. Both the ice and the vessel – after foundering – would drift, so that the exact location of the collision is not known. Interesting books on the Titanic tragedy are *A Night to Remember* by Walter Lord and *The Discovery of the Titanic* by Robert Ballard.

2.4   Conditional probability and Bayes' theorem

All probabilities are conditional upon the information in our possession when we make the assignment. The seamen on the bridge of the Titanic must have known of the presence of sea ice and icebergs ahead of the vessel. There must have been in their minds some consideration as to the chance of encountering ice. At the same time, there was a misplaced faith in the strength of the vessel: plausible arguments can be made as to the construction of the vessel – riveted plates – and as to the nature of ice loads – highly concentrated in small zones – that

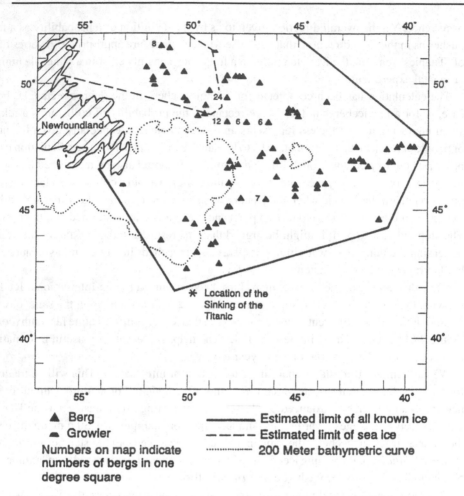

Figure 2.8   Iceberg distribution based on International Ice Patrol map, April 15, 1989

lead to the conclusion that the vessel was vulnerable to damage in a collision with an iceberg. This is, of course, knowledge by hindsight. We are emphasizing how judgement of probability depends on our information at the point in time when we make our judgement.

Let us consider a population that is evenly divided into blue-eyed and brown-eyed people. If a person is chosen at random, we consider the proposition:

A = *'the person's right eye is brown'*.

We would accept that $\mathbf{Pr}(A) = 0.5$. Let us consider the additional proposition:

B = *'the person's left eye is blue'*.

We now consider the conditional probability

$\mathbf{Pr}(A|B)$,

which is understood as the probability of A **conditional on** B or **given that** B is true.

In the present problem, our probability that the person in question has a brown right eye would decrease considerably upon receiving the information 'B', that the left eye is blue. There are a small number of people in the population with eyes of different colour, so that $\mathbf{Pr}(A|B)$ would decrease to some very small value. This illustrates the fact that information is all-important in the assignment of probability. (Probabilistic reasoning and coincidence are taken to a bizarre extreme in Eugène Ionesco's play 'The Bald Soprano'. Both Donald and Elizabeth have a daughter Alice, pretty and two years old with a white eye and a red eye. But Donald's child has a white right eye and a red left eye, whereas Elizabeth's child has a red right eye and a white left eye . . .).

The Bald Soprano

In a similar way, we could let

$A =$ '*the wave height at an offshore location is ≤ 2 m*'.

The probability of this event is quite large; even on the Grand Banks it is about 25%. We are now given the information

B = '*the wind speed has been 50 knots for the last three days*'.

We would clearly want to reduce $\mathbf{Pr}(A|B)$ from the value for $\mathbf{Pr}(A)$ as a result of this information. In fact, information is used in all evaluations of probability. Strictly, one should list the information used and state for example $\mathbf{Pr}$(event E|listed information which is available). This is usually condensed to the notation $\mathbf{Pr}(E)$, with the convention that the background information is understood by those involved in the evaluation. If we obtain **new information** H, we write $\mathbf{Pr}(E|H)$.

Introductory urn models; Raiffa's urns

Urn models, introduced in Section 1.4, provide good illustrations of conditional probability. Figure 2.9 shows an urn containing two balls, labelled '0' and '1'. It is also traditional to denote the drawing of one colour – arbitrarily – as a success. Let us denote the drawing of a red ball as a success: this will correspond to the ball labelled '1' in Figure 2.9. The possible drawings of balls from the urn are illustrated in the 'random walk' shown in the lower part of Figure 2.9 for the case where the balls are not replaced after drawing. There can then be only two drawings. The horizontal axis represents the number of 'trials' or drawings from the urn and the successes and failures are shown on the sloping axes. The vertical axis shows the gain or loss if a reward of $+1$ is paid for each success and -1 for each failure. It should be emphasized that the terminology of 'successes' and 'failures' is conventional in probability theory and the terms can be assigned at our convenience. A 'success' could be 'failure of a system' if we wish in an application relating (for example) to the loading of a structural or electrical system.

Example 2.1 If there is no replacement of balls taken out, the probabilities of successive drawings from an urn will change. For the urn of Figure 2.9, the probability of red is 0.5 on the first draw, and then changes to 0 or 1 on the second, depending on whether the first draw is red or pink respectively. This is stochastic dependence; knowledge of the result of the first draw affects our assessment of the probability of the second (more detail in Section 2.5). If we replace the first ball after drawing and shuffle, our probability of red on the second draw remains at 0.5; knowledge of the results of the first draw does not affect our probabilistic assessment. This is stochastic independence (see Section 2.5).

Raiffa (1968) used two urns, shown in Figure 2.10, in his introduction to decision theory. These are denoted θ_1 and θ_2, and each contains ten balls. Urn θ_1 contains four red and six pink balls, while θ_2 contains nine that are red and one pink. The figure also shows the random walk diagrams for drawing without replacement.

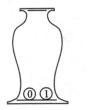

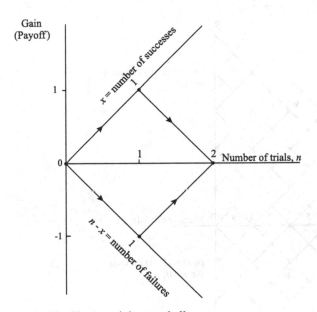

Figure 2.9 Urn containing two balls

Example 2.2 Consider the θ_1 urn and drawing without replacement. If we are told that the first draw yields a red, the probability of drawing a pink on the second is 6/9; if the first is a pink, this becomes 5/9. The probability of drawing a pink on the second draw is 6/10 if no other information is provided. One can think of this as (6/9) weighted by (4/10) – probability of red on first draw – plus (5/9) weighted by (6/10) – the probability of pink on the first draw:

$$\frac{4}{10} \cdot \frac{6}{9} + \frac{6}{10} \cdot \frac{5}{9} = \frac{6}{10}.$$

Example 2.3 We are presented with an urn, and we do not know whether it is of type θ_1 or θ_2. In the case of certain experiments, one can move to a position of certainty: for example, if one were to draw five or more red balls from the urn we would know that it is of type θ_2. This results from the fact that there are only four red balls in the θ_1 urn.

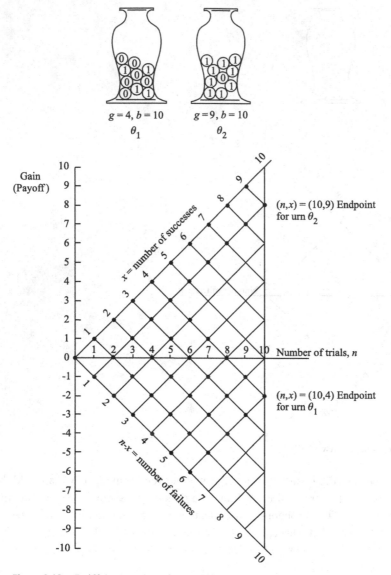

Figure 2.10   Raiffa's urns (based on Raiffa, 1968) with b balls, g of which are red

2.4.1   Bayes' theorem

Introduction with urns

In Raiffa's problem of Figure 2.10, there are 800 urns of type θ_1 and 200 of type θ_2 and one is selected at random and presented to us. The key question is: which type of urn has been presented? The experiment of sampling one ball, or more, from the urn is permitted.

One doesn't know which type of urn is being sampled, but one has the prior information of a probability of 0.80 for θ_1 and 0.20 for θ_2. This problem encapsulates the subject of probabilistic inference. The type of urn (θ_1 or θ_2) represents the uncertain entities which are often referred to as 'states of nature' (traffic in a network, number of flaws in a weld, type of urn, ...). One uses an experiment (measurements, drawing from the urn, ...) to improve one's probabilistic estimate of the state of nature.

The urns can be used to provide an intuitive introduction to Bayes' theorem. We sample a single ball from the urn and find that it is red. What are our amended probabilities of θ_1 and θ_2? We know that there are ten balls per urn, and there are therefore 10 000 balls in all since there are 1000 urns. Now there are $4 \cdot 800 = 3200$ red balls in the θ_1 urns and $9 \cdot 200 = 1800$ red balls in the θ_2 urns. Consequently there are 5000 red balls in all, half of the total of 10 000. The probability of a θ_1 urn given the fact that a red ball has been drawn is then $(3200)/(5000) = 0.64$, and of a θ_2 urn is $(1800)/(5000) = 0.36$. We have changed our probabilities in the light of the new information (red ball). In effect we have changed our space of consideration from urns to balls.

Formal introduction to Bayes' theorem

To focus ideas, let us consider the problem of non-destructive testing of flaws such as cracks or other imperfections in a weld.[6] A length of weld in the construction of an important steel structure – a ship, an offshore rig, a pressure vessel, a major bridge or building component, a gas pipeline – is subjected to testing by a device. Typically this might be based on ultrasonic or electromagnetic methods. The testing device has been calibrated by taking weld samples, some with flaws and some without. We shall use the following notation.

F = 'flaw present in weld'.
$\tilde{F}$ = 'flaw not present in weld'.
D = 'flaw detected by device'.
$\tilde{D}$ = 'flaw not detected by device'.

The machine is tested in the simulated field conditions to see whether or not it can detect the flaw. Repeated checks will yield information that can be used in assessing probabilities. This calibration will yield the following information:

(1) $\mathbf{Pr}(D|F)$, and
(2) $\mathbf{Pr}(D|\tilde{F})$.

If the device were perfect, these values would be 1 and 0, respectively. In general, they will reflect the effectiveness of the testing machine. Suppose that, for the specimens with a flaw, 80% of a long series of readings indicate the presence of a flaw, whereas 30% indicate this when

---

[6] The importance of fracture in engineering is highlighted by disasters such as the Liberty ships, the Comet aircraft and the offshore rig 'Alexander Kielland'.

a flaw is absent. A reasonable conclusion is that values of $\mathbf{Pr}(D|F) = 0.8$ and $\mathbf{Pr}(D|\tilde{F}) = 0.3$ are appropriate.

Probabilities must normalize conditional upon the same information; thus

$$\mathbf{Pr}(D|F) + \mathbf{Pr}(\tilde{D}|F) = 1, \tag{2.19}$$

and

$$\mathbf{Pr}(D|\tilde{F}) + \mathbf{Pr}(\tilde{D}|\tilde{F}) = 1. \tag{2.20}$$

Therefore, $\mathbf{Pr}(\tilde{D}|F) = 0.2$ and $\mathbf{Pr}(\tilde{D}|\tilde{F}) = 0.7$, using the values above. Equations (2.19) and (2.20) follow common sense in interpreting the calibration results. If we had tested 100 samples with a flaw present, and obtained 80 results in which the device indicated a positive result, then there would be 20 results in which the device did not. More generally, they are coherent estimates of probability.

The flaw is either present (F) or absent ($\tilde{F}$). The test device either detects (D) or does not detect ($\tilde{D}$) the flaw. We can interpret this in terms of the Venn diagram of Figure 2.5(c). The individual sample points could consist of possible readings on the device, combined with possible flaw sizes. These could be plotted, for example, on orthogonal axes in a cartesian system. The events F and D are clearly defined in terms of (say) a critical flaw size above which F is considered to have occurred, and a range of readings on the ultrasonic device that indicate the occurrence of D. We can think of the probabilities as being weights attached to the areas, or more generally attached to points within the areas of Figure 2.5(c). Events F and D would correspond to points within the areas F and D shown; points in the overlapping area correspond to event DF.

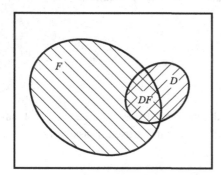

Figure 2.5(c)   Logical products (intersection) and sum (union). The total hatched area is $D \vee F$; the area with double hatching is $D \wedge F$ (or DF)

Let us consider the event $D|F$. If event F is true, the only weights of probability that matter are those within F of Figure 2.5(c); those corresponding to $\tilde{F}$ can be ignored. The weights corresponding to $D|F$ must be in the area (set) DF. It seems natural and consistent to write $\mathbf{Pr}(D|F)$ as the ratio of weights in DF and in F:

$$\mathbf{Pr}(D|F) = \frac{\mathbf{Pr}(DF)}{\mathbf{Pr}(F)}. \tag{2.21}$$

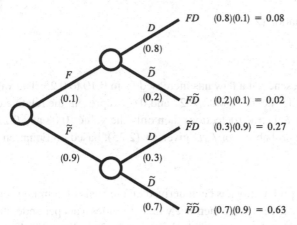

Figure 2.11 Extensive form: Bayes' calculation tree

In deriving Equation (2.21), we considered the probability of detection given that a flaw is present in the weld, $\mathbf{Pr}(D|F)$. This is a useful 'calibration'. In the actual field use of the device, we are interested instead in the **probability that a flaw is present given a positive indication by the test device, $\mathbf{Pr}(F|D)$**. We shall show how to derive this using the calibration. By reasoning analogous to that used in deriving Equation (2.21), we could equally well have deduced that

$$\mathbf{Pr}(F|D) = \frac{\mathbf{Pr}(FD)}{\mathbf{Pr}(D)}. \tag{2.22}$$

Using this result, and Equation (2.21), we have the symmetric form for the probability of $DF(\equiv FD)$:

$$\mathbf{Pr}(DF) = \mathbf{Pr}(D|F)\mathbf{Pr}(F) = \mathbf{Pr}(F|D)\mathbf{Pr}(D),$$

from which we may deduce that

$$\mathbf{Pr}(F|D) = \frac{\mathbf{Pr}(D|F)\mathbf{Pr}(F)}{\mathbf{Pr}(D)}. \tag{2.23}$$

This is **Bayes' theorem**. It is the essential means of changing one's opinion in the light of new evidence. It converts the prior assessment, $\mathbf{Pr}(F)$, into the so-called 'posterior' assessment $\mathbf{Pr}(F|D)$. It is natural that we should be prepared to alter our opinion in the light of new facts, and Bayes' theorem tells us how to do this.

Example 2.4 We now complete a numerical computation to illustrate Equation (2.23). It is useful to sketch the possible results in the 'extensive form' of Figure 2.11. We also refer to this as a 'Bayes calculation tree' as it assists in the use of Bayes' theorem. We are given that $\mathbf{Pr}(D|F) = 0.8$ and $\mathbf{Pr}(D|\tilde{F}) = 0.3$. Let us assume that, in the absence of any result of tests using our device, the prior probability $\mathbf{Pr}(F)$ is equal to 0.10. We can first deduce that

$$\mathbf{Pr}(D) = \mathbf{Pr}(D|F)\mathbf{Pr}(F) + \mathbf{Pr}(D|\tilde{F})\mathbf{Pr}(\tilde{F}) \tag{2.24}$$
$$= (0.8)(0.10) + (0.3)(0.90) = 0.35.$$

Then, by Bayes' theorem, Equation (2.23),

$$\mathbf{Pr}(F|D) = \frac{0.08}{0.35} \cong 0.23. \tag{2.25}$$

Thus, our probability of the presence of a flaw has increased from 0.10 to 0.23. The values on the right hand side of Figure 2.11, i.e. 0.08, 0.02, 0.27 and 0.63, represent the relative odds of the outcomes. If we know that D is true (detection) then only the values 0.08 and 0.27 (odds of 8 : 27) are relevant with the probability of $F|D$ given by (2.25), based on Equation (2.4).

Coherence

The rule (2.21) for conditional probability has been derived on the basis of common sense and consistency. More precisely, it derives from coherence which provides a deeper understanding. We return to the concept of probability as a 'fair bet'. Figure 2.12(a) shows all the possible outcomes given uncertainty regarding both F and D. We place a bet of $(p_1 s)$ on DF, and receive 's' if it occurs. We now consider this same bet as a two-stage process, as shown in Figure 2.12(b). Here, we first bet $(p_2 t)$ on F and receive t if it is true, in other words, given that there is a flaw in the weld. Thus, p_2 represents $\mathbf{Pr}(F)$. We now reinvest this amount in a bet on detection D **in the knowledge that there is a flaw present**. This would correspond to our calibration on a weld with a flaw present. The amount t is any small amount of money so that we can set the amount 't' in order that we receive the reward 's' if D occurs and are content to accept the series of two bets.

Thus, we may write $t = p_3 s$ where $p_3 = \mathbf{Pr}(D|F)$. Now the rewards for the two betting scenarios of Figure 2.12(a) and (b) are identical so that we would expect that

$$p_1 s = p_2 t = p_2 p_3 s, \text{ i.e.}$$

$$p_1 = p_2 p_3, \text{ or}$$

$$\mathbf{Pr}(DF) = \mathbf{Pr}(D|F)\mathbf{Pr}(F),$$

which is identical to Equation (2.21).

Failure to follow this rule for conditional probabilities will result in the possibility of a combination of bets that one is sure to lose – incoherence, again! Let us assume that one assigns probabilities such that

$$\mathbf{Pr}(D|F)\mathbf{Pr}(F) > \mathbf{Pr}(DF), \text{ or } p_2 p_3 > p_1. \tag{2.26}$$

The assignor is indifferent between the status quo and the bets shown in Figure 2.12(a) and (b) and would therefore exchange either bet, buying or selling, for the status quo. In order to make a sum of money for sure, we merely have to sell lottery (b) and to buy (a). The net payout is zero, no matter which outcome occurs; if DF occurs we receive 's' for lottery (a) and have to pay out the same amount for lottery (b). Other outcomes result in no exchange of money. Yet we have received $(p_2 p_3 s)$ for (b) and paid $(p_1 s)$ for (a). We have therefore gained $(p_2 p_3 - p_1)s$ by the inequality (2.26) above. For other incoherent assignments, the reader might wish to work out the appropriate combinations of bets.

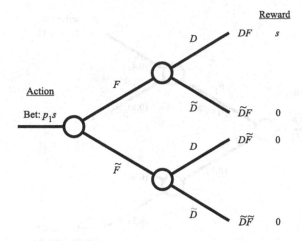

(a) Bet on DF; $p_1 = \mathbf{Pr}(DF)$

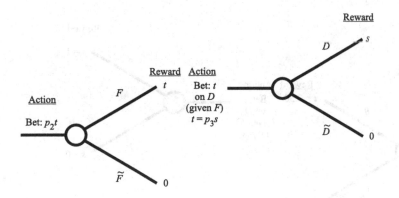

(b) Successive bets on F and then D given F. The amount 't' is adjusted to give the final reward of 's', if D occurs. The value $p_2 = \mathbf{Pr}(F)$ and $p_3 = \mathbf{Pr}(D|F)$

Figure 2.12   Coherence and conditional probability

Example 2.5 We can combine our thoughts on bets, odds and urns by considering Raiffa's urns of Figure 2.10. Figure 2.13(a) shows reasonable odds on the outcome $(\theta_1 R)$, as against $(\theta_2 R)$, given the probabilities shown, whereas (b) shows an incoherent set. If the assignor agrees to these two system of bets, we could buy the first and sell the second, making 8 cents for sure.

Extending the conversation

We can expand on the way of writing $\mathbf{Pr}(D)$ given by Equation (2.24) to the case where there are several Fs. Rather than a single event F in Equation (2.23), we consider a member, say

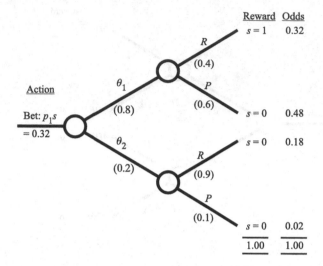

(a) Odds on $\theta_1 R$ and other events in accordance with the probabilities (extensive form)

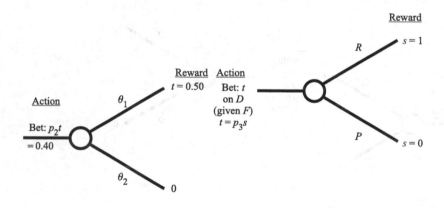

(b) Incoherent bet on $R|\theta_1$

Figure 2.13   Bets regarding Raiffa's urns

F_i, of the partition $\{F_1, F_2, \ldots, F_i, \ldots, F_n\}$, where the F_i represent n possible crack lengths of a flaw in our weld. Each F_i corresponds now to a specific interval of crack lengths (e.g. 1 to 10 mm, 11 to 20 mm, and so on). Then, D might be the information 'the reading on the ultrasonic testing device is x (specified value)'. We have obtained enough information using the calibration process to assess to our satisfaction the probabilities $\mathbf{Pr}(D|F_i)$ for $i = 1, 2, \ldots, n$. Figure 2.14 illustrates the situation.

It often suits us to write D as

$$D = DF_1 + DF_2 + \cdots + DF_n. \tag{2.27}$$

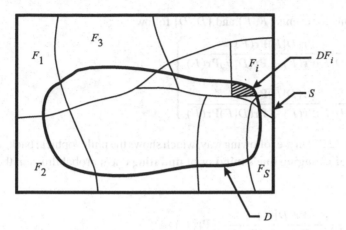

Figure 2.14　Venn diagram for Bayes' theorem

There must be no possible occurrence of D outside the partition $\{F_1, \ldots, F_n\}$. Equation (2.27) is used to write

$$\mathbf{Pr}(D) = \mathbf{Pr}(DF_1) + \mathbf{Pr}(DF_2) + \cdots + \mathbf{Pr}(DF_n), \tag{2.28}$$

using the same arguments, via coherence – Equation (2.10) – as before. Equation (2.21) shows that we can write, for any event F_i in the partition,

$$\mathbf{Pr}(DF_i) = \mathbf{Pr}(D|F_i)\mathbf{Pr}(F_i), \tag{2.29}$$

and therefore that

$$\mathbf{Pr}(D) = \sum_{i=1}^{n} \mathbf{Pr}(D|F_i)\mathbf{Pr}(F_i). \tag{2.30}$$

This equation generalizes (2.24). It has been termed 'extending the conversation' by Myron Tribus (1969). In the particular case of a partition consisting of two mutually exclusive events $\{F, \tilde{F}\}$,

$$\mathbf{Pr}(D) = \mathbf{Pr}(D|F)\mathbf{Pr}(F) + \mathbf{Pr}(D|\tilde{F})\mathbf{Pr}(\tilde{F}), \tag{2.31}$$

which recaptures Equation (2.24). We used this in the example above in deducing $\mathbf{Pr}(F|D) = 0.23$. Substituting Equation (2.30) into Bayes' equation (2.23), we have the following useful form:

$$\mathbf{Pr}(F_i|D) = \frac{\mathbf{Pr}(D|F_i)\mathbf{Pr}(F_i)}{\sum_{j=1}^{n} \mathbf{Pr}(D|F_j)\mathbf{Pr}(F_j)}, \tag{2.32}$$

where we have used 'j' in the denominator to avoid confusion with 'i' in the numerator. The following simplifications for events $\{F, \tilde{F}\}$ and $\{D, \tilde{D}\}$ follow:

$$
\left.\begin{array}{l}
\mathbf{Pr}(F|D) = \dfrac{\mathbf{Pr}(D|F)\mathbf{Pr}(F)}{\mathbf{Pr}(D|F)\mathbf{Pr}(F) + \mathbf{Pr}(D|\tilde{F})\mathbf{Pr}(\tilde{F})} \\[4mm]
\text{and} \\[2mm]
\mathbf{Pr}(\tilde{F}|D) = \dfrac{\mathbf{Pr}(D|\tilde{F})\mathbf{Pr}(\tilde{F})}{\mathbf{Pr}(D|F)\mathbf{Pr}(F) + \mathbf{Pr}(D|\tilde{F})\mathbf{Pr}(\tilde{F})}
\end{array}\right\} \tag{2.33}
$$

We can cast Equation (2.32) in the following way, which shows the philosophical background to the equation – **that of changing one's mind** or of **updating** one's probabilities in the face of new evidence.

$$
\mathbf{Pr}(F_i|D) = \frac{\mathbf{Pr}(D|F_i)}{\sum_{j=1}^{n} \mathbf{Pr}(D|F_j)\mathbf{Pr}(F_j)} \cdot \mathbf{Pr}(F_i)
$$

Revised (posterior) = (Updating factor) $\quad \cdot$ (Initial (prior)

probability of F_i $\qquad\qquad\qquad\qquad$ probability of F_i)

given information D $\hfill$ (2.34)

The denominator of the term denoted the 'updating factor' is the summation which gives $\mathbf{Pr}(D)$; it is, in fact, a normalizing constant which ensures that

$$
\sum_i \mathbf{Pr}(F_i|D) = 1.
$$

It is natural to consider the term $\mathbf{Pr}(D|F_i)$ as the probability of the new information, or evidence, D, as a function of all the states of nature F_i. For example, we have obtained a certain reading on our ultrasonic testing device, denoted D. The first thing that occurs to us is: how likely is this reading D as a function of the possible flaw lengths $F_1, F_2, \ldots, F_n$? Given evidence in a trial that a murderer is lefthanded with 90% probability, we would immediately investigate whether our suspects are lefthanded or not. (This example is discussed in the next subsection). The term $\mathbf{Pr}(D|F_i)$ considered as a function of the F_i has come to be known as the **likelihood**, denoted ℓ. We can then generally write Bayes' theorem as:

(Revised probability of event) $\propto$ (likelihood) $\cdot$ (initial probability of event),

or in the case above

$\mathbf{Pr}(F_i|D) \propto \ell \cdot \mathbf{Pr}(F_i)$. $\hfill$ (2.35)

The likelihood ensures that the revised probabilities of each event are changed in proportion to the probability of the evidence on the condition that the event is, in fact, true. Successive applications of Bayes' theorem lead generally to a 'sharpening' of probability distributions. The more relevant information we obtain and incorporate, the closer we move to certainty.

Example 2.6 *Communication system.* In any communication system, there is no assurance that the signal transmitted will be received correctly. In the system under consideration using smoke signals, three symbols are transmitted. These are $\{t_1, t_2, t_3\}$. The corresponding symbols $\{r_1, r_2, r_3\}$ are received. Past experience shows that $\{t_1, t_2, t_3\}$ are transmitted with probabilities $\{0.2, 0.3, 0.5\}$ respectively. Trial runs using the system have yielded the following values of $\mathbf{Pr}\left(R = r_i | T = t_j\right)$:

	t_1	t_2	t_3
r_1	0.85	0.05	0.10
r_2	0.10	0.80	0.25
r_3	0.05	0.15	0.65
$\Sigma_i \, \mathbf{Pr}(R = r_i | T = t_j)$	1.00	1.00	1.00

The problem of determining the probabilities $\mathbf{Pr}\left(T = t_j | R = r_i\right)$ is at the centre of any communication system: given that we have received a particular symbol, what is the probability that the various symbols were sent? The analysis is presented in the tree of Figure 2.15. Given

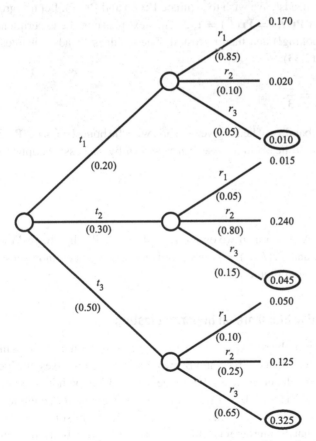

Figure 2.15   Extensive form for communication problem. Ovals indicate relative odds given r_3

that r_3, for example, was received, we can work out the probabilities of $\{t_1, t_2, t_3\}$ having been sent. The relative odds are enclosed in the ovals of Figure 2.15, and we find (rounded to three decimal places) that

$$\mathbf{Pr}\,(T = t_1 | R = r_3) = \frac{0.010}{0.010 + 0.045 + 0.325} = 0.026,$$

$$\mathbf{Pr}\,(T = t_2 | R = r_3) = \frac{0.045}{0.010 + 0.045 + 0.325} = 0.118, \text{ and}$$

$$\mathbf{Pr}\,(T = t_3 | R = r_3) = \frac{0.325}{0.010 + 0.045 + 0.325} = 0.855.$$

These can determined directly from Equation (2.32) or (2.34).

Example 2.7 Consider the case of two boxes which are indistinguishable in appearance: one box contains a two-headed coin, while the other contains an 'honest' coin. The two boxes are mixed up and one is presented to us. Let E be the event 'the box in our hands is the one containing the coin with two heads'. We wish to estimate $\mathbf{Pr}(E)$ and $\mathbf{Pr}(\tilde{E})$. Let us agree that our state of mind is such that $\mathbf{Pr}(E) = \mathbf{Pr}(\tilde{E}) = 1/2$. The next phase in the experiment is to remove the coin (without looking!) and then to toss it. The result is 'heads', denoted 'H'. Using the first of Equations (2.33):

$$\mathbf{Pr}(E|H) = \frac{(1)(\frac{1}{2})}{(1)(\frac{1}{2}) + (\frac{1}{2})(\frac{1}{2})} = \frac{2}{3},$$

where we have judged the probability of heads given the box with the honest coin, i.e. $\mathbf{Pr}(H|\tilde{E})$, to be $1/2$. If we obtain $H_n = $ 'n heads in n tosses', it is shown by successive application of Bayes' theorem that

$$\mathbf{Pr}(E|H_n) = \frac{1}{1 + (\frac{1}{2})^n} \tag{2.36}$$

which tends to 1 as $n \to \infty$. At any time if we obtain $\tilde{H} = T = $ 'a tail', then Bayes' theorem shows us that $\mathbf{Pr}(E|T) = 0$, and $\mathbf{Pr}(\tilde{E}|T) = 1$, in accordance with our common sense. The Bayes calculation tree is shown in Figure 2.16.

2.4.2   Bayes' theorem and everyday life: more on imperfect testing

The logic of Bayes' theorem is at the root of everyday thinking. Consider the case of a murder in which there are three suspects: Jones, Green and Brown. Evidence on motives, the location of the individuals at the time of the crime and other evidence, leads to the initial assignment $\mathbf{Pr}(J) = 0.1$, $\mathbf{Pr}(G) = 0.4$ and $\mathbf{Pr}(B) = 0.5$, where $J = $ '*Jones is guilty of the murder*', and $G = $ '*Green is guilty*', and $B = $ '*Brown is guilty*'. Now, a forensic scientist investigating the crime reaches the conclusion that the murderer is lefthanded with 90% probability, in contrast to

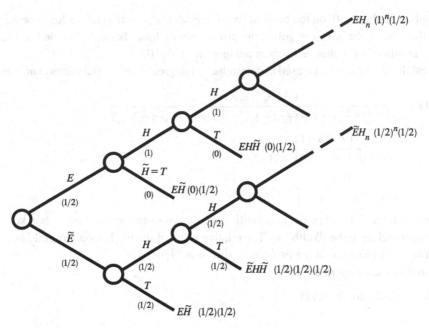

Figure 2.16   Extensive forms of coin tossing problem; $H_n =$ heads in n tosses

the fact that lefthandedness in the population is quite rare, estimated at one in twenty persons for the population of interest. All of this is now treated as background information in the problem. We cannot change our probability assignment until more information is obtained. Tests are done on Jones, Green and Brown and it is found that Jones is lefthanded while Green and Brown are righthanded. The new information on lefthandedness – after the prior assignment – is denoted 'I'. In other words,

I = 'Jones is lefthanded while Green and Brown are righthanded.'

We would as amateur sleuths feel a sudden surge of suspicion that Jones is guilty. Bayes' theorem offers the best route to change our probability assignment:

$$\mathbf{Pr}(J|I) = \frac{\mathbf{Pr}(I|J)\mathbf{Pr}(J)}{\mathbf{Pr}(I|J)\mathbf{Pr}(J) + \mathbf{Pr}(I|G)\mathbf{Pr}(G) + \mathbf{Pr}(I|B)\mathbf{Pr}(B)}. \tag{2.37}$$

Let us consider first $\mathbf{Pr}(I|J)$ on the numerator of the right hand side of (2.37). Given that Jones is guilty of the murder, what is the probability that Jones is lefthanded, and Green and Brown righthanded? The value is

$$\mathbf{Pr}(I|J) = (9/10) \cdot (19/20) \cdot (19/20).$$

Note that in this probabilistic statement we do not **know** that Jones is lefthanded: we are assessing the **probability** of I which includes the fact that Jones is lefthanded and also that Green and Brown are righthanded, all given J = 'Jones is guilty'. The probability of Jones being lefthanded is 9/10, on the basis of forensic evidence, conditional on Jones being guilty. Given that one of the others is guilty, the probability of Jones being lefthanded is 1/20, and the probability of a righthanded person being guilty is 1/10.

Bayes' theorem can now be used to determine our degree of belief that Jones *et al.* are guilty.

$$\mathbf{Pr}(J|I) = \frac{(\frac{9}{10})(\frac{19}{20})(\frac{19}{20})(\frac{1}{10})}{(\frac{9}{10})(\frac{19}{20})(\frac{19}{20})(\frac{1}{10}) + (\frac{1}{10})(\frac{1}{20})(\frac{19}{20})(\frac{4}{10}) + (\frac{1}{10})(\frac{19}{20})(\frac{1}{20})(\frac{5}{10})}$$

$$= \frac{(9)(19)(1)}{(9)(19)(1) + (1)(4) + (1)(5)}$$

$$= \frac{171 \cdot (1)}{171 \cdot (1) + 1 \cdot (4) + 1 \cdot (5)} = \frac{19}{20} = 0.95.$$

The factors related to the prior probabilities are shown in in the last line in brackets (); the other factors relate to the likelihood. The relative value of the likelihood terms, 171 : 1 : 1, is vital. The extensive form of the problem is shown in Figure 2.17.

By similar reasoning to the above,

$$\mathbf{Pr}(G|I) = 0.02 \text{ (strictly } 0.0\dot{2})$$

and

$$\mathbf{Pr}(B|I) = 0.03 \text{ (strictly } 0.02\dot{7}).$$

The analysis above shows that one can carry out a numerical calculation of our instinctive reaction to 'suspicious' circumstances. These are common in fiction: rare or unusual traits,

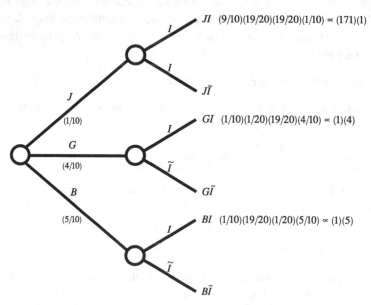

Figure 2.17   Extensive form of murderer problem and relative odds given '*I*'

scar-faced suspects, men with peg legs form the basis of many a mystery story. But such factors and their likelihood also are common in real life. There are subtleties in the presentation of the forensic evidence (see Lindley (1987) who gives an analysis of the problem above in some depth).

Another common problem is that of imperfect testing. We have already dealt with this problem in our weld-testing example with which we introduced Bayes' theorem. The use of imperfect testing devices is common in medical and other fields. One example is lie-detection, which according to Paulos (1988) is notoriously inaccurate. Let us take a medical example, also discussed by Paulos and others.

Example 2.8 The presence of a relatively rare disease (cancer, Aids, ...) is tested by a proce-dure that is 99% accurate. That is, a positive indication (P) will accurately reflect the presence of the disease (D) 99% of the time with 1% of false positive indications when the disease is in fact absent ($\tilde{D}$). Thus, $\mathbf{Pr}(P|D) = 0.99$ and $\mathbf{Pr}(\tilde{P}|D) = 0.01$. Similarly, a negative reading ($\tilde{P}$) is correct 99% of the time, but in 1% of cases, it fails to detect the presence of the disease. Thus, $\mathbf{Pr}(\tilde{P}|\tilde{D}) = 0.99$ and $\mathbf{Pr}(P|\tilde{D}) = 0.01$. We consider a person who has no prior indica-tion of the disease, so that the average for the population (one per thousand is the statistic used in the present example) is a reasonable representation of the probability that the person has the disease. A surprising result is that a single test with a positive result will yield only a 9% probability that the person has the disease, since

$$\mathbf{Pr}(D|P) = \frac{(\mathbf{Pr}(P|D)\mathbf{Pr}(D)}{\mathbf{Pr}(P|D)\mathbf{Pr}(D) + \mathbf{Pr}(P|\tilde{D})\mathbf{Pr}(\tilde{D})}$$

$$= \frac{(0.99)(0.001)}{(0.99)(0.001) + (0.01)(0.999)} = 0.09.$$

Also

$$\mathbf{Pr}(\tilde{D}|P) = 0.91, \mathbf{Pr}(D|\tilde{P}) = 10^{-5} \text{ and } \mathbf{Pr}(\tilde{D}|\tilde{P}) = 1 - 10^{-5}.$$

As Paulos puts it, the person could feel 'cautiously optimistic' upon receiving the positive indication! Paulos mentions results for cervical cancer based on the Pap test that are only 75% accurate. Note that a calculation similar to the above can be carried out mentally by considering 1000 persons: one person on average has the disease. If they are all tested, there will be close to ten false positives, and almost one true positive. The probability of having the disease is then about one in eleven, given a positive indication.

An engineering application of this problem is the question of indicator lights (or similar devices) for rare events. There is a recorded case of an aircraft crash in 1972 which occurred when the crew were distracted while trying to verify whether the nose landing gear was in the down position. Both bulbs in the indicator system were burnt out. At the same time, malfunction of the landing gear is relatively rare. See Exercise (2.23).

2.4.3 Bayes' theorem, calibration and common sense

Application of Bayes' theorem often shows in an uncanny manner what data are needed to change our opinion, or improve a probability estimate. The introductory example, in which we wished to obtain the probability of flaws in a field situation, $\mathbf{Pr}(F_i|D)$, showed us that data on $\mathbf{Pr}(D|F_i)$ were needed in this calculation. It is entirely natural that the calibration of the testing device should be aimed at checking the test device on welds with flaws of **known** sizes F_i, yielding values of $\mathbf{Pr}(D|F_i)$, as required for inference. The characterization of forensic evidence as given above is also a natural way to interpret the situation. The scientist would be likely to say 'on the basis of the evidence, I am $x\%$ sure that the guilty person has a certain characteristic, such as lefthandedness, light weight, great height, or a certain sex, . . . '. Lindley (1987) discusses this aspect: a forensic interpretation in the inverse manner (given the characteristic, to try to estimate the probability of guilt) does not work well. The assessment should proceed on the basis of the correct forensic interpretation and, of course, other information regarding the person such as availability at the time of the crime, other suspects, and so on.

In the assessment of probability of being infected by a certain disease, it is also natural to use the Bayesian procedure. Let us take the case of Aids, and assume that the person under consideration is of a certain sex, with a defined pattern of behaviour (sexual disposition, possible intravenous drug use, . . .). The latter information is denoted 'B'. Then

$$\mathbf{Pr}(\text{Aids}|B) \propto \mathbf{Pr}(B|\text{Aids})\mathbf{Pr}(\text{Aids}).$$

The term $\mathbf{Pr}(B|\text{Aids})$ is commonly obtained by investigation of the behavioural pattern of persons known to have Aids, e.g. $y\%$ of people with Aids are intravenous drug users, $z\%$ are males, and so on.

The classical case of measurement error also leads to a natural assessment using Bayesian methods. A device for measuring weight, or apparatus for measuring distance, would be calibrated against known weights or distances. We would place a known weight repeatedly on the weighing device and take repeated readings, with a similar procedure for distance, or other quantity of interest. Often the procedure leads to the well-known Gaussian error curve, in which the readings are scattered around the accurate weight value. Details on fitting of curves to data is given in Chapter 11, but for the present we note the importance of the calibration to obtain $\mathbf{Pr}(R|W = w)$, where R = reading and w = actual weight (known). This in turn can be used to make inferences regarding an unknown weight W:

$$\mathbf{Pr}(W = w|R = r) \propto \mathbf{Pr}(R = r|W = w)\mathbf{Pr}(W = w).$$

The calibration of the device for various known weights will give information on $\mathbf{Pr}(R|W = w)$. Prior information on the weight W, as represented by $\mathbf{Pr}(W)$, might be vague, or diffuse (see Chapter 8) so that, in this case, the result in terms of $\mathbf{Pr}(W|R = r)$, is dispersed around the reading $R = r$, where r is the particular reading obtained for W with the dispersion given by the previous calibration.

2.4.4   Induction

Deductive logic deals with situations where the truth of the conclusion follows from the truth of the premises. For example, if X is in $[1, 5]$ and in $[5, 10]$, where $[\ldots]$ indicates a closed interval, then $X = 5$. When the arguments do not lead to such a firm conclusion, but rather give good evidence for the conclusion, then the arguments have inductive strength. We can see probability theory as being the basis for inductive reasoning. Bayes' theorem is the principal means of absorbing new information in a rational manner. Methods derived from this theorem have far-reaching implications. Bruno de Finetti (1974), in the context of induction, states:

'The vital element in the inductive process, and the key to every constructive activity of the human mind, is then seen to be Bayes' theorem.'

2.5   Stochastic dependence and independence; urns and exchangeability

There are several kinds of dependence in mathematics and logic. Equations may be linearly dependent on each other. There is a logical dependence which we shall show to be a special case of stochastic dependence. Let us define the concept of stochastic dependence, which is our concern in this work. Two events E and H are said to be **stochastically independent** if and only if we make the judgement that

$$\mathbf{Pr}(E|H) = \mathbf{Pr}(E). \tag{2.38}$$

Knowledge of the truth (certainty) of H does not lead us to change our probability of E. In effect, we are saying that any information we obtain about H is irrelevant to our degree of belief, in terms of the odds we would lay, regarding E. We shall often use the word 'independent', for instance when we refer to independent events, and we mean stochastic independence.

Most of us would be prepared to accept that the rainfall in Lesotho last year is irrelevant to our assignment of probability to various possible rates of inflation in Canada next year. This is an extreme example of independence. We generally have to make a judgement based on information. For example, we might judge that the fact that it rains today is relevant to the probability that it will rain tomorrow. When considering the probability of rain, two, three, $\ldots$, ten days hence, the relevance of 'rain today' on the judgement declines. At some point, we would judge that the events are independent (data on persistence of rain can be obtained).

It is most important to learn and to appreciate that stochastic independence relates to a **judgement regarding probability**. It is not uncommon to read statements like 'two events are independent if one event has no influence on the outcome of the other'. This is an incomplete interpretation of stochastic independence. It applies in the case where we have an 'honest' coin; we might decide that our assignment of probability for 'heads on the next toss' will not change upon the knowledge that we have obtained heads in the preceding three. The events are not 'influencing' each other. Yet if we suspected that the coin might be two-headed, we would judge the events to be dependent – see Equation (2.36) and related example. In both cases, we might judge that one toss of the coin does not 'affect' or 'influence' the next. This judgement therefore does not necessarily imply independence. There are cases where judgement of events 'not influencing each other' leads to a valid use of independence. Wind and earthquakes are phenomena that do not influence each other and we would judge the loads caused by them to be stochastically independent. Wind and waves or tsunami and earthquakes are dependent pairs.

At the same time it must be recognized that stochastic independence is not physical but a function of information and judgement. Jensen (1996) gives the example of Holmes who is concerned about the state of his lawn in the morning – is it wet or not? This could result from two causes: rain overnight or forgetting to turn the sprinkler off. These events are in Holmes' opinion independent. But he arises in the morning to find his lawn wet. He looks across at Watson's lawn – Watson lives next door – and finds that it is also wet. Knowledge of this fact increases the likelihood of rain and reduces the probability that Holmes has left his sprinkler on. The events 'rain overnight' and 'forgetting to turn the sprinkler off' become dependent once Holmes finds Watson's lawn as well as his own to be wet. Independence is conditional on our state of knowledge.

The absence of independence implies dependence. In this case, we use Bayes' theorem – Equations (2.23) and (2.32) – to change our opinion. If the evidence F causes us to increase our probability assignment regarding E, i.e. $\mathbf{Pr}(E|F) > \mathbf{Pr}(E)$, then we say that E and F are **positively correlated**. Conversely, if knowledge of F causes us to decrease the probability of E, i.e. $\mathbf{Pr}(E|F) < \mathbf{Pr}(E)$, then we have **negative correlation.** These are both cases of dependence. The increase in probability of a flaw, as a result of a positive indication on the non-destructive test device, is an example of positive correlation; the decrease of our probability of rain as a result of high atmospheric pressure would indicate negative correlation.

We have already alluded to a misunderstanding regarding events 'influencing each other'. It is often said that successive tosses of a coin are 'independent'. What is generally meant is that $\mathbf{Pr}(H)$, where $H =$ 'heads on the next toss', does not change as we toss the coin, as if probability were a property of the coin. (Assuming that we toss the coin a hundred times, and obtain 98 heads, we would naturally become somewhat suspicious had we assigned **a priori** the value of $1/2$ to $\mathbf{Pr}(H)$.) More generally, if we are trying to **estimate** the probability of a repetitive event, we use the records of past events to **change** our probability assignment and **independence cannot apply**. The results of our tosses of a coin would be used to 'change our mind', via Bayes' theorem, regarding $\mathbf{Pr}(H)$. If Mr and Mrs Jones have had no children, we might estimate $\mathbf{Pr}$(daughter in the next 'trial') as 0.5, but if the Jones' proceed to have seven sons (out of seven 'trials') we would alter our assignment of probabilities, taking into account this new information by using Bayesian inference. Again, independence cannot be said to apply. This class of problem is analysed in detail in Chapter 8.

A consequence of Equation (2.38) is that if we substitute $\mathbf{Pr}(E)$ for $\mathbf{Pr}(E|H)$ in Equation (2.21), and noting that D and F of (2.21) are replaced by E and H in (2.38), we find that

$$\mathbf{Pr}(EH) = \mathbf{Pr}(E)\mathbf{Pr}(H) \tag{2.39}$$

which is the familiar multiplication rule for probabilities – only to be applied if we consider the events E and H to be independent.

Example 2.9 We could take two successive tosses of a coin to be independent if we were sure, as a result of past experience, that $\mathbf{Pr}$ (Heads) $= 1/2$. Then the probability of two heads in two successive tosses is $(1/2)(1/2) = 1/4$.

Example 2.10 We might well consider the events $E_1 =$ 'there is an earthquake of intensity (specified) at a given location and time' and $E_2 =$ 'there is a hurricane wind of speed greater than (specified value) at the same location and time as E_1' to be independent. Then we can apply Equation (2.39); suppose that $\mathbf{Pr}(E_1)$ and $\mathbf{Pr}(E_2)$ are estimated at 10^{-5} and 10^{-4} respectively. The probability $\mathbf{Pr}(E_1 E_2)$ is then the very small value of 10^{-9}. We see why it is common practice **not** to design for all combinations of extreme load at once, if independence is a reasonable assumption.

If we consider $E_1, E_2, \ldots, E_n$ to be mutually independent,

$$\mathbf{Pr}(E_1 E_2 \cdots E_n) = \mathbf{Pr}(E_1 \wedge E_2 \wedge \cdots \wedge E_n)$$
$$= \mathbf{Pr}(E_1)\mathbf{Pr}(E_2) \cdots \mathbf{Pr}(E_n), \tag{2.40}$$

by applying Equation (2.39) successively. It is also worth noting that $\mathbf{Pr}(E|H) = \mathbf{Pr}(E)$ implies that

$$\mathbf{Pr}(EH) = \mathbf{Pr}(E)\mathbf{Pr}(H) = \mathbf{Pr}(H|E)\mathbf{Pr}(E).$$

Provided that $\mathbf{Pr}(E) \neq 0$, we have

$$\mathbf{Pr}(H|E) = \mathbf{Pr}(H) \tag{2.41}$$

as a consequence of $\mathbf{Pr}(E|H) = \mathbf{Pr}(E)$. Thus, independence of E and H implies both $\mathbf{Pr}(E|H) = \mathbf{Pr}(E)$ and $\mathbf{Pr}(H|E) = \mathbf{Pr}(H)$.

The difference between the concepts of incompatibility, when two events cannot both happen, and independence should be noted. For two events to be stochastically independent, it **must** be possible for both of them to occur. As a result they cannot be incompatible. This points to another kind of independence – that of logical independence. We treat it as a special case of stochastic independence very much in the same way as we find the end points 0 and 1, falsity and truth, to be special cases of a probability assignment, which corresponds to any point on the line [0, 1].

Logical independence

Logical independence is concerned with questions of truth and falsity, and not directly with probability. Two events A and B are logically independent if knowledge of the truth of, say, A does not affect the possibility (our knowledge of the truth or falsity) of B. If we have logical dependence, knowledge of the truth or falsity of A would lead us to a definite conclusion regarding the truth or falsity of B. Therefore, incompatibility implies logical dependence. Consider as an example a partition, such as the set of possible results of a test on a transistor with $E_1, E_2, \ldots, E_n$ representing results in different voltage ranges. One and only one of the E_i can occur and we have $\sum E_i = 1$. Knowledge of the truth of, say, E_5, i.e. $E_5 = 1$, implies that all the other E_i are 0, or false. Thus, we could also say that $\mathbf{Pr}(E_i, i \neq 5 | E_5 = 1) = 0$. All the E_i are logically dependent on each other in this case.

For n events $\{E_1, E_2, \ldots, E_n\}$ to be logically independent, it must be possible for all of the 2^n constituents to be possible. If any are not, it implies that one of the events is determined by the occurrence (or non-occurrence) of the others – logical dependence. Finally, it should be clear that logical dependence implies stochastic dependence. It is the extreme case, in the sense that '$E_1 = 1$ implies $E_2 = 0$' is a limiting case of $\mathbf{Pr}(E_2 | E_1) < \mathbf{Pr}(E_2)$.

2.5.1   Urns, independence and exchangeability

Paulos (1988) describes a strategy, ascribed to the mathematician John von Neumann, for making a fair bet using a coin that is possibly biased. What is required is a value of probability equal to $1/2$ for each of two possible outcomes. The coin is tossed twice. If either two heads or two tails occur, the result is ignored. (This also eliminates from consideration a two-headed or two-tailed coin, which cannot be used in this method.) The two other results, heads–tails and tails–heads, are equally likely whether the coin is biased or not, and can be used to decide the outcome between the two parties to the bet. The basis for this method is that the two sequences of results (heads–tails and tails–heads) are judged to be equally likely, regardless of any opinions or inferences regarding the probability of heads in each toss.

The method then consists of a series of 'pairs of tosses' until heads–tails or tails–heads occurs. We would also accept that, **given** that the probability of heads is equal to 0.8, say (and therefore of tails equal to 0.2), the two sequences heads–tails and tails–heads have the same probability $(0.8)(0.2)$ or $(0.2)(0.8)$ on the assumption of independence. Ignoring any heads–heads or tails–tails leads to the equal probability of tails–heads and heads–tails. But this case of known probability and independence is not the most general way of analysing the problem.

Sequences such as the above, where the order of results is considered to be unimportant in making a probability assignment for the sequence of results as a whole, are termed **exchangeable** sequences. It is fundamental to this concept that independence is **not** assumed: in fact independence can be thought of as a special case of exchangeability. We can convert the coin problem to an equivalent urn problem; this idealization has been introduced in Figures 2.9 and 2.10. When a ball is drawn, it can be replaced, or not replaced, or an additional ball or balls of the same or different colour could be added to the urn depending on the result of the draw.

For the case of non-replacement, illustrated in the lower part of Figures 2.9 and 2.10, it is interesting and important to note that the successive drawings are not independent: in Figure 2.9 if we draw a red (1) ball, we know that the remaining one is pink (0) and *vice versa*. Yet we would judge the sequences $\{1, 0\}$ and $\{0, 1\}$ to be equivalent: we would assign the same probability to each sequence. These are therefore judged to be exchangeable. In fact, if we replace each ball after drawing, and mix them up before drawing again, we would regain independence with a probability of $1/2$ on each draw for red or pink. Remember that independence is a special case of exchangeability.

Von Neumann's solution to the biased coin-tossing would correspond to an urn model, with replacement, where the proportion of red balls (say) corresponds to the probability of heads. This proportion might not be known, yet we can still use von Neumann's method for obtaining a 50–50 bet. Or we could draw the balls without replacement and still have the series of $\{0, 1\}$ and $\{1, 0\}$ which are equally probable. In the case of Raiffa's urn (Figure 2.10), repeated sampling without replacement would not result in independent events, but they would be exchangeable.

2.6 Combinatorial analysis; more on urns; occupancy problems; Bose–Einstein statistics; birthdays

The world about us is complex, yet we have seen that relatively simple physical experiments can be used to reflect fundamental ideas in probability theory. Coin tossing and urn models represent good examples; they can lead to basic formulations of probability distributions, exchangeability and statistical mechanics. Combinations and permutations are usually dealt with at an early stage of our formal education. There are several ways of interpreting the experiments and these will first be reviewed briefly. Reduction to a single model is possible in many cases.

An intuitive procedure is embodied in the **multiplication principle**:

If an operation can be carried out in n_1 possible ways, and if for each of these a second operation can be carried out in n_2 ways, and if for each of the first two a third can be carried out in n_3 ways, and so on for r operations, then the sequence of r operations can be carried out in $n_1 \cdot n_2 \cdot n_3 \cdot \ldots \cdot n_r$ ways.

Example 2.11 A pair of dice thrown in succession yields $6 \cdot 6 = 36$ possible results if one counts results with different orders, e.g. $\{1, 3\}$ and $\{3, 1\}$, as separate results. Or consider an electronic unit that is made of three components, the first of which can be manufactured in three different factories, the second in two and the third in five factories. A particular assembled unit could correspond to $3 \cdot 2 \cdot 5 = 30$ sets of origins.

A common application of the multiplication principle is in the study of permutations of objects. We must first recognize the difference between distinguishable and indistinguishable objects. Both of these are used in statistical mechanics in dealing with particle physics, and there are many more commonplace examples. The simplest ones that come to mind are sets of numbers: sets of distinguishable numbers such as $\{1, 2, 3, 4, 5\}$, and sets of indistinguishable numbers such as $\{1, 1, 1\}$, $\{7, 7, 7, 7\}$ or mixtures of the two $\{1, 1, 1, 3, 3, 6, 6\}$.

The idea is important also in analyzing sets of data; for instance, in analysing the distribution of traffic accidents among weekdays one might ignore the type of accident and treat each accident as indistinguishable from the others. A similar treatment of accidents may be appropriate in studying incidents on offshore rigs. All fire-related accidents may be treated as indistinguishable, with a similar approach for blowouts, collisions with supply vessels, and so on. Consequences, for instance, damage due to a collision with repair costs within a certain range, might be treated as indistinguishable.

In considering permutations, we consider initially a set of distinguishable (distinct) objects. A **permutation** or **ordering** is an arrangement of all or part of the set of objects. For instance, the six permutations of the letters $\{a, b, c\}$ are $\{a, b, c\}, \{a, c, b\}, \{b, a, c\}, \{b, c, a\}, \{c, a, b\}$ and $\{c, b, a\}$. To calculate the total number of ways of forming a permutation, we note that there are n possible ways of arranging the first place, $(n-1)$ ways of arranging the second since one object has been used in the first place, $(n-2)$ ways of arranging the third, and so on. One candidate is used up in each successive arrangement. For a **complete permutation**, in which all of the set of n objects is permuted, the total number of permutations is

$$n \cdot (n-1) \cdot (n-2) \cdot \ldots \cdot 3 \cdot 2 \cdot 1 = n! \tag{2.42}$$

in which we have applied the multiplication principle. It should be recalled that $0! = 1$, by definition (and for convenience).

If we choose only r of the n objects, the number of permutations is often written as $_nP_r$, or with the abbreviation $(n)_r$, and is given by:

$$_nP_r \equiv (n)_r = \prod_{i=0}^{r-1}(n-i) = \frac{n!}{(n-r)!}$$
$$= n(n-1)(n-2)\cdots(n-r+1). \tag{2.43}$$

These r terms, multiplied together, represent respectively the number of ways of selecting the first, second, third, $\ldots$, up to the rth object. We shall use the notations $(n)_r$ or $_nP_r$ hereafter. Note that $_nP_0 = (n)_0 = 1$.

In some way or another, all objects are different, but the differences might be small enough to neglect for some purpose, as discussed above. In some cases they might be impossible to distinguish. We do not 'tag' the molecules in a fluid: the balls (of the same colour) in our urn are treated as indistinguishable. A natural way to think of indistinguishable objects is that their order, say when arranged in a line, is of no significance.

There are two combinatorial problems that turn out to be identical. These are as follows.

(1) The number of ways that n **distinguishable** objects can be arranged into k groups (or piles, or boxes), with n_1 in the first group, n_2 in the second, n_3 in the third, $\ldots$, and n_k in the kth group. The order of objects within each pile does not matter; as a result rearrangements within a pile do not constitute a new 'way' of arranging, or combination.

(2) The number of ways that k distinguishable groups of **indistinguishable** objects, totalling n in all, can be arranged in a line. For example, there might be n_1 balls of colour 1 (red for

example), n_2 of colour 2 (pink for example), ..., and n_k of colour k. Apart from colour, the balls are indistinguishable. This is essentially equivalent to an urn model, with k groups of coloured balls, which are removed in turn. Or the indistinguishable objects might be given a colour or identity by means of a subsidiary experiment, such as tossing a coin or spinning a roulette wheel.

In both cases, (1) and (2) above, $\sum_{i=1}^{k} n_i = n$. The equivalence between (1) and (2) will now be demonstrated. We consider initially only two groups or colours in the above, i.e. $k = 2$, as this suits our demonstration in Figure 2.18. But this is readily extended to the case $k > 2$ as shown below. Since $k = 2$, we can consider two piles, A and B (case 1 above) with n_1 and $(n - n_1)$ objects in the first and second piles, respectively. This represents the classical problem of **combinations** of distinct objects, in which we are choosing r objects from n; in the previous notation, $n_1 = r$, and $n_2 = n - r$, with $n_1 + n_2 = n$.

In case 1 above, we use numbers to make this identification – in real problems, the numbers could correspond to people's names, or individual wave heights, for example. The numbers are attached to the balls before they are put into piles – the choice of pile could be on the basis of a subsidiary experiment such as throwing a die or on the basis of a measurement, such as a wave height. In case 2 above, in the series of trials we number the trials, not the objects. The objects are either successes or failures: this could also be decided on the basis of a subsidiary experiment, or on the basis of the colour of the ball, the sex or height of a person, the height of a wave.

The equivalence of the urn model, that is, a series of trials, and the act of placing distinguishable objects into two piles is illustrated in Figure 2.18. There is a one-to-one correspondence between piles of distinguishable balls and ordered trials of sets of indistinguishable ones. The numbers attached to the balls in pile A (case 1) correspond to the numbers of the trials in which successes occur (case 2). For example, ball '3' and the third trial correspond. Furthermore, if we make any alteration in the piles, for example, if we exchange ball 1 in pile A with ball 8 in pile B, this changes the order of successes and failures in case 2. **All of these orders correspond to distinct piles**.

The number of combinations is

$$_nC_r = \binom{n}{r} = \frac{(n)_r}{r!} = \frac{n!}{r!(n-r)!},$$
(2.44)

in which there are r terms. A useful interpretation is the following, for the case of the two piles. The numerator gives the number of ordered arrangements (permutations) of all of the n objects. But these are divided into two piles or groups, say the first r of them, and the remaining $(n - r)$; within these two groups the order of the distinguishable objects is not important. The number of ordered arrangements in the two groups is $r!$ and $(n - r)!$ respectively. If we divide the total number of permutations by these two numbers, we obtain the number of cases in which there are different objects in each group without regard to order within the groups.

The drawing of r things from a group of n is equivalent to forming two piles with r things in one and $(n - r)$ in the other. Since $\binom{n}{r} = \binom{n}{n-r}$, the formula (2.44) gives the

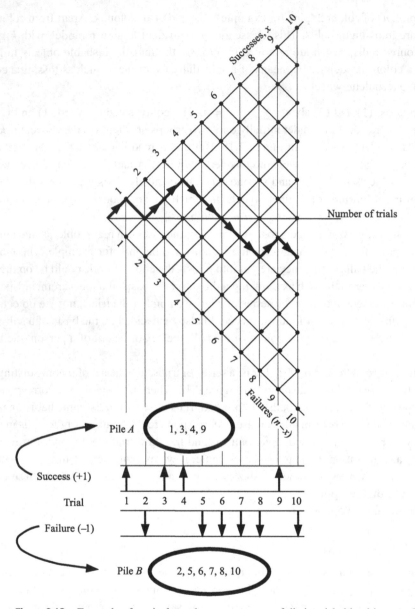

Figure 2.18 Example of equivalence between groups of distinguishable objects and orders of groups of indistinguishable objects

number of combinations of n things taken either r or $(n - r)$ at a time. It can be thought of as: (trials)!/[(successes)!(failures)!]. Equation (2.44) is used in many applications; a popular one is the number of ways that a committee can be formed from a given number of women and men.

Example 2.12 The number of possible hands of poker is $52!/(5!47!) = 2598\,960$. The chance of being dealt four kings is $48/2598\,960 \simeq 1/54\,000$ since there are 48 ways of combining the remaining cards with the four kings.

Example 2.13 The number of possible orders of removing the balls from Raiffa's θ_1 urn is $10!/(4!6!) = 210$.

The logic used just after Equation (2.44) is readily extended into a k- rather than a two-dimensional problem. For the first interpretation, one considers the distinguishable objects. The numerator $n!$ of Equation (2.44) again gives the number of ordered arrangements, but to eliminate orders within groups we must divide as follows:

$$\binom{n}{n_1 n_2 \cdots n_k} = \frac{n!}{n_1! n_2! \cdots n_k!}, \tag{2.45}$$

where the notation of Equation (2.44) has been extended into k dimensions, from the binomial to the multinomial case, and $\sum_{i=1}^{k} n_i = n$. In terms of the second of the two interpretations of combinatorial problems, one has only to think of an urn with balls of k colours, rather than two.

2.6.1 Urn model, Pascal's triangle and binomial formula

A sequence of binary events such as drawing balls of two colours from an urn can be illustrated as in Figure 2.19. Imagine that one starts at the origin, and then proceeds somewhat erratically along a pathway, or random walk, going through a set of nodes in the rectangular grid. The two diagonal axes show the numbers of successes (northeast), denoted x, and failures (southeast), that is $(n - x)$, for n trials. The vertical axis shows the gain: say, one dollar gained for each success, and one loss for each failure. The total gain is thus

$$x - (n - x) = 2x - n.$$

The number of paths leading to each node can be obtained by summing the number of paths to the two preceding nodes giving 'access' to the node in question. Examples are given in the upper part of Figure 2.19. One obtains in this way Pascal's triangle, part of which is now shown in the more familiar vertical array of Table 2.1. Each number is the sum of the two above it (apart from the first one!). There are 2^n possible paths (orders) for the first n trials. The coefficients are in fact the same as those appearing in the binomial expansion:

$$(p + q)^n = p^n + np^{n-1}q + n(n-1)p^{n-2}q^2$$
$$+ \cdots + \binom{n}{x} p^{n-x}q^x + \cdots + q^n. \tag{2.46}$$

This may be expressed in a form more consistent with the usual expression of the binomial

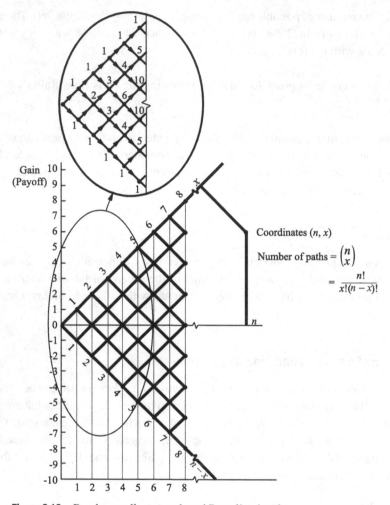

Figure 2.19 Random walk, network and Pascal's triangle

distribution, merely by reversing the order of the terms in Equation (2.46),

$$(p+q)^n = q^n + npq^{n-1} + n(n-1)p^2q^{n-2}$$
$$+ \cdots + \binom{n}{x}p^x q^{n-x} + \cdots + p^n,$$

or

$$(p+q)^n = \sum_{x=0}^{n} \binom{n}{x}p^x q^{n-x}. \tag{2.47}$$

In this case the terms on the right hand side correspond to the probabilities of $0, 1, 2, \ldots, x, \ldots, n$ successes in n trials, with p = probability of success in a single trial,

Table 2.1 *Pascal's triangle*

n	The binomial coefficients	2^n possible outcomes
0	1	1
1	1　1	2
2	1　2　1	4
3	1　3　3　1	8
4	1　4　6　4　1	16
5	1　5　10　10　5　1	32
6	1　6　15　20　15　6　1	64
7	1　7　21　35　35　21　7　1	128
8	1　8　28　56　70　56　28　8　1	256

and $q = 1 - p = \tilde{p}$. For the use of this formula the trials must be stochastically independent. The binomial distribution is considered further in Chapter 3.

Although the appearance of the factor $\binom{n}{x}$ in Equation (2.47) may now be obvious, the following serves to confirm the result. Consider each term in the expansion of $(p + q)^n$ to be numbered. Thus

$$(p + q)^n = (p + q)\,(p + q)\,(p + q) \ldots (p + q)$$
$$\text{Number} \quad\quad 1 \quad\quad 2 \quad\quad 3 \quad\quad\quad n$$

Each 'p' in p^x is chosen from x of these n factors and each 'q' in q^{n-x} is chosen from the remaining $(n - x)$. There are $\binom{n}{x}$ ways of putting the n factors in two piles of size x and $(n - x)$ respectively.

In exchangeable sequences, order is not important so that all paths are equally likely. For any node (n, x) where $n =$ number of trials and $x =$ number of successes, exchangeability would dictate that

$$\omega_{x,n} = (\text{number of paths}) \cdot p_{x,n} \tag{2.48}$$

where $\omega_{x,n} =$ probability of getting to the node in question by all paths, and $p_{x,n} =$ probability of getting to the node in question by a single specified path. The evaluation of $p_{x,n}$ could, for example, follow from the content of the urn: the number of red and pink balls and the strategy for replacement after a draw. The number of paths is given by Equation (2.44); here x replaces r.

The use of Pascal's triangle goes back to the early history of probability theory (Bernstein, 1996), and, indeed, was known to a Chinese mathematician in the fourteenth century (Chu's precious mirror). An early application was to the fair odds in an unfinished game – balla – discussed by Paccioli. We shall apply it to the finals in the National Hockey League or the World Series, in both of which the 'best of seven' wins. Figure 2.20 shows the possible results in such a series, and we can work out the probability that our team will win if it has lost the first game. It is assumed that the teams are evenly matched and that they have equal probabilities

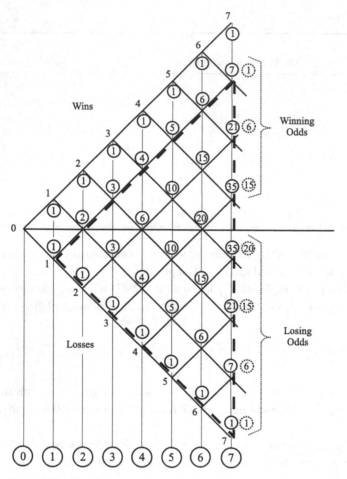

Figure 2.20 Possible results in a 'best of seven' series. The broken line shows the possible results if our team has lost the first game. The probability then of winning is $(1 + 6 + 15)/2^6 \simeq 1/3$

(50%) of winning each game. To use this method, one assumes all games are played; for example, if a team has won four games and wins the series, it nevertheless goes on to complete the remaining three games. The resulting odds are correct since all permutations of results of the unplayed games are included.

2.6.2 Further combinatorial results

We use our random walk diagram to prove an important result. Figure 2.21 shows a sequence of two random walks, the first from A to B with m trials and r successes, and the second from B to C with n trials and $(t - r)$ successes. If one went from A to C through any path, there would be $(m + n)$ trials with t successes. The number of paths from A to B is $\binom{m}{r}$. One

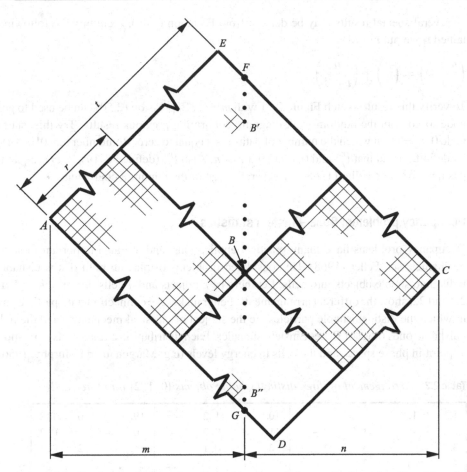

Figure 2.21 Random walk from A to C considered as random walks from A to B and then from B to C. All paths are covered if B takes all the dotted positions shown, from F to G

then starts at B and finishes at C; the number of paths to C through B are $\binom{m}{r} \cdot \binom{n}{t-r}$ by the multiplication principle, and since the path from B to C involves $(t-r)$ successes in n trials. In order to arrive at C by any path, it is necessary then to consider all the possible positions for B. Thus,

$$\binom{m+n}{t} = \sum_{r=0}^{t} \binom{m}{r}\binom{n}{t-r}. \tag{2.49}$$

This result is the addition theorem for binomial coefficients.

There are many more paths from A to C in Figure 2.21 when B is near the centre of the rectangle $ADCE$ than when it is near the corners D or E. We shall see later that winning lotteries corresponds to taking one particular extreme path, for example AEC, when all the paths through the random 'maze' are equally likely (see Chapter 3 under 'hypergeometric distribution' for a probabilistic solution).

Several special results may be derived from Equation (2.49), including the following, obtained if we put $m = 1$:

$$\binom{n+1}{t} = \binom{n}{t} + \binom{n}{t-1}. \tag{2.50}$$

To verify this result, sketch Figure 2.21 with $m = 1$. Expression (2.50) can be used to go from node to node in the random walk diagram, confirming previous results. Try this, starting at node 0 ($n = 0$), at which the number of paths is set equal to zero. Remember that $0! = (n)_0 = 1$ by definition and that $\binom{n}{r}$ and $(n)_r$ are 0 if $r > n$. Also $\binom{n}{r}$ (defined as $(n)_r / r!$) is equal to 0 if r is negative. See Feller (1968) for generalization of the binomial coefficients.

2.6.3   Occupancy problems; Bose–Einstein statistics

Occupancy problems have many practical applications, and are also important in statistical mechanics; see Feller (1968). He considered problems of placing both distinguishable and indistinguishable objects into cells. Let there be r objects and n cells. For $n = r = 3$, Tables 2.2 and 2.3 show the different arrangements. Feller makes the connection with particle physics, in which the distinguishable particles are the subject of classical mechanics, and the indistinguishable ones those of Bose–Einstein statistics. Each distribution among cells corresponds to a 'point in phase space' and the cells to energy levels (e.g. Margenau and Murphy, 1956). For

Table 2.2 *Arrangement of three distinguishable objects {0, 1, 2} into three cells*

1.	0, 1, 2	–	–	10.	0	1, 2	–	19.	–	0	1, 2
2.	–	0, 1, 2	–	11.	1	0, 2	–	20.	–	1	0, 2
3.	–	–	0, 1, 2	12.	2	0, 1	–	21.	–	2	0, 1
4.	0, 1	2	–	13.	0	–	1, 2	22.	0	1	2
5.	0, 2	1	–	14.	1	–	0, 2	23.	0	2	1
6.	1, 2	0	–	15.	2	–	0, 1	24.	1	0	2
7.	0, 1	–	2	16.	–	0, 1	2	25.	1	2	0
8.	0, 2	–	1	17.	–	0, 2	1	26.	2	0	1
9.	1, 2	–	0	18.	–	1, 2	0	27.	2	1	0

Table 2.3 *Arrangement of three indistinguishable objects {0, 0, 0} into three cells*

1.	000	–	–	6.	0	00	–
2.	–	000	–	7.	0	–	00
3.	–	–	000	8.	–	00	0
4.	00	0	–	9.	–	0	00
5.	00	–	0	10.	0	0	0

some time, probabilities were assigned on the assumption that the 27 different possibilities of Table 2.2 were equally likely, giving the basis for Maxwell–Boltzmann statistics ('statistic' is the word used for this kind of inference by physicists).

It was later accepted that some of the possibilities could not in fact be distinguished, for example arrangements 4, 5, and 6 in Table 2.2 collapse into arrangement 4 in Table 2.3, as a result of failure on our part to recognize an underlying reality. The question then arises: should one assign a probability of $3/27 = 1/9$ to the arrangement just discussed (Table 2.2), or $1/10$ (Table 2.3)? Bose and Einstein showed that, for some particles, the second assignment is appropriate. This problem will be discussed in more depth in Chapter 6. It will be seen that urn models can be used to obtain the distributions used in particle physics.

The number of arrangements of indistinguishable objects (e.g. Table 2.3) will now be derived. Let us take $n = 10$, and $r = 8$ (r objects and n cells). A typical arrangement would be of the following kind: $|00||0|0||||000|0||$ in which the vertical lines $|$ delineate the boundaries of the cells. If we omit the first and last of these barriers we are left with $(n - 1)$ barriers. Combining these with the objects, we have $(n + r - 1)$ entities which can be divided into two groups ($|$ or 0). The number of arrangements corresponds to the number of ways that two distinguishable groups of indistinguishable objects can be arranged in a line (see Equation (2.44) and the text preceding it). Therefore, the number of arrangements in the present case is

$$a_{r,n} = {}_{n+r-1}C_{n-1} = \binom{n+r-1}{n-1} = \binom{n+r-1}{r}. \tag{2.51}$$

If no cell is to remain empty, we consider the spaces between the r objects. There are $(r - 1)$ of these spaces. There are again $(n - 1)$ barriers omitting the first and last ones. If $(n - 1)$ of the $(r - 1)$ spaces are occupied by barriers then no two barriers can be adjacent to each other. (Note that $r \geq n$.) There are

$$_{r-1}C_{n-1} = \binom{r-1}{n-1}$$

ways to choose the $(n - 1)$ spaces from the $(r - 1)$ ones available, and this is therefore the number of arrangements leaving no space empty. Try the arrangements with, for example, $n = 3, r = 5$.

Replacement of objects

We have seen that replacement of objects taken from a pile, or urn, has implications with regard to the probability for the next draw. If we have coloured balls (say red and pink) in an urn, the same probability will apply in two successive draws if we replace the ball after the first draw and then mix the balls up. The same applies in the case of a deck of cards: if we replace a card and shuffle the deck, we judge that it has the same probability of reappearing.

Replacement also has an effect on the number of ways a group of distinguishable objects can be arranged. Consider a group of n such objects, from which r are sampled. If there is no replacement, the number of distinct samples is $(n)_r$, following the usual rule for permutations.

If the objects are replaced after each sample is taken, then there are $n \cdot n \cdot n \cdots \cdot n$ (r terms) $= n^r$ different samples. In this case, there can be repetitions of the same object in the sample; without replacement, each object can appear only once. Both calculations count different orderings. Given replacement after each draw, the probability of no repetition in the sample is

$$\frac{(n)_r}{n^r} \tag{2.52}$$

If n balls are placed at random in n cells, the probability that each cell is occupied is $n!/n^n$. For $n = 6$, this equals $0.0154\ldots$; it is thus rather unlikely that in six throws of a die, all faces appear. The famous birthday problem is also solved by the same methods. As in all the cases considered here, in evaluating probabilities it is assumed that all distributions are equally likely.

2.6.4   The birthday problem

In a group of r people, what is the probability that at least two have the same birthday? The results have been found to be counter-intuitive and indeed, they may be for some. Let us calculate, in particular, the number r for which the probability of at least two common birthdays is greater than $1/2$. Let us first work out the probability of **no** common birthdays, then the probability of at least one shared birthday is $1 - \mathbf{Pr}(\text{none})$. We assume in the following that a person has the same probability ($1/365$) of having a birthday on any particular day.

The total number of ways that the birthdays can be arranged is 365^r; 365 days for the first person, similarly for the second, and so on, assuming of course that we don't have a leap year! If there are to be no shared birthdays, the first person still has the choice of 365 days, but the second is constrained to 364 to avoid whichever day is the birthday of the first, the third has 363 possibilities and so on. Therefore, the number of ways of avoiding any shared birthdays is

$$365 \times 364 \times 363 \times (365 - r + 1) = 365!/(365 - r)! = (365)_r,$$

and the probability of no shared birthdays is thus a special case of Equation (2.52):

$$p = \frac{(365)_r}{365^r}, \tag{2.53}$$

and the probability of at least two shared birthdays is

$$\tilde{p} = 1 - p = 1 - \frac{(365)_r}{365^r}. \tag{2.54}$$

Evaluation of Equation (2.53) involves rather large numbers: for expedient solutions Stirling's approximation (Section 6.4) is often used. For present purposes, we rather first note that Equation (2.53) can be written as

$$p = \left(1 - \frac{1}{365}\right)\left(1 - \frac{2}{365}\right)\cdots\left(1 - \frac{r-1}{365}\right). \tag{2.55}$$

Table 2.4 *Results for birthday problem*

	Approximate: Equation (2.56)		Exact: Equation (2.55)	
r	p	$\tilde{p}$	p	$\tilde{p}$
5	0.973	0.027	0.973	0.027
10	0.877	0.123	0.883	0.117
15	0.712	0.288	0.747	0.253
20	0.479	0.521	0.589	0.411
25	0.178	0.822	0.431	0.569
30	–	–	0.294	0.706

A first approximation can be made by assuming that r is small, i.e. $r \ll 365$. Then all cross products in the formula can be neglected; thus

$$p \simeq 1 - \frac{1 + 2 + \cdots + (r-1)}{365}$$

$$\text{and } \tilde{p} \simeq \frac{1 + 2 + \cdots + (r-1)}{365} \tag{2.56}$$

The results from this calculation are given in Table 2.4, together with accurate results calculated by Equation (2.55). Equation (2.56) can be solved mentally (at least as a fraction) but is only accurate up to about $r = 10$. The accurate results show that a surprisingly small number of people are required for $\tilde{p} = $ (probability of at least two birthdays) to exceed 0.5. This happens between $r = 22$ ($\tilde{p} = 0.476$) and $r = 23$ ($\tilde{p} = 0.507$).

The problem can be used nicely in a classroom with, say, 30 students. A piece of paper is passed around upon which the students enter their birthdays. While this happens, the problem is solved. With a bit of luck (there is a 70% probability for $r = 30$), there will be at least two common birthdays. See Feller (1968, Vol. 1) for another approximate solution.

2.6.5   Urns with balls of many colours; Pólya's urn: relation to occupancy

We complete our study of combinatorial analysis using now urns with b balls, of k different colours. The number of balls initially is

$$
\begin{array}{lll}
g_1 & \text{of colour} & 1, \\
g_2 & \text{of colour} & 2, \\
\vdots & & \\
g_i & \text{of colour} & i, \\
\vdots & & \\
\underline{g_k} & \text{of colour} & k, \\
b & \text{in total.} &
\end{array} \tag{2.57}
$$

There are various strategies for drawing balls from the urn. We consider the following basic ones.

(a) No replacement; eventually all of the balls will be used up. This leads to the hypergeometric distribution (Sections 3.2 and 3.4.1).
(b) Replacement followed by mixing of the balls; the composition then stays the same. This leads to the binomial or multinomial distributions (Sections 3.2 and 3.4.1).
(c) Double replacement, with both balls being of the same colour as that just drawn, followed by mixing of the balls; this is Pólya's urn model.

We draw n balls in the experiment and find that there are

$$
\begin{array}{lll}
x_1 & \text{of colour} & 1, \\
x_2 & \text{of colour} & 2, \\
\vdots & & \\
x_i & \text{of colour} & i, \\
\vdots & & \\
x_k & \text{of colour} & k, \\
\hline
n & \text{in total.} &
\end{array}
\tag{2.58}
$$

For any of the above sampling strategies, the **order of results** is not of significance in probabilistic analysis – exchangeability applies. An example will illustrate this. We take Pólya's urn as the example. Let $k = 3$, so that there are three colours. Let $g_1 = 2$, $g_2 = 3$ and $g_3 = 5$; consequently, $b = 10$. If we draw six balls and obtain two of colour 1, one of colour 2 and three of colour 3, let us say in the order 3-1-1-2-3-3, then the probability is

$$
\frac{5}{10} \cdot \frac{2}{11} \cdot \frac{3}{12} \cdot \frac{3}{13} \cdot \frac{6}{14} \cdot \frac{7}{15}.
$$

Any other order of results corresponds to a permutation of the terms in the numerator with no change in the denominator. For example, the order 1-1-2-3-3-3 results in the probability

$$
\frac{2}{10} \cdot \frac{3}{11} \cdot \frac{3}{12} \cdot \frac{5}{13} \cdot \frac{6}{14} \cdot \frac{7}{15},
$$

which is the same as the previous result. It will be found generally that the probability of obtaining a particular set of balls (2.58) drawn from the Pólya urn model (2.57), given a particular order, is the same as for any other order. The reader might wish to check that this is also the case for the other cases introduced above, with no replacement or single replacement.

We shall pursue the general case for the Pólya urn, since this is relevant to the occupancy problem, and to statistical mechanics: see Chapter 3 for more on the basic distributions, Chapter 6 for more on statistical mechanics, and Chapter 8 for the use of the distributions, particularly the hypergeometric, in exchangeability theory. We have seen that the same probability is obtained for any order of drawing. We may then assume that we draw all balls of colour 1 first, then those of colour 2, and so on, and that the result will be the same for any other order. The probability for this order is then

$$
\frac{(g_1)}{(b)} \cdot \frac{(g_1 + 1)}{(b + 1)} \cdots \frac{(g_1 + x_1 - 1)}{(b + x_1 - 1)} \cdot \frac{(g_2)}{(b + x_1)} \cdots \frac{(g_2 + x_2 - 1)}{(b + x_1 + x_2 - 1)} \cdots
$$

$$
\cdot \frac{(g_k)}{(b + x_1 + x_2 + \cdots + x_{k-1})} \cdots \frac{(g_k + x_k - 1)}{(b + n - 1)}. \tag{2.59}
$$

In the denominator of the last term, $n = \sum x_i$. The equation (2.59) can be expressed as

$$
\frac{\prod_{i=1}^{k} g_i(g_i + 1)(g_i + 2) \cdots (g_i + x_i - 1)}{b(b + 1)(b + 2) \cdots (b + n - 1)},
$$

or

$$
\frac{\prod_{i=1}^{k} (g_i + x_i - 1)_{x_i}}{(b + n - 1)_n}. \tag{2.60}
$$

This probability corresponds to a single path in a (multidimensional) random walk from the origin to node $(x_1, x_2, \ldots, x_k)$. There are

$$
\frac{n!}{x_1! x_2! \cdots x_k!} = \frac{n!}{\prod_{i=1}^{k} x_i!} \tag{2.61}
$$

paths to the node in question. Therefore the probability of getting to node $(x_1, x_2, \ldots, x_k)$ by any path is the product of Equations (2.60) and (2.61). This can also be written as

$$
\frac{\prod_{i=1}^{k} \binom{g_i + x_i - 1}{x_i}}{\binom{b + n - 1}{n}}. \tag{2.62}
$$

In this equation we have just solved our first multidimensional probability distribution; the set of random quantities $\{X_1, X_2, \ldots, X_k\}$ with particular values $\{x_1, x_2, \ldots, x_k\}$ constitutes a multidimensional set. The formal methods for representing probability distributions are given in Chapter 3.

It is interesting to note that the model for the number of ways of distributing r balls among n cells in Equation (2.51) is in fact identical to an urn model. The term

$$
\binom{g_i + x_i - 1}{x_i} \tag{2.63}
$$

represents the number of ways of distributing x_i indistinguishable objects amongst g_i cells; this is clear by reference to Equation (2.51). This is used in statistical mechanics. Indeed, the principal distributions for fundamental particles are derived in Chapter 6 by a demon who plays various games using roulette wheels and urns. The values x_i and g_i represent the number of particles and degeneracy associated with state i.

We note in concluding that in the case where all the g_i are equal to unity, i.e. $g_i = 1$, for $i = 1, \ldots, k$, then each term in the product term of the numerator of Equation (2.62) is equal to $[x_i!/(x_i!0!)] = 1$, and the resulting probability is

$$\frac{1}{\binom{b+n-1}{n}}. \tag{2.64}$$

This is an urn with single balls of the various colours. Pólya's urn is also discussed in de Finetti (1974), Feller (1968) and Blom *et al.* (1994).

2.7   Exercises

Basic exercises are included to revise odds, set theory and counting.

2.1 An engineer states that the odds of a beam surviving a load of 3 MN for 25 years is '5 to 1 on'. What is the engineer's probability for this event? Draw a decision tree as in Figure 2.1(a) to illustrate the assignment of probability.

2.2 The odds of a prototype microchip surviving a current of $3\mu A$ for 2 hours are estimated as 4 to 3 against. What is the estimated probability of this event? Draw a decision tree as in Figure 2.1(a) to illustrate the estimator's assignment of probability.

2.3 The event $E = $ (the score on a roll of a fair die is a composite number; that is, a non-prime number greater than 1). Calculate the odds on E, expressed as a ratio reduced to its lowest terms.

2.4 *Odds vs. bookie's odds.* In a five-horse race, a bookmaker sets a stake of $240. The bookie quotes odds of $r_1 = 5$ to 3 on, $r_2 = 2$ to 1 against, $r_3 = 5$ to 1 against, $r_4 = 19$ to 1 against and $r_5 = 19$ to 1 against.[7] Calculate 'probabilities' q_i ($i = 1, 2, \ldots, 5$) based on these odds. You may place a bet of $(240 \cdot q_i)$ on any of the five horses and if event E_i ($=$ horse i wins) occurs, then you win the bookie's stake of $240.

   (a) From your point of view, do these odds constitute fair bets? Are the q_i acceptable as probabilities? Justify your answer by determining whether or not the probabilities associated with these odds are coherent.

   (b) What is the bookie's guaranteed profit if you place one bet on each of all five horses?

2.5 *Fair vs. unfair odds.* Six teams are competing in a design competition. An engineer offers a stake of $100 to be won by a colleague who guesses the winner. He offers odds of 19 to 1 against team A, 9 to 1 against team B, 9 to 1 against C, 3 to 1 on team D, 4 to 1 against E, and 7 to 3 against F. Calculate 'probabilities' based on these odds. Are they coherent? If a person bets on each team, what is the engineer's guaranteed profit? Assume that the odds reflect well the relative likelihood of each team winning, so that for example teams B and C are considered equally likely to win. Rescale the probabilities and odds so that the bets become fair bets and the probabilities become coherent.

---

[7] Usually odds in horse racing are expressed as 'against'.

2.6 Derive the condition of coherence for a bookmaker by using Equation (2.7) without making the assumption that $s_1 = s_2 = \cdots = s_n = s$ (de Finetti, 1937).

2.7 Verify the following logical relationships. Set relationships are similar with S replacing 1 and $\emptyset$ replacing 0.

    1. $A \wedge A = A = A \vee A$.
    2. $\sim (\tilde{A}) = A$.
    3. $A \wedge B = B \wedge A; A \vee B = B \vee A$ (Commutative rules).
    4. $(A \wedge B) \wedge C = A \wedge (B \wedge C)$; similarly for $\vee$ (Associative rules).
    5. $A \wedge (B \vee C) = (A \wedge B) \vee (A \wedge C); A \vee (B \wedge C) = (A \vee B) \wedge (A \vee C)$ (Distributive rules).
    6. $A \wedge 1 = A; A \vee 0 = A$.
    7. $A \wedge \tilde{A} = 0; A \vee \tilde{A} = 1$.
    8. $A \wedge 0 = 0$.
    9. $\tilde{0} = 1$.
    10. $\sim (A \vee B) = \tilde{A} \wedge \tilde{B}$ (De Morgan's theorem).
    11. $\sim (A \wedge B) = \tilde{A} \vee \tilde{B}$ (De Morgan's theorem).
    12. $A \wedge (A \vee B) = A$.
    13. $A \wedge (\tilde{A} \vee B) = A \wedge B$.

2.8 The union and intersection of a finite collection of sets $A_i (i = 1, 2, \ldots, n)$ are written as $\cup_i A_i$ and $\cap_i A_i$ respectively. Verify for, say, $i = 3$ the following relationships.

    1. $\sim (\cup_i A_i) = \cap_i (\tilde{A}_i)$ (general version of De Morgan's theorem).
    2. $\sim (\cap_i A_i) = \cup_i (\tilde{A}_i)$ (general version of De Morgan's theorem).
    3. $A \cup (\cap_i B_i) = \cap_i (A \cup B_i)$.
    4. $A \cap (\cap_i B_i) = \cap_i (A \cap B_i)$.
    5. $A \subset \cap_i B_i \leftrightarrow A \subset B_i$ for every i.

2.9 Demonstrate the associative, commutative and distributive rules for sets using Venn diagrams.

2.10 Verify the following relationships.

    1. $A \subset A; \emptyset \subset A \subset S$.
    2. $A = B \leftrightarrow A \subset B$ and $B \subset A$.
    3. $A \subset B \leftrightarrow A \cap \tilde{B} = \emptyset$.
    4. $A \subset B \leftrightarrow A = A \cap B$.

2.11 Verify the following duality relationship in a few cases. If we interchange

    $\subset$ and $\supset$

    $\emptyset$ and S

    $\cup$ and $\cap$

then the new relationship also holds.

2.12 Expand Equation (2.17) for $n = 3$. Write the corresponding probabilistic equation using the principle in Equations (2.11) and (2.12).

2.13 For the sample space S which consists of the first ten natural numbers $\{1, 2, \ldots, 10\}$, events A, B, C are defined as: $A = \{$even numbers in $S\}$, $B = \{$prime numbers in $S\}$, and $C = \{3, 5\}$.

(a) Represent these sets in a Venn diagram.
(b) Express the set $D = \{$even prime numbers in $S\}$ as a set function of the above sets. How many elements are in this set?
(c) Simplify the expression $B \cap C$.
(d) Simplify the expression $B \cup C$.
(e) Simplify the expression $A \cup C$.
(f) What are the elements in the set $\tilde{A} \cap \tilde{B} \cap \tilde{C}$?

2.14 There are three possible causes for failure of a computer system. One or more of the causes may be associated with a failure. Let $A_i = \{$cause i is associated with a failure$\}$, for $i = 1, 2, 3$ and suppose that $\mathbf{Pr}(A_1) = 0.30$, $\mathbf{Pr}(A_2) = 0.39$, $\mathbf{Pr}(A_3) = 0.32$, $\mathbf{Pr}(A_1 \wedge A_2) = 0.16$, $\mathbf{Pr}(A_2 \wedge A_3) = 0.12$, $\mathbf{Pr}(A_3 \wedge A_1) = 0.12$, and $\mathbf{Pr}(A_1 \wedge A_2 \wedge A_3) = 0.02$. Draw a Venn diagram that represents this situation. Calculate the probability of each of the following events:

(a) $A_1 \vee A_2$
(b) $\tilde{A}_1 \wedge \tilde{A}_2$
(c) $A_1 \vee A_2 \vee A_3$
(d) $\tilde{A}_1 \wedge \tilde{A}_2 \wedge \tilde{A}_3$
(e) $\tilde{A}_1 \wedge \tilde{A}_2 \wedge A_3$.

2.15 The quality of mass-produced ceramic components for an electricity supply system is checked by an inspection system. The products are unacceptable if they are fractured, oversize or warped. From past experience, the probability that a component is warped is 0.0015, the probability that it is fractured is 0.0020, the probability that it is oversize is 0.0014, the probability that it is either warped or fractured or both is 0.0028, the probability that it is either fractured or oversize or both is 0.0029, the probability that it is either oversize or warped or both is 0.0025 and finally the probability that it has at least one of these imperfections is 0.0034.

(a) What is the probability that a component is free of all imperfections?
(b) What is the probability that a component is oversize but is free of other imperfections?
(c) What is the probability that a component has all of the specified imperfections?

2.16 In a component failure, there are three possible causes $E_i, i = 1, 2, 3$. At failure, a reliability analysis indicates that

$$E_1 + 2E_1 E_2 + E_3 = 1.$$

What can be said regarding the values $\mathbf{Pr}(E_i)$?

2.17 The six faces of a cube of side 10 cm have been painted purple. The cube is then cut into 1000 smaller cubes, each of side 1 cm, and these smaller cubes are placed in an urn. One of these cubes is selected at random from the urn. What is the probability that this cube will have two painted faces?

2.18 Consider Raiffa's urns of Figure 2.10. An urn is chosen at random from a room containing 800 of type θ_1 and 200 of type θ_2. Two balls are sampled without replacement. What are the probabilities that the urn chosen is a θ_1 or θ_2 urn if the sampled balls are

(a) both red;
(b) both pink; and
(c) red and pink?

2.19 *Everyday life 1*.

(a) What kind of bets for the future are manifested in 'puts and calls' in security trading? Other bets, on sporting events or in insurance might be pondered.
(b) A Canadian automobile insurance company threatens to double the insurance rate for a client if she puts in a claim after a minor accident. But they will leave the rate the same if the claim is not entered even though they know about the accident. Bayes' theorem applies equally well as regards the probability of a future accident whether the claim is entered or not since the company knows of the accident. The client's future risk is the same in either case. Could it be that the insurance company is attempting to force the client to pay for her repairs herself? Comment on this quite common practice.

2.20 *Everyday life 2*. An elderly public figure is accused by several women of his age of having committed serious improprieties in his youth involving them individually. The women do not know each other and appear reluctant to give evidence. They are ordinary people with no apparent motive falsely to accuse the public figure. Comment on the probability of guilt.

2.21 *Everyday life 3*. Three door problem. This problem was a common topic of discussion some years ago. The host in a television show has a set with three doors. Behind one is a prize, say a certificate entitling the contestant to a brand new Maserati. The contestant may choose one door at random, giving a one-in-three chance of winning the prize. He selects a door. At this point the host intervenes and offers to open one of the two remaining doors. She does so, and then offers the contestant the option of changing his selection to the remaining closed door. Should he do so? It may be assumed that the host will not open a door that contains the prize.

2.22 *Warning light 1*. Warning lights and devices are used in many applications, for example fire alarms, lights for oil level in an automobile and devices for methane gas detection in coal mines. In automobile dashboards, warning lights are used rather than the more expensive gauges. Consider for example, a device to indicate event T = 'oil level below a specified level'. A warning light comes on – event L – to indicate that T might have occurred. Tests can be conducted to obtain $\mathbf{Pr}(L|T)$ and $\mathbf{Pr}(\tilde{L}|\tilde{T})$. Let these values be denoted p and q respectively, and let $\mathbf{Pr}(T) = t$. Find $\mathbf{Pr}(T|L)$, $\mathbf{Pr}(\tilde{T}|L)$, $\mathbf{Pr}(T|\tilde{L})$ and $\mathbf{Pr}(\tilde{T}|\tilde{L})$ in terms of p, q and t.

2.23 *Warning light 2*. The devices in Problem 2.22 are designed to indicate the occurrence of a rare event. This can pose difficulties in the design of the system. Consider the following problem. A warning device on an aircraft includes a light that is 'on' when the landing gear is correctly positioned. It accurately reflects the landing gear correctly in position 99.9% of the time. The remaining 0.1% corresponds to failure of the lighting system in the device, incorrectly indicating failure of the landing gear to position itself. Given that failure of the landing gear occurs on average once in 10^6 landings, what is the probability that the landing gear is incorrectly positioned if the light indicates 'failure'? Assume that the warning device works perfectly when the landing gear has failed. Comment.

This problem might be analysed with a real-world application in mind; see *Aviation Weekly and Space Technology*, July 30 and August 6, 1973. A Lockheed L-1011 aircraft crashed with many fatalities after there was doubt whether the front landing gear was in the down position. The bulbs in the system indicating the position of the landing gear were found after the crash to have burned out, and the probable cause of the accident was preoccupation of the crew regarding the landing gear position, resulting in distraction from the instruments and altitude of the aircraft.

2.24 An electronic company 'A' manufactures microprocessors. An important competitor, 'B', has introduced a new product which is selling well. Company A has to decide whether to introduce a competing microprocessor. There are concerns regarding the response of company B, which could expand production, keep it at the same level, or reduce. From past experience, the chances of the three actions are respectively 60%, 30%, and 10%. One source of information to company A is to study B's advertising campaign. From past experience, they have assigned probabilities to the rival reducing, maintaining, and increasing their advertising given that it is their intention to cut, keep at same level, or expand production as in the table below.

		Advertising is		
		decreased	maintained	increased
	cut back	0.60	0.30	0.10
Company B is going to	keep same level	0.30	0.50	0.20
	expand	0.10	0.30	0.60

It is then observed that advertising is in fact increased. What are the revised probabilities for the rival's course of action?

2.25 In an industrial plant, a flow meter is used to measure the direction of flow in a pipe, with an electrical signal being sent back to the control room. Under ideal conditions, a positive signal should arrive when the flow is in one direction – positive flow – with a negative signal level when the flow is in the opposite, or negative, direction. In reality, a positive signal level can be observed even if the flow is negative and *vice versa*, as a result of turbulence and system noise. The possible received signal levels X are $-2, -1, 0, +1, +2$. The probabilities of the value of X conditional on the direction of the flow are as follows.

x	$\mathbf{Pr}(X = x\,|\,\text{positive flow})$	$\mathbf{Pr}(X = x\,|\,\text{negative flow})$
+2	0.40	0.00
+1	0.30	0.15
0	0.20	0.15
−1	0.05	0.30
−2	0.05	0.40

It is also known that positive flow is 1.5 times as likely as negative flow. The operator in the control room must make a decision based on the received signal level. She assumes that the more likely flow direction indicates the actual flow direction.[8] In other words, upon receiving signal level $X = x$, if **Pr**(positive flow | signal level $X = x$) $\geq$ **Pr**(negative flow | signal level $X = x$), the operator takes the flow direction to be positive. Otherwise the flow direction is taken to be negative.

(a) Find the signal levels which cause the operator to assume that the flow is positive.
(b) What is the probability that the operator's assumption of flow direction is incorrect?

2.26 In the construction of a gas pipeline, the number of flaws due to welding at joints in the metal is a question of major importance since the flaws are associated with possible subsequent failure of the pipeline. The engineer in charge first studies previous pipelines of similar construction to estimate the number of major flaws, S, per section. On this basis, the engineer decides that $S = 0$, 1, or 2 **major flaws** are possible, with probabilities of 0.92, 0.05 and 0.03 respectively. An expensive X-ray testing procedure can be used to detect the presence and depth of the flaws; the results are embodied in the values $I = 0$, 1 or 2, representing the number of major flaws **indicated** by the X-ray machine. Non-destructive testing methods of this kind are not perfect, and previous use of similar equipment yields the following conditional probabilities **Pr**($I = i | S = s$).

		I	
S	0	1	2
0	0.80	0.15	0.05
1	0.10	0.70	0.20
2	0.10	0.10	0.80

Calculate the matrix of values **Pr**($S = s | I = i$).

2.27 A message is sent from a transmitter to a receiver via three relays. Bits (0 or 1) are transmitted. The relays may commit a reversal error in which case a '1' is sent when a '0' is received and *vice versa*. The probability of a reversal error in each relay is 0.03. Stochastic independence from one relay to another can be assumed.

(a) Calculate the probability that the received bit is identical to the transmitted bit.
(b) From the past, it is known that 60% of all bits transmitted are '1'. Find the probability that the received bit is a '1' and the probability that a '1' was transmitted given that a '1' has been received.

---

[8] One must beware of converting uncertainty to certainty. A decision could be based upon the signal but this does not mean that flow is positive or negative – in general only a probability can be estimated.

2.28 In the two systems below, the components have a reliability of 98%. This means that the probability of failure is 0.02. Find the system reliability in each case. Stochastic independence should be assumed. The systems are operational if there is a working path from A to B.

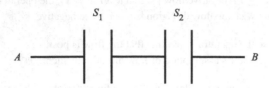

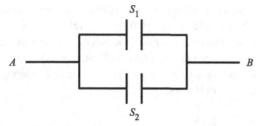

2.29 Describe an example of dependence that is conditional on our state of knowledge.

2.30 A batch of ten cables contains three defective cables. A random sample of four cables is taken from this batch and tested. Find

(a) the number of ways there can be exactly two defective cables in the sample, and
(b) the number of ways there can be at least two defective cables in the sample.

2.31 A water diviner claims that he can locate hidden sources of water using a carved whalebone. A test is conducted by asking him to separate ten covered cans into two groups of five, one consisting of cans containing water. It is arranged beforehand that five cans contain water. What is the probability that the diviner correctly puts at least three cans into the 'water group', entirely by chance?

2.32 At the 'Group of Eight' meetings in 2002, what is the probability that George Bush and Tony Blair stand next to each other at the photo-op, purely by chance? How many ways can the eight persons be seated around a circular table?

2.33 An urn contains seven red balls and three green balls. A second urn contains four red balls and six green balls. A ball is chosen randomly from the first urn and placed in the second urn. Then a ball is chosen randomly from the second urn and placed in the first urn.

(a) What is the probability that a red ball is selected both from the first urn and then from the second urn?
(b) What is the probability that a red ball is selected from the second urn?
(c) At the conclusion of the selection process, what is the probability that the numbers of red and green balls in each urn are identical to the initial numbers?

Now consider the generalization to the case where there are $n(> 1)$ balls initially in each urn, with a (such that $0 < a < n$) red balls in urn 1 and b (with $0 < b < n$) red balls in urn 2 initially.

(d) Show that the probability that a red ball is selected from the second urn during this process is, in general, different from the probability of drawing a red ball directly from the second urn, without transferring a ball from the first urn.

(e) Find a necessary and sufficient condition on the values of a and b for the two probabilities in part (d) above to be equal.

2.34 Find a proof of Equation (2.49) based on the expansion of Newton's formula for $(p + q)^{m+n}$.

2.35 *Two age-old problems.*

(a) Find the number of distinct ways of arranging the letters in the word 'STATISTICS'.
(b) A hand of five cards is dealt from a standard deck of 52 cards. What are the odds against the event that all five cards are of the same suit?

2.36 *Hat problem; also an old problem.*

'I'm a decent man I am I am and I don't want to shout, but I had a hat when I came in and I'll have a hat when I go out!' (Old Newfoundland song)

At a club for men on a wet and windy night all are wearing hats which they check in upon arrival. The person checking the hats becomes confused and returns them at random. What is the probability that no man receives his hat? Consider n men and the change in value of probability as n increases.

2.37 *General model of drawing from urn.* In an urn with balls of two colours, g_1 are of colour 1, and g_2 of colour 2. A ball is drawn, and then replaced. Then h_1 of the colour drawn and h_2 of the other colour are added to the urn and mixed. A ball is now drawn from the urn, which contains $g_1 + g_2 + h_1 + h_2$ balls. The values h_1 and h_2 are arbitrary and could be negative. See Feller (1968). We consider now the case $h_2 = 0$ and $h_1 > 0$, which constitutes a general Pólya urn. Show that the probability of drawing a ball of colour 2 at any drawing is $g_2 / (g_1 + g_2)$.

2.38 Eight passengers share a compartment in a train which stops at eleven stations. What is the probability that no two passengers leave at the same station?

3 Probability distributions, expectation and prevision

Bring out number, weight and measure in a year of dearth.

William Blake, *Proverbs of Hell*

3.1 Probability distributions

We are now in the following situation. We have a probability mass of unity, and we wish to distribute this over the possible outcomes of events and quantities of interest. We also wish to manipulate the results to assist in decision-making. We have so far considered mainly random events, which may take the value 0 or 1. A common idealized example that we have given is that of red and pink balls in an urn, which we have labelled 1 and 0, respectively. Thus, the statement $E =$ 'the ball drawn from the urn is red' becomes $E = 0$ or $E = 1$, depending on whether we draw a pink or red ball. The event E either happens or it does not, and there are only two possible outcomes. **Random quantities** can have more than two outcomes. For the case of drawings from an urn, the random quantity could be the number of red balls in ten (or any other number) of drawings, for example from Raiffa's urns of Figure 2.10. A further example is the throwing of a die which can result in six possible outcomes; the drawing of a card from a deck can result in 52 possible outcomes, and so on.

The possible outcomes often take discrete values, as in the examples just given; in the case of a die these are $\{1, 2, 3, 4, 5, 6\}$. Or in the case of the tests on the concrete cylinder of the last chapter (Section 2.1), we might assign (arbitrarily if we wish) the values $1, 2, 3, \ldots$, to intervals of strength, e.g. 0–20 MPa, 20–40 MPa, and so on. The number of vehicles that can queue in the turning lane at a traffic light takes a discrete set of possible values. We have in all of these cases a **discrete random quantity**. This should be contrasted with a **continuous** one, which can take on any value in a range, for instance any strength in the interval 0–100 MPa. Such **continuous random quantities** will have their mass of probability spread out or distributed over an interval.

A random quantity will always be indicated by an upper case letter, for example X, Y, A. The event E above is random; an event is a special quantity with only two possible outcomes. It becomes tedious to insist that entities represented by an upper case letter are *always* random.

Often we have a deterministic quantity which we wish to denote with an upper case letter. Consequently an upper case letter does not necessarily indicate that a quantity is random but a random quantity will always have an upper case. **Probability distributions** are defined over the possible values of a random quantity. We shall introduce some common distributions in this chapter; Appendix 1 contains a listing of probability distributions and their properties, for easy reference.

The possible outcomes that are being considered might be limited to a finite discrete set such as the results of die-tossing or drawing from an urn with specified numbers of 'trials'. If the space in the turning lane is limited so that the maximum number of vehicles is fixed, then again we may set an upper limit to the number of vehicles, which may then be modelled as random and finite. Or there may be a set that we might wish to regard as unbounded. For example, if we are considering the distribution of ages of human beings in a certain population, we might consider the possible values as corresponding to the countably infinite set $\{1, 2, 3, \ldots\}$. No doubt, we could devise an upper bound. But this is often unnecessary, provided that the distribution fitted to the 'tail' assigns negligible probabilities to very great ages. For example, if the probability of a person being greater than, say, 140 years is very small, there might be no consequence to the analysis by including an infinite set.

Another example might be the set of extreme wave heights in considering the design of an offshore structure. There are no doubt physical bounds that could be set on the possible wave heights. One has then to assess whether there is any advantage to such a procedure. The probability distribution fitted to the extreme wave heights might well have negligible probability mass corresponding to 'unrealistically high' values of wave height. This can always be checked after the analysis is complete to see whether the design situation is realistic. There is a danger of bounding a parameter at an inappropriate value and excluding valid events of small probability from consideration.

3.2   Discrete probability distributions; cumulative distribution functions

The probability distribution of the random quantity X is the set of values of probability associated with the possible outcomes. This is written $p_X(x)$ for the discrete case. For the random quantity Y the distribution is denoted $p_Y(y)$, and so on for other quantities. The advantage of this notation is that we can see immediately from say $p_W(\cdot)$ that we are referring to the probability distribution of W. The lower case value of the random quantity indicates that the random quantity is taking on a particular value. For example, for the case of a fair die, with $X =$ outcome of throw,

$$p_X(x) = \mathbf{Pr}(X = x) = \frac{1}{6}, \text{ for } x = 1, 2, \ldots, 6.$$

A further example might be

$$p_Y(y) = \mathbf{Pr}(Y = y) = \frac{y^2}{14}, \text{ for } x = 1, 2, \text{ or } 3.$$

In these cases we may also write

$$p_X(5) = \frac{1}{6}; p_Y(3) = \frac{9}{14}.$$

Probability distributions are defined over a partition, so that by Equation (2.9), we must have that

$$\sum_X p_X(x) = 1 \text{ and } \sum_Y p_Y(y) = 1. \tag{3.1}$$

3.2.1   Repeated trials

A common case of practical importance is that of 'repeated trials of the same phenomenon'. Examples are repeated tosses of a coin, or drawings of a ball from an urn. These are the basis for a set of distributions such as the binomial one, to be explained in the following. Instead of coins or coloured balls, consider the following events, which have a more practical flavour.

1. *'The strength of the steel sample is greater than 400 MPa.'*
2. *'The component failed under the given voltage.'*
3. *'This item is defective.'*

Each of these events has a complement and can be considered as having outcomes of 0 or 1. Repeated trials correspond to a series of steel specimens, to a set of components, or to a series of tests to tell whether a manufactured item is defective or not.

3.2.2   Bernoulli and binomial distributions

When we consider the probability of success ($E = 1$ for event E) to be constant and known for each trial we have the **binomial** case. A single trial is termed a **Bernoulli trial** and the series is a binomial process. In case 1 above, if we have substantial data from samples of similar steel, the probability might reasonably be considered constant and known; this might also be true in the other cases. In the idealized example of the urn, the case of constant and known probability corresponds to the case where the composition is known and the ball is replaced after drawing, with the balls subsequently mixed up by shaking before the next draw. Alternatively we may consider an urn with a very large number of balls with the proportion of red to pink known. Then if the number of balls is much greater than the number of draws, the probability of a red or pink in a draw is approximately constant. Each draw is a Bernoulli trial: the probability remains the same on each draw. The judgement of stochastic independence of the trials is required for the probability to remain the same. This situation prevails in coin-tossing, die-throwing or in drawing cards from a deck if the cards are returned and the deck reshuffled.

If the Bernoulli trial is repeated n times, the binomial process results. Let X be the total number of successes in these n trials;

$$X = \sum_{i=1}^{n} E_i,$$ (3.2)

where E_i denotes the ith Bernoulli (0:1) trial. The probability distribution of X is the **binomial distribution**, i.e.

$$\Pr(X = x) = p_X(x) = \binom{n}{x} p^x (1 - p)^{n-x} = \binom{n}{x} p^x q^{n-x},$$ (3.3)

where $\Pr(E_i) = p$ and $q = 1 - p = \tilde{p}$ for each trial i. We can model the process as a special case of Figure 2.19, in which we take successive trials as being independent with constant probability p. For a particular path in Figure 2.19, the probability of x successes is p^x and $(n - x)$ failures is $(1 - p)^{n-x}$. The binomial coefficient $\binom{n}{x}$ adds up all the paths to the node (n, x) in this random walk diagram, counting all the orders of successes and failures. The binomial distribution of Equation (3.3) is often abbreviated to $b(x, n, p)$.

There are $\sum_x \binom{n}{x} = 2^n$ possible paths to all nodes for given n; see Table 2.1. We can verify this using Equation (2.46) by substituting $p = q = 1$ (note that these cannot be probabilities!). If all of the paths are equally likely, Equation (3.3) becomes $\binom{n}{x} / 2^n$. This corresponds to Bernoulli trials with a probability of $1/2$ at each trial. The binomial distribution is symmetrical if $p = 1/2$. We see also that Equation (3.3) corresponds to terms of Newton's expansion $(p + \tilde{p})^n$; see Equation (2.47).

Probability distributions for discrete random quantities represent masses of probability attached to the various possible outcomes, and are denoted **probability mass functions** (*pmf*s).

Example 3.1 Examples of Bernoulli ($n = 1$) and binomial ($n = 5$) distributions are given in the upper of the pairs of the diagrams of Figure 3.1, showing the *pmf*s for $p = 0.5$ and $p = 0.8$. The results could correspond to the probability of success in one or five trials on devices with known reliability of 0.5 and 0.8. In five trials, the number of successes could be any value from 0 to 5. The individual probability values are calculated using Equation (3.3).

Cumulative distribution functions (*cdf*s)

We now introduce the **cumulative distribution function** or *cdf*, $F_X(x)$. This is defined as

$$\Pr(X \leq x) = F_X(x) = \sum_{X \leq x} p_X(x).$$ (3.4)

Examples are shown in Figure 3.1 and other figures in the following. It is seen that one simply accumulates the probability as the value of the quantity increases. The value of $F_X(x)$ jumps in value as each new probability is encountered as x increases. Finally the *cdf* is equal to unity.

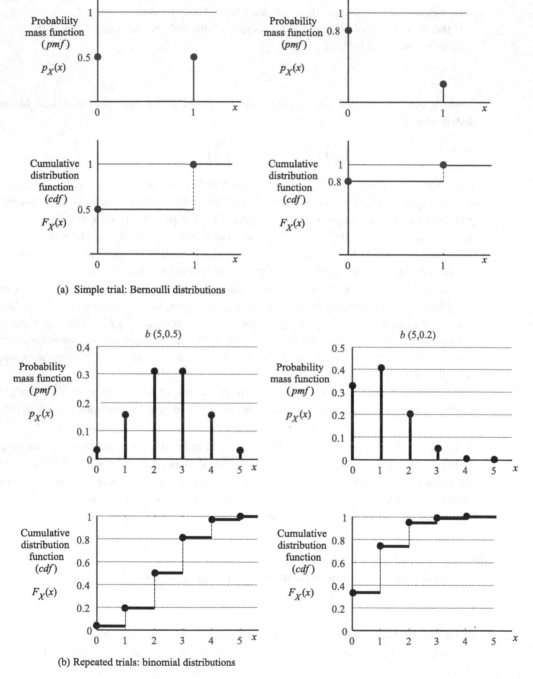

(a) Simple trial: Bernoulli distributions

(b) Repeated trials: binomial distributions

Figure 3.1   Bernoulli and binomial distributions

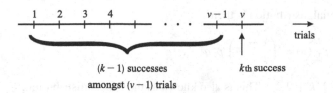

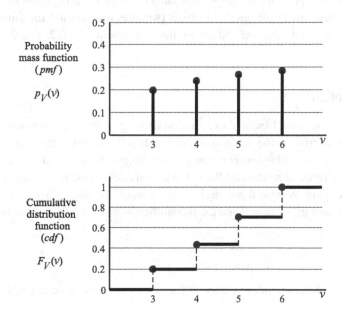

(a) Basis for derivation

(b) Example distribution

Figure 3.2   Negative binomial distribution

3.2.3   Negative binomial distribution

Rather than the number of successes in n trials, we might be interested in the number of trials, say V, to the kth success; V is random. It is assumed that the kth success occurs when $V = v$, and that $(k - 1)$ successes are dispersed amongst the preceding $(v - 1)$ trials; see Figure 3.2(a). The probability of the second of these conditions is given by the binomial distribution

$$\binom{v - 1}{k - 1} p^{k-1} q^{v-k}. \tag{3.5}$$

Since the probability of the final success on the vth trial is p, we multiply (3.5) by p. The negative binomial distribution is then

$$\mathbf{Pr}\,(V = v) = p_V\,(v) = \binom{v-1}{k-1}\,p^k q^{v-k}, \tag{3.6}$$

for $v = k, k+1, k+2, \ldots$ This is also known as the Pascal distribution.

Example 3.2 We are sampling for defectives in a batch in which the probability of a defective is 0.2, and we are interested in the number of trials (samples tested) until the third defective is found. Figure 3.2(b) shows the *pmf* and *cdf* for the case where $p = 0.2$, $k = 3$, calculated using Equation (3.6).

3.2.4   Geometric distribution

A special case of the negative binomial distribution is the geometric distribution, where we consider the number of trials to the first success, rather than the general kth value. Thus $k = 1$ in the negative binomial distribution. For example, we might be interested in the number of years to the first occurrence of an extreme flood. Let us consider a flood level with a probability of exceedance p per year. A typical practical situation would be the level with a 1% probability of exceedance in a given year. Then the probability of $v - 1$ years without the extreme flood is

$$(1 - p)^{v-1},$$

assuming, as before, stochastic independence. If the first extreme flood occurs in the vth year, then

$$\mathbf{Pr}\,(V = v) = p_V\,(v) = (1 - p)^{v-1}p = pq^{v-1}, \text{ for } v = 1, 2, \ldots, \tag{3.7}$$

since the extreme event is preceded by $(v - 1)$ years without an extreme occurrence (there is only one path to this result in Figure 2.19). Equation (3.7) represents the **geometric** distribution.

Example 3.3 We wish to obtain the probability of 100-year design storm being first exceeded in a specified year. Then $p = 0.01$. The probability varies from 0.01 for year 1 ($v = 1$) to 0.0083 for year 20 ($v = 20$), to 0.0037 for year 100 ($v = 100$).

3.2.5   Hypergeometric distribution

This can be introduced using an urn of known composition. It contains b balls, g of which are red and $(b - g)$ pink. Drawing is carried out without replacement. The probability of obtaining X 'successes' (red balls) in n trials (drawing of balls from the urn) is given by the

hypergeometric distribution. Writing the result for this in full:

$$\Pr(X = x|b, n, g) = \frac{\dfrac{g!}{x!(g-x)!} \cdot \dfrac{(b-g)!}{(n-x)!(b-g-n+x)!}}{\dfrac{b!}{n!(b-n)!}}$$

$$= \frac{\dbinom{g}{x}\dbinom{b-g}{n-x}}{\dbinom{b}{n}},$$

(3.8)

for $x = 0, 1, 2, \ldots$, up to the lesser of g, n. Equation (3.8) gives the number of ways, or combinations, of choosing x red balls from g and $(n-x)$ pink balls from $(b-g)$, divided by the number of ways of choosing n from b. All of these choices are without regard to order: see Equation (2.44).

The right hand side of Equation (3.8) may be written as

$$\frac{n!}{x!(n-x)!} \cdot \frac{(g)_x(b-g)_{n-x}}{(b)_n}.$$

(3.9)

The expression $\binom{n}{x} = n!/[x!(n-x)!]$ isolated on the left hand side of expression (3.9) is again the number of ways of getting to the node (n, x) in our random walk diagrams (Figure 2.19). Since all paths are equally likely, the probability of a single path is simply

$$\frac{(g)_x(b-g)_{n-x}}{(b)_n}.$$

(3.10)

This can be understood as an application of the multiplication principle (Section 2.6): there are $(g)_x$ ways (permutations) of choosing the x red balls, $(b-g)_{n-x}$ ways of choosing the $(n-x)$ pink balls and $(b)_n$ ways of choosing the n balls regardless of colour. Therefore Equation (3.10) gives the probability of getting to node (n, x) by a particular path since the order is specified. Note that if $b, g \to \infty$ with g/b constant and equal to p, then Equation (3.10) tends to $p^x(1-p)^{n-x}$, the probability of a single path in the binomial distribution. Also, if $n = b, x = g$, Equation (3.9) is equal to unity since we have sampled every ball in the urn. Equation (3.10) is then equal to the inverse of $\binom{b}{g}$, the number of paths to the endpoint. Each is equally likely.

It should be noted the probability of red (success) in each trial is constant unless one has the results of the preceding trials: see Exercise 2.37 and Example 3.5 below.

There are four forms of the hypergeometric distribution that give different insights; see Exercise 3.6.

Example 3.4 Typical probability distributions for the Raiffa urn problem of Figure 2.10 are shown in Figure 3.3.

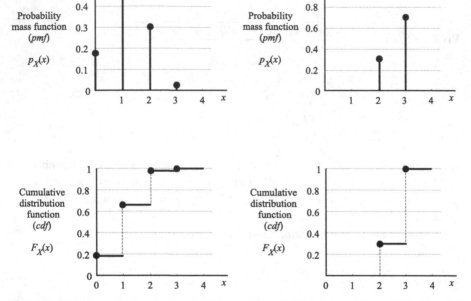

Figure 3.3   Hypergeometric solutions with three trials to Raiffa's urns of Figure 2.10; θ_1 urn left, θ_2 urn right

Example 3.5 Consider Raiffa's θ_1 urn (Figure 2.10). The probability of drawing red on the first draw is $4/10$. On the second, without knowledge of the first result, the probability is

$$\frac{4}{10} \cdot \frac{3}{9} + \frac{6}{10} \cdot \frac{4}{9} = \frac{36}{9 \cdot 10} = \frac{4}{10}.$$

Example 3.6 Consider the game Lotto 6–49, most popular in Canada in recent years. One buys a ticket in the lottery for a small sum and enters six different numbers from the set of 49 consecutive numbers $\{1, 2, 3, \ldots, 47, 48, 49\}$. The lottery organization also chooses six, presumably at random, and if our six numbers match theirs, regardless of order, then we win first prize – on occasions, millions of dollars.

We may regard the six winning numbers as 'special' and the remaining 43 numbers as 'ordinary'. The probabilistic solution to the problem is the standard hypergeometric distribution, in which we choose six numbers, which may be of two kinds, special or ordinary. The probability of drawing x special numbers (specified by the lottery organization) in six trials, each trial being our choice of a number in entering the lottery, is as follows, from Equation (3.8):

$$\mathbf{Pr}(X = x) = \frac{\binom{6}{x}\binom{43}{6-x}}{\binom{49}{6}}, x = 0, 1, 2, \ldots, 6.$$

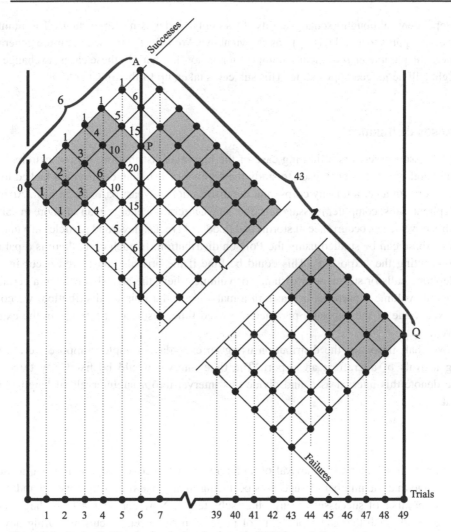

Figure 3.4 Lotto 6–49: winning is the single path *OAQ* through the maze!

For $x = 6$, we deduce the probability of winning as 1:13 983 816. The result is shown graphically in Figure 3.4; in the figure, if $x = 4$, the result given by the equation above can also be seen as

$$\frac{(\text{No. of paths from } O \text{ to } P) \times (\text{No. of paths } P \text{ to } Q)}{\text{No. of paths from } O \text{ to } Q}$$

The winning result ($x = 6$) corresponds to the single path *OAQ* divided by the number of paths from *O* to *Q*. The end point, *Q* in Figure 3.4, represents always the composition of the urn.

The hypergeometric distribution is of fundamental importance in studying exchangeability and has many other practical as well as theoretical applications. The successive trials are not independent, although exchangeability does apply (order is not important). The number of orders, or paths to a node, is $\binom{n}{x}$, as shown above. We have mentioned the more general case where the parameter p or the urn composition is not 'known'. In these cases, exchangeability might still be judged appropriate. This subject is taken up further in Chapter 8.

3.2.6   Poisson distribution

The Poisson process is a limiting case of the binomial distribution. It is of extremely wide applicability and has been used to study the number of Prussian cavalrymen kicked to death by a horse in a certain army corps, and the distribution of yeast particles in beer. Arrivals of telephone calls, computer messages, traffic, ships, aircraft, icebergs at an offshore installation, radioactive decay, occurrence of storms and floods, of cracks, flaws and particles in a material – all of these can be studied using the Poisson distribution. In each case, there is a parameter representing the 'exposure'. This could be time if the arrivals considered occur in time – telephone calls or storms – or distance or volume where the arrivals occur in a certain distance or volume – particles or flaws in a material. In the Poisson distribution, we consider a fixed value of the exposure parameter: a fixed time, distance or volume in the examples given.

We shall introduce the distribution using an everyday example. Suppose we are studying arrivals of aircraft at an airport. The time interval would be fixed, say three hours. We denote this interval as t, and divide this interval into n subintervals of length Δt such that

$$t = n\Delta t.$$

Within each subinterval, an arrival of an aircraft is a 'success', and non-arrival a 'failure'. We judge that the number of successes occurring in any one time subinterval is independent of the number of successes in any other subinterval. Now consider n to be large, and Δt to be small, such that the probability of more than one event occurring during any Δt is negligible. Let the rate of events – number per unit time – be λ, on average. For example, extensive records of past arrivals of aircraft might indicate that $\lambda = 5.6$ aircraft per hour arrive on average during the interval of three hours under consideration. Then, for each small Δt, the probability of an event is $p = \lambda \Delta t$. We consider λ to be constant for the time period t. For any subinterval

$$p = \lambda \Delta t = \frac{\lambda t}{n} = \frac{\nu}{n}, \tag{3.11}$$

where $\nu = \lambda t$ is the expected (mean) number of successes during the time period t. The process that we are describing is a binomial one: a series of n independent trials with a probability of success equal to p on each trial.

If X is the number of successes in the n trials, then

$$\Pr(X = x) = p_X(x) = \binom{n}{x} p^x (1 - p)^{n-x}$$

$$= \frac{n!}{x!\,(n-x)!} \left(\frac{\nu}{n}\right)^x \left(1 - \frac{\nu}{n}\right)^{n-x}$$

$$= \frac{\nu^x}{x!} \left(1 - \frac{\nu}{n}\right)^n \left\{ \frac{\overbrace{n\,(n-1)(n-2)\cdots(n-x+1)}^{x \text{ terms}}}{\left[n\left(1 - \frac{\nu}{n}\right)\right]^x} \right\}, \tag{3.12}$$

in which we have rearranged the terms. We now let $n \to \infty$ and $\Delta t \to 0$, while keeping t constant. As a result, $\nu = \lambda t$ is also constant since λ is as well. The ratio of the x terms in the numerator and the corresponding ones in the denominator within the braces {} of Equation (3.12) tends to unity, while

$$\left(1 - \frac{\nu}{n}\right)^n = 1 - n\left(\frac{\nu}{n}\right) + \frac{n!}{2!\,(n-2)!}\left(\frac{\nu}{n}\right)^2 - \frac{n!}{3!\,(n-3)!}\left(\frac{\nu}{n}\right)^3 + \cdots$$

$$\to 1 - \nu + \frac{\nu^2}{2!} - \frac{\nu^3}{3!} + \cdots$$

$$= \exp(-\nu).$$

Then

$$\Pr(X = x) = p_X(x) = p(x, \nu) = \frac{\exp(-\nu)\,\nu^x}{x!} \tag{3.13}$$

for $x = 0, 1, 2, 3, \ldots$.

This is the Poisson distribution often associated with a stochastic process (Section 3.6).

Example 3.7 Long distance calls at a particular exchange arrive on average at a rate of one every four minutes. This rate can be considered to be constant for the time periods under consideration. Suggest probability distributions for the number of calls in (a) five minutes and (b) an hour. In case (b), what is the probability that the number of calls exceeds 15? 20? 30? An appropriate distribution is the Poisson distribution with (a) $\nu = 5 \cdot \frac{1}{4} = 1.25$ and (b) $\nu = 60 \cdot \frac{1}{4} = 15$ calls. The distributions are plotted in Figure 3.5. In case (b), if X is the number of calls,

$$\Pr(X > 15) = 1 - F_X(15) = 1 - 0.5681 = 0.4319,$$

$$\Pr(X > 20) = 1 - F_X(20) = 1 - 0.9170 = 0.0830,$$

$$\Pr(X > 30) = 1 - F_X(30) = 1 - 0.9998 = 0.0002.$$

The values of F_X for the calculations can be obtained from standard computer software, from statistical tables or they can be computed from first principles.

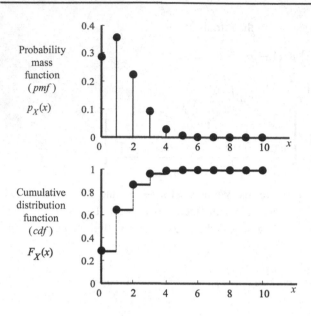

(a) Poisson distributions with $\nu = 1.25$

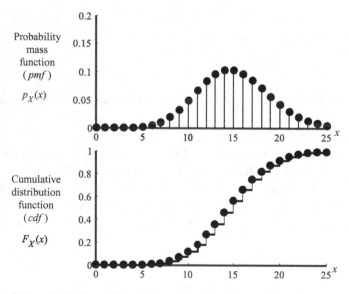

(b) Poisson distributions with $\nu = 15$

Figure 3.5   Poisson distribution with different values of ν

In the derivation above and in the examples, λ was taken as a constant. The main result, Equation (3.13), can be used if λ is not constant provided that one uses the correct expected value. In this case

$$\nu = \int_0^t \lambda(\tau) \, d\tau \tag{3.14}$$

where λ varies over the interval from 0 to t. One can think of 'pseudotime' if one wishes such that the rate is constant with respect to this parameter rather than ordinary time (or distance, volume, and so on).

3.3   Continuous random quantities; probability density; distribution functions for the combined discrete and continuous case

In the previous section, 'weights' of probability were assigned to the discrete points corresponding to the possible outcomes of the random quantity. In some limiting distributions, we might be interested in a countably infinite set of discrete outcomes, for example in the Poisson distribution just discussed. The random quantity can then take any value in the infinite sequence $\{0, 1, 2, 3, \ldots\}$. The approach is the same as for a finite set of results: the sum of the 'weights' of probability must approach unity as the number of outcomes tends to infinity.

An entirely different case, but one which also involves infinite sets, arises when we have a quantity which is continuous over some range. For instance, a voltage across a circuit could take any value in a range, say 0 to 5 Volts. For such a **continuous random quantity**, we 'smear' out the probability into a density, much as we smear out a body's mass when considering its density, or as we smear out a load in considering a distributed load. In the present case, we smear out our degree of belief over a continuous range.

An amusing initial example will illustrate the approach. This is Lewis Carroll's 'equally frangible rod'.[1] By this is meant a rod, or let us say a specimen of steel of uniform cross section, which is subjected to a tensile stress in a test machine. The probability of failure at any point along the length of the rod is assumed to be equal to that at any other point. In reality, there may be an increased probability of failure near the grips which attach the specimen to the test machine, as a result of stress concentrations. But let us accept the idealization of equal probability for the present illustration.

Figure 3.6(a) shows Lewis Carroll's rod. The length is 'a', and the distance from the left hand end to the breaking point is x. Since the value of x is not known, we denote it in general as a random quantity X. We denote the **probability density** of X as $f_X(x)$ and this is defined such that

$$\Pr(x \leq X \leq x + dx) = f_X(x) \, dx. \tag{3.15}$$

---

[1] Lewis Carroll (Charles Dodgson) *Pillow Problems*. Reprinted by Dover, 1976.

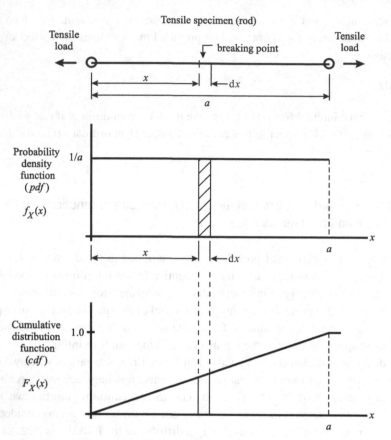

Figure 3.6   (a) Lewis Carroll's 'equally frangible rod'. The *pdf* is a uniform function of *x* and the *cdf* increases linearly with distance *x*

The density is illustrated in Figure 3.6(a) for the equally frangible rod and in Figure 3.6(b) for a density where the probability is not uniformly distributed: certain values of X are more likely than others. The probabilities of all the possible values of X must normalize so that

$$\int_{\text{all } X} f_X(x)\, dx = 1. \tag{3.16}$$

In the case of the rod with uniformly distributed probability of the breaking point,

$$f_X(x) = \frac{1}{a},\ 0 \le x \le a$$
$$= 0\ \text{otherwise.} \tag{3.17}$$

This probability density normalizes in accordance with our desire for coherence expressed in Equation (3.16) above. Equation (3.17) is our first example of a **probability density function**.

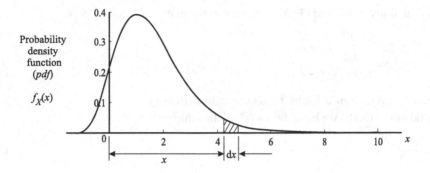

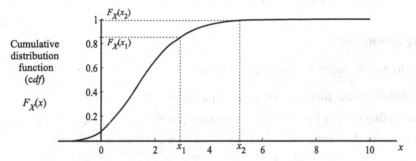

Figure 3.6   (b) Schematic illustration of a *pdf* whose density varies non-uniformly, with its associated *cdf*

We shall abbreviate this term to '*pdf*'. Figure 3.6(a) shows the *pdf* for the breaking point, and (b) shows that for a more general non-uniform distribution.

The cumulative distribution function (*cdf*) is defined for a continuous random quantity as

$$F_X(x) = \mathbf{Pr}(X \leq x)$$
$$= \int_{-\infty}^{x} f_X(u)\,du, \tag{3.18}$$

in which u is a dummy variable. This is analogous to the summation in Equation (3.4). In the case of the breaking point, (3.18) becomes

$$F_X(x) = \int_{-\infty}^{x} f_X(u)\,du = \int_{0}^{x} \frac{1}{a}\,du = \frac{x}{a} \text{ for } 0 \leq x \leq a. \tag{3.19}$$

For example, the probability of the rod breaking in the first third of its length is $1/3$ corresponding to $0 \leq x \leq a/3$. We might at this point raise Lewis Carroll's original question: 'If an infinite number of rods be broken: find the chance that one at least is broken in the middle'. His answer is $0.632\,12\ldots$, and the method is given in Appendix 2. But there are certain implications of his method as discussed there.

From Equations (3.15) and (3.18), we can also see that

$$\mathbf{Pr}(x_1 < X \leq x_2) = \int_{x_1}^{x_2} f_X(x)\,\mathrm{d}x$$
$$= F_X(x_2) - F_X(x_1). \tag{3.20}$$

This is illustrated in Figure 3.6(b), for a non-uniform density.

From Equation (3.18) we have the useful relationship

$$f_X(x) = \frac{\mathrm{d}F_X(x)}{\mathrm{d}x}. \tag{3.21}$$

For example, if $F_X(x) = 1 - \mathrm{e}^{-\lambda x}$, $f_X(x) = \lambda \mathrm{e}^{-\lambda x}$. This is the exponential distribution, which we shall now derive for a particular application.

3.3.1 Summary of notation

We have adopted the following shorthand notation:

pmf = probability mass function, for discrete quantities,

pdf = probability density function, for continuous quantities,

cdf = cumulative distribution functions, for discrete or continuous quantities.

3.3.2 Exponential distribution

We introduced probability density functions by means of a simple uniform density on Lewis Carroll's rod, and illustrated a non-uniform density in Figure 3.6(b). We now consider a specific example of a non-uniform density. This is the waiting time to an event. The event could be the failure of a component, for example an electrical or mechanical component in an assembly or the warning lights in Exercises 2.22 and 2.23, or the arrival of a packet of information in a computer network, or a vehicle in a traffic problem, or a customer in a financial enterprise. Let T be the (random) time to the first event; there could be several failures, arrivals, events, and we are interested in the time to the first. We make the assumption that the probability of an event occurring in the time interval Δt is $(\lambda \Delta t)$ where λ, the rate of arrival per unit time, is a constant. This is the assumption in the Poisson distribution of Section 3.2.6 above.

Since λ is constant, the probability $(\lambda \Delta t)$ is the same for any interval Δt in T. The interval Δt is small enough that the probability of more than one event in any Δt is negligible. Let $n \Delta t = t$. Then the probability of no events in the first n intervals is $(1 - \lambda \Delta t)^n$. This follows from the binomial distribution, using independence. If this period of time $t = n \Delta t$ with no event is followed directly by an event, then

$$\mathbf{Pr}(t < T \leq t + \Delta t) = f_T(t)\Delta t$$
$$= (1 - \lambda \Delta t)^n \lambda \Delta t, \tag{3.22}$$

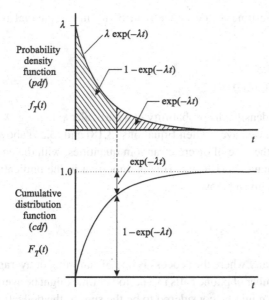

Figure 3.7   Distribution of waiting times and exponential 'tail'

where $f_T(t)$ is the probability density function of T. As noted, we have used in the above the binomial process with n failures followed by one success – a single-path route to the first success (Figure 2.19). Now Equation (3.22) can be written as

$$f_T(t) = \lambda(1 - \lambda\Delta t)^n$$
$$= \lambda(1 - \frac{\lambda t}{n})^n.$$

In the limit of large n,

$$f_T(t) = \lambda \exp(-\lambda t), t \geq 0$$
$$= 0 \text{ for } t < 0. \tag{3.23}$$

This is the **exponential distribution of waiting times**, an example of which is given in Figure 3.7.

The *cdf* for the exponential distribution is as follows, by Equation (3.18), for $t \geq 0$.

$$\mathbf{Pr}(T \leq t) = F_T(t) = 1 - \exp(-\lambda t). \tag{3.24}$$

Figure 3.7 illustrates the exponential distribution and also the fact that the 'tail' area is given by the simple exponential function $\exp(-\lambda t)$. We shall encounter such 'exponential tails' later on in extremal analysis.

If we require the probability of the time to the first arrival being in the interval t_1 to t_2 seconds, this is given by

$$\mathbf{Pr}(t_1 < T \leq t_2) = F_T(t_2) - F_T(t_1)$$
$$= \exp(-\lambda t_1) - \exp(-\lambda t_2).$$

When probabilities are specified by a density, the probability of a particular value, e.g. $X = x_1$ or $T = t_2$, is zero by Equation (3.15); we have written Equations (3.18) and (3.24) above in a form equivalent to Equation (3.4) in the case of discrete random quantities, with the equality included in $\mathbf{Pr}(X \leq x)$ so as to maintain consistency. This is important in the unification of discrete and continuous distributions, given below.

3.3.3   Normal (Gaussian) distribution

This is a common distribution, especially where the process is one of summing or averaging a number of random quantities. The number of phone calls in a period of time might be averaged; the deflection of a structural member might be considered to be the sum of the deflections of a number of elements; we might find the average of a set of data points measuring distance, mass or any quantity of interest. Many observations have been found naturally to follow the distribution, including early ones on heights of human beings and other measurements, for example chest size (Bernstein, 1996). These measurements are doubtless the result of many additive random factors including genetic ones. The process of summing and averaging is studied in detail in Chapter 7, where the normal distribution is found as a limiting distribution to these processes. The normal distribution also appears naturally as that which maximizes entropy subject to the constraints that the mean and standard deviation are fixed (Chapter 6). At the same time, it must be emphasized that many quantities of engineering interest, for example extreme loads, are generally not normal distributed. In Chapter 9 we see that such processes tend to the different class of **extremal** distributions. The conclusion is to choose distributions carefully in applications.

The **normal** distribution takes the following form:

$$f_X(x) = \frac{1}{\sqrt{2\pi}\sigma} \exp\left[-\frac{1}{2}\left(\frac{x-\mu}{\sigma}\right)^2\right], \tag{3.25}$$

where $\mu =$ mean and $\sigma =$ standard deviation (see Section 3.5). It is the familiar bell-shaped curve shown in Figure 3.8(a). The *cdf* cannot be obtained in closed form integration, but the standard form (see below) is widely tabulated and available in computer programs. It is illustrated in Figure 3.8(b).

Changing μ has the effect of moving the distribution to a new location while σ dictates the spread, with small σ resulting in peaked distributions and large σ more flat ones; see Figure 3.8(c). As a result μ and σ are known as **location** and **scale** parameters. We shall adopt the shorthand notation $N(\mu, \sigma^2)$ for the distribution (3.25). The normal distribution has the

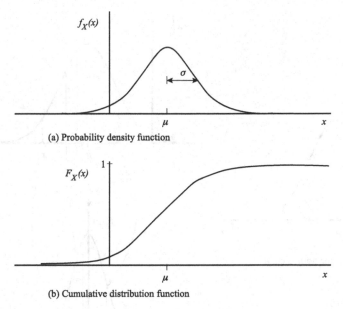

(a) Probability density function

(b) Cumulative distribution function

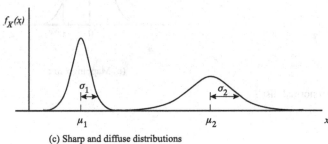

(c) Sharp and diffuse distributions

Figure 3.8   Normal distributions

following properties.

1. It is symmetrical about its mode at $x = \mu$.
2. Points of inflexion occur at $x = \mu \pm \sigma$.
3. The curve approaches the x-axis asymptotically as $x \to \pm\infty$.
4. As with all coherent *pdf*s, the area under the curve is unity. The mean is μ and the variance is σ^2 (Section 3.5).

It is most convenient to consider a linear transformation of the random quantity X, by changing the location and scale. We use the relationship

$$Z = \frac{X - \mu}{\sigma}.$$

(3.26)

Then we ensure that

$$f_Z(z)\,dz = f_X(x)\,dx;$$

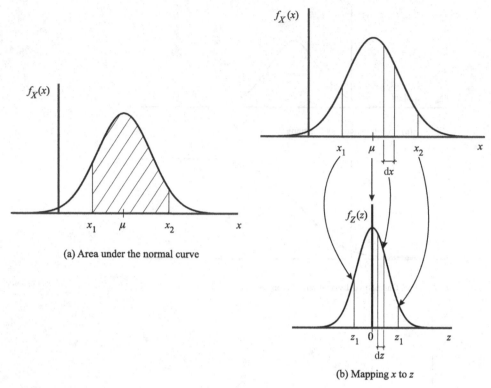

(a) Area under the normal curve

(b) Mapping x to z

Figure 3.9   Use of standard normal distribution

noting that

$$dz = \frac{dx}{\sigma}$$

we obtain the result

$$f_Z(z) = \frac{1}{\sqrt{2\pi}} \exp\left(-\frac{z^2}{2}\right). \tag{3.27}$$

The quantity Z is the **standardized** normal random quantity and the distribution (3.27) is the **standard** normal distribution. Comparison of Equations (3.25) and (3.27) shows that Z has a mean of zero and standard deviation of unity. (The transformation of random quantities must be done according to proper mathematical rules, as outlined in Chapter 7.)

Equation (3.27) is most useful in calculating areas under the normal curve. Consider Figure 3.9; the area under the normal curve between x_1 and x_2 is

$$\mathbf{Pr}\,(x_1 < X \le x_2) = \int_{x_1}^{x_2} \frac{1}{\sqrt{2\pi}\sigma} \exp\left[-\frac{1}{2}\left(\frac{x - \mu}{\sigma}\right)^2\right] dx$$

$$= F_X(x_2) - F_X(x_1).$$

The areas under this curve can be obtained from the standard normal distribution by making the substitution (3.26) and noting that $dz = dx/\sigma$, with the result that

$$\mathbf{Pr}\,(x_1 < X \leq x_2) = \int_{z_1}^{z_2} \frac{1}{\sqrt{2\pi}} \exp\left(-\frac{z^2}{2}\right) dz$$

$$= \mathbf{Pr}\,(z_1 < Z \leq z_2) = F_Z(z_2) - F_Z(z_1). \tag{3.28}$$

The transformation (mapping) from Z to X is shown in Figure 3.9 with z_1 and z_2 corresponding to x_1 and x_2. The standard normal distribution is the same for any problem so that areas under the curve can be tabulated and used in any problem under consideration. Also areas under the normal distribution are widely available in computer software programs, for example MATLAB®. Use of the method is shown in the following examples.

Example 3.8 The strength of masonry for a project can be reasonably approximated using a normal distribution with mean of 18 MPa and standard deviation of 2.5 MPa. Find

(a) $\mathbf{Pr}(X \leq 15)$ MPa,
(b) k such that $\mathbf{Pr}(X \leq k) = 0.2578$,
(c) $\mathbf{Pr}(17 < X \leq 21)$, and
(d) k such that $\mathbf{Pr}(X > k) = 0.1539$.

In case (a),

$$\mathbf{Pr}\,(X \leq 15) = \mathbf{Pr}\left(Z \leq \frac{15 - \mu}{\sigma}\right)$$

$$= \mathbf{Pr}\left(Z \leq \frac{15 - 18}{2.5}\right)$$

$$= \mathbf{Pr}\,(Z \leq -1.20) = 0.1151.$$

In case (b), the z-value corresponding to $k = 0.2578$ is $z = -0.65$ (e.g. from tables). Then, using

$$x = \mu + \sigma z$$

from (3.26),

$$k = 18 + (2.5)(-0.65) = 16.38.$$

For case (c),

$$z_1 = \frac{17 - 18}{2.5} = -0.40 \text{ and } z_2 = \frac{21 - 18}{2.5} = 1.2;$$

consequently

$$\mathbf{Pr}\,(17 < X \leq 21) = \mathbf{Pr}\,(-0.40 < Z \leq 1.20) = 0.5403.$$

The z-value in (d) corresponding to an area of $(1 - 0.1539) = 0.8461$ is $z = 1.02$. Then

$$k = 18 + (2.5)(1.02) = 20.55.$$

3.3.4 __Summary of properties of *pmf*s, *pdf*s, and *cdf*s__

The following summarizes the properties of *pmf*s and *pdf*s.

pmf	*pdf*

1. $0 \le p_X(x) \le 1$ $f_X(x) \ge 0$,
   but not necessarily ≤ 1
2. $\sum_{\text{all } x} p_X(x) = 1$ $\int_{-\infty}^{\infty} f_X(x)\,dx = 1$
3. $\mathbf{Pr}(X = x) = p_X(x)$ $\mathbf{Pr}(x_1 < x \le x_2) = \int_{x_1}^{x_2} f_X(x)\,dx$

It is interesting to note that, in the case of both *pmf*s and *pdf*s, we can write the properties of the *cdf* in a common manner, as follows.

1. $F_X(x) = \mathbf{Pr}(X \le x)$.
2. $0 \le F_X(x) \le 1$ since $F_X(x)$ is a probability.
3. $F_X(-\infty) = 0$.
4. $F_X(\infty) = 1$.
5. If a jump in value occurs, $F_X(x)$ always takes the upper value, as in *cdf*s for discrete random quantities.
6. $\mathbf{Pr}(x_1 < X \le x_2) = F_X(x_2) - F_X(x_1)$.

3.3.5 __Unification of notation__

Some unification arises from this, but it would be a step forward if we could treat the summation over probability mass functions and the integration over the density from a common standpoint. The expressions that we have encountered so far are as follows, for mass and density respectively.

$$F_X(x) = \sum_{x_i \le x} p_X(x_i)$$

$$F_X(x) = \int_{u \le x} f_X(u)\,du. \qquad (3.29)$$

We shall soon encounter summations and integrals of the kind

$$\sum_{x_i} g(x_i) p_X(x_i),$$

$$\int g(x) f_X(x)\,dx. \qquad (3.30)$$

when we calculate expected values.

It would be advantageous to unify these pairs of integrals since they occur frequently, and since they contain the same concept, that is weighting outcomes either by point masses, or by density (smeared probability), in both cases over the possible values of the random quantity. There are two ways of achieving the required unification. One is by the use of Stieltjes integrals

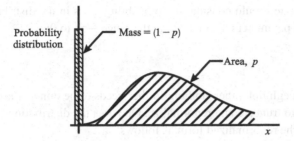

Figure 3.10   Random quantity X with combined discrete and continuous distribution

and the second is by the use of Dirac delta and Heaviside step functions. In both cases the integration includes the jumps over masses as well as the gradual variation associated with distributed masses: see Appendix 2. In the case of Stieltjes integrals, we note the main result, the inclusion within the same integral of both concentrated and distributed masses using one expression. This is given in the right hand side of the following (corresponding to sets of Equations (3.29) and (3.30) above).

$$\left.\begin{aligned} F_X(x) &= \sum_{x_i} p_X(x_i) \\ F_X(x) &= \int f_X(u)\mathrm{d}u \end{aligned}\right\} \rightarrow F_X(x) = \int \mathrm{d}F_X(x) \tag{3.31}$$

$$\left.\begin{aligned} \sum_{x_i} g(x_i)p_X(x_i) \\ \int g(x)f_X(x)\mathrm{d}x \end{aligned}\right\} \rightarrow \int g(x)\,\mathrm{d}F_X(x). \tag{3.32}$$

The term $\mathrm{d}F_X(x)$ in the expressions above should be interpreted either as a sum over a step or a smooth integration over a continuous function. This is the Stieltjes integral. The mathematical background is briefly outlined for the interested reader in Appendix 2.

3.3.6   Mixed discrete and continuous distributions

Occasionally mixed discrete and continuous distributions are needed; for instance, the strength X of small units of a heterogeneous body may be modelled by a continuous distribution together with a 'spike' at $x = 0$, the latter corresponding to the possibility of a flaw traversing the small part. Another instance is the maximum loading X in a year caused by a rare event, such as an earthquake. The spike at $x = 0$ corresponds to the case where the event does not occur within the time period, giving a discrete probability mass at $x = 0$. A diffuse spread of opinion regarding the earthquake severity and consequent load for the case where the value is not zero is shown by the density function adjacent to the spike at zero. In order to be coherent, the sum of the discrete mass and the area under the curve must be unity (see Figure 3.10).

3.3.7   Truncation and renormalization

Occasionally we encounter the case where the random quantity of interest is limited to the range $(0, \infty)$ but we wish nevertheless to model it using a distribution such as the normal, which ranges from $-\infty$ to $+\infty$. First, one should consider the probability mass in the distribution, denoted $f_X(x)$, with the chosen parameters, that occurs in the range $-\infty$ to 0:

$$F_X(0) = \int_{-\infty}^{0} f_X(x)\,dx.$$

If this quantity is deemed to be negligible, then nothing further needs to be done. If not, one way to deal with the problem is to truncate and renormalize. One uses the distribution $f_X(x)$ only in the range 0 to ∞ and in the renormalized form as follows:

$$f_X^*(x) = \frac{f_X(x)}{1 - F_X(0)}.$$

Whether this is appropriate or not is a question of probabilistic judgement.

3.4   Problems involving two or more dimensions

Many, if not most, problems require consideration of more than one random quantity. In a study of the response of a structure, we might be considering wind, wave and earthquake loading on the structure, all of which are random quantities. In an economic study, we might have to consider future interest rates as well as the rate of inflation. These might both be considered as uncertain. In both of these studies, we must consider not only the various random quantities noted, but also their possible dependence, in the stochastic sense, upon each other. Knowing the history of interest rates, we might be able to reduce our uncertainty about the rate of inflation, or *vice versa*. Similarly, knowledge of the wind speed would enable us to estimate, with reduced uncertainty, the wave field.

We shall consider a straightforward example to introduce the approach. This is a model for the movement of people, goods, money, packets of information, or other quantity of interest from one defined area of a city or region to another. The subject will be taken up again in Chapter 6 (Section 6.10) which deals with the subject of entropy maximization. Figure 3.11 illustrates a region divided into n geographical subregions. As an example, we shall focus our discussion onto the movement of people: in particular, we wish to estimate the probability of a person residing in region x working in region y. This could be important in a transportation study. The random quantity X denotes the origin and Y the destination of the person in question (let us say chosen randomly from amongst the population). We are therefore interested in the event $[(X = x)(Y = y)]$. Note that this is a logical product, that is, the event that both $(X = x)$ and $(Y = y)$ occur. As is the case for any event, it takes the values 1 or 0 depending on its truth or falsity.

Example 3.9 The result of the probability assignment can be expressed in terms of a two-dimensional probability mass function $p_{X,Y}(x, y)$, shown in tabular and graphical form in Table 3.1 and Figure 3.12 respectively. The actual assignment in a particular case would be based on past statistics combined with modelling such as the gravity model (Chapter 6).

Table 3.1   *Values of $p_{X,Y}(x, y)$*

| | X = x | | | |
Y = y	1	2	3	4
1	0.15	0.10	0.05	0.0
2	0.10	0.20	0.10	0.0
3	0.05	0.05	0.15	0.05

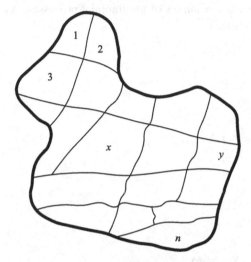

Figure 3.11   Geographical region for transportation study

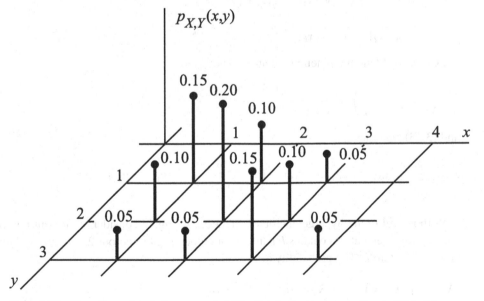

Figure 3.12   Two-dimensional discrete probability distribution corresponding to Table 3.1

3.4.1   Multinomial and hypergeometric distributions

The previous derivation of the binomial and hypergeometric distributions generalizes quite naturally into several dimensions. In the multinomial process, rather than two possible outcomes of each trial, there are k in number. We might consider an urn with balls of k different colours; again if we replace the balls, the probability of drawing each colour remains the same on each draw. Tossing a die several times, taking cards from a deck, again with replacement, sampling from a distribution with known parameters (for example the normal with known μ and σ^2) and putting the results into 'bins', are examples of multinomial processes. As in the case of the binomial process, we sample n times.

If the possible outcomes

$$\{E_1, E_2, \ldots, E_i, \ldots, E_k\}$$

occur

$$\{X_1 = x_1, X_2 = x_2, \ldots, X_i = x_i, \ldots, X_k = x_k\}$$

times with probabilities

$$\{p_1, p_2, \ldots, p_i, \ldots, p_k\},$$

then

$$
\begin{aligned}
\mathbf{Pr}\,(X_1 &= x_1, X_2 = x_2, \ldots, X_i = x_i, \ldots, X_k = x_k) \\
&= p_{X_1, X_2, \ldots, X_i, \ldots, X_k}\,(x_1, x_2, \ldots, x_i, \ldots, x_k) \\
&= \frac{n!}{x_1! x_2! \cdots x_i! \cdots x_k!} p_1^{x_1} p_2^{x_2} \cdots p_i^{x_i} \cdots p_k^{x_k}.
\end{aligned}
\tag{3.33}
$$

We can also extend the binomial notation using

$$
\begin{pmatrix} n \\ x_1 x_2 \cdots x_i \cdots x_k \end{pmatrix} = \frac{n!}{x_1! x_2! \cdots x_i! \cdots x_k!},
\tag{3.34}
$$

and (3.33) becomes

$$
p_{X_1, \ldots, X_i, \ldots, X_k}\,(x_1, \ldots, x_i, \ldots, x_k) = \begin{pmatrix} n \\ x_1 \cdots x_i \cdots x_k \end{pmatrix} p_1^{x_1} \cdots p_i^{x_i} \cdots p_k^{x_k}.
\tag{3.35}
$$

With regard to the hypergeometric distribution, we again consider an urn but now sample without replacement. It contains b balls, g_1 of colour 1, g_2 of colour 2,..., g_i of colour i,..., and g_k of colour k. The probability of obtaining

$$\{X_1 = x_1, X_2 = x_2, \ldots, X_i = x_i, \ldots, X_k = x_k\}$$

of colours $1, 2, \ldots, i, \ldots, n$ respectively in n drawings without replacement is the hypergeometric distribution:

$$\mathbf{Pr}\{X_1 = x_1, X_2 = x_2, \ldots, X_i = x_i, \ldots, X_k = x_k\}$$

$$= p_{X_1, X_2, \ldots, X_i, \ldots, X_k}(x_1, x_2, \ldots, x_i, \ldots, x_k)$$

$$= \frac{\dfrac{g_1!}{x_1!(g_1 - x_1)!} \cdot \dfrac{g_2!}{x_2!(g_2 - x_2)!} \cdots \dfrac{g_i!}{x_i!(g_i - x_i)!} \cdots \dfrac{g_k!}{x_k!(g_k - x_k)!}}{\dfrac{b!}{n!(b - n)!}}. \tag{3.36}$$

There are four forms of the multidimensional hypergeometric distribution, as was the case in Section 3.2.5. See also Exercise 3.6.

3.4.2 Continuous random quantities

Instead of a density $f_X(x)$ on a single random quantity, we consider a density in more than one dimension. We start with two dimensions, the random quantities being X and Y, say. Figure 3.13 illustrates the idea. The probability of the logical product

$$(x \leq X \leq x + dx)(y \leq Y \leq y + dy)$$

is given by the volume

$$\mathbf{Pr}[x \leq X \leq x + dx)(y \leq Y \leq y + dy)] = f_{X,Y}(x, y)\, dx dy. \tag{3.37}$$

Therefore the probability of the random quantities X and Y being in the interval noted is the appropriate volume under the two-dimensional density $f_{X,Y}(x, y)$. Figure 3.13 illustrates Equation (3.37).

To illustrate, we can extend the exponential distribution of (3.23), which has been used as a distribution of waiting times in Section 3.2. These could be the lifetime of devices, whether they be light bulbs or electronic components. Equation (3.38) below could then represent the joint distribution of lifetimes $\{X, Y\}$ of two such components of an assembly, or the waiting times for two transportation modes.

$$f_{X,Y}(x, y) = ab \exp[-(ax + by)] \text{ for } x, y \geq 0 \tag{3.38}$$
$$= 0 \text{ elsewhere.}$$

3.4.3 Coherence and *cdf*s

In order to be coherent (Section 2.2) we have to ensure that our probabilities add up to unity. This must also be the case for two- or n-dimensional random quantities. It is easy to check that the 'spikes' of probability in Figure 3.12 normalize so that this is a coherent evaluation. The condition can be written as

$$\sum_{\text{all } x_i} \sum_{\text{all } y_i} p_{X,Y}(x_i, y_i) = 1. \tag{3.39}$$

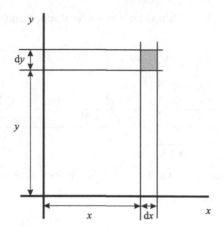

(a) X - Y plane of intergration

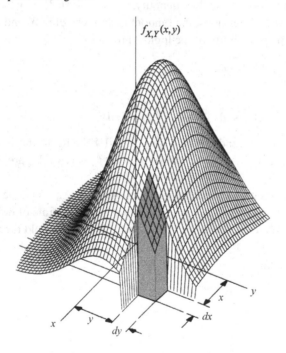

(b) Function $f_{X,Y}(x,y)$ truncated at (x,y) with volume of Equation (3.37) illustrated

Figure 3.13   Two-dimensional continuous random quantity $\{X,Y\}$

Table 3.2 *Values of $F_{X,Y}(x, y)$*

$Y = y$	X = x			
	1	2	3	4
1	0.15	0.25	0.30	0.30
2	0.25	0.55	0.70	0.70
3	0.30	0.65	0.95	1.00

For continuous random quantities,

$$\int_{-\infty}^{\infty} \int_{-\infty}^{\infty} f_{X,Y}(x, y) \, dx \, dy = 1. \tag{3.40}$$

This can be confirmed for Equation (3.38):

$$\int_{0}^{\infty} \int_{0}^{\infty} ab \exp[-(ax + by)] \, dx dy = 1.$$

The cumulative distribution function (*cdf*) is defined by

$$F_{X,Y}(x, y) = \mathbf{Pr}[(X \le x)(Y \le y)]$$
$$= \sum_{x_i \le x} \sum_{y_i \le y} p_{X,Y}(x_i, y_i) \tag{3.41}$$

for the discrete case, and by

$$F_{X,Y}(x, y) = \mathbf{Pr}[(X \le x)(Y \le y)]$$
$$= \int_{-\infty}^{y} \int_{-\infty}^{x} f_{X,Y}(u, v) \, du \, dv \tag{3.42}$$

for the continuous case.

Example 3.10 The *cdf* $F_{X,Y}(x, y)$ for the distribution of Figure 3.12 is given in Table 3.2.

We can use the *cdf*s to deduce that

$$\mathbf{Pr}[(x_1 < x \le x_2)(y_1 < Y \le y_2)]$$
$$= \sum_{x_1 < x_i \le x_2} \sum_{y_1 < y_i \le y_2} p_{X,Y}(x_i, y_i) \text{ for discrete random quantities, and}$$
$$= \int_{y_1}^{y_2} \int_{x_1}^{x_2} f_{X,Y}(u, v) du dv \text{ for continuous random quantities.} \tag{3.43}$$

Example 3.11 Referring to Figure 3.12, $\mathbf{Pr}[(1 < X \le 3)(Y = 2)] = 0.30$.

Example 3.12 Consider the continuous probability distribution $f_{X,Y}(x, y)$:

$$\left. \begin{aligned} f_{X,Y}(x, y) &= x \exp[-x(y + 1)] \text{ for } 0 \le x \le \infty, 0 \le y \le \infty \\ &= 0 \text{ otherwise.} \end{aligned} \right\} \tag{3.44}$$

First, it is to be noted that

$$\int_0^\infty \int_0^\infty x \exp\left[-x(y+1)\right] dydx = \int_0^\infty \left[-\exp\left(-x(y+1)\right)\right]_0^\infty dx$$

$$= \int_0^\infty \exp\left(-x\right) dx = \left[-\exp\left(-x\right)\right]_0^\infty = 1.$$

Also

$$\mathbf{Pr}[(0 < X \leq 1)(0 < Y \leq 2)] = \int_0^1 \int_1^2 x \exp\left[-x(y+1)\right] dydx$$

$$= \int_0^1 \left[-\exp\left(-3x\right) + \exp\left(-2x\right)\right] dx$$

$$= \left[\frac{\exp\left(-3x\right)}{3} - \frac{\exp\left(-2x\right)}{2}\right]_0^1$$

$$= \frac{\exp\left(-3\right)}{3} - \frac{\exp\left(-2\right)}{2} - \frac{1}{3} + \frac{1}{2} = 0.116,$$

and

$$F_{X,Y}(x, y) = \int_0^x \int_0^y u \exp\left[-u(v+1)\right] dvdu$$

$$= \int_0^x \{-\exp\left[-u(y+1)\right] + \exp\left(-u\right)\} du$$

$$= 1 - \exp\left(-x\right) + \frac{1}{y+1} \{\exp\left[-x(y+1)\right] - 1\}.$$

Summary of properties

The following summarizes the properties of two-dimensional distributions.

1. $0 \leq p_{X,Y}(x, y) \leq 1$ for all x, y
2. $\sum_{\text{all } x} \sum_{\text{all } y} p_{X,Y}(x, y) = 1$
3. $\mathbf{Pr}[(X, Y) \in A]$
   $= \sum \sum_A p_{X,Y}(x, y)$
   for any region A in the xy plane.

1. $f_{X,Y}(x, y) \geq 0$ for all x, y
2. $\int_{-\infty}^\infty \int_{-\infty}^\infty f_{X,Y}(x, y) dxdy = 1$
3. $\mathbf{Pr}[(X, Y) \in A]$
   $= \int \int_A f_{X,Y}(x, y) dxdy$
   for any region A in the xy plane.
4. $f_{X,Y}(x, y) = \dfrac{\partial^2}{\partial x \partial y} F_{X,Y}(x, y).$

The discrete and continuous cases can be combined for conditions 2 and 3 using Stieltjes integrals as discussed in Section 3.3.5 above.

3.4.4 Marginal distributions

It is possible to obtain from the joint distribution of X and Y the individual distributions of X and Y. This is done simply by 'summing' over the other quantity. For instance, if the distribution of Y is required, we sum over X. The 'summing' is a summation or integration depending

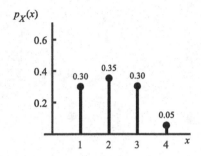

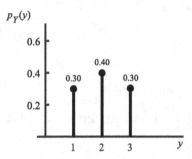

Figure 3.14   Discrete marginal distributions

on whether we are dealing with discrete or continuous random quantities, respectively. The marginal distributions $p_X(x)$ and $p_Y(y)$ in the discrete case are given by

$$p_X(x) = \sum_{\text{all } y} p_{X,Y}(x, y)$$

$$\text{and } p_Y(y) = \sum_{\text{all } x} p_{X,Y}(x, y). \tag{3.45}$$

In the case of continuous distributions,

$$f_X(x) = \int_{-\infty}^{\infty} f_{X,Y}(x, y)\,dy$$

$$f_Y(y) = \int_{-\infty}^{\infty} f_{X,Y}(x, y)\,dx. \tag{3.46}$$

Example 3.13 Consider the example of a discrete distribution given earlier in Figure 3.12; the marginal distributions, obtained using Equations (3.45), are shown in Figure 3.14.

Example 3.14 In Example 3.12 given for the continuous case, using the first of Equations (3.46), we have

$$f_X(x) = \int_0^{\infty} x \exp\left[-x(y + 1)\right] dy$$
$$= \exp\left(-x\right),$$

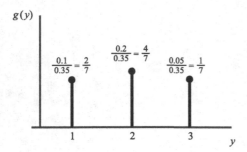

Figure 3.15 Discrete conditional probability distribution $p_{Y|X=2}$

and using the second,

$$f_Y(y) = \int_0^\infty x \exp\left[-x(y+1)\right] dx = \frac{1}{(y+1)^2}.$$

Example 3.15 The probability of Y occurring in any range, e.g. $\mathbf{Pr}(5 < Y \leq 10)$, can be obtained in the usual way:

$$\mathbf{Pr}(5 < Y \leq 10) = \int_5^{10} \frac{1}{(y+1)^2} \, dy = -\left[\frac{1}{y+1}\right]_5^{10}$$
$$= \frac{1}{6} - \frac{1}{11} = 0.0758.$$

3.4.5 Conditional probability distributions

The **conditional** probability distributions are defined as follows, for the discrete case:

$$p_{Y|X}(x, y) = \frac{p_{X,Y}(x, y)}{p_X(x)}$$

$$p_{X|Y}(x, y) = \frac{p_{X,Y}(x, y)}{p_Y(y)}. \tag{3.47}$$

The meaning of these equations is best understood by considering a typical application using the discrete illustration used previously (Table 3.1). Consider for example $p_{Y|X=2}(2, y) = g(y)$, say; this is given in Figure 3.15. The process of obtaining the conditional probability amounts to normalizing those probabilities corresponding to $X = 2$, by ensuring that the sum of the probabilities is unity. The same reasoning applies for any given value of X (or for any Y in the case of $p_{X|Y}$).

Example 3.16 The complete set of conditional probabilities in our discrete example are given below in Tables 3.3.

Table 3.3(a) *Conditional discrete probabilities* $p_{Y|X}(x, y)$

		$X = x$		
$Y = y$	1	2	3	4
1	1/2	2/7	1/6	0
2	1/3	4/7	1/3	0
3	1/6	1/7	1/2	1
Σ	1	1	1	1

Table 3.3(b) *Conditional discrete probabilities* $p_{X|Y}(x, y)$

			$X = x$		
$Y = y$	1	2	3	4	Σ
1	1/2	1/3	1/6	0	1
2	1/4	1/2	1/4	0	1
3	1/6	1/6	1/2	1/6	1

In the case of continuous distributions, conditional probability densities are defined in a similar manner to Equations (3.47):

$$f_{Y|X}(x, y) = \frac{f_{X,Y}(x, y)}{f_X(x)}$$

$$f_{X|Y}(x, y) = \frac{f_{X,Y}(x, y)}{f_Y(y)}. \tag{3.48}$$

Example 3.17 In Example 3.12 considered previously we have

$$f_{Y|X}(x, y) = \frac{x \exp[-x(y + 1)]}{\exp(-x)} = x \exp(-xy) \text{ for } x, y > 0.$$

$$f_{X|Y}(x, y) = \frac{x \exp[-x(y + 1)]}{1/(y + 1)^2}$$

$$= (y + 1)^2 x \exp[-x(y + 1)] \text{ for } x, y > 0.$$

These may be interpreted in a similar way as in the discrete case; the functions represented by Equations (3.48) may be viewed as $f_{X,Y}$ with X or Y considered fixed (respectively) and normalized by division by f_X and f_Y respectively. In the example taken, let us consider $f_{Y|X=0.5}$, that is $0.5 \exp(-0.5y)$. This function, and the parent distribution

$$f_{X=0.5,Y} = 0.5 \exp[-0.5(y + 1)]$$

$$= [0.5 \exp(-0.5)] \exp(-0.5y)$$

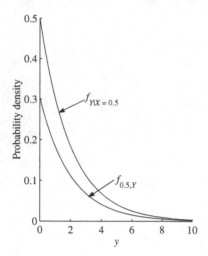

Figure 3.16   Continuous conditional probability distribution

are plotted in Figure 3.16. The ratio between the ordinates, for any given value of y, is a constant:

$$\frac{f_{Y|X=0.5}}{f_{0.5,Y}} = \exp{(0.5)} = 1.649,$$

so that the distribution $f_{Y|X=0.5}$ is simply scaled from the parent $f_{0.5,Y}$.

Conditional probability distributions can be manipulated in the same manner as ordinary probability distributions, for example

$$\mathbf{Pr}[(a < x \leq b)|Y = y] = \int_a^b f_{X|Y}(x, y)\mathrm{d}x.$$

Other conditional distributions can also be derived using the same principles as those above, for example $f_{Y|(X \geq 5)}$, $p_{X|(Y \geq y)}$, and so on.   ■

Example 3.18 Consider the lifetime T of a component, introduced in Section 3.3.2, given by

$$f_T(t) = \lambda \exp{(-\lambda t)}, t \geq 0.$$

We wish to find the probability density for the lifetime, given that failure has not occurred in the interval $[0, t_1]$. The probability of failure in the interval $[0, t_1]$ is

$$\int_0^{t_1} \lambda \exp{(-\lambda t)}\,\mathrm{d}t = 1 - \exp{(-\lambda t_1)},$$

and the probability that the lifetime exceeds t_1 is $\exp{(-\lambda t_1)}$. The required density is then

$$f_{T|t>t_1}(t) = \frac{\lambda \exp{(-\lambda t)}}{\exp{(-\lambda t_1)}} = \lambda \exp{[-\lambda (t - t_1)]},$$
$$\text{for } t > t_1.$$

If we write $\tau = (t - t_1)$, we see that the new density is $\lambda \exp(-\lambda \tau)$, which is identical in form to the original density (termed the memoryless property of the exponential distribution).

3.4.6 Independent random quantities

In Section 2.5, we defined stochastic independence: events E and H are stochastically independent if we judge that

$$\Pr(E|H) = \Pr(E).$$

We also showed that, as a consequence,

$$\Pr(H|E) = \Pr(H),$$

subject to the trivial condition that $\Pr(E) \neq 0$. The idea is readily extended to random quantities. The random quantities X and Y are stochastically independent if

$$\Pr(X = x|Y = y) = \Pr(X = x) \tag{3.49}$$

and

$$\Pr(Y = y|X = x) = \Pr(Y = y). \tag{3.50}$$

In the case of discrete random quantities, these equations become

$$p_{X|Y}(x, y) = p_X(x) \tag{3.51}$$

and

$$p_{Y|X}(x, y) = p_Y(y). \tag{3.52}$$

Now $p_{X,Y}(x, y) = p_{X|Y}(x, y)p_Y(y)$ from the second of Equations (3.47) and therefore

$$p_{X,Y}(x, y) = p_X(x)p_Y(y). \tag{3.53}$$

In the continuous case, it can correspondingly be shown that

$$f_{X,Y}(x, y) = f_X(x)f_Y(y) \tag{3.54}$$

by considering the small intervals dx and dy (Figure 3.13). Two random quantities X and Y are stochastically independent if Equations (3.53) or (3.54) hold for all x and y.

Example 3.19 In the discrete case, Figure 3.17 shows an example of independent X and Y, for which (3.53) is satisfied.

Example 3.20 In the continuous case, Equation (3.38),

$$f_{X,Y}(x, y) = ab \exp[-(ax + by)] \text{ for } x, y \geq 0,$$

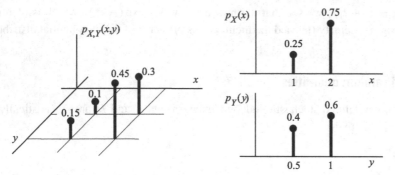

Figure 3.17 Discrete independent random quantities

models independent random quantities X and Y since

$$f_X(x) = ab \int_0^\infty \exp\left[-(ax + by)\right] \mathrm{d}y = a \exp\left(-ax\right),$$

$$f_Y(y) = ab \int_0^\infty \exp\left[-(ax + by)\right] \mathrm{d}x = b \exp\left(-by\right).$$

As a consequence Equation (3.54) is satisfied and X and Y are stochastically independent. For examples where X and Y are **not** independent, see Figure 3.12 (discrete case) and Equation (3.44) (continuous case).

3.4.7 More than two random quantities

In this case the joint *pmf*s or *pdf*s would be of the kind

$$p_{X_1, X_2, \ldots, X_n}(x_1, x_2, \ldots, x_n) \text{ or } f_{X_1, X_2, \ldots, X_n}(x_1, x_2, \ldots, x_n)$$

for random quantities $X_1, X_2, \ldots, X_n$. Examples have been given already in the case of the multinomial and hypergeometric distributions (Section 3.4.1). The cumulative, marginal and conditional distributions are defined in a manner similar to those for two random quantities; thus

$$F_{X_1, X_2, \ldots, X_n}(x_1, x_2, \ldots, x_n) = \mathbf{Pr}[(X_1 \le x_1)(X_2 \le x_2) \cdots (X_n \le x_n)].$$

In the discrete case,

$$
\begin{aligned}
&F_{X_1, X_2, \ldots, X_n}(x_1, x_2, \ldots, x_n) \\
&= \sum\nolimits_{u_1 \le x_1} \sum\nolimits_{u_2 \le x_2} \cdots \sum\nolimits_{u_n \le x_n} p_{X_1, X_2, \ldots, X_n}(u_1, u_2, \ldots, u_n),
\end{aligned}
\tag{3.55}
$$

with an integral replacing the summation for the continuous case. A typical joint marginal distribution for the continuous case would be

$$F_{X_1, X_2}(x_1, x_2) = \int_{-\infty}^\infty \int_{-\infty}^\infty \cdots \int_{-\infty}^\infty f_{X_1, X_2, \ldots, X_n}(x_1, x_2, \ldots, x_n) \, \mathrm{d}x_3 \mathrm{d}x_4 \ldots \mathrm{d}x_n \tag{3.56}$$

and a typical conditional distribution for the discrete case is

$$p_{X_2, X_3 | X_1, X_4, X_5, \ldots, X_n}(x_1, x_2, \ldots, x_n) = \frac{p_{X_1, X_2, \ldots, X_n}(x_1, x_2, \ldots, x_n)}{p_{X_1, X_4, X_5, \ldots, X_n}(x_1, x_4, x_5, \ldots, x_n)}, \tag{3.57}$$

where $p_{X_1, X_4, X_5, \ldots, X_n}(x_1, x_4, x_5, \ldots, x_n)$ is the joint marginal distribution, given by

$$p_{X_1, X_4, X_5, \ldots, X_n}(x_1, x_4, x_5, \ldots, x_n)$$
$$= \sum_{\text{all } x_2} \sum_{\text{all } x_3} p_{X_1 X_2, X_3, X_4, \ldots, X_n}(x_1, x_2, x_3, x_4, \ldots, x_n). \tag{3.58}$$

3.5 Means, expected values, prevision, centres of mass, and moments

The most natural way for engineers to introduce the subject of this section is to consider the probability as a 'mass' of unity which we distribute over the possible outcomes. In this interpretation, the mass represents our degree of belief regarding the random quantity under consideration. Then the mean or 'expected value' is nothing but the centre of gravity of the mass, as was discussed also in Section 1.6. This implies the imposition of a gravity field to produce a force, acting at right angles to the quantity under consideration.

The product of a force and a distance gives a first moment, or simply, a moment. If a body is suspended in a gravitational field, the point at which there is a zero resultant moment is the centre of gravity. This can be seen in a simple experiment by attaching a length of cord to the body, for example a plate in the gravitational field; if the weight is attached at its centre of gravity, it will balance without tilting. The balancing of a scale and its use in measuring weights works on this principle. Thinking of probability as a set of 'weights' attached to particular values, and the balance as our 'balance of judgement', we arrive at the expected value of the random quantity as the point at which the weights balance, the centre of gravity. We now derive the appropriate equations.

3.5.1 Mean (expected value)

Let us take the example of waiting times. These could be for the arrival of a bus, or of data in a network, or of a customer in business. Figure 3.18 shows schematically the probability distribution of waiting times, T: this represents our distribution of 'degree of belief' regarding the various waiting times. The **mean** or 'expected value' of the distribution is the 'centre of mass' of the distribution, given by

$$\langle T \rangle = E(T) = \mu_T = \int_{\text{all } t} t f_T(t) \mathrm{d}t. \tag{3.59}$$

Here, we have used the notations $\langle \cdot \rangle$ and $E(\cdot)$ interchangeably; the first will be our standard version while the second stands for 'expected value'.[2] The third notation (μ) is the Greek '*m*'

---

[2] De Finetti uses $\mathbf{P}(\cdot)$ denoting 'prevision'; if we consider 0:1 random quantities (events), this becomes $\mathbf{Pr}(\cdot)$ or probability. Thus probability can be seen as a special prevision (expected value) for events.

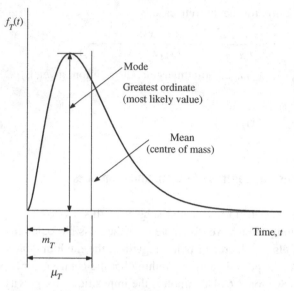

Figure 3.18 Waiting time distribution

for mean. We discussed in Chapter 1 (Section 1.6) the use of the word 'prevision', suggested by de Finetti; we shall use this occasionally as an equivalent to 'mean' or 'expected value'.

The physical interpretation of the mean as the centre of mass is most appropriate and accurate; the (smeared) weights of probability in Figure 3.18 indicate the relative degree to which each value is 'expected'; no particular one is actually expected to the exclusion of the others. In this sense, the term 'expected value' is not appropriate and even misleading. One interpretation for which 'expected value' does make sense would be what is expected in the long run. It is common to talk about 'life expectation': this is only expected on average. In a series of bets, with say 50% chance of winning $10 and 50% chance of no money changing hands, the expected value, or mean, is $5. This would be the expected average gain in a long series of bets but not in a single bet. If one had to take the average of many waiting times, one would expect this to be close to the mean of the distribution. The word 'mean' is neutral in regard to what is expected in a single trial and will often be used in the sequel. 'Expected value' will also be used in conformance with usual usage. But the comments just made should be borne in mind. The **mode**, the value with the greatest ordinate (m_T in Figure 3.18), is the most likely or the most probable. (But we do not 'expect' it to occur in a single trial, either.)

In the case of discrete random quantities, we consider the probability masses as being concentrated at discrete points x_i, and the definition of mean for the random quantity X becomes

$$\langle X \rangle = E(X) = \mu_X = \sum\nolimits_{\text{all } x_i} x_i \, p_X(x_i). \tag{3.60}$$

As we pointed out in Section 3.3.5, integrals involving discrete and continuous random quantities can be written as Stieltjes integrals. The definition of mean for discrete or continuous random quantities X can then be expressed as

$$\langle X \rangle = E(X) = \mu_X = \int_{\text{all } x} x \, dF_X(x). \tag{3.61}$$

Some examples will now be given.

Example 3.21 If we toss an honest coin three times, the mean number of heads is 1.5; this can be obtained by symmetry but can be checked by using Equation (3.60) or (3.61) above.

Example 3.22 Find the mean of the general binomial distribution of Equation (3.3). Applying (3.60), the mean is

$$\langle X \rangle = \mu_X = \sum_{x=0}^{n} x \cdot \binom{n}{x} \cdot p^x (1-p)^{n-x}$$

$$= \sum_{x=1}^{n} \frac{n!}{(x-1)!\,(n-x)!} p^x (1-p)^{n-x}$$

$$= (np) \sum_{x=1}^{n} \frac{(n-1)!}{(x-1)!\,(n-x)!} p^{x-1} (1-p)^{n-x}$$

$$= (np) \sum_{y=0}^{m} \frac{m!}{y!\,(m-y)!} p^y (1-p)^{m-y} \text{ with } y = x-1, m = n-1; \text{ then}$$

$$\langle X \rangle = np,$$

noting that the summation in the second last line is merely the sum of all members $b(y, m, p)$ of a binomial distribution with m trials. The value np makes sense since one would expect on average this number of successes in n trials. For example, one would expect on average one success in ten trials with probability of success per trial equal to 0.1. (But one would not expect one success in every set of ten trials.)

Example 3.23 For the exponential distribution $f_X(x) = \lambda \exp(-\lambda x)$ for $x \geq 0$, and $f_X(x) = 0$ for $x < 0$,

$$\langle X \rangle = \int_{-\infty}^{\infty} x \cdot f_X(x) \, dx$$

$$= \int_{0}^{\infty} x \lambda \exp(-\lambda x) \, dx.$$

Integrating by parts, with $x = u$, $\lambda \exp(-\lambda x) \, dx = dv$,

$$\langle X \rangle = \left[x(-\exp(-\lambda x)) \right]_0^{\infty} - \int_{0}^{\infty} \exp(-\lambda x) \, dx.$$

Noting that $\lim_{x \to \infty} \left[x \exp(-\lambda x) \right] = 0$, by L'Hôpital's rule,

$$\langle X \rangle = \frac{1}{\lambda}. \tag{3.62}$$

3.5.2 Expected values and decisions

We shall see in the next chapter that the expected value of utility is the fundamental basis of decision-making. A special case of this is 'expected monetary value', abbreviated to *EMV*. We introduce this with an example. In Raiffa's urns, we pose the following problem from Raiffa (1968). The urns are as in Figure 2.10.

Example 3.24 An urn is chosen at random from a room containing 800 of type θ_1 and 200 of type θ_2. Two contracts are offered to you at no cost:

α_1 : You receive \$40 if the urn is of type θ_1 and pay me \$20 if it is of type θ_2.
α_2 : You pay \$5 if it is of type θ_1 and receive \$100 if it is of type θ_2.

You wish to investigate the merits of these two decisions; the first favours the outcome being a θ_1 urn whereas the second favours θ_2. A first method might be to weight the outcomes by their probabilities. Let V be the (random) payoff in the contracts. The choice is illustrated in the decision tree of Figure 3.19(a). Then

$$\langle V|\alpha_1\rangle = (0.8)(40) + (0.2)(-20) = \$28, \text{ and}$$
$$\langle V|\alpha_2\rangle = (0.8)(-5) + (0.2)(100) = \$16.$$

The quantity $\langle V|\alpha_i\rangle$ is the 'expected monetary value' (*EMV*) for each contract. These are the amounts that one would expect 'in the long run'. There is no guarantee of gain in a single trial. But since the amounts are both positive, the contracts are attractive since they potentially have value for the person accepting the contract. With larger amounts of money, aversion to the risk of losing will become more important, as explained in Chapter 4.

The expected values are shown in Figure 3.19(a). Of the two choices α_1 is the one with the highest *EMV*, so that this choice is made. The decision crossed out ($\neq$) in Figure 3.19(a) is the choice not taken.

Example 3.25 *Bayesian expected value.* We now offer the same two contracts specified in the preceding example, but include the option of sampling a ball from the urn, at a cost of \$8. This will allow us to change our probabilities of the urn being θ_1 and θ_2. The new decision tree is shown in Figure 3.19(b). The initial choice is between conducting the experiment – sampling a ball – in which the choice is labelled e_1 and not conducting an experiment, labelled e_0. The result of experiment is either a red ball (R) or a pink one (P). The main decision (α_1 versus α_2) is based on the result of the experiment – a very natural procedure. The endpoints in the decision tree reflect the cost of the experiment ($-\$8$), for example, \$32 = \$40 − \$8.

We saw in Section 2.4.1 how to update the probabilities of $\theta_1|R$, $\theta_1|P$, $\theta_2|R$ and $\theta_2|P$. These were done using the number of balls in Section 2.4.1. A typical formal calculation using Bayes'

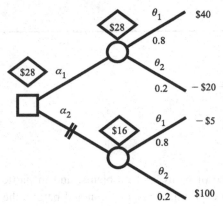

(a) *EMV*s for basic decision tree

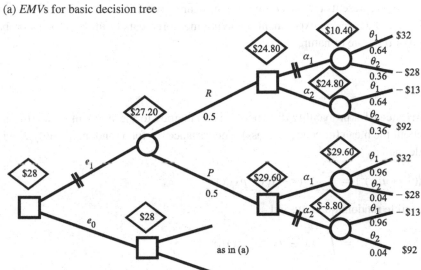

(b) *EMV*s for the Bayesian problem

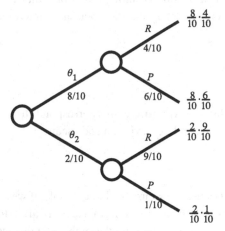

(c) Bayes' calculation tree

Figure 3.19 Optimal decisions

theorem is

$$\mathbf{Pr}(\theta_2|P) = \frac{\mathbf{Pr}(P|\theta_2)\mathbf{Pr}(\theta_2)}{\mathbf{Pr}(P|\theta_2)\mathbf{Pr}(\theta_2) + \mathbf{Pr}(P|\theta_1)\mathbf{Pr}(\theta_1)}$$

$$= \frac{\frac{1}{10} \cdot \frac{2}{10}}{\frac{1}{10} \cdot \frac{2}{10} + \frac{6}{10} \cdot \frac{8}{10}} = \frac{1}{25} = 0.04.$$

The Bayes' tree is shown in Figure 3.19(c).

The Raiffa problem presents an analogue of many real problems, and in particular the fundamental choice between conducting an experiment or not. As noted before, the type of urn (θ_1 or θ_2) represents the uncertain state of nature (traffic in a network, number of flaws in a weld, ...). One uses an experiment involving measurements to improve one's probabilistic estimate of the state of nature.

3.5.3   Variance

The **variance** of a probability distribution is analogous to the moment of inertia of a body taken about the mean (or centre of mass). The variance σ_X^2 for a random quantity X with mean μ_X is defined as

$$\sigma_X^2 = \langle (X - \mu_X)^2 \rangle = \sum_{\text{all } x}(x - \mu_X)^2 p_X(x) \tag{3.63}$$

for a discrete random quantity, and

$$\sigma_X^2 = \langle (X - \mu_X)^2 \rangle = \int_{\text{all } x}(x - \mu_X)^2 f_X(x)\,\mathrm{d}x \tag{3.64}$$

for a continuous one. The total (probability) mass is unity; Equations (3.63) and (3.64) are the same as those for the moment of inertia of a body of unit mass. For either a discrete or continuous distribution with the *cdf* $F_X(x)$ specifying the distribution of mass (Stieltjes convention), we have

$$\sigma_X^2 = \langle (X - \mu_X)^2 \rangle = \int_{\text{all } x}(x - \mu_X)^2\mathrm{d}F_X(x). \tag{3.65}$$

The square root of the variance is the **standard deviation**, σ_X; this corresponds to the radius of gyration in mechanics. The ratio between the standard deviation and mean

$$\gamma \equiv \frac{\sigma_X}{\mu_X} \tag{3.66}$$

is known as the **coefficient of variation**, and is a useful parameter. For example, if we consider the various loadings on a structure, it is important from the point of view of safety to know how much variation there is about the mean; the more there is, the higher the chance of an extreme loading.

Example 3.26 An example of the calculation of variance for the discrete distribution of Figure 3.14 (top), using Equation (3.63). First

$\mu_X = 2.1$, then

$$\sigma_X^2 = (1 - 2.1)^2(0.3) + (2 - 2.1)^2(0.35) + (3 - 2.1)^2(0.3) + (4 - 2.1)^2(0.05)$$
$$= 0.79 \text{ and } \sigma_X = 0.8888, \text{ and}$$

$\gamma = 42\%$ (coefficient of variation).

Example 3.27 For the exponential distribution, $f_X(x) = \lambda \exp(-\lambda x)$, for $x \geq 0$, we recall that the mean is $1/\lambda$ from Equation (3.62). Then using Equation (3.64):

$$\sigma_X^2 = \int_0^\infty \left(x - \frac{1}{\lambda}\right)^2 \lambda \exp(-\lambda x)\, dx,$$

which may be evaluated by parts, giving

$$\sigma_X^2 = \frac{1}{\lambda^2} \text{ or } \sigma_X = \frac{1}{\lambda}.$$

The ratio between the standard deviation and mean, the coefficient of variation $\gamma = \sigma_X/\mu_X$, is then unity.

The interpretation above – as a moment of inertia – leads to some interesting conclusions. We recall first the statement in the second paragraph of this section, that the probability 'balances' at the centre of mass of the distribution: this is equivalent to saying that the first moment of mass is zero about the centre of mass. The variance is the second moment of mass about the centre of mass, and we can show by the 'parallel axis' theorem, well known in mechanics, that the variance is a minimum about the axis through the centre of mass.

Minimum variance: analogy to mechanics   Consider the body shown in the illustration; we calculate the moment of inertia about an axis $B-B$ as

$$\int (r - d)^2 \, dF_R(r), \tag{3.67}$$

where r is the distance from the axis $A-A$ through the centre of mass G to a particle of mass $dF_R(r)$, and d is the distance between the axis $B-B$ and the parallel axis $A-A$ through the centre of mass. All distances are perpendicular to the parallel axes. Equation (3.67) can be written as

$$\int r^2 dF_R(r) - 2d \int r dF_R(r) + d^2. \tag{3.68}$$

(Recall that the total mass in the case of probability is unity by coherence.) Since the first

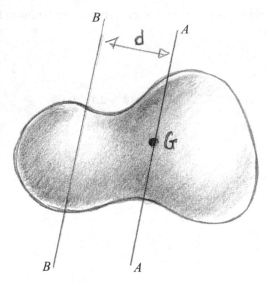

Body of unit mass with parallel axes

moment about the centre of mass is zero by definition,

$$\int r \, dF_R(r) = 0,$$

Equation (3.68) becomes

$$\sigma_G^2 + d^2, \tag{3.69}$$

where σ_G^2 is the moment of inertia about the centre of mass.

Equation (3.69) is at a minimum when $d = 0$, so that the minimum moment of inertia (variance) σ_G^2 occurs when the axis is through G. Equation (3.69) can also be used to deduce the following. The variance of Equation (3.63) or (3.64) was defined as the second moment about the mean (or centre of gravity). Let the origin of coordinates be on the axis B–B. If the distances are measured in the x-direction, we can apply (3.69), and we see that the second moment about the origin, $\langle X^2 \rangle$, is given by

$$\langle X^2 \rangle = \sigma_X^2 + \mu_X^2, \text{ or}$$
$$\sigma_X^2 = \langle X^2 \rangle - \mu_X^2. \tag{3.70}$$

Equation (3.70) gives a useful 'shortcut' formula for the variance. This equation is introduced in more detail below; see Equation (3.87).

To conclude the analogy and to summarize, there are two interesting things about the centre of mass: the first moment of mass about this point is zero, and the moment of inertia is at a minimum. The first of these is identical to the condition we used to define probability, as a 'price' that we would pay for a bet on an outcome with a reward of unity if the outcome happens (Section 2.2). We can associate the prices with the distributed masses above. Consequently we see that the mean is at the centre of mass of these prices. In the case of events there are only two outcomes: the definition of the mean then coincides with the definition of probability. (See the Guide for Life on page 25.) For an event F,

$$\langle F \rangle = E(F) = P(F) \equiv \mathbf{Pr}(F);$$

see also de Finetti (1974). The second interesting point, that the moment of inertia is a minimum at the centre of mass, leads to the following. In order to define the centre of mass, or prevision-point – or the probability in the case of events – assume that a quadratic loss is incurred, corresponding to the term d^2 in Equation (3.69), if we make an estimate away from the centre of mass of our belief. This would ensure that we place our expected value (prevision), or probability, at the centre of mass to minimize this loss. Again see de Finetti (1974).

3.5.4   Mathematical manipulation of mean values

The mean or expected value of a **function**, $g(X)$, of a random quantity X is defined in a manner analogous to the above:

$$\langle g(X) \rangle = \sum_{\text{all } x_i} g(x_i) p_X(x_i), \quad \text{for discrete } X \tag{3.71}$$

or

$$\langle g(X) \rangle = \int_{-\infty}^{\infty} g(x) f_X(x) \, dx, \quad \text{for continuous } X. \tag{3.72}$$

Example 3.28 Take $g(X) = X^2$, and use the *pmf* of Figure 3.14 (top). Then

$$\langle X^2 \rangle = (1)(0.3) + (4)(0.35) + (9)(0.3) + (16)(0.05)$$
$$= 5.20.$$

Example 3.29 Take $g(X) = \exp(-X)$ and $f_X(x) = \lambda \exp(-\lambda x), x \geq 0$, i.e. the exponential distribution used previously. Then

$$\langle \exp(-X) \rangle = \int_0^{\infty} \exp(-x) \lambda \exp(-\lambda x) \, dx = \int_0^{\infty} \lambda \exp[-(\lambda + 1)x] \, dx$$
$$= \frac{\lambda}{\lambda + 1}.$$

The expectation of a function of two random quantities, $g(X, Y)$, is defined in a similar manner to the above:

$$\langle g(X, Y) \rangle = \sum_{\text{all } x_i} \sum_{\text{all } y_j} p_{X,Y}(x_i, y_j) g(x_i, y_j) \text{ (discrete case)} \tag{3.73}$$

$$\langle g(X, Y) \rangle = \int_{-\infty}^{\infty} \int_{-\infty}^{\infty} f_{X,Y}(x, y) g(x, y) \, dx \, dy \text{ (continuous case).} \tag{3.74}$$

Example 3.30 For the discrete case, consider the distribution of Figure 3.12. If $g(X, Y) = XY^2$, then

$$\begin{aligned}
\langle XY^2 \rangle =\ & (1)(1)(0.15) + (1)(4)(0.10) + (1)(9)(0.05) + (20)(1)(0.1) \\
& + (2)(4)(0.2) + (2)(9)(0.05) + (3)(1)(0.05) + (3)(4)(0.10) \\
& + (3)(9)(0.15) + (4)(9)(0.05) \\
=\ & 10.9.
\end{aligned}$$

Example 3.31 As an example of the continuous case, let

$$f_{X,Y}(x, y) = \frac{x(1 + 3y^2)}{4}, \text{ for } 0 \le x \le 2, 0 \le y \le 1,$$
$$= 0 \text{ otherwise.}$$

Then

$$\begin{aligned}
\left\langle \frac{Y}{X} \right\rangle &= \int_0^1 \int_0^2 \frac{y(1 + 3y^2)}{4} \, dx \, dy \\
&= \int_0^1 \frac{y(1 + 3y^2)}{2} \, dy \\
&= \left[\frac{y^2}{4} + \frac{3y^2}{8} \right]_0^1 = \frac{1}{4} + \frac{3}{8} = \frac{5}{8}.
\end{aligned}$$

Several important properties of expectation should be noted.

1. $\langle c \rangle = c$,

   where c is a constant. Using the continuous form as an example,

   $$\langle c \rangle = \int_{-\infty}^{\infty} c f_X(x) \, dx = c \int_{-\infty}^{\infty} f_X(x) \, dx = c.$$

2. $\langle a_1 g_1(X) + a_2 g_2(X) \rangle = a_1 \langle g_1(X) \rangle + a_2 \langle g_2(X) \rangle.$ (3.75)

   This equation may be derived by substituting in the definition of expectation, either discrete or continuous. A special case of Equation (3.75) is

$$\langle a + bX \rangle = a + b \langle X \rangle. \tag{3.76}$$

3. $\langle a_1 g_1(X, Y) + a_2 g_2(X, Y) \rangle = a_1 \langle g_1(X, Y) \rangle + a_2 \langle g_2(X, Y) \rangle. \tag{3.77}$

This again may be proved by substituting in the definition, for example in the discrete case,

$$\langle a_1 g_1(X, Y) + a_2 g_2(X, Y) \rangle$$
$$= \sum_{x_i} \sum_{y_j} [a_1 g_1(x_i, y_j) + a_2 g_2(x_i, y_j)] p_{X,Y}(x_i, y_j)$$
$$= a_1 \sum_{x_i} \sum_{y_j} [g_1(x_i, y_j) p_{X,Y}(x_i, y_j)] + a_2 \sum_{x_i} \sum_{y_j} [g_2(x_i, y_j) p_{X,Y}(x_i, y_j)]$$
$$= a_1 \langle g_1(X, Y) \rangle + a_2 \langle g_2(X, Y) \rangle.$$

A useful special case is obtained if $g_1(X, Y) = X$ and $g_2(X, Y) = Y$; then

$$\langle X + Y \rangle = \langle X \rangle + \langle Y \rangle. \tag{3.78}$$

From Equations (3.75) and (3.77) it is seen that the operation of obtaining an expected value is a linear operation.

2. For two independent random quantities,

$$\langle XY \rangle = \langle X \rangle \langle Y \rangle, \tag{3.79}$$

since (taking the continuous case as an example) $f_{X,Y}(x, y) = f_X(x) f_Y(y)$ and therefore

$$\langle XY \rangle = \int_{-\infty}^{\infty} \int_{-\infty}^{\infty} xy f_X(x) f_Y(y) \mathrm{d}x \mathrm{d}y$$
$$= \int_{-\infty}^{\infty} x f_X(x) \mathrm{d}x \int_{-\infty}^{\infty} y f_Y(y) \mathrm{d}y$$
$$= \langle X \rangle \langle Y \rangle.$$

Example 3.32 The last point may be verified in the case of the distribution in Figure 3.17:

$$\langle XY \rangle = (10)(0.1) + (20)(0.15) + (20)(0.3) + (40)(0.45)$$
$$= 28,$$

and

$$\langle X \rangle = (10)(0.25) + (20)(0.75) = 17.5$$
$$\langle Y \rangle = (1)(0.4) + (2)(0.6) = 1.6$$

Therefore $\langle X \rangle \langle Y \rangle = 28 = \langle XY \rangle$. This would be expected because of the independence of X and Y.

3.5.5   Conditional expectation

Since conditional probability distributions such as $f_{Y|X}$, $p_{X|Y>y}$, ..., are valid normalized probability distributions, they may be used to calculate conditional expectations. For example,

$$\langle X|Y \rangle = \int_{-\infty}^{\infty} x f_{X|Y}(x, y) \mathrm{d}x \tag{3.80}$$

for a continuous distribution and

$$\langle Y|X \rangle = \sum_{\text{all } y_j} y p_{Y|X}(x_i, y_j) \tag{3.81}$$

for a discrete distribution.

3.5.6   Moments

We consider first a single random quantity X. To define moments, we let the function $g(X) = X^k$ in Equations (3.71) and (3.72), where k is a constant. Then the expectations are

$$\langle X^k \rangle = \mu_X^{(k)} = \sum_{\text{all } x_i} x_i^k p_X(x_i), \text{ for discrete } X, \text{ and} \tag{3.82}$$

$$\langle X^k \rangle = \mu_X^{(k)} = \int_{-\infty}^{\infty} x^k f_X(x) \, \mathrm{d}x, \text{ for continuous } X. \tag{3.83}$$

These expectations define the kth **moments about the origin**. In particular, for $k = 1$, $\langle X \rangle = \mu_X^{(1)} = \mu_X$, the mean of the random quantity. The analogy to moments in mechanics is complete. In the discrete case, the probabilities of the *pmf* at values $x_1, x_2, \ldots$, are treated as concentrated masses, and in the continuous case, the *pdf* is treated as a distributed mass, for example a thin uniform plate bounded by the *pdf* and the x-axis, in both cases with total mass $= 1$.

The moments about the mean, or **central moments**, are defined by letting $g(X) = (X - \mu_X)^k$. Then $\langle g(X) \rangle$ is the kth moment about the mean, which we write as $m_X^{(k)}$; thus

$$m_X^{(k)} = \sum_{\text{all } x} (x - \mu_X)^k p_X(x), \text{ for discrete } X, \text{ and} \tag{3.84}$$

$$m_X^{(k)} = \int_{-\infty}^{\infty} (x - \mu_X)^k f_X(x) \, \mathrm{d}x, \text{ for continuous } X. \tag{3.85}$$

Note that the first central moment $m_X^{(1)}$ is equal to zero. A particular case of great practical importance is $m_X^{(2)}$, the variance σ_X^2, as described earlier:

$$m_X^{(2)} = \sigma_X^2 = \langle (X - \mu_X)^2 \rangle; \tag{3.86}$$

σ_X is the standard deviation. Both σ_X^2 and σ_X are, as noted earlier, measures of spread of the

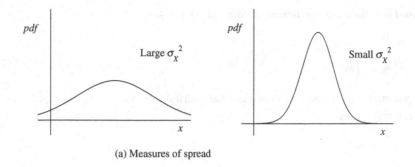

(a) Measures of spread

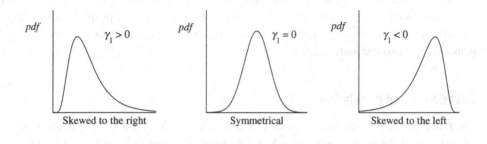

(b) Skewness

Figure 3.20   Practical interpretation of moments

distribution. The following result is useful:

$$\langle (X - \mu_X)^2 \rangle = \langle X^2 - 2\mu_X X + \mu_X^2 \rangle$$
$$= \langle X^2 \rangle - 2\mu_X \langle X \rangle + \mu_X^2$$

i.e. $\langle (X - \mu_X)^2 \rangle = \langle X^2 \rangle - \mu_X^2$ since $\langle X \rangle = \mu_X$

$$\text{or } \sigma_X^2 = \langle X^2 \rangle - \mu_X^2. \tag{3.87}$$

This is the 'shortcut' formula noted before in Equation (3.70). Examples of the calculation of the variance have been given in Section 3.5.3.

The interpretation of $m_X^{(2)}$ as a measure of spread is illustrated in Figure 3.20(a). If $k = 3$, the third central moment, $m_X^{(3)}$, is obtained. The quantity

$$\gamma_1 = \frac{m_X^{(3)}}{\sigma_X^3} \tag{3.88}$$

is the **coefficient of skewness** and varies in value with the skewness of the distribution, as indicated in Figure 3.20(b). Note that if $\gamma_1 = 0$, the distribution is not necessarily symmetrical, but if the distribution is symmetrical, then $\gamma_1 = 0$. For $k = 4$, the fourth central moment can be used to obtain the **coefficient of kurtosis** (flatness):

$$\gamma_2 = \frac{m_X^{(4)}}{\sigma_X^4} = \frac{m_X^{(4)}}{\left(m_X^{(2)}\right)^2}. \tag{3.89}$$

The parameter γ_2 is equal to 3 for a normal distribution. As a result, the coefficient of kurtosis is often defined as

$$\gamma_2' = \gamma_2 - 3. \tag{3.90}$$

The kurtosis is also associated with **tail heaviness**; the more probability mass in the tails of the distribution, the greater the kurtosis. The normal distribution is the standard, termed mesokurtic with $\gamma_2 = 3$, as noted. Then $\gamma_2' = 0$. Distributions with positive values of γ_2' are termed leptokurtic (more peaked); those with negative values are platykurtic (flat, or less peaked). See also Example 7.23.

3.5.7   Covariance and correlation

For functions of two random quantities, if $g(X, Y) = X^k Y^\ell$, then $\langle X^k Y^\ell \rangle$ is the moment of order $k + \ell$ of the random quantities X and Y. Similarly, if $g(X, Y) = (X - \mu_X)^k (Y - \mu_Y)^\ell$, then $\langle (X - \mu_X)^k (Y - \mu_Y)^\ell \rangle$ is the central moment of order $k + \ell$ of the random quantities X and Y. Note that σ_X^2 is obtained if $k = 2$ and $\ell = 0$; similarly σ_Y^2 is obtained if $k = 0$ and $\ell = 2$. If $k = \ell = 1$, the **covariance** $\sigma_{X,Y}$ is obtained:

$$\sigma_{X,Y} = \langle (X - \mu_X)(Y - \mu_Y) \rangle$$
$$= \sum_{\text{all } x_i} \sum_{\text{all } y_j} (x_i - \mu_X)(y_j - \mu_Y) p_{X,Y}(x_i, y_j) \tag{3.91}$$

in the discrete case, and

$$\sigma_{X,Y} = \langle (X - \mu_X)(Y - \mu_Y) \rangle = \int_{-\infty}^{\infty} \int_{-\infty}^{\infty} (x - \mu_X)(y - \mu_Y) f_{X,Y}(x, y) \mathrm{d}x \mathrm{d}y \tag{3.92}$$

in the continuous case.

The **correlation coefficient** ρ_{XY} is defined by

$$\rho_{X,Y} = \frac{\sigma_{X,Y}}{\sigma_X \sigma_Y} \tag{3.93}$$

and it should be noted that $-1 \le \rho_{X,Y} \le 1$, always. (See Exercise 3.25.) The covariance and in particular the correlation coefficient measure the extent of the **linear** relation between X and Y. Thus a value of $\rho_{X,Y}$ close to $+1$ would be associated with probabilities in the X–Y plane being concentrated along a line with positive slope (in the X–Y plane); a value close to -1 with probabilities along a negative slope. See Figure 3.21 for examples in the discrete case.

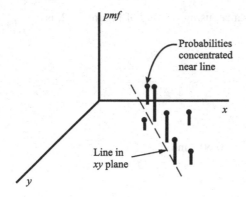

(a) $\rho_{X,Y}$ close to +1

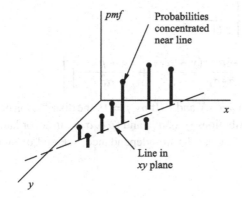

(b) $\rho_{X,Y}$ close to -1

Figure 3.21   Positive and negative correlation coefficients $\rho_{X,Y}$

The following form of the covariance is useful:

$$\langle(X - \mu_X)(Y - \mu_Y)\rangle = \langle XY - \mu_Y X - \mu_X Y + \mu_X \mu_Y\rangle$$
$$= \langle XY\rangle - \mu_X \langle Y\rangle - \mu_Y \langle X\rangle + \mu_X \mu_Y,$$

or

$$\sigma_{X,Y} = \langle XY\rangle - \mu_X \mu_Y. \tag{3.94}$$

Since $\langle XY\rangle = \mu_X \mu_Y$ if X and Y are independent, in this case $\sigma_{X,Y} = \rho_{X,Y} = 0$. But the fact that $\sigma_{X,Y} = \rho_{X,Y} = 0$ does not imply independence. It is important to register clearly the fact that $\sigma_{X,Y} = \rho_{X,Y} = 0$ does not imply independence yet independence does imply $\sigma_{X,Y} = \rho_{X,Y} = 0$. Equations (3.53) and (3.54) are needed for all x and y for independence. See Exercise 3.23.

Example 3.33 The calculation of covariance, using the *pmf* of Figure 3.12, is

$$\mu_X = 2.1, \mu_Y = 2.0, \langle XY \rangle = 4.5$$

and therefore

$$\sigma_{X,Y} = 0.3.$$

Also

$$\langle X^2 \rangle = 5.20, \langle Y^2 \rangle = 4.60, \sigma_X^2 = 0.79, \sigma_Y^2 = 0.60, \text{ and}$$

$$\rho_{X,Y} = 0.44.$$

Example 3.34 The two-dimensional form of the normal distribution for two random quantities X and Y is

$$f_{X,Y}(x, y) = \frac{1}{2\pi \sigma_X \sigma_Y \sqrt{1 - \rho_{X,Y}^2}} \exp - \left\{ \frac{1}{2(1 - \rho_{X,Y}^2)} \right.$$

$$\left. \times \left[\frac{(x - \mu_X)^2}{\sigma_X^2} - \frac{2\rho_{X,Y}(x - \mu_X)(y - \mu_Y)}{\sigma_X \sigma_Y} + \frac{(y - \mu_Y)^2}{\sigma_Y^2} \right] \right\}, \tag{3.95}$$

where $\rho_{X,Y}$ is the correlation coefficient between X and Y. If $\rho_{X,Y} = 0$, the distribution becomes the (separable) product of two normal distributions (μ_X, σ_X^2) and (μ_Y, σ_Y^2), in accordance with (3.54). It is not generally the case that $\rho = 0$ results in independance; see end of preceding page and Example 7.9.

3.5.8 Linear functions of random quantities

Let us now consider a linear function of the random quantity X,

$$U = aX + b;$$

U must also be a random quantity. It has already been shown in Equation (3.75) and sequel that

$$\mu_U = a\mu_X + b. \tag{3.96}$$

In addition,

$$\sigma_U^2 = \langle (U - \mu_U)^2 \rangle$$
$$= \langle (aX + b - a\mu_X - b)^2 \rangle$$
$$= \langle a^2 (X - \mu_X)^2 \rangle = a^2 \langle (X - \mu_X)^2 \rangle. \tag{3.97}$$

In short

$$\sigma_U^2 = a^2 \sigma_X^2. \tag{3.98}$$

If $U = aX + bY$, where X, Y have a joint probability distribution, discrete or continuous, then

$$\mu_U = a\mu_X + b\mu_Y,$$ (3.99)

which is also a consequence of Equation (3.75). The variance is

$$\begin{aligned}
\sigma_U^2 &= \langle (aX + bY - a\mu_X - b\mu_Y)^2 \rangle \\
&= \langle a^2(X - \mu_X)^2 + b^2(Y - \mu_Y)^2 + 2ab(X - \mu_X)(Y - \mu_Y) \rangle \\
&= a^2 \langle (X - \mu_X)^2 \rangle + b^2 \langle (Y - \mu_Y)^2 \rangle \\
&\quad + 2ab \langle (X - \mu_X)(Y - \mu_Y) \rangle,
\end{aligned}$$ (3.100)

or

$$\sigma_U^2 = a^2\sigma_X^2 + b^2\sigma_Y^2 + 2ab\sigma_{X,Y}.$$ (3.101)

For independent random quantities $\sigma_{X,Y} = 0$ and in this case

$$\sigma_U^2 = a^2\sigma_X^2 + b^2\sigma_Y^2.$$ (3.102)

If

$$Y = X_1 + X_2 + \cdots + X_i + \cdots + X_n,$$

where the X_i are mutually independent quantities, with variances

$$\sigma_1, \ldots, \sigma_i, \ldots, \sigma_n,$$

then

$$\sigma_Y^2 = \sum_{i=1}^n \sigma_i^2.$$ (3.103)

Since $\sigma_U^2 = a^2\sigma_X^2$, where $U = aX$, if

$$Y = \frac{X_1 + X_2 + \cdots + X_i + \cdots + X_n}{n},$$ (3.104)

the average of the Xs, then

$$\sigma_Y^2 = \frac{1}{n^2} \sum_{i=1}^n \sigma_i^2.$$ (3.105)

This is the case of independent random quantities; if the σ_i^2 are the same for each X_i, and equal to σ^2, then

$$\sigma_Y^2 = \frac{n\sigma^2}{n^2} - \frac{\sigma^2}{n}.$$ (3.106)

This would be true for *iid* (independent and identically distributed) random quantities. But random quantities need not be identical to have the same σ_i^2, and they could still be independent so that Equation (3.106) has wider application than the *iid* case. Also, by extending (3.75) we find for the relationship (3.104) that

$$\mu_Y = \mu_X \tag{3.107}$$

if the μ_Xs are the same.

3.6   Random processes

We shall introduce briefly the subject of processes. We have already alluded to one: the Poisson process. This was introduced by means of examples of occurrences in time, for example arrivals of computer messages or vehicular traffic. Usually these occur in time, but there are processes, as indicated earlier, which might involve distance or volume, for example. As a first illustration, consider a series of trials such as the random walk in the diagram of Figure 2.19. The process could be modelled as the gain $X(h) = 2E_h - 1$ per trial h, where the gain is $+1$ for a success and -1 for a failure. Or the total gain $Y(n) = \sum_{h=1}^{n} X_h$ could be considered. Examples of these are shown in Figure 3.22(a). If the events happen regularly in time, this could replace n as the abscissa and we would consider $X(t)$ and $Y(t)$. We shall use time in the following for convenience, but as noted other characterizations are possible.

Random processes are also termed stochastic processes. They can be continuous in time, as in Brownian motion where microscopic particles collide with molecules of a fluid. A further example is given in Figure 3.22(b), which shows some regular and some random features. These random quantities – Brownian motion and Figure 3.22(b) – are continuous in time. In addition to considerations of time, the random quantity itself can be discrete or continuous – four combinations of possible kinds of processes. See Exercise 3.33. Processes that are discrete in time generally have outcomes corresponding to times at the integers, as in Figure 3.22(a). They can be thought of as sequences of random quantities.

Papoulis (1965) and Ochi (1990) provide good introductions to the subject from an engineering standpoint. De Finetti (1974) provides excellent insights into the basis and meaning of different processes.

3.6.1   Processes with independent increments

In Figure 3.22(a) we introduced a process which could, for example, consist of successive tosses of a coin leading to $E_h =$ heads and $\tilde{E}_h =$ tails. These events would be judged to be independent. We consider first such processes in discrete time. The process under consideration consists of increments X_h in $Y(n) = \sum_{h=1}^{n} X_h$ as described in the preceding section. These 'jumps' then occur at the discrete points corresponding to the tosses of a coin. The tosses could be conducted at discrete times indexed $h = 1, 2, 3, \ldots$, and then time could be used as the parameter of the process. This would then be described as in the lower part of Figure 3.22(a).

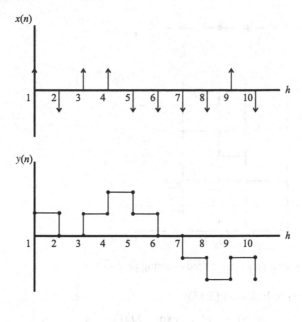

(a) Discrete time processes

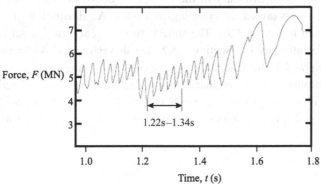

(b) Continuous time process; force during indentation
of ice mass (Jordaan, 2000)

Figure 3.22   Random processes

In **point processes**, events occur at very local points in time. Electron emissions, lightning discharges, and earthquakes are examples of such events (Snyder, 1975). The Poisson process is an example of such a process with the 'jumps' occurring at discrete points in continuous time. These points are random, governed by the assumptions in deriving the Poisson distribution (Section 3.2.6). Associated with the point process is a **counting process**. If the increments, or jumps, are of size 1, then the total number of events in the time interval t is given by X in the

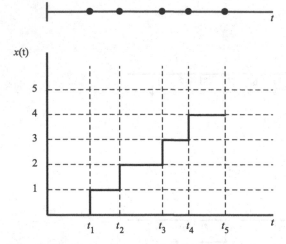

Figure 3.23 Point process (top) and related counting process

Poisson distribution of Equation (3.13):

$$\mathbf{Pr}\,(X = x) = p_X\,(x) = \frac{\exp\,(-\nu)\,\nu^x}{x!} = \frac{\exp\,(-\lambda t)\,(\lambda t)^x}{x!}, \text{ for } x = 0, 1, 2, 3, \ldots,$$

where $\nu = \lambda t$ and λ is the rate or **intensity** of the process. Processes of this kind in which we count the number of events are called counting processes. An illustration of point and counting processes is given in Figure 3.23. The time between events was modelled by the exponential distribution as introduced in Section 3.3.2. See also Exercise 3.34 for the related Erlang distribution, which models the time to the first, second or nth arrival.

The case where λ is constant leads to stationary increments. We pointed out in Section 3.2.6 that the idea behind the Poisson distribution is readily extended to the situation where the rate λ varies with time, i.e. $\lambda = \lambda\,(t)$. In Equation (3.14) we noted that the expected number of arrivals is given by

$$\nu = \int_0^t \lambda\,(\tau)\,d\tau,$$

and (3.13) applies. But the interarrival times are no longer modelled by the exponential or Erlang distributions.

In the preceding the increments have been discrete and of unit size. The increments could be discrete or continuous random quantities such that the increments over disjoint time intervals are independent. The process is termed **homogeneous** if the increments over disjoint time intervals of the same length have the same distribution. In the **compound** Poisson process, the increments (jumps) in time follow the Poisson point process but the quantity is modelled by a further random quantity. For instance, a series of telephone calls might follow the Poisson process and the length of the calls might follow an exponential distribution. We might be interested in the total length of the telephone calls in a certain time interval. The auxiliary random quantity (length of calls) is sometimes called a 'mark' (Snyder, 1975). If the increments occur in continuous time, not at discrete points, we can divide the time period under consideration into small intervals, and we can think of small increments occurring in these intervals. The result is a very large number of small independent increments. These follow the central limit theorem and the result is that the distribution is normal. This is the Wiener–Lévy process. In general the decomposition of continuous processes requires the property of infinite divisibility, discussed briefly in Section 7.4.4.

3.6.2   Markov processes

We might consider that the probability of rain tomorrow depends on whether it is raining today but not on the weather yesterday. This idea of dependence of one step in a process on the preceding but not on events further in the past embodies the idea of a Markov process. Sometimes this is summarized in the statement:

'Knowledge of the present makes the future independent of the past.'

We shall consider the discrete case in which there is again a sequence of values

$$X_1, X_2, X_3, \ldots, X_n, \ldots, \tag{3.108}$$

corresponding to states i, j taking the values 0,1,2,3,... In other words, the Xs will take these discrete values. The probabilities

$$\mathbf{Pr}\,(X_n = i | X_{n-1} = j) = p_{i|j}\,(n) = p_{ij}\,(n) \tag{3.109}$$

are the (one-step) transition probabilities from j to i in moving from one step to another in time.

In the homogeneous case the transition probabilities are independent of time and we write $p_{ij}\,(n) = p_{ij}$:

$$p_{ij} = \begin{bmatrix} p_{00} & p_{01} & p_{02} & \cdots \\ p_{10} & p_{11} & p_{12} & \cdots \\ p_{20} & p_{21} & p_{22} & \cdots \\ \vdots & \vdots & \vdots & \ddots \end{bmatrix}. \tag{3.110}$$

The sums of the columns must normalize for coherence:

$$\sum_i p_{ij} = 1 \text{ for each } j. \tag{3.111}$$

Let the column vector $q_i\,(n)$ represent the probability of state i at step (time) n. The probabilities of states 0,1,2,3,... at time 1 are derived from the preceding set by means of

$$q_i\,(1) = \left[p_{ij}\,(1) \right] q_j\,(0). \tag{3.112}$$

The two-step change is given by

$$q_i(2) = \left[p_{ij}(2) \right] \left[p_{ij}(1) \right] q_j(0), \tag{3.113}$$

and generally for n steps

$$q_i(n) = \left[p_{ij}(n) \right] \cdots \left[p_{ij}(2) \right] \left[p_{ij}(1) \right] q_j(0). \tag{3.114}$$

For homogeneous chains,

$$q_i(n) = \left[p_{ij} \right]^n q_j(0). \tag{3.115}$$

If at any stage in the process $p_{ij} = 1$ for $i = j$ and 0 for $i \neq j$, then the state j is a trapping or absorbing state: transitions are possible into the state but not out of it.

Example 3.35 Counting processes are a special case of the Markov process. Consider the total number of heads in a sequence of tosses with an honest coin. Each step is dependent on the preceding since any increment is added to it. This is a rather trivial example but helps in understanding the procedure. The one-step transition matrix is below. There are no zeros above the diagonal since the process cannot go back to a state once it has been left.

$$p_{ij} = \begin{bmatrix} 0.5 & 0 & 0 & \cdots \\ 0.5 & 0.5 & 0 & \cdots \\ 0 & 0.5 & 0.5 & \cdots \\ \vdots & \vdots & \vdots & \ddots \end{bmatrix}. \tag{3.116}$$

The initial state is

$$q_i(0) = \begin{Bmatrix} 1 \\ 0 \\ 0 \\ \vdots \end{Bmatrix}. \tag{3.117}$$

One can implement Equations (3.112) to (3.115) to obtain the probabilities of states 0, 1, 2, 3, ..., at times 1, 2, 3, ... The reader might wish to attempt this.

Example 3.36 We wish to model a queue – people, packets of data, aircraft. We make the rule that a step is defined by an arrival in or a departure from the queue. Further, we assume that the process is homogeneous, with p = probability of a departure and $1 - p$ = probability of an arrival. The state is the number in the queue. Then

$$p_{ij} = \begin{bmatrix} 0 & p & 0 & 0 & \cdots \\ 1 & 0 & p & 0 & \cdots \\ 0 & 1-p & 0 & p & \cdots \\ 0 & 0 & 1-p & 0 & \cdots \\ \vdots & \vdots & \vdots & \vdots & \ddots \end{bmatrix}, \tag{3.118}$$

and the first four $q_i(t)$ are

$$\begin{Bmatrix} 1 \\ 0 \\ 0 \\ 0 \\ 0 \\ \vdots \end{Bmatrix}, \begin{Bmatrix} 0 \\ 1 \\ 0 \\ 0 \\ 0 \\ \vdots \end{Bmatrix}, \begin{Bmatrix} p \\ 0 \\ 1-p \\ 0 \\ 0 \\ \vdots \end{Bmatrix} \text{ and } \begin{Bmatrix} 0 \\ 2p - p^2 \\ 0 \\ (1-p)^2 \\ 0 \\ \vdots \end{Bmatrix}. \tag{3.119}$$

Ergodic chains

In some cases a Markov chain evolves to a steady state, independent of the initial state. This is termed an ergodic chain. We can analyse this situation by supposing that q_i^* is the steady state. Then, if this exists,

$$q_i(n+1) = q_i(n) = q_i^*.$$

For homogeneous chains

$$q_i(n+1) = [p_{ij}]q_j(n) = [p_{ij}]q_j^*,$$

and consequently

$$q_i^* = [p_{ij}]q_j^*, \tag{3.120}$$

or

$$[p_{ij} - \delta_{ij}]q_j^* = 0. \tag{3.121}$$

The term $[\delta_{ij}]$ is the identity matrix; $\delta_{ij} = 1$ if $i = j$, $\delta_{ij} = 0$ otherwise.

The solution of (3.121) can be carried out as in the example below, or with further insights using eigenvalues and eigenvectors. If the chain is in a steady state, it is said to possess **stationarity**.

Example 3.37 In a queuing problem, three states are considered: $0 =$ no customers, $1 =$ number of customers $\leq n_0$ (specified number) and $2 =$ number of customers $> n_0$. The homogeneous system is modelled by the transition matrix

$$
[p_{ij}] = \begin{bmatrix} 0.8 & 0.2 & 0.6 \\ 0.1 & 0.6 & 0 \\ 0.1 & 0.2 & 0.4 \end{bmatrix}.
\tag{3.122}
$$

For a steady state (3.120) and (3.111) apply. Then

$$
\begin{aligned}
-0.2q_0^* & +0.2q_1^* & +0.6q_2^* & = 0 \\
0.1q_0^* & -0.4q_1^* & & = 0 \\
0.1q_0^* & +0.2q_1^* & -0.6q_2^* & = 0 \\
q_0^* & +q_1^* & +q_2^* & = 1.
\end{aligned}
\tag{3.123}
$$

Note that the third of these equations can be obtained from the first two (add and change sign) so that we need the fourth to solve. Then we find

$$
q_0^* = \frac{2}{3}, q_1^* = \frac{1}{6}, \text{ and } q_2^* = \frac{1}{6}.
\tag{3.124}
$$

3.6.3   Distributions and moments

Some authors consider 'ensembles' of outcomes as representing the random quantity, and then model the behaviour of these as vectors. Each ensemble would constitute a different time series measure of the process – a different realization. It is simpler and more natural to consider merely the random quantity under consideration, which is our approach. The realizations are data. For illustration, we shall concentrate on continuous random quantities; discrete random quantities can be treated in a similar manner. As with other random quantities, those in a stochastic process are modelled using a probability density. Here it is a function of time:

$$
f_{X,t}(x,t).
$$

This will have an associated *cdf*

$$
F_{X,t}(x,t).
$$

These distributions can be conceived of as a multidimensional distribution, with different times corresponding to different random quantities. For example, we could think of the sequence of values $\{X_1, X_2, X_3, \ldots\}$ corresponding to times $\{1, 2, 3, \ldots\}$ as having a *pdf* $f_{X_1,X_2,X_3,\ldots}(x_1, x_2, x_3, \ldots)$. To make progress we must search for simplifications. The calculation of moments assists in capturing essential features of a process.

The mean of $X(t)$ is defined in a manner similar to (3.59) and in Section 3.5.6:

$$
\mu_X(t) = \langle X(t) \rangle = \int_{-\infty}^{\infty} x f_{X,t}(x,t)\,dx.
\tag{3.125}
$$

In order to study the behaviour of the process in time, we consider times t_1 and t_2 and the associated random quantities, $X(t_1)$ and $X(t_2)$. These will where convenient be abbreviated as X_1 and X_2. The mean values are

$$\mu_X(t_1) = \langle X(t_1) \rangle = \langle X_1 \rangle = \int_{-\infty}^{\infty} x_1 f_{X,t_1}(x, t_1) \, dx_1$$

$$= \int_{-\infty}^{\infty} x_1 f_{X_1}(x_1) \, dx_1, \text{ and} \tag{3.126}$$

$$\mu_X(t_2) = \langle X(t_2) \rangle = \langle X_2 \rangle = \int_{-\infty}^{\infty} x_2 f_{X,t_2}(x, t_2) \, dx_2$$

$$= \int_{-\infty}^{\infty} x_2 f_{X_2}(x_2) \, dx_2. \tag{3.127}$$

The **autocorrelation function** is defined as

$$\varrho_{X,X}(t_1, t_2) = \langle X(t_1) X(t_2) \rangle = \langle X_1 X_2 \rangle$$

$$= \int_{-\infty}^{\infty} \int_{-\infty}^{\infty} x_1 x_2 f_{X_1,X_2}(x_1, x_2) \, dx_1 dx_2. \tag{3.128}$$

This is a second order moment, as explained in Section 3.5.6. It is not a correlation coefficient, which is normalized as in (3.130) below. It is helpful to subtract the contribution of the means, giving the **autocovariance**

$$\sigma_{X,X}(t_1, t_2) = \langle X(t_1) X(t_2) \rangle - \mu_X(t_1) \mu_X(t_2) \tag{3.129}$$

$$= \varrho_{X,X}(t_1, t_2) - \mu_X(t_1) \mu_X(t_2)$$

$$= \langle (X_1 - \mu_{X_1})(X_2 - \mu_{X_2}) \rangle.$$

This central moment of order two is analogous to the covariance of Equation (3.94). It is equal to the variance of $X(t)$ if $t_1 = t_2 = t$.

The **autocorrelation coefficient** is defined as the normalized quantity

$$\rho_{X,X}(t_1, t_2) = \frac{\sigma_{X,X}(t_1, t_2)}{\sigma_X(t_1) \sigma_X(t_2)}, \tag{3.130}$$

analogously to (3.93).

A warning note: usage with regard to the above terms varies; sometimes 'covariance function' is used for what we have termed the autocorrelation function. Other usages also exist.

The functions above depend on t_1 and t_2. The autocorrelation and the autocovariance may depend only on the difference $t_1 - t_2 = \tau$, and not on the values t_1 and t_2. Then

$$\varrho_{X,X}(t, t+\tau) = \varrho_{X,X}(\tau) \text{ and} \tag{3.131}$$

$$\sigma_{X,X}(t, t+\tau) = \sigma_{X,X}(\tau) \tag{3.132}$$

are a function of τ only. This is true for stationary and weakly stationary processes. In the latter case the mean value is constant in time and the autocovariance depends on τ, not on t.

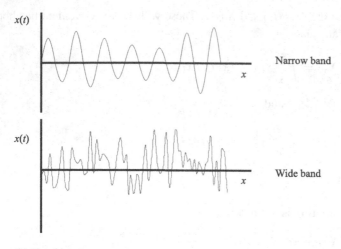

(a) Time histories

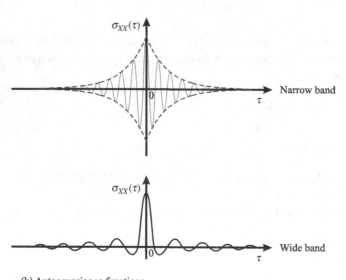

(b) Autocovariance functions

Figure 3.24 Narrow and wide band processes

Example 3.38 Consider the process given by

$$X(t) = a \cos(\omega t + \Xi), \tag{3.133}$$

in which a and ω are constants and Ξ is random with a uniform distribution in $[-\pi, \pi]$. Then

$$\langle X(t) \rangle = \int_{-\pi}^{\pi} a \cos(\omega t + \xi) \frac{1}{2\pi} d\xi = 0$$

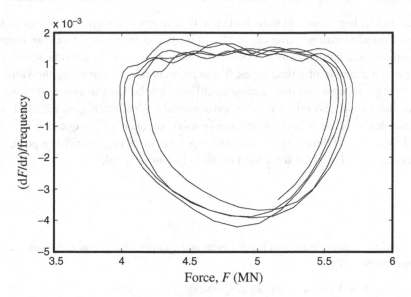

Figure 3.25 Phase-plane diagram for segment 1.22–1.34s of Figure 3.22(b). See Jordaan (2000)

and

$$\sigma_{X,X}(\tau) = \left\langle (X_1 - \mu_{X_1})(X_2 - \mu_{X_2}) \right\rangle$$

$$= a^2 \int_{-\pi}^{\pi} \{\cos(\omega t + \xi)\cos[\omega(t + \tau) + \xi]\} \frac{1}{2\pi} d\xi$$

$$= \frac{a^2}{4\pi} \int_{-\pi}^{\pi} [\cos(2\omega t + \omega\tau + 2\xi) + \cos(\omega\tau)] d\xi$$

$$= \frac{a^2}{2} \cos(\omega\tau), \text{ with} \tag{3.134}$$

$$\rho_{X,X}(\tau) = \cos(\omega\tau). \tag{3.135}$$

Example 3.39 *Narrow and wide band processes.* In narrow band processes, the frequency of the process is concentrated in a band near some frequency. We deal with spectral densities in Chapter 7; see Figure 7.11. Figure 3.24 illustrates the nature of the processes in terms of time series and autocovariance functions. Figure 3.25 shows a phase-plane diagram, based on the data from Figure 3.22(b). This represents a useful way of interpreting narrow band processes; see also Ochi (1990).

As noted above, ensembles represent particular sets of results of a time series, with each member of the ensemble representing a particular set of values in time; in effect, data. The question arises in analysing and interpreting data: should one calculate expected values over time or at a particular time over different ensembles? Ergodicity is a property that is often used as a basis. We alluded to this for Markov chains. The ergodic property requires stationarity of the time series. The implication in statistics is that time averages for ergodic processes from a sample are identical to averages over ensembles. It is often convenient to work out means and other moments from a particular time series rather than from a set of such time series. Figure 3.22(b) shows an example of a time series. There are methods for removing the variation of the mean during a process and time-scaling could possibly be used to ensure stationarity of the process. But the process might well be non-stationary in nature. Figure 3.22(b) is of this kind. The physical process does not suggest or support stationarity. The question is in reality one of modelling and inference. De Finetti (1974) points out that ergodicity is a property that may or may not apply but is not some kind of 'autonomous principle'.

3.7   Exercises

It is a good idea to develop the habit of making plots and sketches of pdfs and cdfs of all probability distributions encountered.

3.1 A random quantity X is modelled using the probability mass function

$$p_X(x) = k(4-x), \text{ for } x = 0, 1, 2, 3, \text{ and}$$
$$= 0, \text{ otherwise.}$$

(a) Find the constant k and the *pmf*. Sketch the distribution.
(b) Find and sketch the *cdf*.
(c) Find $\mathbf{Pr}(X < 2)$.

3.2 For the following probability densities, determine the constant k, sketch the density, and find and sketch the *cdf*.

$$f_X(x) = k\left(1 - x^2\right) \text{ for } 0 < x < 1,$$
$$= 0, \text{ elsewhere.}$$
$$f_X(x) = \exp\left(-k\,|x|\right) \text{ for all real } x.$$
$$f_X(x) = k\sin(kx) \text{ for } 0 < x < \frac{\pi}{4},$$
$$= 0, \text{ elsewhere.}$$

For the first *pdf*, find the $\mathbf{Pr}(X > 0.5)$, for the second find $\mathbf{Pr}(X > 1)$ and for the third, find $\mathbf{Pr}\left(X > \frac{\pi}{6}\right)$.

For an application of the second distribution, see Example 7.7.

3.3 A random quantity is modelled with a *cdf*

$$F_T(t) = \begin{cases} 0, & \text{for } t < 1 \\ t^2 - 2t - 1, & \text{for } 1 \leq t \leq 2 \\ 1, & \text{for } t > 2. \end{cases}$$

Determine the *pdf* $f_T(t)$, and $\mathbf{Pr}(T \leq 1.5)$.

3.4 In planning the use of a route for emergency vehicles in a city, there are ten signalized intersections. At each one, the traffic light is green for 40 seconds, yellow (amber) for 5 and red for 30 seconds. The emergency vehicle will stop at an intersection only if it is red. The signalling system is not coordinated. What probability distribution models the number of stops? Determine the probability that the emergency vehicle will have to stop at

   (a) the first intersection;
   (b) none of the intersections;
   (c) the first four intersections;
   (d) the second, fourth, ..., (even numbered) intersections but not at the others;
   (e) more than five intersections.
   (f) It is desired to reduce the probability in (e) to 0.02 seconds. What should the 'red' time be so that this is obtained, keeping the amber and total cycle times constant?

3.5 A lot of 1000 transistors contains 900 that pass an acceptance test. The transistors are used in groups of five to construct a component. It is decided to mix the failed transistors with those that have passed the test and to select five randomly from the lot. The number of failed transistors in the group of five is of interest, denoted X.

   (a) What is the distribution of X? Can the binomial distribution be used as an approximation?
   (b) Find the probability that not all the transistors have passed the test, using an exact method and a binomial approximation.
   (c) What is the probability that more than two transistors in the component have failed the test?
   (d) What are the mean and the standard deviation? Use the approximate and exact distributions.

3.6 Summarize the four forms of the hypergeometric distribution as described and emphasized by de Finetti (1974).

3.7 In a telephone system, long distance telephone calls from a certain origin arrive at a rate of 12 per five minutes. This rate can be considered as constant for the times under consideration.

   (a) What models should be used to describe the time T between successive calls and for the number of calls X in a minute?
   (b) If a time τ elapses after a call, during which time no call is received, what is the distribution to the next call?
   (c) What is the probability that two calls arrive less than 30 seconds apart? What is the probability that the next call will arrive after a minute elapses?
   (d) Find the probability that two successive calls are separated by an interval that is within two standard deviations of the mean? Sketch the *pdf* showing the calculation.

3.8 The number of flaws X per square metre of the surface of a plate is a random quantity. It is modelled using a Poisson distribution with parameter $\lambda = 1.67$ per square metre.

   (a) What are the mean and standard deviation of X?
   (b) What is the probability that a square metre of plate has no flaws?

   If a square metre of plate has no flaws on its surface, it fails at a pressure of 5.6 kPa. If there are one or more flaws, it fails at 2.8 kPa.
   (c) Find the mean pressure at which a square metre of plate fails.

(d) Show that the standard deviation in kPa of the pressure for the square metre of plate is

$$2.8 \exp(-1.67) \sqrt{\exp(1.67 - 1)}.$$

3.9 Given that an earthquake has occurred, the magnitude R on a Richter scale is modelled as the exponential distribution

$$f_R(r) = 2.35 \exp(-2.35r) \text{ for } r > 0,$$
$$= 0, \text{ otherwise.}$$

Find the *cdf* and the probability that the next earthquake will have a magnitude greater than 6.0.

3.10 The pressure X acting on a ship's hull during a collision is modelled by an exponential distribution with a mean of 2 MPa. The number of impacts per year is modelled by a Poisson distribution with a mean equal to 10. Consider now only those impacts with $X > x$ MPa.

(a) What model would you use for the number of these impacts $(X > x)$ per year? Find the probability that in a year the number of such impacts is equal to zero for the case $x = 6$ MPa.
(b) Consider now the maximum pressure in a year, written as Z. Using the result just obtained, derive the *cdf* $F_Z(z)$.
(c) If the rate of impacts varies with time, with the same expected value, would the results derived still apply?

3.11 The concentration X of pollutant in a lake has been modelled using a normal distribution based on past measurements, with a mean of 2.21 and a standard deviation of 0.46, in parts per million (ppm). Find the following.

(a) $\mathbf{Pr}(X < 1.5)$.
(b) $\mathbf{Pr}(1.5 < X < 2.5)$.
(c) $\mathbf{Pr}(X > 2.5)$.
(d) The probability that in six samples, (a) occurs three times, (b) twice and (c) once.

3.12 Two rebars used in the construction of a gravity-based concrete offshore structure are connected by a coupling. The strength of each rebar under the cyclic loading expected is normally distributed with a mean of 300 MPa and a standard deviation of 30 MPa.

(a) Let X denote the strength of a rebar. What is the value of strength $X = x_f$ corresponding to $F_X(x) = 0.025$?
(b) It is judged that the strengths of the two rebars are stochastically independent. What is the probability that exactly one of the two rebars will fail at a strength less than x_f?
(c) An assembly of two rebars is to be tested. Denote the strength of the assembly as Y. What is the value of $F_Y(x_f)$ given that $F_X(x_f)$ is 0.025?

3.13 Find the mean, variance and standard deviation for Problem 3.1 and the first of 3.2.

It is a good idea to work through the derivations of the means and variances of commonly encountered distributions (Appendix 1).

3.14 The Cauchy *pdf* with parameter a,

$$f_X(x) = \frac{k}{x^2 + a^2}, \quad \text{for } -\infty < x < \infty, \tag{3.136}$$

is a bell-shaped curve resembling in a general way the normal distribution. But it has much thicker tails.

(a) Find the value of k so that the *pdf* is well defined.

(b) Find the *cdf* $F_X(x)$.

(c) Find the inter-quartile range, the difference in x between $F_X(x_L) = 1/4$ and $F_X(x_U) = 3/4$.

(d) Find the mean and standard deviation.

3.15 The Pareto *pdf* is given by

$$f_X(x) = \frac{kt^k}{x^{k+1}}, \text{ for } x \geq t \tag{3.137}$$
$$= 0, \text{ for } x < t.$$

(a) Find necessary conditions on k to ensure that $\langle X^n \rangle$ is finite.

(b) Find the range of values for which the variance is finite. Find an expression for this variance.

3.16 An object (possibly an iceberg) is drifting on the ocean surface with a speed X that is random and modelled by a gamma distribution:

$$f_X(x) = \frac{x^{\alpha-1} \exp(-x/\beta)}{\beta^\alpha \Gamma(\alpha)}.$$

This distribution is based on a survey of all objects in a large area. Consider now the event that the object collides with an offshore installation. The probability of this event is proportional to the speed X. Derive an updated velocity distribution using Bayes' theorem given the event and using the *pdf* noted as a prior distribution. Show that this updated distribution is a gamma distribution with shape parameter changed from α to $\alpha + 1$. Plot the prior and updated distributions for $\alpha = 2.5$, and with β taking the two values 0.11 and 0.38 ms^{-1}.

3.17 An interesting example arises in arctic engineering. The area of an ice floe (assuming it to be circular of area A and radius R, approximately) is given by $A = \pi R^2$. If R is random, then so is A. What is the mean area?

3.18 *Geometric probability 1.* A space probe is planned to land at a point P on the surface of a planet. A local coordinate system is constructed in the tangent to the surface at P. The probability of landing outside a circle of radius 5 km with P as centre is considered to be zero, and the probability within is judged to be uniformly distributed over the circular area with the density proportional to the area. Find the probability that the space probe will land

(a) within 2 km of P,

(b) between 2 and 3 km from P,

(c) north of the aiming point,

(d) south of the aiming point and more than 2 km from P, and

(e) exactly due north of P.

3.19 *Geometric probability 2.* Two signals are sent to a receiver, and arrive within the time interval t. The instants at which the signals arrive are denoted X_1 and X_2. If the signals arrive within the small interval τ of each other, interference and deterioration of the signals result. Find the probability of this event, on the assumption that X_1 and X_2 are independent and equally likely at any point in the interval $[0, t]$.

3.20 *Geometric probability 3.*

(a) Two ships berth at the same quay in a harbour. The arrival times for the ships are independent and uniformly distributed (equally probable) during any 24 hour period. If, during the time w_i (needed for mooring, loading and unloading) between arrival and departure of either ship, the other ship arrives, then the other ship has to wait to use the berth. For the first ship this waiting time is $w_1 = 1$ hour. For the second ship the waiting time is $w_2 = 2$ hours. Estimate the probability that one of the ships will have to wait to use the berth during the next 24 hour period.

(b) A ferry carries freight continually from one bank of a river to another. This takes one hour. A ship moving along the river at right angles to the direction of travel of the ferry can be seen from the ferry if it passes the trajectory of the ferry less than 20 minutes away from its position. All times and places of intersection are judged to be equally likely. What is the probability that the ship will be seen from the ferry?

3.21 *Geometric probability 4.* On the surface of the earth the area bounded by two degrees of longitude and by the line of latitude θ to $\theta + 1$ degrees is to a good approximation, a rectangle of sides 111.12 km (north–south) and $111.12 \cos(\theta + 0.5)$ km (east–west). A ship of beam (width) b is crossing the degree-rectangle on a path of total length d. A stationary iceberg (or other object) of width w is placed at a random position in the degree-rectangle.

(a) Find the probability that the ship will collide with the object during the crossing of the degree-rectangle and evaluate the probability given $\theta = 41°$, $b = 28$ m, $w = 110$ m and $d = 83.2$ km (these are plausible values for the RMS Titanic in 1912), or using some other values of interest.

(b) Using the values obtained in (a) find the probability that there are no collisions after 100 crossings, and the number of crossings above which the probability of a collision just exceeds 0.50.

3.22 Find the skewness of the exponential distribution

$$f_X(x) = \frac{1}{\beta} \exp(-\beta x) \text{ for } x \geq 0$$
$$= 0 \text{ otherwise.}$$

3.23 The joint probability distribution $p_{X,Y}(x, y)$ of the random quantities X and Y is as follows.

		x			
		0	1	2	3
	1	$\frac{2}{27}$	0	0	$\frac{1}{27}$
y	2	$\frac{6}{27}$	$\frac{6}{27}$	$\frac{6}{27}$	0
	3	0	$\frac{6}{27}$	0	0

(a) Check for coherence and determine the marginal distributions $p_X(x)$ and $p_Y(y)$.

(b) Determine $p_{X|Y}(x, y)$, $p_{Y|X}(x, y)$ and $p_{Y|X=0}(y)$.

(c) Find the means μ_X, μ_Y, the standard deviations σ_X, σ_Y as well as $\langle XY \rangle$.

(d) Determine the covariance $\sigma_{X,Y}$.

(e) Are X and Y independent?

3.24 Given the *pdf*

$$f_{X,Y}(x, y) = 2\exp(-x - y), \text{ for } 0 \le x \le y \text{ and } 0 \le y < \infty$$
$$= 0 \text{ otherwise,}$$

find the marginal distributions and check for independence.

3.25 Prove that the correlation coefficient ρ_{XY} is such that

$$-1 \le \rho_{X,Y} \le 1,$$

always. Comment on the meaning of $\rho_{XY} = 0$.

3.26 *Raiffa decision problem.* Consider an additional choice in Examples 3.24 and 3.25 of Section 3.5.2. You are offered the option of sampling two balls for a fee of $12 rather than the single ball at $8 in Example 3.25. The contracts offered remain the same as before. Note that the updated probabilities are the subject of Exercise 2.18. Is the new option an attractive one based on *EMV* compared to the others?

3.27 A long distance telephone company is planning to increase the capacity of its networking system. A team set up to study the future demand is unable to reach a firm decision and concludes that the annual growth rate will be 1, 2, or 3 per cent with equal probability. Three capacities will be adequate for these three growth rates for the design period. The respective capacities are 5, 10 and 15, in arbitrary units. Economics of scale given by the equation

$$\text{cost} = k\,(\text{capacity})^{0.6}$$

can be applied to the cost of the three capacity units. Draw a decision tree for this problem and indicate what decision you would recommend on the basis of maximum expected monetary value. Complete the calculation in terms of present monetary units, i.e. ignore interest on money and the effects of inflation.

3.28 Aggregates for a highway pavement are extracted from a gravel pit. Based on experience with the material from this area, it is estimated that the probabilities are $\mathbf{Pr}(G) = \mathbf{Pr}(\text{good-quality aggregate})$ $= 0.70$, and $\mathbf{Pr}(\tilde{G}) = \mathbf{Pr}(\text{poor-quality aggregate}) = 0.30$. In order to improve this prior information, the engineer considers a testing program to be conducted on a sample of the aggregate. The test method is not perfectly reliable – the probability that a perfectly good-quality aggregate will pass the test is 80%, whereas the probability of a poor-quality aggregate passing the test is 10%.

(a) Draw both Bayes' calculation and decision trees for the problem. The decision tree should include the options '*do not implement test programme*' and '*implement test programme*', as

well as *'accept the aggregate'* and *'reject the aggregate'*. Determine the optimal decision
on the basis of minimum-expected-cost, given the following monetary values (on a nominal
scale). Accept, good aggregate, 15 units; accept, poor aggregate, −8 units; reject, good or poor
aggregate, 5 units; cost of testing, 2 units.

(b) If a sample is tested, and if it passes, what is the probability of good-quality aggregate?

(c) If a second test is performed, and the sample again passes, what is the probability of good-quality
aggregate? One can use the result of part (b) as a prior probability in this case.

3.29 In the design of a dam for flood control, the height of the dam depends on two factors. The
first is the level of water in the dam, denoted X, and the second is the increase caused by flood
waters coming into the reservoir, denoted Y. The level X takes the values {30, 40, 50 metres} with
probabilities {0.2, 0.5, 0.3} respectively, while Y takes values {10, 20, 30 metres} with probabilities
{0.3, 0.4, 0.3} respectively. It is decided to use a decision analysis to determine the dam height. Two
values are under consideration, 65 and 75 m, with costs of 12 and 15 units respectively. The
consequences of overtopping are flooding downstream with costs of 15 units. (All costs can be
treated as additive.) A hydrological study is under consideration in order to obtain a better estimate
of the probabilities of the flood levels noted. For the purposes of the present analysis, the results
of the study can be summarized by the values of $R = 1$ (low), $R = 2$ (medium) and $R = 3$ (high).
The method has the reliabilities indicated in the following table.

	Pr(R|Y)		
	$R = 1$	$R = 2$	$R = 3$
$Y = 10$	0.70	0.20	0.10
$Y = 20$	0.10	0.80	0.10
$Y = 30$	0.05	0.05	0.90

Update the probabilities of Y, i.e. determine the values **Pr(Y\|R)**. Complete a decision analy-
sis with tree, including the option: (complete study, do not complete study) and the
option (dam height= 65 m, dam height = 75 m). The cost of the study is 2 units.

3.30 Referring to Exercise 2.26, the engineer has to decide whether or not to close down a section of
the pipeline to conduct major repairs. The following values of utility are assigned. Subtract 2 units
as a 'cost' if tests are undertaken. Draw a decision tree and determine the optimal course of action
based on the criterion of maximum expected utility. The symbol R denotes 'repairs undertaken'.

S	Given R	Given $\tilde{R}$
0	7	10
1	6	5
2	5	0

3.31 The speeds of vehicles on a highway are modelled by a normal distribution with a mean of 60 kmh^{-1}
and a standard deviation of 15. Determine the *pdf* of the relative speed of two vehicles moving in

opposite directions. Assume now that the drivers, having seen a police car in the area, obey a speed limit of 75 kmh^{-1}; those exceeding the limit now drive at the limit, with the other speeds remaining unchanged. Sketch and describe the new distribution of speeds (in one direction).

3.32 Consider a sequence of four waves in a storm. The wave heights are $\{X_1, X_2, X_3, X_4\}$. These quantities are judged to be exchangeable, so that all 24 permutations such as $\{X_1, X_3, X_2, X_4\}$ or $\{X_4, X_2, X_3, X_1\}$ are considered equivalent. For quantities that are not exchangeable there are four values $\langle X_h \rangle$, six $\langle X_h X_i \rangle$ for $(h < i)$, and four $\langle X_h X_i X_j \rangle$ for $(h < i < j)$. What can be said of these in the case of exchangeable random quantities?

3.33 Sketch random processes that illustrate the four combinations of discrete or continuous in time and discrete or continuous random quantities.

3.34 In a Poisson process with constant intensity λ, show that the time T to the kth event is modelled by the Erlang distribution

$$f_T(t) = \frac{\lambda(\lambda t)^{k-1}\exp(-\lambda t)}{(k-1)!}, \text{ for } t > 0, k \geq 1. \tag{3.138}$$

Verify that this is a special case of the gamma distribution.
If accidents occur at the rate of 0.2 per month on a highway, describe the distribution of the time in months to the third accident.

3.35 Aircraft are serviced at a loading bay. An engineer is studying the waiting times and usage of space and decides to model the situation by dividing time into a discrete set of small time intervals during which one arrival is expected. During one such time step, the crew at the loading bay can service $\{0, 1, 2\}$ aircraft with probabilities $\{3/8, 1/2, 1/8\}$. The following strategy is adopted. One aircraft is allowed to wait to enter the bay if one is being serviced there. Consequently, when the bay has zero or one aircraft being serviced, an additional aircraft is then permitted to enter or to wait. In the case where one is being serviced, and one is waiting, any arriving aircraft are routed elsewhere. The states under consideration are $\{0, 1, 2\}$, corresponding to the number of aircraft at the bay (waiting or being serviced) at the end of each time step. The end of a time step is defined such that any completed aircraft have moved on and before a new one is accepted.

(a) Derive the one-step transition matrix.
(b) Find the long-run stationary state, if there is one.
(c) If the bay starts out the day empty, so that an aircraft is admitted at the beginning of the next step, write down the first three probability vectors q_i.

3.36 Find the autocovariance of

(a) $X(t) = A\cos(\omega t + \Xi)$,
   where A and Ξ are independent random quantities, the latter uniformly distributed in the interval $[-\pi, \pi]$.
(b) $X(t) = \sum_{i=1}^{n}[A_i\cos(\omega_i t) + B_i\sin(\omega_i t)]$,
   where all of the A_i and the B_i are independent random quantities with $\langle A_i \rangle = \langle B_i \rangle = 0$ and $\sigma_{A_i}^2 = \sigma_{B_i}^2 = \sigma_i^2$.

4 The concept of utility

'Your act was unwise,' I cried, 'as you see
By the outcome.' He calmly eyed me:
'When choosing the course of my action,' said he,
'I had not the outcome to guide me.'

Ambrose Bierce

4.1 Consequences and attributes; introductory ideas

We have shown that two ingredients are combined in our decision-making: probability and utility. But utility has not been well defined. Up to this point, we have given detail of the probabilistic part of the reasoning. We now present the basic theory with regard to utility. Probabilities represent our opinions on the relative likelihood of what is possible, but not certain; utilities encode our feelings regarding the relative desirability of the various consequences of our decisions. Examples of where utilities fit into the decision-making process were given in Section 1.2. A special case is expected monetary value (*EMV*), dealt with in Section 3.5.2.

There can be various components of utility, depending on the application. These components will be referred to as '**attributes**'. For example, in a study of airport development for Mexico City, de Neufville and Keeney (1972) considered attributes of the following kind.

- cost
- airport capacity
- safety
- access time
- noise
- displacement of people.

There will usually be a set of such attributes, which depends on the problem at hand. Attributes may have

- **increasing value**, for example money, profit, volume of sales, time saved in travelling, and wilderness areas available for enjoyment
- **decreasing value**, for example size of an oil spill, level of noise pollution, and deaths and injuries in an accident.

Cost and safety are common attributes in engineering, and there are often tradeoffs between the two. It is relatively easy to design a structure that is safe enough by excessive size of its members; it is the art of engineering to provide a structure that is adequately safe (and hopefully elegant) with appropriate economy.

The essence of utility is to transform the attributes into a single scale which measures the desirability of the outcome. We shall introduce the concept by summarizing the transformation of a single attribute, money. The basic idea is contained in the following analysis that was given by Daniel Bernoulli in the eighteenth century. He considered the utility u corresponding to money, x. If a person possesses a sum x, Bernoulli reasoned that a small change in the person's utility, du, is proportional to a change dx in their capital, and also inversely proportional to their working capital x. For example, \$5 will be of higher utility to a man under poverty-stricken circumstances than under relative affluence.[1] Thus, $du = (k_1/x)dx$ where k_1 is a constant, and

$$u = (k_1 \ln x + k_2),$$ (4.1)

where k_2 is a constant of integration.

The logarithmic relationship of (4.1) is an example of the non-linearity of utility with an attribute, money. Figure 4.1 illustrates such a relationship. As a 'summary' value we often use a mean value; the mean of two amounts of money A and B are shown as $\bar{x}$ in the figure. This would be our expected value corresponding to a bet with even odds on A and B. Imagine probabilistic weights of 0.50 attached to the points A and B. Their centre of gravity is at point G and the mean utility is $\bar{u}$. Utility theory tells us to use $\hat{x}$, the x-coordinate value corresponding to $\bar{u}$, obtained from the nonlinear relationship, instead of $\bar{x}$ in decision-making. The germ of the concept of utility is 'curving it'! If all of this seems too complicated, it is explained in much more detail in the following. We hope merely to whet your appetite.

In using utility theory we begin with an attribute measured on some scale, a prominent example being money as in the preceding example. Many decisions concern attributes which might at best be ordered. For example, flooding of an area in dam construction and consequent effect on wildlife might be difficult to measure on a scale. This is not to say that these consequences are not important: many would rank them as being amongst the highest in importance. The point is that placing these consequences on a numerical scale might present difficulties since measurement of the attributes might not be possible. The final decision would then have to be made on the basis of a qualitative assessment of all the consequences with numerical support where possible. Exercise 4.1 is intended to alert the reader to these difficulties.

---

[1] We shall see that utility is a measure of an **individual's** preferences: to speak of a beggar and a rich man is not quite what we mean. Rather, this applies for the **same** person under the two different circumstances.

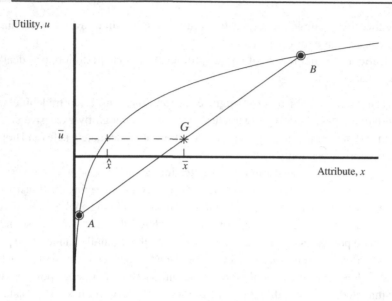

Figure 4.1   Utility and money: 'weights' of probability act upwards at A and B with their equilibrating unit weight acting downwards at G

The primary focus of this chapter is the **quantitative** analysis of attributes. This should be carried out where possible in important decisions and serves to assist in understanding risk aversion and in dealing with qualitative measures which should be given their proper weight in the decision. A further area of concern is that of the use of benefit–cost ratios. These can provide a reasonable basis for certain comparisons but contain arbitrary elements, for example in dealing with attributes measured in a qualitative manner. Risk aversion is not included and, furthermore, when considering the time value of money, the choice of interest rates can have a decisive effect on the results. The latter aspect applies to all methods unless the economic aspects are carefully modelled – and they need to include uncertainty. Exercise 4.2 points in this direction using very simple models.

An excellent introduction to utility is the simple yet logically convincing treatment by Raiffa (1968) which is much recommended. More detail is given in Raiffa and Schlaifer (1961) and in Keeney and Raiffa (1976, reprinted 1993). The latter is a tremendously useful book, motivated very much by practice. Much of the development of utility theory has taken place during the twentieth century, a prominent example being the work by von Neumann and Morgenstern (1947).

4.2   Psychological issues; farmer's dilemma

The attributes in any particular problem have to be carefully considered and 'massaged' before they can be considered as utilities. How do we weigh our feelings in making decisions? The

Table 4.1 *Payoff matrix v_{ij}*

	θ_1 : perfect	θ_2 : fair	θ_3 : bad
α_1 : crop A	11	1	−3
α_2 : crop B	7	5	0
α_3 : crop C	2	2	2

need for a unique method of ranking decisions will be introduced by the 'farmer's dilemma' (Rubinstein, 1975). This concerns a farmer who has to choose between three crops at the beginning of the planting season. The three crops will behave differently in different weather. For simplicity, we consider only three possible weather conditions: perfect, fair and bad. These are literally our states of nature, and the farmer is unsure as to which will prevail. The possible crops are designated A, B and C. Crop A is potentially bountiful and a good money-maker, but will do well only if the weather is very good. If not, it will likely fail. Crop B is more hardy, but will not have the same potential in good weather. Crop C is most hardy, and is little affected by the weather, within the limits of the region.

Table 4.1 shows the potential payoffs to the farmer for the set of planting strategies, combined with the states of nature. The payoffs can be considered to represent money to some arbitrary scale. We shall use the notation α_i to represent the possible actions (or strategies) available to the farmer, and θ_j to represent the states of nature. For some combination of these two, the payoff is v_{ij}. The farmer's problem can also be viewed in the form of a decision tree, as in Figure 4.2.

Lest it be thought that the farmer's dilemma does not encapsulate a real problem in the engineering or economic world, Table 4.2 illustrates the dilemma facing the regulatory authority in Canada when a major new bridge was being designed for harsh environmental loadings. There is often considerable uncertainty in loadings under such circumstances, and, indeed, differences of opinion amongst experts, with load estimates ranging from high (extremely conservative) to low (less conservative, but possibly realistic). Table 4.2 illustrates the payoff matrix that might be in the regulator's mind when attempting to decide upon a decision amongst the α_i. As in all decisions, it might appear as a risky choice versus a safe one.

4.2.1   Pessimist's solution: maximin

The first possible solution to the dilemma is to make the assumption that 'the worst will happen'. The farmer inspects the payoffs for each strategy, and chooses the minimum. This amounts to taking a minimum for each row, as shown in Table 4.3. The farmer then makes the best of the situation by choosing the 'best of the worst', the maximum of the minima. This would result in the choice of crop C in Table 4.3 (action α_3). This is termed the 'maximin' solution. There are three possible interpretations of the farmer's behaviour.

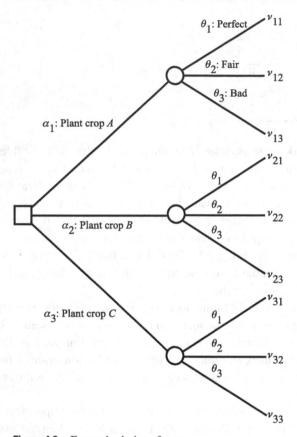

Figure 4.2   Farmer's choice of crop

Table 4.2   *Three load estimates for bridge design*

	Actual load (state of nature)		
	θ_1 : high	θ_2 : medium	θ_3 : low
α_1 : high estimate	2	2	2
α_2 : medium estimate	1	3	4
α_3 : low estimate	−1	2	6

Table 4.3   *Pessimist's solution*

	θ_1	θ_2	θ_3	Row minimum
α_1	11	1	−3	−3
α_2	7	5	0	0
α_3	2	2	2	$\boxed{2}$

Table 4.4 *Optimist's solution*

	θ_1	θ_2	θ_3	Row maximum
α_1	11	1	-3	$\boxed{11}$
α_2	7	5	0	7
α_3	2	2	2	2

1. He or she considers the probability of bad weather to be unity, in other words a certainty. This may be justified, but in general is unlikely to be a realistic judgement.
2. The farmer is very averse to taking risks and chooses a safe solution which is assured. This may well be related to the farmer's present working capital.
3. If the farmer considers the 'opponent' – nature in this case – to be out to 'get' the farmer, the reaction is a reasonable one. We shall see in Chapter 5 that this attitude is a natural one in situations of conflict; in the present case, it may indicate poor judgement.

4.2.2   Optimist's solution: maximax

The optimist assumes that the best always happens. Table 4.4 shows this solution. For each possible action, the maximum value of the payoff is identified as the value to choose. The 'best of the best', the maximum of the maxima is then chosen, with the result that the optimist chooses action α_1, crop A. There are three interpretations of the optimist's behaviour, mirroring the interpretations of the pessimist above. The probability of perfect weather could in fact be judged to be unity, or the decision-maker is willing to take a risk, or the judgement of the decision-maker could be faulty, indicating an unrealistically optimistic attitude.

4.2.3   The 'in-betweenist'

The optimist and the pessimist represent extreme points of view which many of us would wish to avoid, by taking a compromise between them. This criterion (due to Hurwicz) attempts to find common ground between the optimist and the pessimist. A weight β is placed on the pessimist's solution, with the weight $(1 - \beta)$ on the optimist's. The value β, such that $0 \le \beta \le 1$, is used as follows, using the above solutions for the optimist and the pessimist. We assume that $\beta = 0.6$.

Action α_1 :   weighted value $= (0.6)(-3) + (0.4)(11) = 2.6$
Action α_2 :   weighted value $= (0.6)(0) + (0.4)(7) = 2.8$
Action α_3 :   weighted value $= (0.6)(2) + (0.4)(2) = 2.$

The in-betweenist would choose action α_2, crop B, since this gives the greatest of the weighted values. The 'index of pessimism' β is chosen subjectively, with $\beta = 1$ indicating complete pessimism, and $\beta = 0$ complete optimism.

Table 4.5 *Regret matrix r_{ij}*

	θ_1	θ_2	θ_3	Row maximum
α_1	0	4	5	5
α_2	4	0	2	$\boxed{4}$
α_3	9	3	0	9

Table 4.6 *Payoff matrix illustrating inconsistency*

	States			
Actions	θ_1	θ_2	θ_3	Row minimum
α_1	5	4	2	2
α_2	4	1	3	1
α_3	3	6	7	$\boxed{3}$

4.2.4 Regrets

Another attempt to find a decision-making criterion is the 'regrettist' solution. The regrettist always regrets decisions if the result could have been better. In this method, the decision-maker transforms the original matrix v_{ij} of Table 4.1 into a 'regret' matrix r_{ij}. Given any particular state of nature θ_j, the column is inspected for the highest value. In the regret matrix, this is assigned the value 0 (zero regret). For example, taking state of nature θ_2 in Table 4.1, the value $v_{22} = 5$ is the highest (action α_2) and is assigned the value 0 in Table 4.5. The values above and below this value, v_{12} and v_{32}, are the differences $(5 - 1)$ and $(5 - 2)$ respectively, and represent the amounts one would 'regret' if the state is θ_2 (fair) and if one had undertaken actions α_1 and α_3 respectively, rather than α_2. Having completed the table, the regrettist now obtains the row maxima, and chooses the action which minimizes these values, action α_2 in the present case. The criterion is also termed the 'minimax regret' criterion.

4.2.5 Inconsistencies; tentative way out

There are inconsistencies associated with all of the methods above, if we attempt to make a probabilistic interpretation. Consider the payoff matrix of Table 4.6. Let us apply the pessimist's solution, and try to interpret this as a rational assignment of a high probability $\simeq 1$ to the worst consequence. We see that the row minimum corresponds to a different state of nature for each action. For α_1, the state of nature giving the minimum is θ_3; for α_2, it is θ_2; while for α_3 it is θ_1. This can only mean that the decision-maker assumes different states of nature to have a high probability, depending on the decision made. In other words, the assumption is that if one plants crop A, it will hail, if crop B, snow will appear, and crop C will somehow cause frost. Nature does not generally behave in this way. The only reasonable interpretation is total inconsistency, perhaps with an element of paranoia! The one situation where the reaction of the decision-maker might reasonably follow this pattern is where there is a genuine situation of conflict. This would not constitute a 'game against nature' but rather a game against an intelligent opponent, who is out to win. This situation will be dealt with in Chapter 5.

The other methods above also suffer from the problem of inconsistency outlined in the preceding paragraph. Further, the regrettist solution suffers from a further drawback. Addition of a new alternative, which we would in any event not choose on the regrettist's criterion, can alter our original choice (Hall, 1962).

Farmer's tentative way out

Although the in-betweenist has attempted to make a compromise between the pessimist and the optimist, probabilities have not been properly introduced into the problem. There needs to be a clear separation of the probabilities and the consequences in the decision-making process. The farmer realizes this, and consults a friend who knows probability theory, and who happens also to be a meteorologist. They work together on the problem, addressing Table 4.1 and come up with the following probabilities.

$\mathbf{Pr}(\theta_1) = 0.1,$

$\mathbf{Pr}(\theta_2) = 0.5,$ and

$\mathbf{Pr}(\theta_3) = 0.4.$

The farmer decides to weight the monetary values in Table 4.1 by the corresponding probabilities, conditional on the action taken. Remembering that v_{ij} represents the payoff, we denote it V_{ij} to emphasize that it is random.

$\langle V_{1j}|\alpha_1\rangle = (0.1)(11) + (0.5)(1) + (0.4)(-3) = 0.4,$

$\langle V_{2j}|\alpha_2\rangle = (0.1)(7) + (0.5)(5) + (0.4)(0) = 3.2,$ and

$\langle V_{3j}|\alpha_3\rangle = (0.1)(2) + (0.5)(2) + (0.4)(2) = 2.0.$

These are termed the farmer's expected monetary values (*EMV*s), as introduced in Section 3.5.2. We now consider the following criterion.

Tentative criterion: *choose that action which maximizes the expected monetary value.*

Accordingly, the farmer decides to follow action α_2, plant crop *B*. At the same time, a farmer who has a lot of money in the bank might wonder whether crop *A* might not have been a better choice whereas a farmer struggling financially might consider crop *C* a better option, since the prospect of a gain of zero (crop *B*) or a loss of (-3) in the case of crop *A* if the weather is poor are not very attractive alternatives (imagine the amounts to be in $1 000 000).

4.3   Expected monetary value (*EMV*)

In this section we shall continue to deal with that universal attribute, money, and explore the usefulness of expected monetary value *(EMV)*. We shall therefore explore further the concept of *EMV*, but first introduce an interesting set of problems that will be of use in this chapter and elsewhere.

Brief digression: games with two dice   Problems concerning two dice occur frequently in probability theory. One of the early problems in probability theory considered by Pascal was posed to him by de Méré; this is used to introduce extremes in Chapter 9. We use two-dice problems twice in this chapter, once in this section, and also in discussing transitivity of choices (Section 4.6). If we toss two dice, the possible results can be shown in a matrix $\{i, j\}$ in which i denotes the number on the first die, and j the number on the second. All results are equally likely, with probability 1/36.

11 12 13 14 15 16
21 22 23 24 25 26
31 32 33 34 35 36
41 42 43 44 45 46
51 52 53 54 55 56
61 62 63 64 65 66

In the game of craps, two dice are thrown and the sum of the two scores is the deciding factor, so that calculation of the probabilities of $(i + j)$, the total of the values on the two dice, is needed. There are 11 possible results, from 2 to 12, with probabilities corresponding to the diagonal values in the matrix. The probabilities increase in equal steps of 1/36, from 1/36 for a result of 2, to 6/36 for 7, and then decrease in the same way to 1/36 for 12. In craps, a first throw of seven or eleven wins, with a probability of $8/36 = 2/9$. A first throw of two, three or twelve loses. This has a probability of $4/36 = 1/9$. Any other first throw, to win, must be repeated before a seven is thrown in the game.

For the amusement of the reader, we pose the following problem: two dice are thrown. We happen to observe that one of them shows a five. What is the probability that the other shows a four? (Answer: 2/11. Why?)

Back to *EMV* A decision-maker is considering whether to join in a simplified version of the game of craps. We shall consider small stakes, to assist in making our point that *EMV* is acceptable in this case. In our simplified game, two dice are thrown, and three states of nature, as above, are of interest, but we terminate the game after one throw. The states are $\theta_1 =$ 'the total is 7 or 11', $\theta_2 =$ 'the total is 2, 3 or 12' and $\theta_3 =$ 'any other result'. In a particular throw of the dice, we gain \$9.00 in the case that θ_1 happens, lose \$9.00 if θ_2 happens, with no exchange of money if θ_3 happens. We judge the dice to be 'fair' and therefore assign $\mathbf{Pr}(\theta_1) = 2/9$, $\mathbf{Pr}(\theta_2) = 1/9$ and $\mathbf{Pr}(\theta_3) = 2/3$.

To find the *EMV* associated with a throw of the dice we weigh the various monetary values by the corresponding probabilities. The expected monetary value is

$$(9)(2/9) + (-9)(1/9) + (0)(2/3) = \$1.00.$$

This would be the amount that we would expect 'on average' in a long series of throws; in a single throw we would expect $+9$, -9 or 0 with the probabilities noted.

We are pondering an invitation to join the game just described. We have a few hundred dollars in our pocket, and since the amounts involved in the game are trivial compared with the amount in our pocket, and since the *EMV* is positive, we expect to gain 'in the long run', and agree to play the game. Were the *EMV* negative, we should decline to join. We conclude that the *EMV* criterion is acceptable in this case as a guide in decision-making. But many interesting problems in life consist of single trials with large consequences which we do not have the opportunity to repeat.

Such a stark contrast is provided if we return to the problem of the oil wildcatter of Section 1.2. Monetary values for the wildcatter's problem are shown in Figure 4.3. Suppose that the wildcatter has consulted a geologist and that she subsequently assigns the probabilities $\mathbf{Pr}(\theta_1) = 0.15$ and $\mathbf{Pr}(\theta_2) = 0.85$. To implement the tentative criterion, we need to calculate the *EMV* for each decision:

$$\langle V|\alpha_1 \rangle = (0.15)(10\,000\,000) + (0.85)(-1\,500\,000) = \$225\,000$$

whereas $\langle V|\alpha_2 \rangle = \$1\,000\,000$, where V stands for monetary value.

On the basis of our tentative criterion of maximizing expected monetary value, the wildcatter should make the decision not to drill for oil. This seems reasonable. But the upper and lower decision branches of Figure 4.3 become equivalent in terms of expected monetary value when $\mathbf{Pr}(\theta_1) \simeq 0.22$, as can be easily verified. This means that the wildcatter, on the *EMV* criterion,

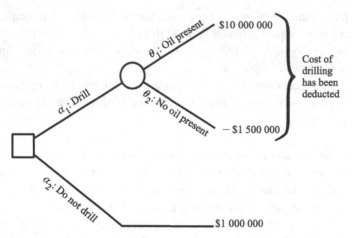

θ_1: Oil present — $10 000 000

α_1: Drill

θ_2: No oil present — $1 500 000

Cost of drilling has been deducted

α_2: Do not drill — $1 000 000

Figure 4.3 Monetary values for oil wildcatter's decision

would decide to drill if the probability of success was 25%. Is it worth risking $1 000 000 for sure, as against a 25% chance at a gain of ten million dollars with a 75% chance at losing $1 500 000? Shouldn't the wildcatter's decision depend on her working capital? This is the subject of the following section.

4.4 Is money everything? Utility of an attribute

We shall demonstrate in this section that maximizing our expected monetary value constitutes a reasonable decision criterion only as a special case of the more general criterion of maximizing our expected **utility**. Again, we should emphasize that we are using money as an example of an attribute, and that we could just as well be discussing some other attribute, such as the number of injuries in an explosion, or the level of noise pollution.

Consider the following two alternative choices, α_1 and α_2, involving chance, which I offer to you. In order to have reasonable agreement on the assignment of probabilities, I present you with a die. We wish to assign probabilities of 5/6 (say faces 1 to 5) and 1/6 (say face 6) to events which we shall specify. Let us assume that you have satisfied yourself as to the honesty of the die. Now to describe the two 'gambles'. In the first (α_1), there is a 5/6 chance of gaining $2000 and a 1/6 chance of losing $4000, whereas in the second (α_2) there is a 5/6 chance of no exchange of money, the status quo, and a 1/6 chance of gaining $6000. Imagine that I am offering them both to you, and you have to choose only one of them. In other words, you must choose between α_1 and α_2 in Figure 4.4.

Most people would consider α_2 to be 'safer' than α_1; to be sure, in accepting α_2 your probability of obtaining no reward is 5/6 but the probability of losing money is zero. On the other hand, α_1 involves a high probability of 'winning' but there is a clear risk, a probability of 1/6, of being 'in the red' to the tune of $4000. It is likely that you prefer α_2; this is termed

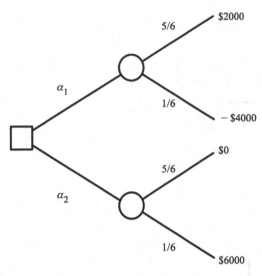

Figure 4.4   Choose a gamble: α_1 or α_2?

risk averseness. If you prefer α_1, the reason might be that you are attracted to the 5/6 chance of a gain of $2000 and you might have enough money at hand to cover the 1/6 contingency of losing $4000. If this is the case, let us try again by multiplying each value in Figure 4.4 by 100, and consider sums of $200 000, $400 000, $0, $600 000 instead of $2000, $4000, $0, $6000 respectively. If you still do not prefer α_2, it is probable that you are not risk averse (or you have a large amount of money at your disposal). We could multiply the amounts by a further factor, say 10; then these would become $2 million, $4 million, $0, and $6 million respectively. It is unlikely that you still prefer α_1; if you do, this could imply that you are **risk prone**, that you are attracted by risk, essentially the characteristic of a chronic gambler! In any event, it is rare that a person will find the two gambles in Figure 4.4 equally attractive. We assume risk-averseness in the following.

Let us try the expected-monetary-value criterion using the values in Figure 4.4. It is readily checked that the *EMV*s are the same in the two cases:

$$\langle V|\alpha_1\rangle = \langle V|\alpha_2\rangle = \$1000$$

(again V stands for monetary value). The *EMV* fails dismally as a criterion for discriminating between α_1 and α_2. How can we express our preference for α_2?

Possible measure using mean and standard deviation   One approach that has been proposed in the past is to determine a 'measure of spread' of the outcomes, in addition to the mean. If the possible outcomes are widely dispersed, then high and low values would be expected to be more likely to occur, the lower values in our case of monetary value involving greater 'loss' and therefore more risk. If the distribution is such that the probabilities are clustered together, there is less chance of an unexpected big loss. Measures based on the use

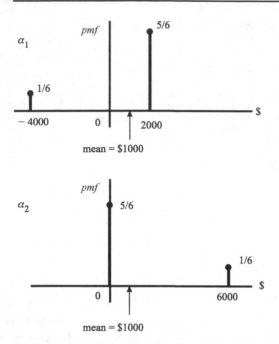

Figure 4.5   Probability mass functions for gambles α_1 and α_2

of both mean and standard deviation have been proposed. For example, Raiffa (1968) notes a common suggestion that one can use a single value to represent the lottery. This is equal to the mean less a constant times the standard deviation. The measure would comprise

$$\mu_V - k\sigma_V, \tag{4.2}$$

where μ_V is the mean $\langle V \rangle$, σ_V is the standard deviation of V, and k is a constant depending on our degree of risk-aversion. The value of expression (4.2) would be calculated separately for each action, α_1 and α_2.

The distributions for the present case are shown in Figure 4.5 for the two cases under consideration. The variance σ_Z^2 is, as we have noted in Section 3.5, the moment of inertia of the probability masses about the mean ($1000). In both cases α_1 and α_2 of Figure 4.5 these must by inspection be the same, and equal to the moment of inertia of two probability masses of (5/6) and (1/6) at 'distances' 1000 and 5000 (dollars) respectively from the centre of mass (1000 in both cases). The variance is calculated as 5×10^6 (dollar)^2. The square root, the standard deviation (or radius of gyration as explained in Section 3.5) is $2236 in both cases. The criterion expressed by (4.2) simply does not work since μ_V and σ_V are the same for each action. The mean and standard deviation are not adequate descriptions: we cannot choose between α_1 and α_2 on the basis of these parameters since they are identical. Yet we prefer α_2. The answer to the problem is to elicit our utilities for the various outcomes. To do this we use *BRLT*s, which will be taken up in the next section.

The idea contained in expression (4.2) has also been used in safety studies. In brief outline, first a random quantity representing risk is derived. This is the **safety margin**, given by the resistance minus the load such that a negative value indicates failure. If this quantity is denoted M, the parameter

$$\beta = \frac{\text{mean of } M}{\text{standard deviation of } M} = \frac{\mu_M}{\sigma_M} \tag{4.3}$$

is often used to describe safety levels; β is denoted the **reliability index**. As just demonstrated, cases can be found where the use of mean and standard deviation together fail to serve as an adequate criterion for choices under uncertainty. The question of safety is taken up in more detail in Chapter 10, Section 10.3.1.

4.4.1   Basic reference lottery tickets (*BRLTs*)

This device was suggested by Raiffa (1968). One first isolates the best, or most desirable consequence (B) and the worst, or least desirable, consequence (W). In the example of the previous section (Figure 4.4), these are $B = \$6000$, $W = -\$4000$. One constructs the basic reference lottery ticket ($BRLT$) as shown in Figure 4.6. The value of π in Figure 4.6 represents the probability of obtaining B and $(1 - \pi)$ is the probability of obtaining W. Since $\pi = 1$ implies receiving B with probability one and W with probability zero, it is equivalent to B. Similarly, $\pi = 0$ (or $(1 - \pi) = \tilde{\pi} = 1$) is equivalent to W.

Let us work again on the assumption that YOU are the subject performing the utility analysis. Then you have to study the lottery of Figure 4.6 carefully and try to find values of π corresponding to various values of the attribute in the interval between B and W, such that you are indifferent between any given value of the attribute and the $BRLT$. Let us list the attributes, as follows, which are in the interval [$\$6000, -\$4000$]. It can be seen that two of the π values are already assigned.

$$\$6000(B) \quad \pi = 1$$
$$\left.\begin{array}{c}\$2000\\\$0\end{array}\right\} \quad \pi \text{ to be determined}$$
$$-\$4000(W) \quad \pi = 0.$$

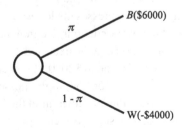

Figure 4.6   Basic reference lottery ticket (*BRLT*) for decision of Figure 4.4

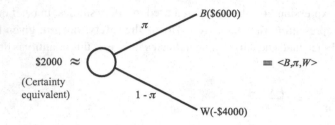

(a) *BRLT* for certainty equivalent of $2000

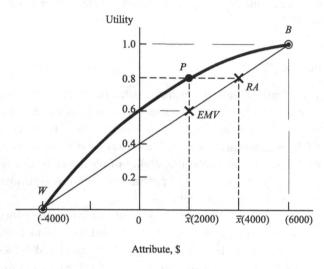

(b) Mean values

Figure 4.7   Deciding on utility for $2000, with trial value of $\pi = 0.8$

Consider for example the value of $2000. You must reflect carefully and suggest a value of π such that, for you, the situation in Figure 4.7(a) is true. In other words, you must be indifferent between the two alternatives on either side of the $\approx$ sign, where $\approx$ indicates indifference, as indicated before in Sections 1.4 and 2.1. The amount of $2000 used in this example is termed the **certainty equivalent,** and is denoted $\hat{x}$. It is the amount on the left hand side of the indifference expression of Figure 4.7(a). The right hand side expresses the lottery in which there is an uncertain outcome with some element of risk. We now ponder our assignment of π. Let us consider for a moment the relation between the value $2000 and the expected monetary value (EMV) of the $BRLT$. If $\pi = 0.6$, then the EMV would be equal to $2000. If you choose this π-value, you would be saying that the lottery was equivalent, for you, to the possession of $2000 for sure. The clear implication of this is that the 'risk element' inherent in the lottery is of no consequence to you. This indicates a neutral attitude to risk. If you choose $\pi > 0.6$, you would then be risk averse, since you would be prepared to exchange $2000 for the lottery **only if** the EMV of the lottery is greater than $2000.

Suppose that you decide that $\pi = 0.8$ is acceptable to you. Then the expected value of the lottery is $(0.8)(6000) + (0.2)(-4000) = 4000\ (>2000)$. The difference between the expected value of the lottery (since we are using money, this is the $EMV = \$4000$) and the certainty equivalent ($\hat{x} = \$2000$) is termed the **risk premium**. This is $2000 in the present instance. The risk premium is the amount that would have to be added to the certainty equivalent as an incentive, in units of monetary value in the present case, to induce you to take the risk involved in the lottery. We have also introduced an abbreviated notation $\langle B, \pi, W \rangle$ for the $BRLT$ as shown in Figure 4.7(a). Your value of π is such that you are indifferent between the exchange of $2000 for $\langle \$6000, \pi, -\$4000 \rangle$ or *vice versa*. The notation $\langle B, \pi, W \rangle$ is similar to our notation for expected value, except that there are three entities inside the brackets. The π-value will be shown actually to represent a prevision so that there is in fact common ground.

We can view the procedure with new insights by means of Figure 4.7(b). The attribute – money, in this case – is shown as the abscissa, with the π-value – the utility, as it turns out – as the ordinate. The $BRLT$ can be idealized using the straight line in Figure 4.7(b), which represents the 'balance of probability'. Initially the EMV corresponds to the case where 'weights' are placed at B and W equal to 0.6 and 0.4 respectively: consider a weight of unity at the point EMV balanced by the 'reaction points' of 0.6 and 0.4 at B and W in equilibrium. The certainty equivalent $\hat{x}$ is also shown on the abscissa of Figure 4.7(b). Risk aversion can be visualized using analogies in mechanics in which we think of the probabilities as masses in a gravitational field perpendicular to the plane of the paper. The two steps are as follows.

- We move our unit mass from point EMV to point RA. This results in an increase in the reaction (probability) from 0.6 (EMV) to 0.8 (RA) on point B and a reduction of the reaction from 0.4 to 0.2 on point W. The higher probability of B expresses our risk aversion. The mean $\bar{x} = \$4000$ is obtained by taking moments of the masses about the utility axis.
- Two lines join points B and W, a straight one and a curved one as shown in Figure 4.7(b). The curve represents the utility (equal to π as we shall soon see). The point of intersection of the coordinates $\hat{x}$ and $\pi = 0.8$ define the point P on the utility curve. One can think of two steps: first, moving the unit mass from EMV to point RA (as just discussed) and second, moving it from RA to point P. Then we can take moments about either the utility axis, with the unit mass at RA, to give $\bar{x} = \$4000$ as above, or about the attribute axis, with the mass at RA or P, to give the mean value of $\pi = \bar{\pi} = 0.8$ (recall that there is a mass of 0.8 at $\pi = 1.0$ and a mass of 0.2 at $\pi = 0$). The value $\pi = 0.8$ is also the utility of $x = \$2000$; this is obtained from the curve of utility (π) versus x.

To return to our original problem of Figure 4.4, we continue by repeating the process described above for various certainty equivalents, in the range of interest. In the case of any of the values of interest ($0 and $2000 have to be considered in this example), it might help

Table 4.7 *Author's π-values*

Certainty equivalent	π-value of lottery (mine!)
−$4000	0.00
−$2000	0.55
0	0.80
$2000	0.92
$4000	0.97
$6000	1.00

to imagine that you have an IOU for this amount, and expect to receive a lottery in return. One has next to reflect on the $BRLT$, assigning different π-values until one is satisfied that the certainty amount of the IOU is indeed equivalent to the lottery. It is very useful initially in the case of each attribute to calculate the π-value such that the EMV of the lottery is equal to the certainty equivalent; for the certainty equivalents of $0 and $2000 the values are $\pi = 0.40$ and 0.60. In each case the lottery is the same, $\langle \$6000, \pi, -\$4000 \rangle$. One can express one's risk aversion by adjusting one's value of π; a higher value of π would indicate more risk aversion than a lower value.

In Table 4.7 I have entered my own π-values as an illustration, using additional certainty equivalents to complete the curve. As already noted, two of the π-values are immediately obtainable and will be shared by all, corresponding to B ($6000) and W (−$4000), that is, $\pi = 1$ and $\pi = 0$ respectively. The values of π for the intermediate values are assessed subjectively. My values of π in Table 4.7 attest to my risk-averseness; the value for a certainty equivalent of $2000 is given as $\pi = 0.92$ (higher than the value of 0.8 used above as a discussion point). The key point is that the π values represent our utility and permit us to make the choice between α_1 and α_2 in Figure 4.4. This is explained in the next section. The values given in Table 4.7 are plotted in Figure 4.8. In the case where utility increases with the attribute value, risk averse behaviour corresponds to a **concave** utility function, as in this figure. Sharper curves would indicate more risk-averseness, flatter curves less, and straight lines none.

4.5   Why do the π-values represent our utility? Expected utility

Our original motivation was to find a criterion that enables us to express a preference for one of the choices α_1 and α_2 in the gamble of Figure 4.4. The next step is merely to replace the monetary amounts in Figure 4.4 with the corresponding π-$BRLT$, i.e. the $BRLT$ that is equivalent – for you or for me, depending on who is doing the analysis! – to the amount in question. Using as an example the values in Table 4.7 and Figure 4.8 (my π-values), the amount $2000 is replaced by $\langle \$6000, 0.92, -\$4000 \rangle$, and so on. The problem of Figure 4.4 is therefore entirely equivalent (for me) to that shown in Figure 4.9(a), since the monetary values of Figure 4.4 have all been replaced by their π-$BRLT$s, which I have by definition agreed are equivalent.

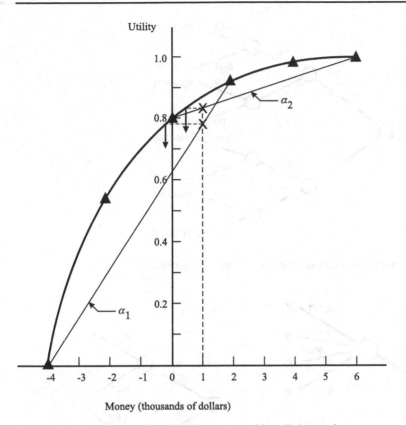

Figure 4.8 The author's utility for illustrative problem. Points × denotes centres of mass and arrows their certainty equivalents. The actions α_1 and α_2 are indicated by lines joining the possible outcomes

The next step is to infer the tree shown in Figure 4.9(b); the important point to note is that the π-values are treated as probabilities **conditional** upon the events associated with the probabilities 5/6 or 1/6. Thus the values 0.767 and 0.067 in Figure 4.9(b), for example, are equal to $(5/6 \times 0.92)$ and $(5/6 \times 0.08)$ respectively, and the values 0.000 and 0.167 are equal to $(1/6 \times 0.00)$ and $(1/6 \times 1.00)$ respectively. The other values are obtained in a similar manner. The values in Figure 4.9(c) are found by collecting together the 'B' and the 'W' values of probability; for instance, the value of 0.77 is $(0.767 + 0.000)$, rounded, whereas 0.23 is $(0.067 + 0.167)$, also rounded. This is clearly permissible since the 'B' and 'W' events for each decision branch are mutually exclusive (incompatible).

Having arrived at Figure 4.9(c), one finally has a basis on which to make a decision; one chooses that decision with the higher probability of 'B'. Actually, I should say 'I have a basis for making a decision' since the π-values are mine; you should enter yours in order to determine your optimal course of action. It is evident that, for me, decision α_2 is optimal since in this case the probability of B (i.e. 0.83) is greater than the probability of B were I to follow action α_1 (i.e. 0.77).

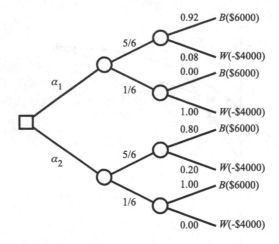

(a) Original consequences of Figure 4.4 replaced by their *BRLT*s

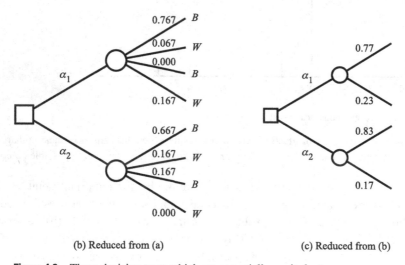

(b) Reduced from (a)   (c) Reduced from (b)

Figure 4.9   Three decision trees which are essentially equivalent

The result may be simplified by treating the π-values for B as utilities; see Figure 4.10, in which the utilities are denoted by U. It is then immediately apparent that the probabilities of B in Figure 4.9(c) are formally identical to the **expected utilities** in Figure 4.10; the higher the expected utility, the higher the probability of B. Therefore, the following rule becomes our basic decision criterion:

> CHOOSE THAT ACTION α_i WHICH MAXIMIZES
> OUR EXPECTED UTILITY

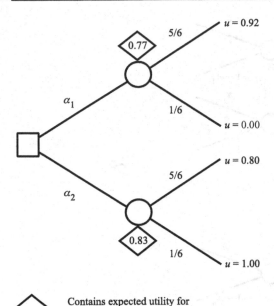

Contains expected utility for
action under consideration

Figure 4.10   Expected utility makes sense as a criterion for optimal decisions

The results above are readily generalized (Raiffa, 1968) to the following basic theorem. The decision-maker is indifferent between the following set of lotteries:

$$\left.\begin{array}{l} \langle B, \pi_1, W \rangle \text{ with probability } p_1 \\ \langle B, \pi_2, W \rangle \text{ with probability } p_2 \\ \vdots \quad \vdots \quad \vdots \quad \vdots \quad \vdots \\ \langle B, \pi_j, W \rangle \text{ with probability } p_j \\ \vdots \quad \vdots \quad \vdots \quad \vdots \quad \vdots \\ \langle B, \pi_n, W \rangle \text{ with probability } p_n \end{array}\right\} \text{ such that } \sum_{j=1}^{n} p_j = 1$$

and the single lottery $\langle B, \overline{\pi}, W \rangle$, where

$$\overline{\pi} = p_1 \pi_1 + \ldots + p_n \pi_n = \sum_{j=1}^{n} p_j \pi_j. \tag{4.4}$$

The steps in the logic involved in deriving this expression are the same as before, and are illustrated in Figure 4.11. The value $\overline{\pi} = \sum p_j \pi_j$ represents the prevision, or expectation, of our utility U (or π). Hence, we may write π as U and

$$\langle U \rangle = \overline{\pi} = \sum_{j=1}^{n} p_j \pi_j. \tag{4.5}$$

We shall use U, rather than π, in the sequel, and also emphasize the convention that the upper

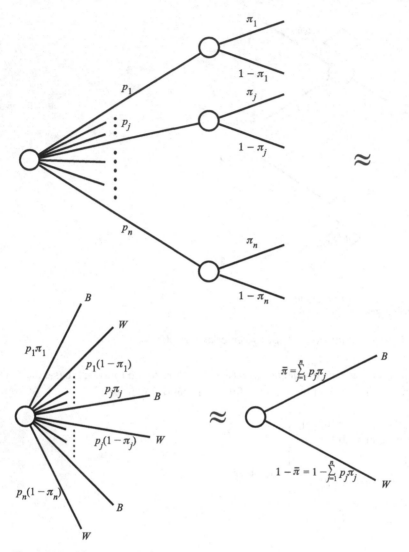

Figure 4.11   The expected-π-value theorem

case letter U represents utility, in general, since it is random, with a lower case, u, indicating a particular value. The notation is the same as that used for random quantities in general.

4.5.1   Positive linear transformations

Having defined utility by means of the $BRLT$, we will have values in the range 0 to 1. Utility, unlike probability, does not have to be confined to this range. Consider the decision tree of Figure 4.12. The problem is to choose one action from the set $\alpha = \{\alpha_1, \alpha_2, \ldots, \alpha_m\}$, each with a discrete set of outcomes $\Theta = \{\theta_1, \theta_2, \ldots, \theta_n\}$. Each of the pairs (α_i, θ_j) has a consequence.

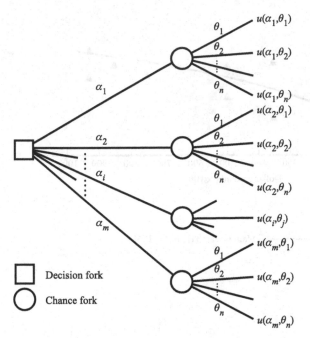

Figure 4.12   Actions α_i and states of nature θ_j with consequences expressed as utilities $u(\alpha_i, \theta_j)$

Imagine that initially all of the utilities $u(a_i, \theta_j)$ are obtained by means of $BRLT$s and are therefore in the range [0, 1].

If we transform all of the values linearly by means of the following equation:

$$u' = a + bu, b > 0, \tag{4.6}$$

where u' is the transformed utility, the **ranking** of the expected utilities

$$\langle U' \rangle = a + b \langle U \rangle \tag{4.7}$$

is the same whether we use U or U'. The optimal decision is therefore unaffected by the linear transformation (4.6), which is therefore permissible provided that the transformation is positive ($b > 0$). This is often expressed as follows.

Utility is determined up to a positive linear transformation.

Two sets of utilities, one of which is a positive linear transformation of the other, are 'strategically equivalent', that is, they will give exactly the same ranking of decisions in terms of optimality.

We shall therefore omit the primes in Equations (4.6) and (4.7) in future consideration of utility, and our fundamental criterion

> CHOOSE THAT ACTION α_i WHICH MAXIMIZES
> OUR EXPECTED UTILITY $\langle U \rangle$

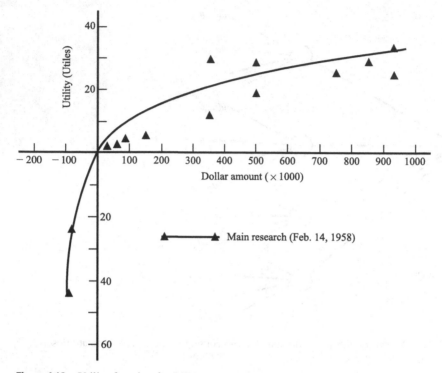

Figure 4.13   Utility function for Bill Beard, adapted from Grayson (1960)

will be taken to refer to utility U on any scale that is a positive linear transformation of any other valid scale, including the *BRLT* π-scale.

There are many examples of the elicitation of utilities in practical situations; a classical case is the work of Grayson (1960) on decisions by oil and gas operators. Figure 4.13 shows the utility function for Bill Beard, its concavity exhibiting a risk-averse behaviour. The units of utility are sometimes termed 'utiles', which is generally a transformation of the π-scale. A logarithmic relationship was found to fit Grayson's data points by Kaufman (1963).

4.6   Attributes and measurement; transitivity

The purpose of decision theory is to construct a rational criterion for the choice of a particular action under conditions of uncertainty. We pointed out in Section 4.1 that an outcome could be described by a set of attributes. In the intervening examples we have used money as the attribute for illustrative purposes, since money is well understood by all. We now consider a wider set of attributes and define rules for their measurement. There is a great variety of attributes, some of which are easier to measure than others. It is easy to measure money in most circumstances, but what of the enjoyment of a wilderness area? Measurement may be defined as the act of assigning a number to the attribute under consideration with a set of rules to guide us. We might rank the level of pleasure associated with enjoyment of nature but it is difficult to define a precise scale. In using probabilistic methods, we need to know the 'rules of the game'.

There are four main scales of measurement:

- Nominal
- Ordinal
- Interval
- Ratio.

One rule concerns transitivity. This is required for ordinal, interval and ratio scales, so we shall define it at the outset. The property of transitivity requires the following, in which we use preference relations; the same definition applies for numerical relations $>$, $<$, $=$ instead of preferences $\succ$, $\prec$, $\approx$.

1. $a_2 \succ a_1, a_1 \succ a_3 \rightarrow a_2 \succ a_3$.
2. $a_2 \approx a_1, a_1 \approx a_3 \rightarrow a_2 \approx a_3$.
3. $a_2 \succ a_1, a_1 \approx a_3 \rightarrow a_2 \succ a_3$.

In this, the a_i are the consequences (entities) under consideration, and the symbol '$\rightarrow$' should be understood as 'implies'. Many real-world events are not transitive; soccer team X might beat team Y, which might beat Z, yet Z beats X. An interesting example of an intransitive choice is given at the end of the present section, using special dice.

In the **nominal** scale, a number is used only as a classification. For instance, in a factory, bolts (or widgets) may be classified into types 1, 2, 3, ... In the nominal scale the number does not indicate size or some other property on an increasing or decreasing scale; it merely distinguishes types, for example, it could indicate the factory to which the widgets are to be sent, or the colour that they are painted for purposes of variety in marketing. We can construct histograms, for instance of the number of the types 1, 2, 3, ... in stock. We can also obtain the mode of the histogram, giving the classification with the most number in stock.

In the **ordinal** scale, we start with a classification, as in the nominal scale, but use a common attribute of the entities under consideration such as cost, weight, or indeed our preferences, to rank them. For example, we might rank the grades of concrete that we are manufacturing, such that grade 3 is stronger than grade 2, which in turn is stronger than grade 1, and so on. In the ordinal scale, grade 2 is not twice as strong as grade 1, merely stronger. The early temperature measurements, an extreme case of which is the use of touch, resulted in a ranking such as 'object A is hotter than object B'. These temperatures could be denoted 2 and 1 respectively, but object A is not necessarily twice as hot as object B. Earthquake severity has often been measured on an ordinal scale by means of an interpretation of resulting damage to buildings (modified Mercalli scale). Rock hardness is also measured on an ordinal scale. In environmental studies, where we might rank consequences into categories of most preferable, acceptable, not preferable, for example, ordinal scales are of great usefulness. Ordinal scales can be subjected to a positive monotone transformation and still preserve the order; the new scale is then equivalent to the old one. The transformation could be non-linear (an essential difference from the interval and ratio scales).

With regard to statistical analysis, care must be taken when using ordinal scales. We may use histograms, obtain the mode (most common occurrence), as for nominal scales. Medians and fractiles can be added to the list of useful statistics; we can for example state that 60% of orders of computers are for grade 5 and higher. An example will be given after dealing with the next two categories, of the possible misuse of an ordinal scale.

An **interval** scale is an ordered set of numbers, but with the additional condition that equal differences at different points on the scale stand for the same difference in the attribute being measured. A good example is the temperature scale, once thermometry had been developed; expansion of fluids as against gases gave the required result. Thus the Fahrenheit and Celsius scales are interval scales; differences of 0 to 5 degrees and 40 to 45 degrees represent the same differences in temperature. Additivity can be invoked if a zero is agreed upon by convention; examples are the temperature scales just noted, or altitude if this is done with reference to sea level (for example). But the zero in the interval scale is arbitrary. All the usual statistics can be used, including the mean, standard deviation, skewness, and so on. The only reservation is that the mean and therefore the coefficient of variation (equal to the standard deviation divided by the mean) are arbitrary since the zero point is agreed only by convention. A positive linear transformation is permissible, for example from Fahrenheit to Celsius scales, but non-linear transformation will destroy the hard-won properties of the interval scales.

In **ratio** scales, an absolute, or natural, zero is added to the interval scale. The most compelling example is again temperature and the development of the Kelvin scale. Quantities such as mass, length and electrical resistance have natural zeros. Additivity applies and the scale can be modified by multiplication by a positive constant, but not by a change in the zero. All statistical measures, including the coefficient of variation, can be used and have meaning. Ratio scales are common in physics and engineering, but less so in the social sciences. We may recall in this context Bridgman's definition of a concept, which is identified with the set of operations which are used in making a measurement of the concept, mentioned at the end of Section 1.3.

Transitivity is required for ordinal scales, and must also apply to interval and ratio scales. This property is required in all our methods; for the use of utility theory, we require further that attributes be measured using interval or ratio scales. These are needed for correct calculation of expected utility or expected value of any kind. This may be a reasonably clear requirement, but the point is emphasized in the following examples.

Example 4.1 *Preferences under certainty*. In this case we can use the transitive property of ordinal, interval or ratio scales to express preferences. If we consider two choices α_1 and α_2 with single-attribute consequences on an ordinal scale that indicate increasing value, the optimal

Table 4.8 *Ordinal ranking and monetary values*

	θ_1 – money	θ_1 – ordinal scale	θ_2 – money	θ_2 – ordinal scale
α_1	\$1 000 000	3	\$0	0
α_2	\$500 000	2	\$250 000	1

choice would correspond to the highest ordinal value. Any positive monotonic transformation of the ordinal ranking could also be used to make choices. Interval or ratio scales can be used in a similar manner.

Example 4.2 *Preferences under uncertainty: misuse of ordinal scales.* Consider Table 4.8 describing payoffs for two actions α_1 and α_2. The consequences are presented in terms of monetary value, and also on an ordinal scale. The probabilities of occurrence of θ_1 and θ_2 are both 0.50. The nominal ranking in the table accords with the monetary values; $\theta_1|\alpha_1 \succ \theta_1|\alpha_2 \succ \theta_2|\alpha_2 \succ \theta_2|\alpha_1$. Let V denote the monetary value and W the ordinal value. It can be verified that $\langle V|\alpha_1 \rangle = \$500\,000$, and $\langle V|\alpha_2 \rangle = \$375\,000$. Yet $\langle W|\alpha_1 \rangle = \langle W|\alpha_2 \rangle = 1.5$, and the clear preference of α_1 (on the basis of *EMV*) over α_2 is lost. It is easy to construct an ordinal scale that reverses the preference indicated by expected monetary value, for example by making the ordinal value for $\theta_2|\alpha_1$ equal to -1.

Although the ordinal scale represents well the individual preferences, it is arbitrary in nature and should not be used with statistical measures such as the mean. It is easy to invent examples that show the same thing in comparisons of expected ordinal measures with maximum expected utility.

Special dice: example of intransitive real-life choice   There are many common situations in life that are intransitive (such as the sporting events noted), but Bradley Efron (Paulos, 1988) has introduced a set of dice that can lead to intransitive choices. This is presented as an interesting example of real-life intransitivity. Now to consider the dice: they are played in pairs by two individuals, with one die tossed by each player. The highest number wins. With normal dice, there would be no reason to choose one die over the other (unless one considered one of them to be weighted, which we exclude in this example). Efron's dice do not have the usual numbering. There are four of them, labelled A, B, C and D. The numbering system is as follows.

Die A Number 4 on four faces, and 0 on two faces,
Die B Number 3 on all six faces,
Die C Number 2 on four faces, and 6 on two faces,
Die D Number 5 on three faces, and 1 on three faces.

Now it can be shown that

- A beats B two-thirds of the time
- B beats C two-thirds of the time
- C beats D two-thirds of the time

yet

- D beats A two-thirds of the time!

This can be shown by means of the two-dice arrays introduced in Section 4.3. Taking the dice C and D for example, the following array represents the possible results, each with equal probability. As before, the possibilities for the first die are represented in the rows, and for the second in columns.

```
25 25 25 21 21 21
25 25 25 21 21 21
25 25 25 21 21 21
25 25 25 21 21 21
65 65 65 61 61 61
65 65 65 61 61 61
```

Given a choice of die to play in a game, the following preferences apply

$$A \succ B \succ C \succ D \succ A \succ \ldots,$$

an intransitive situation. A possible scam is evident: suppose that we make a set of dice as above, and colour them red, blue and so on. We memorize the preference system, and go to our local pub. We offer a friend the choice of a die, and then choose one (obviously, one that will win two-thirds of the time). We then make bets and of course we are successful, unless our opponent sensibly declines first choice.

4.7   Utility of increasing and decreasing attributes; risk proneness

Let us first review the main result achieved so far. Figure 4.14(a) shows an attribute X with increasing value, and a risk-averse utility function. Taking two values of the attribute, x_1 and x_2, with $x_1 > x_2$ as shown, consider the lottery

$$\langle x_1, \pi = 0.5, x_2 \rangle .$$

This special lottery, with a π-value of 0.5, is written compactly as

$$\langle x_1, x_2 \rangle . \tag{4.8}$$

Risk aversion implies that we will exchange the certainty equivalent $\hat{x}$ for this lottery,

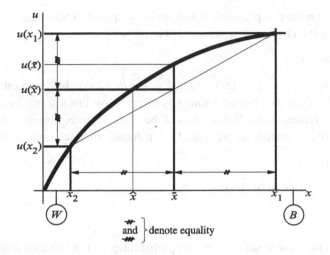

(a) Utility for an attribute of increasing value

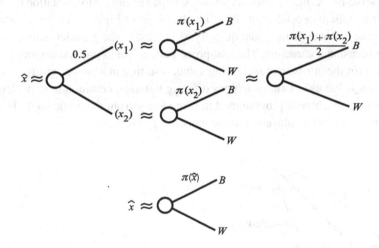

(b) Two ways of analysing x using *BRLT*s. The complementary values $(1 - \pi)$ have been omitted for clarity

Figure 4.14   The essential concavity of utility versus attribute given risk-averseness

that is,

$$\hat{x} \approx \langle x_1, x_2 \rangle , \tag{4.9}$$

such that

$$\hat{x} < \bar{x} = \frac{x_1 + x_2}{2}. \tag{4.10}$$

The difference $(\bar{x} - \hat{x})$ is the risk premium, which we are prepared to 'sacrifice' so as to avoid risk. We now consider the utilities of x_1, x_2 and $\hat{x}$. First,

$$x_1 \approx u(x_1) \text{ and } x_2 \approx u(x_2).$$

This should be clear since $x_1 \approx \langle B, \pi(x_1), W \rangle$ and $x_2 \approx \langle B, \pi(x_2), W \rangle$, as shown in Figure 4.14(b). These π-values can then be transformed by any positive linear transformation to the scale $u(\cdot)$. The utility represents the 'value' to us of the attribute under consideration, and the quantities x and $u(x)$ can therefore be exchanged when considering preferences. Now

$$u(\hat{x}) \approx \langle u(x_1), u(x_2) \rangle, \tag{4.11}$$

from the indifference relation (4.9). Thus

$$u(\hat{x}) = \frac{u(x_1) + u(x_2)}{2}, \tag{4.12}$$

since the relationship (4.11) represents a bet with probabilities of 0.50 on each outcome. This is shown in terms of *BRLT*s in Figure 4.14(b). Since $\hat{x} < \bar{x}$, and in view of Equations (4.10) and (4.12), the curve in Figure 4.14(a) must be concave.

A person who is risk-prone, that is, attracted to risk, will pay to enter into a situation involving risk. Consider for simplicity a 50–50 bet $\langle 0, x \rangle$ as illustrated in Figure 4.15 for an attribute with increasing value – money is the usual quantity of interest to the gambler, although risks to life are also an occasional attraction. The risk-prone person would **pay** the amount $(\hat{x} - \bar{x})$ to enter into the bet for the attribute of increasing value, resulting in a convex utility function. A common situation is the use of slot machines offering lotteries, common in pubs. Even in the knowledge that the machine is programmed to take money from the participant, there are still large numbers of risk-prone players of these games.

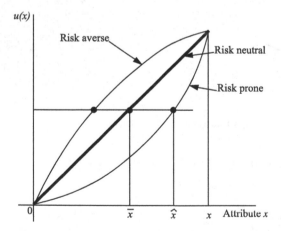

Figure 4.15   Risk proneness represented by a convex curve. Risk neutrality is represented by a straight line; if money is the attribute, this amounts to the *EMV* criterion

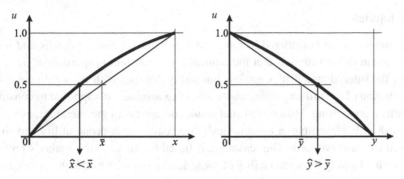

Figure 4.16   Increasing and decreasing utility functions expressing risk aversion

We mentioned in Section 4.1 that attributes may either be of increasing or decreasing value as the size of the attribute increases. Money, profit and the size of wilderness areas are examples of attributes with increasing value. We shall denote attributes with decreasing value by Y. We give some examples: the number of deaths and injuries in an accident, or the volume of oil spilled in a collision at sea. Or imagine that we are subject to a lawsuit that we believe we may lose, say $\$y$, with 50% probability, but we believe that we may possibly suffer no loss, also with 50% probability. Now assume that $y = \$50\,000$. We might be prepared to settle out of court for $\$30\,000$, i.e. $\hat{y} = \$30\,000$, so as to avoid the risk of losing $\$50\,000$. Thus $\hat{y} > \bar{y}$ expresses risk aversion in this case, the converse of the case $\hat{x} < \bar{x}$ for the attribute X with increasing utility. This situation corresponded to a lottery $\langle 0, \$50\,000 \rangle$. But $\hat{y} > \bar{y}$ will express risk aversion in any lottery $\langle y_1, y_2 \rangle$, or $\langle y_1, \pi, y_2 \rangle$ (except for the special cases $\pi = 0, 1$).

The contrast between an attribute Y with decreasing utility and an attribute X with increasing utility is illustrated in Figure 4.16, the right hand side showing the case of decreasing value. Risk aversion is characterized by $\hat{y} > \bar{y}$; for example, we might accept an oil spill of $120\,000$ m^3 for sure to avoid a 50–50 lottery

$$\langle 0, \ 200\,000 \ \text{m}^3 \rangle .$$

The risk-averse utility function is concave, as in the case for attributes of increasing value.

We have considered, in the above, attributes for which the utility is monotonically increasing or decreasing with the attribute value. This is a useful property, which we shall assume in the rest of this work. But many attributes are not of this kind. For example, consider your utility with regard to ambient temperature. For most of us, utility will increase with temperature until some optimal value, and then decline. This non-monotonic behaviour can also occur with attributes such as blood sugar level, which has an optimal value of highest utility. This situation can result in non-unique values of the certainty equivalent (see Keeney and Raiffa, 1976), and requires special consideration.

4.7.1   Deaths and injuries

An attribute that has to be considered in safety studies is the number of deaths and injuries. We shall assume in this discussion that the injuries are a constant proportion of deaths, and only consider the latter, denoted W. Certainly our utility declines with W; we do not consider the bizarre situation of the military commander who, one assumes, is supposed to consider the enemy's deaths a good thing. Various unusual results do appear in the literature, as reviewed by Nessim (1983). For example, a monotonically decreasing risk-prone utility function has been proposed by some workers. The reasoning is based on the decision-maker's preference for a lottery with relatively high probability of some deaths and with no deaths associated with the complementary probability, as against a situation with even one certain death. As a specific example, a 50–50 lottery between 0 and 60 deaths was preferred to 10 certain deaths. While the desire to avoid deaths is commendable, this actually expresses risk proneness. A risk-averse decision-maker would accept more than the mean of the lottery, that is, more that 30 deaths, to avoid the 50% probability of 60 deaths. The risk-prone attitude in the 50–50 lottery is an example of faulty reasoning spurred by a misplaced desire to avoid deaths.

We have emphasized the fact that risk aversion is related to the question of the size of the quantity risked. For example, we would not wish to risk $10 000 in a 50–50 bet if our present capital was only (say) $11 000. The aversion to risk in this case occurs when we are faced with a 'one-off' bet – given the amounts and our present working capital, we would be unlikely to wish to enter into a long-run series of bets. On the other hand, given the same amount of capital, we might be quite content to risk $10 in a series of bets, perhaps for our amusement. There is nothing amusing about death, but a similar situation prevails. In considering repeated incidents involving deaths, the carnage of the roads might be mentioned. Over 40 000 deaths a year occur from this cause in the United States, mostly singly or in small numbers. If we exchanged 10 certain deaths so as to obtain instead a 50–50 lottery between 0 and 60 deaths, and if this were repeated in the long run, many unnecessary deaths would result. Our attitude should be closer to risk neutrality, but it may well be a good idea to include risk aversion into a decision-making strategy where we might be able to bias the actions so as to avoid large accidents and thereby reduce overall risks. Such a risk-averse curve is illustrated in Figure 4.17.

4.8   Geometric representation of expected utility

We shall deal with risk-averse utility functions for an attribute X. (Risk proneness is handled easily as a variation of the present treatment.) We are concerned with decision-making under uncertainty and consequently the attribute X is random. The decision tree is illustrated in Figure 4.12; we use X as the random quantity representing the state of nature rather than Θ in the present discussion. And since the utility is derived from X, it is also random, denoted U. There will often be a function relating these two quantities, for example $U = \ln X$. If we

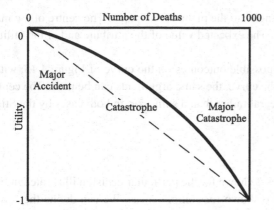

Figure 4.17   Risk-averse utility with regard to large accidents

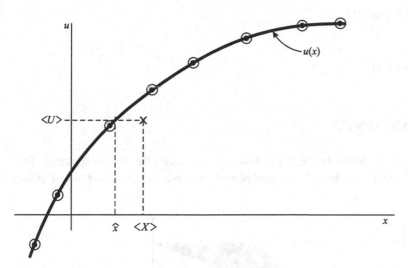

Figure 4.18   Schematic illustration of possible outcomes for a decision α_i of Figure 4.12 indicated by dots with circles. Masses of probability are attached to these points. The expected result is shown by a cross

have the probability distribution of X, the distribution for U can be derived from it; methods for analysing functions of random quantities are given in Chapter 7.

Figure 4.18 shows utility U as a function of the attribute X; the particular values shown by points might correspond to one of the branches of the decision tree of Figure 4.12. The criterion for optimal decision-making is to maximize our expected utility, as in (4.5) or the equivalent transformation (4.7). Imagine that the probabilities of various outcomes are indicated by the dots enclosed by small circles, and treat these as small masses. If the probability masses are placed on these 'consequence points', the point $(\langle X \rangle, \langle U \rangle)$ is at the centre of mass – imagine a gravitational field applied at right angles to the plane of the paper. The centre of the masses is indicated by the cross in the figure. The expected value of the attribute and of the utility are also indicated in the figure.

In the case of a continuous set of possible outcomes on the curve of Figure 4.18, with the probabilistic mass spread out along the curve, the same arguments can be used. The centre of mass would be determined by an integral rather than a sum – or, in both cases by the Stieltjes integral

$$\langle U \rangle = \int u \, \mathrm{d}F_U(u). \tag{4.13}$$

The criterion for ranking the decisions is shown for the particular decision illustrated in Figure 4.18 by $\langle U \rangle$. The value of $\langle X \rangle$ is also shown for completeness. We can define the certainty equivalent $\hat{x}$ in a similar way to that previously used for simple lotteries, as illustrated in the figure. If $u(\cdot)$ denotes the function transforming X into U, then

$$u(\hat{x}) = \langle u(X) \rangle = \langle U \rangle \tag{4.14}$$

or

$$\hat{x} = u^{-1}[\langle u(X) \rangle]. \tag{4.15}$$

4.9   Some rules for utility

It is important to delineate the utility function $u(\cdot)$ carefully. We have already made some progress. Given that we have determined point D between the 'best' and 'worst' points x_B and

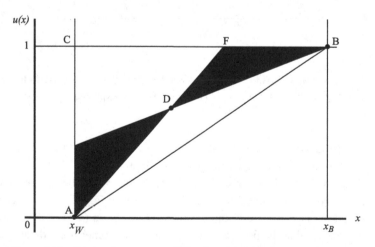

Figure 4.19   Region for concave utility function given points A, D and B

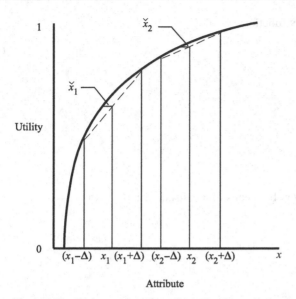

Figure 4.20   Risk premium $\check{x}_2 < \check{x}_1$, indicating a decreasingly risk-averse utility function

x_W respectively, the shaded area in Figure 4.19 shows the possible region in which a concave utility function can be placed. But we need to develop further rules to guide us. The first step is to enquire: how risk-averse are we, and how does this vary over the range of attributes? Figure 4.20 illustrates a utility function that is decreasingly risk-averse. In this, two points x_1 and x_2 are contrasted, using two 50–50 lotteries $\langle x_1 - \triangle, x_1 + \triangle \rangle$ and $\langle x_2 - \triangle, x_2 + \triangle \rangle$. These are lotteries centred on x_1 and x_2 respectively, with the same range $2\triangle$. The risk premium is denoted $\check{x}$ and is defined by

$$\check{x} = \bar{x} - \hat{x}.$$

We have **decreasing** risk aversion if $\check{x}$ decreases with the attribute x; for example, in Figure 4.20 the two values x_1 and x_2 have risk premiums $\check{x}_1$ and $\check{x}_2$ such that $\check{x}_1 > \check{x}_2$. Therefore a good measure of the characteristic of decreasing risk aversion for a given utility curve is the behaviour of $\check{x}$. The fact that the slope of the curve of u versus x decreases as in Figure 4.20 results in a decreasing $\check{x}$; if the slope eventually becomes constant, then we have risk neutrality in that region, with $\check{x} = 0$. This might suggest that the second derivative

$$u'' = \frac{\mathrm{d}^2 u}{\mathrm{d} x^2}$$

be a measure of risk aversion. But this contains pitfalls. We know that we can transform a utility curve linearly and obtain the same ranking of decisions, that is, strategically equivalent solutions. The risk premiums are therefore the same but the values of u'' differ by the constant of proportionality. This may be verified by the following example.

Example 4.3 Consider the utility function

$$u(x) = a\left[1 - b\exp(-cx)\right]; \tag{4.16}$$

then

$$u' = \frac{du}{dx} = abc\exp(-cx)$$

and

$$u'' = \frac{d^2u}{dx^2} = -abc^2\exp(-cx).$$

If we transformed the scale of $u(x)$ by means of a constant $d > 0$, with

$$u_1(x) = ad\left[1 - b\exp(-cx)\right],$$

then

$$u_1'' = du''$$

so that the measure of risk aversion alters even though the function expresses the same utility.

To generalize this idea, we have to normalize u'', and a useful measure was suggested by Keeney and Raiffa. We shall denote the **local risk aversion** measure as $\gamma(x)$, and this is defined as

$$\gamma(x) \equiv -\frac{u''}{u'}. \tag{4.17}$$

It can be shown that $\gamma > 0$ implies concavity and risk averseness, $\gamma = 0$ implies risk neutrality, and $\gamma < 0$ risk proneness. A constant $\gamma(> 0)$ implies constant risk aversion. This means that a 50–50 lottery $\langle x - \Delta, x + \Delta \rangle$ will have the same risk premium for any x. This characteristic is exemplified by exponential utility functions.

Example 4.4 Using the function of Example 4.3, it can be verified that γ is a constant:

$$\gamma(x) = c.$$

The essential part of Equation (4.16) is the following linear transformation:

$$u(x) = -\exp(-cx). \tag{4.18}$$

Example 4.5 We now explore the behaviour of the risk premium $\check{x} = \bar{x} - \hat{x}$ for the exponential relationship of Equation (4.18). For the 50–50 lottery $\langle x - \Delta, x + \Delta \rangle$, $\bar{x} = x$, and

$$E(U) = \langle U \rangle = -\frac{1}{2}\{\exp[-c(x - \Delta)] + \exp[-c(x + \Delta)]\}.$$

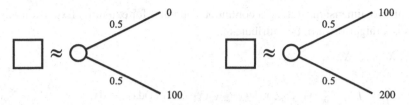

Figure 4.21   Constant risk aversion requires the values in the boxes to have the same difference from their respective means (50 and 150, left and right)

Then using Equation (4.14), that $u(\hat{x}) = \langle U \rangle$, we have

$$\exp(-c\hat{x}) = \exp(-cx)\left\{\frac{1}{2}\left[\exp(c\Delta) + \exp(-c\Delta)\right]\right\}$$

and

$$\hat{x} = x - \frac{1}{c}\ln\left\{\frac{1}{2}\left[\exp(c\Delta) + \exp(-c\Delta)\right]\right\}.$$

Therefore $\check{x} = (x - \hat{x}) = (\bar{x} - \hat{x})$ is a function only of Δ (and c) and we have constant risk aversion, as indicated in Example 4.4. For an attribute in the range of 0 to 200, consider the lotteries of Figure 4.21. If one entered 65 in the box on the left (for a decreasing attribute), then constant risk aversion would require that one would have to enter 165 in the box on the right. The risk premium is the same (15) in each case. Decreasing and increasing risk aversion would imply a reduction or increase in the risk premiums, respectively.

We have introduced several ways of assessing our utility. The various mathematical functions have implications with regard to our attitude to risk. The exponential function represents constant risk aversion (or proneness, depending on the sign of the constant in the exponent). This is a useful base case, and other functions can be assessed using the methods introduced. For example, when we introduced the idea of utility in Section 4.1, a logarithmic relationship was used. This arose from the thinking of Daniel Bernoulli, and can be shown to lead to decreasing risk aversion; this can be checked by calculating the value of γ. See Exercise 4.8.

4.10   Multidimensional utility

When we derived probability distributions in more than one dimension, we considered two or more random quantities, say X and Y (taking as an example the case of two dimensions). In considering utility, the situation is similar, and we wish to obtain a function $u_{X,Y}(x, y)$ which expresses our preferences regarding two attributes, X and Y. The same fundamental rule applies in terms of maximizing one's expected utility; this now becomes

$$\langle U \rangle = \iint_{\text{all } x, y} f_{X,Y}(x, y)\, u_{X,Y}(x, y)\mathrm{d}x\mathrm{d}y \tag{4.19}$$

in the case of two dimensions, taking a continuous quantity for example. Extension to higher dimensions is straightforward. For attributes

$$X_1, X_2, \ldots, X_i, \ldots, X_n,$$

$$\langle U \rangle = \int \cdots \int_{\text{all } x_i} f_{X_1, \ldots, X_n} (x_1, \ldots, x_n) \, u_{X_1, \ldots, X_n}(x_1, \ldots, x_n) \mathrm{d}x_1 \cdots \mathrm{d}x_n. \tag{4.20}$$

The theory of multidimensional utility is analogous in a striking way to the theory for multidimensional probability: probability and utility are dual random quantities in their vector spaces. We deal here with an introduction to preferences with several attributes so as to give a flavour of the theory. Before dealing with utility of several attributes, it is useful to consider preferences under certainty and the use of value functions.

4.10.1 Preferences under certainty

In Example 4.1 we pointed out that any increasing scale expressing value can be used to indicate preferences under certainty. Ordinal scales can be used as well as interval or ratio scales, or positive monotonic transformations thereof. Instead of a single attribute, we are now faced with a set of attributes expressed as the vector

$$\mathbf{x} = x_1, x_2, \ldots, x_i, \ldots, x_n. \tag{4.21}$$

Dominance might be used to eliminate some possible choices. We consider here attributes that are increasing in value. (It is easy to convert an attribute with decreasing value to one with increasing value by altering the sign.) If we are considering two decisions α_1 and α_2 it might happen that the sets of consequences, represented by the vectors $\mathbf{x}^{(1)}$ and $\mathbf{x}^{(2)}$ respectively – of the kind (4.21) – are related in the following way:

$$x_h^{(1)} \geq x_h^{(2)} \text{ for all } h, \text{ and}$$
$$x_h^{(1)} > x_h^{(2)} \text{ for some } h. \tag{4.22}$$

Then we must prefer α_1 to α_2 and we say that $\mathbf{x}^{(1)}$ dominates $\mathbf{x}^{(2)}$. This rule might serve to eliminate some possible decisions.

We might in some cases be able to construct a **value function** of the form[2]

$$v_{\mathbf{X}} (\mathbf{x}) = v_{X_1, X_2, \ldots, X_i, \ldots, X_n} (x_1, x_2, \ldots, x_i, \ldots, x_n) \tag{4.23}$$

which indicates preferences; the higher the value of v, the greater the desirability of the choice. Some attributes could be based on interval or ratio scales, with others on a subjective ranking scale, say in the range 0–10. It is convenient often to use the additive form

$$v_{\mathbf{X}} (\mathbf{x}) = \sum_{i=1}^{n} w_i v_{X_i} (x_i), \tag{4.24}$$

---

[2] We use upper case subscripts in this subsection, anticipating the random attributes of subsection 4.10.2.

and in particular the linear form with appropriate weights

$$v_{\mathbf{X}}(\mathbf{x}) = \sum_{i=1}^{n} w_i x_i = w_1 x_1 + w_2 x_2 + \cdots + w_i x_i + \cdots + w_n x_n. \qquad (4.25)$$

The weights in (4.24) and (4.25) reflect the relative importance of the attributes x_i.

The use of such simplifications can be used in a heuristic manner – seeking to understand the structure and to develop a set of important attributes. In some studies, a large number of attributes are initially considered. For example, in a study of the cleanup of waste in subsurface disposal (Parnell *et al.*, 2001), 21 evaluation measures were considered. It is often necessary and appropriate to reduce the number of attributes so as to study the important ones in more detail. We outline in the following the approach that can be used to justify on a rigorous basis the use of simplified forms (4.24) and (4.25).

Indifference surfaces and rate of substitution

Often decisions will be concerned with attributes which involve a tradeoff, with an improvement in one attainable only through a sacrifice in the other. A common tradeoff in engineering occurs between cost and safety. We continue to discuss attributes with increasing value so that we would consider negative cost, in other words, monetary value. We can use the value function v introduced above in (4.23) to construct **indifference surfaces** which are surfaces of equal value of v. Then

$$v_{X_1,\ldots,X_i,\ldots,X_n}(x_1,\ldots,x_i,\ldots,x_n) = v_1, \qquad (4.26)$$

a constant, would describe such a surface. The higher the value of v, the higher the 'value' to the decision-maker. Figure 4.22 shows a set of indifference curves for the two-dimensional case using attributes x and y, with $v_1 < v_2 < v_3$. The consequences of three decisions α_1, α_2 and α_3 are shown as points on the curves. Inspection of their location leads to the conclusion

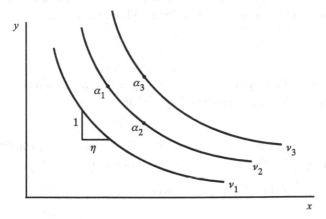

Figure 4.22   Indifference curves $v(x, y) = $ constant

that

$$\alpha_1 \approx \alpha_2 \prec \alpha_3,$$

so that we are indifferent between α_1 and α_2 yet prefer α_3.

An important parameter is the **marginal rate of substitution**, η, which is the negative reciprocal of the slope of an indifference curve, given by

$$\eta = -\frac{1}{dy/dx} \tag{4.27}$$

for $v(x, y) = $ constant. It specifies how much of attribute x the decision-maker is just willing to sacrifice in order to gain one unit of y. This is shown in Figure 4.22. We can also write

$$\eta = \frac{\partial v/\partial y}{\partial v/\partial x} \tag{4.28}$$

at the point in question.

Example 4.6 If $v_{XY} = xy = v_1$,

$$\eta = \frac{x}{y}.$$

Example 4.7 If the marginal rate of substitution is a constant for all (x, y) of interest then v is a linear function and a suitable value function is

$$v_{X,Y}(x, y) = x + \eta y. \tag{4.29}$$

The rule of constant marginal rate of Example 4.7 gives us guidance in determining whether a linear function of attributes is an appropriate model to use. In effect, the amount of one attribute that one would give up in order to obtain a unit of the other does not depend on the current level (x, y). If this is not considered reasonable, we might be able to transform x and y using functions $v_X(x)$ and $v_Y(y)$ so as to achieve a linear additive structure,

$$v_{X,Y}(x, y) = v_X(x) + v_Y(y). \tag{4.30}$$

Example 4.8 If $v_{XY} = xy = v_1$, then $v_X(x) = \ln x$ and $v_Y(y) = \ln y$ would achieve a linear additive structure. The value would then be expressed by $v' = \ln v$.

When three attributes x_1, x_2 and x_3 are under consideration the relationship of **conditional preference** is useful. In this, $\left(x_1^{(1)}, x_2^{(1)}\right)$ given a level x_3 is written as $\left(x_1^{(1)}, x_2^{(1)}|x_3\right)$. For two sets of consequences $\left(x_1^{(1)}, x_2^{(1)}\right)$ and $\left(x_1^{(2)}, x_2^{(2)}\right)$ both given x_3,

$$\left(x_1^{(1)}, x_2^{(1)}|x_3\right) \succ \left(x_1^{(2)}, x_2^{(2)}|x_3\right) \tag{4.31}$$

implies conditional preference (given x_3). If the preferences

$$(x_1, x_2 | x_3)$$

for all (x_1, x_2) do not depend on x_3 for all x_3, there exists **preferential independence**. This implies that the indifference curves for (x_1, x_2) are independent of x_3.

If each pair of attributes in (x_1, x_2, x_3) is preferentially independent of the remaining attribute, then we have **pairwise preferential independence**. This can be used to justify the additive function

$$v_{X,Y}(x, y) = v_{X_1}(x_1) + v_{X_2}(x_2) + v_{X_3}(x_3), \tag{4.32}$$

and extended into higher dimensions if mutual preferential dependence between the attributes is judged appropriate; see Keeney and Raiffa (1976).

4.10.2   Preferences under uncertainty

For decision-making under uncertainty, we need to use utility, for all the reasons given in this chapter, and principally to account for aversion to risk. We now have a vector $\mathbf{x}$ of n attributes $(x_1, \ldots, x_i, \ldots, x_n)$. This set of attributes might be used to construct the entire utility function

$$u_{\mathbf{X}}(\mathbf{x}) = u_{X_1, \ldots, X_i, \ldots, X_n}(x_1, \ldots, x_i, \ldots, x_n) \tag{4.33}$$

over all values $(x_1, \ldots, x_i, \ldots, x_n)$ of interest, perhaps using π-$BRLT$s together with the rules for utility outlined above. Structuring of a such a multiattribute utility function might be assisted by using a value function $v(x_1, \ldots, x_n) = v_{\mathbf{X}}(\mathbf{x})$. It might be possible to enumerate $u_{\mathbf{X}}[v_{\mathbf{X}}(\mathbf{x})]$ over all points of interest. This could be a lengthy process. It is emphasized that in the final decision analysis $\mathbf{X}$ and $U_{\mathbf{X}}(\mathbf{x})$ will be random and we need to weight U as in (4.19) and (4.20). We might wish to pay more attention to those areas of the space with greater probability.

Understanding the structure of the utility function can lead to better appreciation and analysis of the problem. The various functional forms have conceptual meaning, much as the multidimensional forms in probability do. Let us start with two attributes X and Y. Under what conditions is the additive function

$$u_{X,Y}(x, y) = k_X u_X(x) + k_Y u_Y(y), \tag{4.34}$$

where the ks are constants, reasonable? This kind of simplification can be achieved by considering forms of independence that are strikingly similar to those found in studying probabilities where it was termed stochastic independence. This, we have noted, is conditional on our information when making the judgement. Here we begin by defining **utility independence**.

The attribute X is utility independent of attribute Y if preferences regarding lotteries concerning values of X given $Y = y$ are judged not to depend on the level $Y = y$.

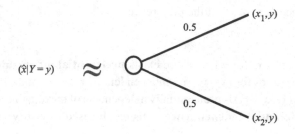

(a) Level of Y in utility independence

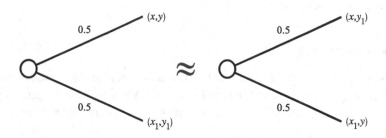

(b) Lotteries for additive independence

Figure 4.23   Inferences regarding independence of attributes in utility functions

This is illustrated by means of Figure 4.23(a) which shows a 50–50 lottery between two values of X, x_1 and x_2, given that $Y = y$. We would exchange the lottery for the certainty equivalent $(\hat{x}|Y = y)$, and utility independence implies that $\hat{x}$ does not depend on the value $Y = y$. In other words, if we are satisfied with a particular certainty equivalent $\hat{x}$ for $Y = y$, we would not wish to change this if we changed y.

Given utility independence, we can deduce that the utility function $u_{X,Y}(x, y)$ must be a positive linear transformation of the utility $u_{X,Y}(x, y_1)$ based on Equation (4.6):

$$u_{X,Y}(x, y) = a(y) + b(y)u_{X,Y}(x, y_1), \tag{4.35}$$

where $b(y) > 0$. The parameters a and b are now functions of y. Positive linear transformations ensure strategic equivalence of the utilities. See Exercise 4.10.

A further simplification results if we invoke **additive independence**. This applies if we are indifferent between the two lotteries shown in Figure 4.23(b) for all (x, y) and arbitrary (x_1, y_1). This condition can be written as

$$\langle (x, y), (x_1, y_1) \rangle \approx \langle (x, y_1), (x_1, y) \rangle. \tag{4.36}$$

As a consequence

$$\frac{1}{2}u_{X,Y}(x, y) + \frac{1}{2}u_{X,Y}(x_1, y_1) = \frac{1}{2}u_{X,Y}(x, y_1) + \frac{1}{2}u_{X,Y}(x_1, y).$$ (4.37)

We then set $u_{X,Y}(x_1, y_1) = 0$, treating (x_1, y_1) as an arbitrary reference point. Then

$$u_{X,Y}(x, y) = u_{X,Y}(x, y_1) + u_{X,Y}(x_1, y),$$

from (4.37). We can write

$$u_{X,Y}(x, y_1) = k_X u_X(x),$$

since x is the only quantity that is changing its value, and similarly

$$u_{X,Y}(x_1, y) = k_Y u_Y(y).$$

We then have the principal result of additive independence, that

$$u_{X,Y}(x, y) = k_X u_X(x) + k_Y u_Y(y).$$ (4.38)

Mutual utility independence occurs if X is utility independent of Y and if Y is utility independent of X. Under these conditions, it can be shown that the utility function can be represented in the following bilinear form:

$$u_{X,Y}(x, y) = k_X u_X(x) + k_Y u_Y(y) + k_{XY} u_X(x) u_Y(y),$$ (4.39)

where the ks are scaling constants. The proof is based on two equations of the kind (4.35), one for each kind of utility independence in the mutual relationship. For details see Keeney and Raiffa (1976). The form (4.38) can be obtained also by mutual utility independence and condition (4.36), the latter at only one set of values (x, y, x_1, y_1).

We have given a flavour of the theory of multidimensional utility. It is important in many projects to have a rational basis for dealing with and combining consequences of various kinds. Multidimensional utility theory gives us guidance in this respect. There are many analogies to the theory of multidimensional probability. An example is the relationship between cross sectional shapes of functions for independent quantities representing probabilities or utilities – in both cases there is a scaling relationship. There is considerable scope for improving our understanding and ability to analyse by studying and applying the rules for structuring multi-dimensional utility functions. Keeney and Raiffa (1976) is the important reference in this area.

4.11   Whether to experiment or not; Bayesian approach; normal form of analysis

Very often the question is asked: do we need to undertake tests or experimentation in order to assist our decision-making? We know that we can incorporate the results of experiments in our probabilistic assessments by means of Bayes' theorem. But we wish to make an assessment of the value of experimentation before we have the results. There is a formal way of doing this which fits neatly into our decision-making procedure. This is sometimes rather clumsily termed 'preposterior' analysis. The 'normal' form of analysis also follows, and this is a good introduction to our next subject, the question of game theory, dealt with in Chapter 5.

I shall introduce the present subject by means of an example. This concerns a building extension. In fact I had experience in practice of a very similar case soon after my graduation, so that this example had real appeal. A structural engineer is considering whether to extend an existing building or to build instead on a separate site. The second option is more expensive, but there is uncertainty regarding the strength of the concrete in the ground floor columns of the existing building. The uncertainty is increased by the fact that the existing building was constructed some fifty years ago, and that records of the strength of the concrete are rather meagre. Remedial action on the existing columns could be considered, but for the present illustration, the simple choice will be made between the two decisions

$\alpha_1 = $ '*extend existing building*',

and

$\alpha_2 = $ '*build on separate site*'.

Similarly, we shall simplify the states of nature into

$E = $ '*the strength of the concrete in the columns is $\geq$ 30 MPa*'

and

$\tilde{E} = $ '*the strength of the concrete in the columns is $<$ 30 MPa*'.

The engineer is considering whether to obtain additional information by means of an experiment on the existing column, for example ultrasonic pulse tests, or the use of impact (hammer) devices, or the taking of small core samples by drilling and testing these. After some research, it is decided to consider taking core samples. But the engineer wishes to estimate the efficacy of the procedure *a priori*, before proceeding with the tests. Therefore two additional decisions are added into the analysis,

$e_0 = $ '*do not experiment*'

and

$e_1 = $ '*take core samples*';

we consider the core sample tests as a single result for simplicity. Two possible results of the core tests are included, as follows.

$T = $ '*successful core tests*'

and

$\tilde{T} = $ '*unsuccessful core tests*',

where T and $\tilde{T}$ are based on the strength results of the core tests.

Since the core tests (of diameter 1.5 cm) will not give a perfect indication of the strength of the existing columns, the engineer studies the literature for records of the frequency distribution of core strengths, given the strength of the larger specimens, and uses these values to assess the probability of core strengths given column strength. A further study of methods of construction leads to an assignment of probabilities of column strength. For the purposes of the present analysis, the following results are pertinent.

$$\mathbf{Pr}\,(E) = 0.7,\;\; \mathbf{Pr}(\tilde{E}) = 0.3,$$

and

$$\mathbf{Pr}\,(T|E) = 0.95,\;\; \mathbf{Pr}(\tilde{T}|E) = 0.05,$$
$$\mathbf{Pr}(\tilde{T}|\tilde{E}) = 0.90,\;\; \mathbf{Pr}(T|\tilde{E}) = 0.10.$$

The decision tree for the problem is illustrated in Figure 4.24. The set of outcomes is different for the two different actions. As we see, this minor complication can be dealt with quite simply and naturally. The problem is an example of probabilistic inference, dealt with in more detail

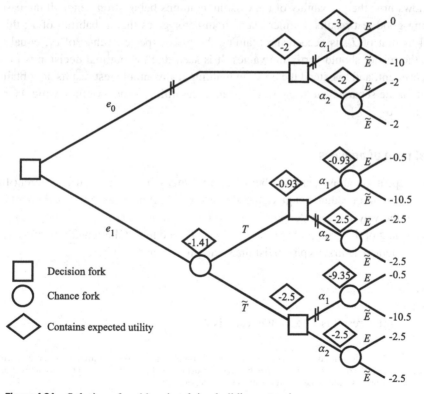

Figure 4.24   Solution of problem involving building extensions

in Chapter 8. The solution is also shown in Figure 4.24. Utilities have been assigned as shown, including a 'cost' of 0.5 units for the core tests.[3]

The decision tree of Figure 4.24 shows a chance fork after the decision e_1, indicating the possible results of the experiment. These results are not known at this stage; it is often most important to conduct an analysis of the potential usefulness of the 'experiment', which could be a major undertaking. For example, it might be considered important to conduct a major field study on the wave regime to determine the loads on an offshore structure. This could be a very costly experiment, and careful consideration must be given to the potential benefits to be gained. The structuring of the problem suggested in Figure 4.24 is aimed at a general approach to such a consideration. The idea is to consider the usefulness of an experiment (in the most general sense) before conducting it.

In deriving the solution, we proceed from the right hand side to the left in the tree of Figure 4.24. The expected values of the upper (e_0) branch are derived from the probabilities of E and of $\tilde{E}$, as stated above. The reader should verify the values. For the lower (e_1) branch, Bayes' theorem is applied. For example, the values of probability of E given the result T are obtained from

$$\mathbf{Pr}\,(E|T) = \frac{(0.95)\,(0.7)}{(0.95)\,(0.7) + (0.1)\,(0.3)} = \frac{0.665}{0.695} = 0.957.$$

This shows how the probability of the existing columns being strong enough increases upon obtaining good test results. Further, the denominator gives the probability of T; this value, as well as that of $\tilde{T}$, is needed in obtaining the final expected value, of e_1, equal to 1.41. Again, the reader should verify the values. It is seen that the optimal decision is to conduct the experiment, and to extend the existing building if favourable test results are obtained, but to build on the separate site if the tests are unsuccessful. In short, we may write the result as $\{e_1; \alpha_1 \text{ if } T, \alpha_2 \text{ if } \tilde{T}\}$.

4.11.1   Normal form of analysis

We now repeat the above analysis, but in **normal form**. This approach is different from the one just given in procedure but gives the identical result. It provides many additional insights, and paves the way for consideration of game theory. It also permits the analyst to postpone the assessment of the prior probabilities of E and $\tilde{E}$, and to test the sensitivity of the results to this assessment. The first step is to list the **strategies** σ_i, as follows.

$\sigma_1 : e_0; \alpha_1$
$\sigma_2 : e_0; \alpha_2$
$\sigma_3 : e_1; \alpha_1$ if the result is T, α_1 if the result is $\tilde{T}$

---

[3] If the cost of the core tests is constant in monetary units, and if utility is non-linear with money, the same cost will be reflected by different amounts of utility depending on the position on the utility curve. The present case of constant 'cost' in utility units is valid if the non-linearity is small or negligible.

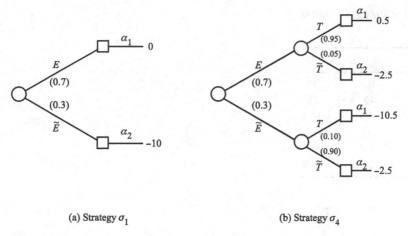

(a) Strategy σ_1 (b) Strategy σ_4

Figure 4.25 Strategies σ_1 (left) and σ_4 (right) in normal form

$\sigma_4 : e_1; \alpha_1$ if the result is T, α_2 if the result is $\tilde{T}$
$\sigma_5 : e_1; \alpha_2$ if the result is T, α_1 if the result is $\tilde{T}$
$\sigma_6 : e_1; \alpha_2$ if the result is T, α_2 if the result is $\tilde{T}$.

Some of these strategies, for example σ_5, are not very logical and could have been omitted. But it is instructive in this first example to consider all possible actions. Now consider each strategy in turn, and calculate the expected utility of the strategy given each state of nature. For example

$$u(\sigma_1|E) = 0,$$
$$u(\sigma_1|\tilde{E}) = -10,$$

and

$$u(\sigma_4|E) = (-0.5)\mathbf{Pr}(T|E) + (-2.5)\mathbf{Pr}(\tilde{T}|E) = -0.6,$$
$$u(\sigma_4|\tilde{E}) = (-10.5)\mathbf{Pr}(T|\tilde{E}) + (-0.5)\mathbf{Pr}(\tilde{T}|\tilde{E}) = -3.3.$$

These calculations are illustrated in Figure 4.25.

The resulting values may be put in tabular form, as in Table 4.9, and also illustrated in the graphical presentation of Figure 4.26. In general, the states of nature could have any dimension, rather than the two in this illustration. But the two-dimensional case of Figure 4.26 is a good aid to understanding. In the figure, the further we are in the 'northeast' direction, the greater the utility. The lines joining σ_2 and σ_4, and σ_1 and σ_4, are termed the 'efficient' set or the 'admissible' set, since they correspond to possible optimal decisions. The optimal decision σ_i corresponds to the maximization of

$$\langle U \rangle = p_1 u(\sigma_i|E) + p_2 u(\sigma_i|\tilde{E}), \tag{4.40}$$

Table 4.9 *Utilities given states of nature*

Strategy σ_i	$u(\sigma_i|E)$	$u(\sigma_i|\tilde{E})$
σ_1	0	−10
σ_2	−2	−2
σ_3	−0.5	−10.5
σ_4	−0.6	−3.3
σ_5	−2.4	−9.7
σ_6	−2.5	−2.5

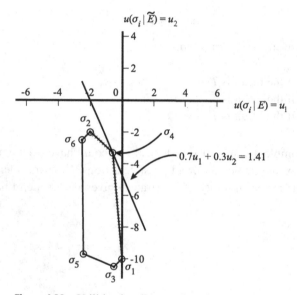

Figure 4.26   Utilities for all strategies, given states of nature

where $p_1 = \mathbf{Pr}(E)$ and $p_2 = \mathbf{Pr}(\tilde{E})$. This corresponds to a straight line in the space of Figure 4.26, the slope of which is dictated by the prior probabilities p_1 and p_2. To illustrate, we use the abbreviated notation

$$u(\sigma_i|E) \equiv u_1, \ u(\sigma_i|\tilde{E}) \equiv u_2.$$

Then (4.40) becomes

$$\langle U \rangle = p_1 u_1 + p_2 u_2 = c, \text{ say.} \tag{4.41}$$

For any value of c, this is the line illustrated in the upper part of Figure 4.27. Its slope is

$$-\frac{p_1}{p_2} = -\frac{p_1}{\tilde{p}_1} = -\frac{p}{\tilde{p}},$$

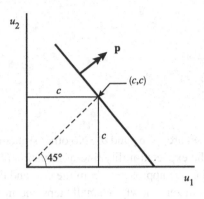

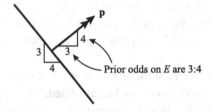

Figure 4.27   Prior probability vector and line representing expected utility with prior odds

writing p_1 simply as p. The slope of the *normal* to the line is then $\tilde{p}/p$ (recall that the product of the slopes of two perpendicular straight lines in two-dimensional space in cartesian coordinates is equal to -1).

The vector normal to the line of expected utility is shown as **p** in Figure 4.27. The ratio $p/\tilde{p}$ represents the odds on the event under consideration as in Equation (2.5), and $\tilde{p}/p$ the odds against. For example if $p = 3/7$ with $\tilde{p} = 4/7$, then the slope of the line is $(-3/4)$ and the normal vector **p** has components $(3/7, 4/7)$ as in the lower part of Figure 4.27. The odds on the event are then 3:4. Note that **p** is not a unit vector; the scalar sum of the components add to unity so that its length is less than unity except where $p = 0$ or $p = 1$. The vector $(3/7, 4/7)$ has length $5/7$. It is advisable to work through one or two examples. See Exercises 4.12, 4.14 and 4.15.

The optimal point is the one which maximizes expected utility. One can then imagine a line with the slope of Equation (4.40) – or its equivalent (4.41) – but free to move in the space of Figure 4.26 while keeping the same slope. This effectively changes $\langle U \rangle$ – or its equivalent c – which is the quantity that we wish to maximize. The further we move in the northeast direction, the larger is $\langle U \rangle$. In our building-extension problem, the furthest point in Figure 4.26 is that corresponding to σ_4. The line corresponding to the assigned prior probability $p = 0.70$ with slope $(-p/\tilde{p})$ is shown, together with the optimal point (σ_4) and value of expected utility (1.41).

We can also deduce the values in Table 4.10. In producing this table, we suppose that we have no knowledge of the prior probabilities, and imagine that the line corresponding to Equation (4.41) can have any slope. The question arises: under what conditions are the different strategies

Table 4.10   *Optimal strategies*

$p = \mathbf{Pr}(E)$	Optimal strategy
0.00 to 0.48	σ_2
0.48 to 0.92	σ_4
0.92 to 1.00	σ_1

of the efficient set optimal? The members of this set are σ_1, σ_2 and σ_4. No other strategies can be optimal. Let us begin with σ_1. First imagine the expected-utility line of Equation (4.41) is vertical in Figure 4.26. This means that $p = 1$; the line approaches from the east and the first point that it touches is σ_1. We now increase p from zero slowly in small steps and move the resulting line each time from the northeast towards the efficient set. It will continue contacting σ_1 first until its slope is the same as the line joining σ_1 to σ_4 in Figure 4.26, at which stage both σ_1 and σ_4 are optimal. Increasing the slope further will result in σ_4 being optimal until the slope of the line coincides with σ_4–σ_2. Finally the line is horizontal ($p = 0$, $\tilde{p} = 1$) and approaches from the north.

To obtain the values of Table 4.10, we need the slopes of σ_1–σ_4 and σ_4–σ_2. Since we have the coordinates in Table 4.9 above, these can easily be calculated. Then to get the values of probability in Table 4.10, we use the fact that the normal vector $\mathbf{p}$ is at right angles to the slope of the lines just noted (σ_1–σ_4 and σ_4–σ_2). We exploit the simple rule that the product of these slopes is -1. The direction of the vector $\mathbf{p}$ gives also the prior odds on E. It is well worth working through the exercise of obtaining the values in Table 4.10; further examples are given in Exercises 4.12, 4.14 and 4.15. Exercise 4.16 deals with the extension to more than two dimensions.

We also note that the strategies can be used to reduce the problem to a decision tree such as that of Figure 4.12, with all the possible actions stated at the start. The ideas above can be readily generalized into several dimensions. This is illustrated in the general decision trees of Figure 4.28. Continuous random quantities can be dealt with using integration rather than summation.

4.12   Concluding remarks

Probability and utility form natural components in a coherent theory of decision. That both of these attach to a person is a natural outcome of the nature of decision and of opinion. Utility is then a personal concept: we cannot specify preferences, especially risk aversion, for others. This is as it should be, since one's attitude to risk, and to the relative desirability of a set of attributes is very much a personal affair. The personalistic aspect should then be seen as a natural outcome of proper modelling of the decision-maker's dilemma.

The theory presents a powerful guide to the decision-maker, for structuring decisions and thinking about risk aversion. It must be adapted to the situation where we wish to develop policies for a company, regulatory body or other organization. We might guide the client to

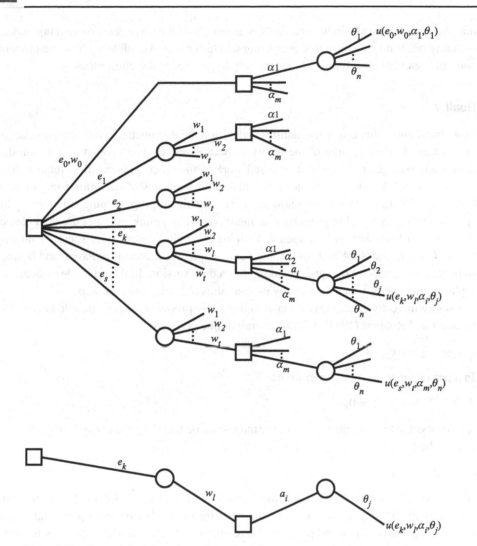

Figure 4.28   General decision tree and condensed tree

policies that reflect a reasonable level of risk aversion. An example might be a design code, or advice to a company on operating policy for a manufacturing process, in which it is desired to achieve a certain level of safety. In these cases, close consultation with the client or the group involved in the decision-making, with the objective of reaching a consensus, is needed. There is the responsibility to ensure that risks to human life are not excessive. Chapter 10 gives guidance on risk levels in society and on target levels.

The quotation by Ambrose Bierce given at the beginning of the chapter underlies a fundamental truth. We might optimize our decision but when reality dawns, the unlikely state of nature might well turn out to be true. The columns in the building might turn out not to be of adequate strength, the desired states of nature might not happen. We can arrange things so that the chance of undesirable outcomes is acceptably small. Yet decision-making under uncertainty holds no guarantee that everything will turn out as we might wish, while probability and utility nevertheless represent the best way to proceed in the circumstances.

4.12.1 Duality

Probability and utility are dual quantities in decision theory, much like force and displacement in mechanics. The structure of the theory in both cases revolves around a scalar product – maximum expected utility in one case and work in the other. Mathematical relationships of duality abound. We saw in Chapter 2 the duality between sets of operations; in particular in Exercise 2.11. In Section 5.4.3 below we have dual linear-programming solutions in linear spaces. Duality is found in projective geometry, whereby points and lines are dual elements; intersection of two lines gives a point and joining two points gives a line. These are dual operations. Courant and Robbins (1941) show, for example, the theorems of Pascal and Brianchon to be dual statements. Desargues' theorem has a dual version. In the principle of duality, one replaces each element in a theorem by its dual, thus obtaining another theorem.

Duality in vector spaces is connected to the scalar product (see for example Greub, 1975). Courant and Robbins (1941) describe a straight line given by

$$ax_1 + bx_2 + cx_3 = 0.$$

In a particular realization, for example

$$4 \cdot 1 + 2 \cdot 6 - 8 \cdot 2 = 0,$$

it is impossible to distinguish between the interpretation that the point $\{x_1 = 1, x_2 = 6, x_3 = 2\}$ lies on the line

$$4x_1 + 2x_2 - 8x_3 = 0,$$

written in short as $\{4, 2, -8\}$, and the interpretation that the point $\{4, 2, -8\}$ lies on the line $\{1, 6, 2\}$. This symmetry provides the basis for the duality between point and line. Any relationship between line and point can be reinterpreted as a relation between point and line by interchanging the coordinates.

We can characterize the maximization of expected utility as the scalar product

$$\langle U \rangle = p \cdot u = \sum_i p_i u_i, \tag{4.42}$$

in which p_i is the probability of event E_i, and u_i is the utility (payoff) associated with its occurrence. If one considers a set of random events E_i, with associated utilities u_i, then

$$U = \sum_i u_i E_i \tag{4.43}$$

represents the gain that one would achieve, given the results $E_i = 1$ or 0 depending on the occurrence or non-occurrence of E_i (respectively). Equation (4.42) can be derived from (4.43) by taking the expected values of both sides. Both Equations (4.42) and (4.43) can be taken to

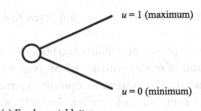

(a) Fundamental lottery

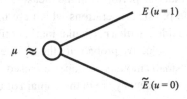

(b) Definition of probability: $\mu \equiv \mathbf{Pr}(E)$

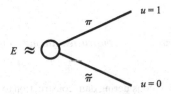

(c) Definition of utility: $\pi \equiv u(E)$

Figure 4.29   Probability and utility

represent a scalar product of two vector quantities in dual vector spaces. This does not change if integrals rather than sums are considered; generally

$$\langle U \rangle = \int u \, dF.$$

One should expect then for there to be a dual relationship between probability and utility. Consider the fundamental lottery shown in Figure 4.29(a). This shows a chance fork in which one obtains maximum utility ($u = 1$) with probability p or π, and minimum utility ($u = 0$) with the complementary value $(1 - p)$ or $(1 - \pi)$. To define probability and utility one can proceed as follows. See Figure 4.29(b) and (c).

Probability p of E. Take a unit of utility and divide it into two parts $(\mu, 1 - \mu)$ such that the certain reward μ is equivalent to uncertain E with unit of utility as reward if E occurs. Then $p \equiv \mu$.
Utility u of E. Take a unit of probability and divide it into two parts $(\pi, 1 - \pi)$ such that the certain event $E = 1$ is equivalent to the uncertain reward of a unit of utility, with probability π. Then $u \equiv \pi$.

The separation of probability and utility is more than a matter of convenience. It is quite natural to want to separate questions of impartial judgement regarding the occurrence of a future event, that is, assignments of probability, and questions of advantage, or gain given that the event occurs, that is, assignment of utility.

It is awkward, as in the above definition, to involve both probability and utility, so in defining probability, we generally start with small amounts of money with no risk aversion and proceed in an iterative manner. A person in the street, if asked to express her opinion in terms of a fair bet, would be forced to an honest and optimal expression of opinion. The role of the bet is much the same as the operations in the *operational definition* of a concept due to Bridgman. In this, the concept is associated with the operations that one performs in the measurement of the concept, for example length is identified with the physical operations which are used in measuring it, laying a measuring rod with one end coincident with a particular mark, noting the mark, taking the reading at the other end, and so on. In subjective probability theory, the bet is the method for making a measurement of uncertainty; small amounts of money are used so that factors such as risk aversion do not enter into the picture since they are in the domain of utility.

4.13   Exercises

There are wide areas of application for utility theory. Projects can be constructed around subjects of interest in particular disciplines.

4.1 *Benefit–cost ratio and qualitative measures.* An authority is considering dam construction to supply clean water to the region. Two alternatives, 1 and 2 below, are being considered for water supply and flood control. The following table shows costs and benefits associated with the two alternatives together with other consequences. The costs and benefits are annual values.

Alternative		Water supply	Flood control
1: one large dam	Costs	20.2×10^6	8.0×10^6
	Benefits	23.2×10^6	8.3×10^6
2: two small dams	Costs	28.0×10^6	9.1×10^6
	Benefits	24.5×10^6	10.0×10^6

Alternative		Other consequences
1: one large dam	Costs	Flooding of 50 km^2 of natural woodland and habitat
		Relocation of people, reduction in wild life
	Benefits	Improved human health for local population
		Recreational facilities for 100 000 persons per year
2: two small dams	Costs	Flooding of 23 km^2 of natural woodland and habitat
		Relocation of people, reduction in wild life
	Benefits	Improved human health for local population
		Recreational facilities for 200 000 persons per year

(a) Which project has the better benefit–cost ratio? Use stated dollar amounts.
(b) Suggest ways to assist the authority to make a decision taking into account the other qualitatively measured consequences listed above.
(c) Comment on the merits of benefit–cost ratios in decision-making in particular in the above.

4.2 *Time value of money.* Commodity prices can be very volatile, and the problem is posed in arbitrary units. An oil company has to make a decision on whether to invest in offshore production from a marginal oil field, the size of which is estimated at 200 MB (million barrels). The development requires capital expenditure of 150 million per year for five years at which point production would commence at the rate of 10 MB per year for 20 years. The production period would also incur operating and maintenance costs of 75 million per year. The costs noted so far are in real (constant-worth) monetary units, with constant purchasing power with regard to year 1 of the project. In a preliminary analysis of the problem, an engineer considers two scenarios as described in the table below. These represent two possible developments of the world economy in average terms, the second corresponding to a period of conflict in the Middle East (rather alarmist but useful for contrast). The values are in actual (inflated) monetary units.

Calculate the net present worth for the two scenarios. For real monetary units, use a real interest rate calculated from the figures in the table. Calculate the *EMV* based on a probability of 0.80 for scenario 1 as against 0.2 for scenario 2.

Scenario 1 – stable			Scenario 2 – volatile		
Oil price	Interest rate	Inflation rate	Oil price	Interest rate	Inflation Rate
15 per bbl increasing at 5% per year	8% per year over 20 year period	4% per year over 20 year period	25 per bbl increasing at 10% per year	15% per year over 20 year period	12% per year over 20 year period

4.3 Consider the following payoff matrix for a fisher, who has to decide which species to fish on in a season. The states of nature are uncertain. Solve the problem using the maximin, maximax, Hurwicz ($\alpha = 0.5$) and minimax regret criteria. Comment on the strategies. What is meant by utility dominance? Give an example.

	State of nature		
	θ_1	θ_2	θ_3
Species A	1	1	1
Species B	1	3	0
Species C	0	4	0
Species D	1	2	0

4.4 Imagine that you are a member of the Board of Directors of a small company, with assets of about $50 000. The table below shows a choice to be made by the board between three contracts (α_1, α_2

and α_3). The entries show the consequences with associated probabilities in brackets. Draw the decision tree. Infer utility values for each outcome using π-basic-reference lottery tickets. Plot the resulting values of utility against money. Iterate if necessary. Complete the decision analysis using expected monetary value and expected utility criteria.

	Consequences in $ (probability)	
α_1	$-80\,000$ (0.2)	$50\,000$ (0.8)
α_2	$20\,000$ (0.6)	4000 (0.4)
α_3	$100\,000$ (0.2)	0 (0.8)

4.5 Consider the statement: 'A decision-maker is risk-averse if she or he prefers the expected consequences of a lottery to the lottery itself.' Take two examples:

 (i) an attribute with increasing preferences, e.g., money,
 (ii) an attribute with decreasing preferences, e.g. size of an oil spill.

Illustrate the statement above with regard to these two cases and sketch the utility functions. Assume that the utilities range between 0 and 1. Use lotteries with 50–50% probabilities.

4.6 An engineer is working with a client in formulating a response to a potential oil spill in an offshore region. The volumes of spill under consideration are in the range 0–50 000 m³. Sketch a utility function expressing risk aversion. Prove that the shape shown must express risk aversion by using the certainty equivalent of $\langle 0, 50\,000 \rangle$. Consider the function

$$u(v) = b[c - \exp(av)],$$

where v = volume of spill, and a, b and c are constants. Discuss whether this might be used as a utility function. Let $u(0) = 1$, and $u(50\,000) = 0$. Show how the certainty equivalent noted can be used with this information in evaluating the constants a, b and c.

4.7 Study the problem of noise, for example in reduction of noise pollution in siting an airport. What would constitute a good attribute in attempting to formulate utility? Consider intensity (energy per unit area per unit time – Wm⁻²). For utility, would decibels serve? (Note that doubling the decibel level 'buys' intensity on a log scale.)

4.8 Contrast the exponential form of Example 4.3 and the logarithmic form

$$u(x) = \ln[a(x + b)]$$

for expressing risk aversion. Obtain the local risk aversion measure $\gamma(x)$ of Equation (4.17). Use examples from your experience (cost, money, time saved in construction or manufacturing, delays, quality of product, ...).

4.9 Find the local risk aversion measure $\gamma(x)$ for the quadratic utility function

$$u(x) = -ax^2 + bx + c$$

with $a, b > 0$, and $x < b/2a$, so that u is increasing. Comment.

4.10 Strategic equivalence of two utility functions is defined such that the same preferences are implied for any lotteries with the same consequences. Show that the relationship between the two utility functions u_A and u_B of the kind given in Equation (4.6), i.e. $u_B = a + bu_A$, $b > 0$ results in strategic equivalence. Use Equation (4.15) to assist in solution. Show also that strategic equivalence results in the positive linear transformation noted.

4.11 Consider a problem of your choice that involves multidimensional utility. Examples are cost and safety, travel distance and noise (siting of airport or similar facility) and service rate provided for incoming data and the queuing delay (in data transmission). Explore the relation between the attributes for utility and additive independence and speculate on the various forms of the utility function, for example those given by Equations (4.38) and (4.39).

4.12 Solve the problem of Example 3.25 in normal form, listing all the strategies and plotting a graph similar to that in Figure 4.26. Find the optimal solution and the range of probabilities for which various strategies are optimal, using the method given for Table 4.10.

4.13 An engineer is advising a city administration on the question of whether or not to recommend that a public transportation system (for example light rail transit) be built in the medium sized North American city. The population of the area that would benefit from the service is well defined at 420 000. The key to the success of the project is the proportion M of the population who will use the facility. This is treated as a random quantity. In the preliminary study being undertaken, three values of M are considered. These are $M = 0.2, 0.4$ and 0.6. Initial best estimates of the probability of these events are

$$\mathbf{Pr}(M = 0.2) = 0.5; \mathbf{Pr}(M = 0.4) = 0.4; \mathbf{Pr}(M = 0.6) = 0.1.$$

The following table gives the utility associated with various outcomes.

Action	State	Utility (0–100)
Build system	$M = 0.2$	30
	$M = 0.4$	60
	$M = 0.6$	90
Do not build		60

The engineer realizes that a survey as to intentions might help, but that the results are unlikely to match subsequent behaviour. A person might indicate aversion to using the system yet subsequently find it a useful form of transport, and *vice versa*. Let Y designate the proportion of the population who indicate that they will use the system; this is also discretized into three possible values, 0.2, 0.4 and 0.6. Psychologists and previous literature are consulted and the following table is the result. It gives values of $\mathbf{Pr}(Y|M)$, the probability of the three values of Y (indications of use of the system), given the three subsequent levels of usage (the values in the table are for illustrative purposes).

		Y		
		0.2	0.4	0.6
	0.2	0.50	0.30	0.2
M	0.4	0.10	0.60	0.3
	0.6	0.15	0.15	0.70

Assume now that a survey has been carried out with the result that $Y = 0.6$. To account for the cost of the survey, subtract two units of utility. What decision maximizes expected utility?

4.14 Reconsider the problem posed in Exercise 4.13.

(a) Analyze the problem in normal form, assuming that the survey has not been conducted. Include the option whether to experiment (survey), at cost of 2 units, or not in the analysis. List the ten strategies and obtain the utilities of the strategies given the states of nature. List these in a table for the three states of nature. Obtain the optimal decision using the given values of $\mathbf{Pr}(M = 0.2, 0.4, 0.6)$.

(b) Obtain optimal strategies for the table of (a) using maximin, maximax, Hurwicz ($\alpha = 0.4$), and minimax regret criteria.

(c) Use again the table of utilities of (a) but ignore the results for $M = 0.4$. Plot a graph of $U(\sigma_i|M = 0.2)$ versus $U(\sigma_i|M = 0.6)$ and find the ranges of probability for which the various σ_i are optimal, considering that only the two states of nature $M = 0.2, 0.6$ are possible.

(d) Discuss consequences with regard to the environment of building the system, and possible social measures to affect and change public opinion regarding use of public transportation over time, for example to raise gasoline prices.

4.15 A company manufacturing microprocessors purchases batches of microchips by contract. To evaluate the quality of the chips a test is conducted. The results are P (pass) or $\tilde{P}$ (fail). It is found that the test method is not perfect and an analysis of performance of microchips purchased in the past indicates the following, where S = satisfactory performance and $\tilde{S}$ = unsatisfactory performance. The values in the table were obtained by conducting tests on microchips found to have either satisfactory or unsatisfactory performance. The utility associated with satisfactory performance is 15 units and that associated with unsatisfactory performance -50 units, while that associated with rejecting a batch is 0 units.

$\mathbf{Pr}\left(P \text{ or } \tilde{P}|S \text{ or } \tilde{S}\right)$		
Test indication	S	$\tilde{S}$
P	0.85	0.30
$\tilde{P}$	0.15	0.70

(a) Draw a decision tree for the problem of deciding whether or not to accept or reject a batch of microchips, conditional on the results of the test.

(b) A contractor with a good track record – 98% of her batches of microchips in the past has been satisfactory – provides a batch for which the test shows the result $\tilde{P}$. Which decision, accept or reject, maximizes expected utility?

(c) Repeat part (b) in the case of a contractor with no track record; assume then that $\mathbf{Pr}(S) = \mathbf{Pr}(\tilde{S}) = 0.50$.

(d) Conduct an analysis in normal form, listing all the strategies and plotting a graph similar to that in Figure 4.26. Find the optimal solution and the range of probabilities for which various strategies are optimal, using the method given for Table 4.10.

4.16 Find a method to evaluate the ranges of probabilities for which various decisions are optimal in the 'normal form' analysis, similar to the method for Table 4.10, but consider the case where there are more than two dimensions.

4.17 *Unfinished games and utility.* In the early days of mathematical probability theory, one area of endeavour related to utility was that of fair odds in an unfinished game; see Figure 2.20 and the related discussion. This was applied to the finals in the National Hockey League or the World Series, in both of which the 'best of seven' wins. We now introduce a more general problem. Three players A, B and C are engaged in a game of chance taking turns in this order. The winner will receive the entire pot of S dollars. When it is A's turn, his probability of winning is p_A. If A does not win it becomes B's turn, whose probability of winning is p_B. If B does not win, it becomes C's turn, and so on, returning to A if nobody has won, and continuing as before. The game is still in progress but suddenly must be abandoned. It is A's turn. Find the fairest way to divide the pot. Apply the result to the following two cases.

(a) The winner obtains first 'doubles' in a throw with a pair of fair dice. The pot is $273.

(b) An archery contest between three archers, in which the weakest (A) goes first. The first to hit the bull's eye receives the pot of $96. The following probabilities have been derived empirically through observations of the performance of the archers:

$$p_A = 0.2, \, p_B = 0.5, \, p_C = 0.9.$$

5 Games and optimization

Intelligence and war are games, perhaps the only meaningful games left.

William Burroughs

We are survival machines – robot vehicles blindly programmed to preserve the selfish molecules known as genes.

Richard Dawkins

A great truth is a truth whose opposite is also a great truth.

Thomas Mann

5.1 Conflict

Up to now, we have considered decision-making in cases where we are the 'protagonist', and where the 'antagonist' is relatively benign, or more precisely, indifferent to our desires. In this case, we characterize the antagonist as 'nature' in which we have we considered nature to consist of those aspects such as weather that do not react to our everyday behaviour. We can then regard the activity of engineering design as a 'game against nature'. In the example of design of an offshore structure to resist wave loading, we might choose the design wave height and 'nature' chooses the actual wave heights which the structure has to resist.

But nature includes creatures other than ourselves! In activities in which other human beings – or indeed, animals – are involved, we cannot rely on any kind of indifference. If we set up a business, our competitors are out to do better than us, and this implies trying to take business from us. Strategies will be worked out to achieve this objective. We can no longer rely on probabilistic estimates of our opponent's strategies based on a study of 'impartial' nature, as we might do in studying a record of wave heights.

The situation where we face an antagonist who is out to beat us in some way or another has strong links to animal behaviour in general. Dawkins (1989) in his book *The Selfish Gene* points out the connections between strategies for survival of animals (or, more correctly, of genes) and game theory. We introduce this briefly, so as to bring into play an important and interesting line of thinking. Two animals in conflict could adopt strategies such as the following.

- Fight always as hard as possible, retreating only if seriously injured.
- Always retreat.
- Attack and pursue an opponent that runs away but retreat if the opponent retaliates.
- Fight back only when attacked.

Most of us know from personal experience that it is not sensible to fight to the finish in every confrontation, so that we might sometimes follow an aggressive approach and at other times retreat, regardless of the stance of our opponent. This is a mixed strategy. The problem originally approached using mathematical methods by von Neumann was of the school playground variety, but it has striking resemblances to conflict between animals as we shall describe. This game was 'match-penny'. Each of two players reveals the face of a coin at the same moment; if the faces are the same (both heads or both tails) one player wins. If they are unlike, the other player wins. A variation of this is the rock–paper–scissors game, in which paper covers rock, scissors cut paper, and rock breaks scissors. The three objects are mimicked by the hands of the two antagonists, who each select one and reveal their selection simultaneously. The first of the pairs in the preceding sentence wins.

A species of lizards shows similar behaviour.[1] Three male types have evolved, distinguished by the colour of their throats: orange, blue and yellow. Mating – and consequently contribution to the gene pool – is the objective of the game. Let us start with the moderate blue-throated specimens. They keep relatively small harems of three females, with small territories. Orange-throated lizards are much more aggressive, full of testosterone, and they easily take over from the blues. Their harems are large, up to seven females, with corresponding large territories, keeping the orange-throated lizards very busy indeed. The yellow-throated lizards infiltrate these large harems by simulating female behaviour and secretly mating with the females within. (The reviewer refers here to the movie *Shampoo,* in which a human follows the same strategy.) Cycling by generation occurs, with the blue-throated lizards taking over from the yellows, the orange from the blues and so on. The ideas of game theory have been applied to problems of biology notably by John Maynard Smith; see Dawkins (1989).

Our objective is to introduce the ideas of game theory so as to demonstrate another important way of assigning probabilities. We wish to show that the allocation of probabilities follows different rules under conditions of conflict, and that it is important to understand the influence of this situation. Certainly we do not aim at an exhaustive treatment. We deal with two-person zero-sum games. The main concepts that we wish to convey are embodied in their solution, which can conveniently be carried out using linear programming. This is part of optimization theory, an important subject for engineers, so that this is also introduced. Lagrange multipliers are introduced as part of this subject; they are used extensively in Chapter 6.

---

[1] See Scientific American of June, 1996.

Table 5.1 *Payoff matrix*

		B			
		τ_1	τ_2	τ_3	τ_4
	σ_1	12	3	9	8
A	σ_2	5	4	6	5
	σ_3	−3	0	6	7

5.2   Definitions and ideas

At the outset, some basic definitions need to be given.[2] There are two-person, three-person, or generally n-person games, depending on the number of participants. We shall consider only two-person games. In previous chapters, we used the symbol α to denote actions in considering decisions in cases where there is no conflict. We now use σ and τ for the strategies in conflict situations. These are the strategies open to the two people, A and B respectively. We introduce the payoff matrix of Table 5.1. After each play of the game, a $\{\sigma_j, \tau_i\}$ pair is identified by virtue of A having played σ_j and player B τ_i. As a consequence, player B pays the amount entered at the $\{\sigma_j, \tau_i\}$ intersection in Table 5.1 to player A. For example, if A plays σ_2 and B plays τ_3, B pays A 6 units. We shall illustrate in Section 5.2.1 how games can be reduced to a matrix of the kind in Table 5.1. (In most real games the matrix is much larger.) The idea is similar to the $\{\alpha, \theta\}$ pairs used previously in Chapter 4 (for instance in Figure 4.12), the new element being that of conflict.

It is a convention in game theory the amounts are paid by the column player (B here) to the row player (A); where the amounts are positive, A receives the amount. If there are negative entries, as in $\{\sigma_3, \tau_1\}$, the amount (3 units) is paid by A to B. Positive entries are negative for B who pays. The payoffs could be utilities; for the present with single payoffs for each $\{\sigma_j, \tau_i\}$ pair, the relationship between utility and monetary value – or other attribute – would have to be shared by both A and B. A further complication is that A's positive values are negative for B, so that the sharing noted would have to be symmetrical about the origin. To avoid these complications, we shall simply regard the amounts as relatively small monetary payoffs with no complications involving factors such as risk aversion.

If A plays σ_2 and B plays τ_3, then B pays A 6 units. Since the sum of the payoffs to each player at each terminal state is zero (+6 for A and −6 for B in the case just noted), the game is termed a **zero-sum game**: what one player pays, the other receives. When interests are diametrically opposed, as in the two-person zero-sum games under consideration, the games are games of **pure conflict** or **antagonistic** games, again in the jargon of game theory. Further usage includes

---

[2] The monumental work on games is the book by von Neumann and Morgenstern (1947); other excellent books such as Jones (1980) and Luce and Raiffa (1957) may be consulted. Owen (1982) introduces games from a mathematical point of view. Bernstein (1996) gives interesting historical perspectives.

Table 5.2 *Minimax solution*

		B				Row
		τ_1	τ_2	τ_3	τ_4	minima
	σ_1	12	3	9	8	3
A	σ_2	5	4	6	5.	4
	σ_3	−3	0	6	7	−3
Column maxima		12	4	9	8	

finite and **infinite** games, depending on the number of strategies. In the end-game in chess, it is possible sometimes to formulate a set of moves which go on infinitely. This can be dealt with by stopping rules, for instance specifying a draw after a certain number of repetitions. Or one can construct infinite coin-tossing games. We shall consider only finite games. **Cooperative** and **non-cooperative** games correspond to instances where there is, or is not, cooperation between the players. We shall deal only with the non-cooperative variety. Moves can include chance elements, for instance in scrabble or poker, where there is uncertainty regarding the tokens or cards received. There will always be uncertainty regarding our opponent's future moves.

Engineering design is sometimes thought of as a 'game against nature'. In Section 4.2 we dealt with a real game against nature involving a farmer. Various approaches were considered that had different psychological implications: the optimist, the pessimist, and others. In the present case of games, we are up against an intelligent antagonist who is out to better us in a situation of conflict. The natural reaction to this situation is that of the pessimist assuming that the worst will happen. This results in a 'maximin' solution, and will be the standard way of dealing with games. We now return to the problem of Table 5.1. Since the payoffs are to A, he selects the minimum in each row as shown in the last column of Table 5.2, and then chooses the strategy that maximizes these minima. Therefore A chooses σ_2. Player B, meanwhile, undergoes the same mental turmoil wondering what A is going to do. B has to make the payoffs so they are negative to her. B therefore selects the column maxima – the worst that can happen, the most that B might have to pay. Then the minimum of these is chosen, leading to the choice of τ_2. It is to be noted that this minimax strategy is in fact identical to A's, the difference – minimax versus maximin – arising from the fact that the payoffs are positive for A, and negative for B.

The solution of the game is shown in Table 5.2, and is $\{\sigma_2, \tau_2\}$. It turns out that the solution leads to the saddle point shown, at which

Pure MAXIMIN = Pure MINIMAX. (5.1)

The value 4 at the saddle point is termed the **value** of the game. When this occurs, we have a solution in **pure strategies**. Both players are satisfied after the game that had they chosen

Table 5.3 *Match-penny*

		B		Row
		τ_1	τ_2	minima
A	σ_1	−1	1	−1
	σ_2	1	−1	−1
Column maxima		1	1	

Table 5.4 *Game with no saddle point*

		B		Row
		τ_1	τ_2	minima
A	σ_1	3	6	3
	σ_2	5	4	4
Column maxima		5	6	

differently the result would have been worse. Neither regrets their decision once the other player's choice is known. A game in which the value is zero is a **fair** game; the present game is not fair as its value is 4.

The question arises: is there always a saddle point? The answer is no. If there were always a saddle point, this would mean that, in the example in the introductory section 'Conflict' above, a single strategy such as 'always attack' or 'always retreat' would be the invariable solution. We shall explore the absence of saddle points by means of examples.

Example 5.1 Consider the game of match-penny: the original game studied by von Neumann. Player A chooses heads or tails on his coin, while B makes the same choice on hers. Neither player knows the other's choice. If the coins match, player B receives a payoff of unity from A. If the coins don't match, A receives a payoff of unity from B. In the matrix of Table 5.3, σ_1 is the choice 'heads' for A, and σ_2 'tails'; for B τ_1 is 'heads and τ_2 is 'tails'. The payoff matrix is shown in Table 5.3. There are no distinct maximin or minimax values, and no saddle point.

Example 5.2 A variation of this game is called 'odds and evens'. The two players point either one or two fingers simultaneously. If the sum of the fingers is odd a designated player 'odd' wins; if even, the designated player 'even' wins.

In the examples, there is neither a unique row maximin nor a column minimax. These can exist yet there may be no saddle point, as in the game in Table 5.4. In this game and other similar ones we have

Pure MAXIMIN ≤ Pure MINIMAX. (5.2)

The solution to match-penny consists of a strategy of not demonstrating any pattern in showing a particular face. For example, if one shows a head more often than a tail, or demonstrates any regularity of choice, our opponent will quickly realize this and adjust their actions accordingly and begin to gain. The essence of the solution is to show heads or tails randomly, without any pattern. One might have a subsidiary device such as a coin, a die, or roulette wheel, which permits us to choose heads or tails with a probability of $1/2$ each time. This is an example of the use of a **mixed strategy**; in Table 5.3, within both pairs $\{\sigma_1, \sigma_2\}$ and $\{\tau_1, \tau_2\}$ the strategy would be chosen with a probability of $1/2$. More details and proofs of this are given in Section 5.5.

5.2.1   Extensive form, normal form and reduction to matrix

We have introduced games using a payoff matrix with a set of strategies for each player. How are real games reduced to this form? Example 5.1 showed how this is done for the game of 'match-penny'. We shall demonstrate further the method of reducing a game to a matrix, using the game of 'Nim', following Jones (1980). This gives a reasonably convincing idea of how any game can be reduced to a matrix game. In Nim, two distinct piles of matches are created. Two players alternately remove matches, one or more, from either of the two piles. They must remove one or more matches from only one of the two piles at each turn. The game terminates when the last match is drawn, and the player who removes the last match loses the game. The game has been analysed in some detail by various persons; see Jones (1980) for specifics. Since both players know the full details at each step, it is a **game of perfect information**.

We can illustrate the extensive form of the game simply using 2–2 Nim. In this, there are only two matches in each pile. Figure 5.1 shows the game in the form of a game tree, in which each possible state and play of the two players A and B is shown successively. At the start, on the left of the tree, it is A's turn. A either removes one or two matches from a pile – this is shown for the lower of the two piles only; a symmetrical situation results with removal from the upper pile. Then it is B's turn; depending on A's move, all of B's possible moves are shown. For example, given that A removed a single match as in the lower of the first choices open to him, B can remove both matches of the upper pile, a single one, or the single match in the lower pile left by A. These three choices are shown in the lower part of Figure 5.1, following B's turn. Then A plays, and so on to the end of the game. Again, if there is a symmetrical choice, only one is shown in the figure. 'Dumb moves' have been included, where a player makes a move that unnecessarily leads to their loss.

The game tree of Figure 5.1 is also known as the **extensive form** of the game, as it shows all the possible states – except for the symmetrically identical moves noted which have been omitted so as to avoid unnecessary complication. (These can easily be added if wanted.) Now we have reduced the game to sets of strategies for each player and the corresponding endpoints. To illustrate, consider A's possible strategies. These are shown in Figure 5.1 and summarized in Table 5.5. Strategy σ_1 consists of removal of both matches in a pile; thereafter A makes no further choices. To assist in following the strategies, the nodes in Figure 5.1 have been

Table 5.5 *Possible strategies for A*

Strategy	First turn	Second turn
σ_1	Remove two matches from a pile; node $1 - 2$	
σ_2	Remove one match from a pile; node $1 - 3$	Remove two matches from the remaining pile; node $8 - 12$
σ_3	Remove one match from a pile; node $1 - 3$	Remove one match from the remaining pile; node $8 - 13$

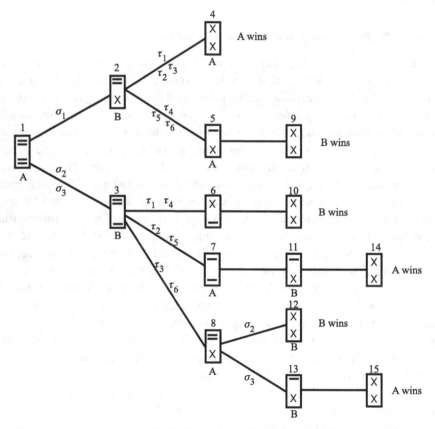

Figure 5.1   Extensive form of 2–2 Nim illustrated in a game tree. 'X' indicates that no matches remain in the pile

numbered from 1 to 15. Strategy σ_1 corresponds to the choice $1 \rightarrow 2$. In strategies σ_2 and σ_3, A removes one match, and has to make a further choice if B removes the other from the same pile; σ_2 results if A removes both remaining matches and σ_3 if a single match is removed on the second turn. These are paths $1 \rightarrow 3 \rightarrow 8 \rightarrow 12$ and $1 \rightarrow 3 \rightarrow 8 \rightarrow 13$ respectively in Figure 5.1. The strategies for B are listed in Table 5.6. B's first choice will depend on A's prior

Table 5.6 *Possible strategies for B*

Strategy	If at	Move to	Path
τ_1	2	4	$1 \to 2 \to 4$
	3	6	$1 \to 3 \to 6$
τ_2	2	4	$1 \to 2 \to 4$
	3	7	$1 \to 3 \to 7$
τ_3	2	4	$1 \to 2 \to 4$
	3	8	$1 \to 3 \to 8$
τ_4	2	5	$1 \to 2 \to 5$
	3	6	$1 \to 3 \to 6$
τ_5	2	5	$1 \to 2 \to 5$
	3	7	$1 \to 3 \to 7$
τ_6	2	5	$1 \to 2 \to 5$
	3	8	$1 \to 3 \to 8$

Table 5.7 *Game of 2–2 Nim in normal form*

		τ_1	τ_2	τ_3	τ_4	τ_5	τ_6
				B			
	σ_1	4(A)	4(A)	4(A)	9(B)	9(B)	9(B)
A	σ_2	10(B)	14(A)	12(B)	10(B)	14(A)	12(B)
	σ_3	10(B)	14(A)	15(A)	10(B)	14(A)	15(A)

Table 5.8 *Game of 2–2 Nim in matrix form*

		τ_1	τ_2	τ_3	τ_4	τ_5	τ_6
				B			
	σ_1	1	1	1	-1	-1	-1
A	σ_2	-1	1	-1	-1	1	-1
	σ_3	-1	1	1	-1	1	1

choice in the first play. She therefore starts at either node 2 or 3 in Figure 5.1. A complete description of a strategy must take into account all possible states, and this is done by specifying a course of action that depends on A's first choice as in Table 5.6. All possible strategies are covered in this set.

Now we can see how a specified course of action by A and B leads to a definite result. For example, if the set of actions $\{\sigma_3, \tau_3\}$ is decided upon, the result is node 15 in Figure 5.1, with A winning; this is handily deduced using Table 5.7. If A and B now agree that \$1 is paid from A to B if B wins, and *vice versa,* then the matrix of payoffs given in Table 5.8 is obtained. The game of 2–2 Nim has been reduced to a matrix game, similar in form to the payoff matrices of the preceding section. At this stage, A and B could sit in their favourite pub – this is after all a pub game! – armed with Tables 5.7 and 5.8. They could then ignore their piles of matches, and simply choose a pair of strategies $\{\sigma_j, \tau_i\}$ independently, and look up the result in one of the tables. Or if necessary pass their choices to an impartial referee to determine the payoff. This of course removes all the fun and interest from the game. But our purpose is to demonstrate that games can be reduced to a simple matrix of results corresponding to vectors of possible strategies and to the matrix form noted. This can be used to show, for example in Table 5.8, that strategy τ_4 guarantees a win for B regardless of A's choice.

In introducing the game of 2–2 Nim, player A started the play. In fact, player B could have started the game, or the starting player could have been decided by chance. For example, a coin could have been tossed to decide between players A and B. This is a simple example of a chance move; note that there will (almost!) always be uncertainty as to our opponent's potential moves. In a chance move, an external device such as the tossing of a coin or die, or the drawing of a card, or of letters in Scrabble, is introduced. We can think of such devices as extra players if we wish, who play certain moves specified by the nature of the device. We can denote these players as 'nature' if we wish, and the plays by nature would follow the rules of the device; for example a die tossed will yield six possible outcomes, each with a probability of $1/6$. Nature's plays would then correspond to mixed strategies, as mentioned at the end of the last section, and again in Section 5.5. The probabilities of nature's strategies would be predetermined, and not optimized as would those of the main players. The latter subject requires knowledge of optimization theory, to be introduced in the next sections.

Example 5.3 The game tree for match-penny, as introduced in Example 5.1 is shown in Figure 5.2. The payoff matrix has been given in Table 5.3.

In non-zero-sum games the sum of the payoffs is not zero. As an example, let us consider the game of match-penny illustrated first in Table 5.3. Suppose now that if A loses, \$2 is paid,

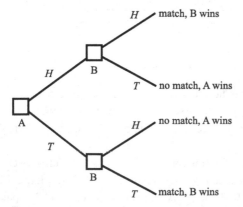

Figure 5.2   Game tree for match-penny with players A and B

Table 5.9 *Bimatrix for match-penny with non-zero-sum payments*

		B	
		τ_1	τ_2
	σ_1	$(-2, 1)$	$(1, -2)$
A	σ_2	$(1, -2)$	$(-2, 1)$

with $1 going to B, and the remaining $1 to the 'banker' running the game. Similarly, if B loses, $2 is paid, but only $1 goes to A. The payoff matrix is then represented by the bimatrix of Table 5.9. Only zero-sum games are dealt with in this chapter, and the bimatrix is introduced only to illustrate the more general situation.

5.3   Optimization

5.3.1   Importance in engineering

In engineering, in science and in economics, we often wish to obtain an optimal solution with regard to some quantity: minimum cost, minimum weight, maximum safety, shortest transmission time for electronic data, maximum profit, shortest route or completion time. We shall consider in this section and the next deterministic systems, which involve decisions under uncertainty, but not related to probability. The uncertainty arises as to the best optimal solution. Yet the methods can be applied to the solution of problems of probability as well. The coverage of the subject of optimization given here is introductory in nature; there are many texts which give a detailed treatment that can be consulted. Non-linear optimization is used in maximizing entropy, the subject of Chapter 6, and the solution to games with mixed strategies is carried out using linear optimization in this chapter.

To introduce ideas, we consider the design of a bridge (Figure 5.3). We are interested in determining the optimal span length x and number of piers n for a total bridge length ℓ. Thus $nx = \ell$ (see Figure 5.3). The bridge, apart from the abutments, is to be made of steel. The cost of the abutments is a, in nominal units that we use for this problem. The cost of a pier is b, and this may be taken for initial design purposes as constant, regardless of the span. The cost per unit weight of steel is d, and m is the cost per unit length of the floor system, cross bracing and roadway. The main variable of interest is w, the weight of steel per unit length of span, which is taken to be a linear function of x:

$$w = kx, \tag{5.3}$$

where k is a constant for the type of steel girder under consideration.

This relationship can be seen to be approximately correct if we consider that bending moment is proportional to the square of the span ($\propto x^2$); section modulus s is proportional to the square

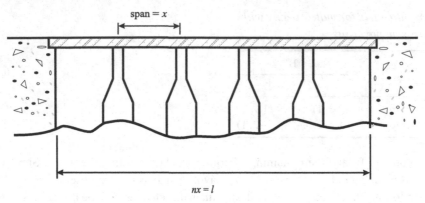

Figure 5.3   Spacing of bridge piers

of girder depth h, (i.e. $s \propto h^2$) assuming constant width. Since section modulus is proportional to bending moment, the span and girder depth – or weight – are linearly related ($h \propto x$). If weight is proportional to h, we have the result in Equation (5.3). Then the total cost c is given by

$$c = a + (n - 1)b + dw\ell + m\ell;$$

this is to be minimized, and we write

$$\text{Min} \left[c = a + \left(\frac{\ell}{x} - 1 \right) b + dkx\ell + m\ell \right]. \tag{5.4}$$

In problems of optimization we use the notation 'Min' and 'Max' rather than 'min' and 'max' to distinguish between the desire to obtain the maximum of a mathematical function – possibly subject to constraints – and the desire to study the maximum of a set of random quantities, as we do in Chapter 9. The relationship (5.4) is the **objective function** in the jargon of optimization theory. The solution to the function (5.4) is

$$\frac{dc}{dx} = -\frac{b\ell}{x^2} + dk\ell = 0, \text{ or}$$

$$x = \sqrt{\frac{b}{dk}}, \text{ or}$$

$$b = dkx^2 = dwx. \tag{5.5}$$

The last line may be interpreted as the rule 'make the cost of one span $\simeq$ the cost of a pier'. This might form an approximate starting point in the spacing of piers and deciding the span of the main girders in the design of the bridge.

We check the second derivative:

$$\frac{d^2c}{dx^2} = \frac{2b\ell}{x^3} > 0.$$

Therefore there is a minimum at $x = \sqrt{b/dk}$.

Real problems will tend to be more complex than this example. Often the introduction of constraints is important. We might want to make the depth of a girder a minimum subject to the condition that its deflection under working loads is not excessive. This might be achieved by specifying a maximum span-to-depth ratio, or by more detailed calculations.

Example 5.4 A product-mix problem: there are many examples of such problems, a famous one being the diet problem, to be dealt with later in Section 5.4.3. Suppose that we have written a contract to supply a mixture of two ores, which come from two sources, 1 and 2. The following have been determined by measurement.

- Ores 1 and 2 contain 5 kg/tonne and 10 kg/tonne of element A, respectively.
- Ores 1 and 2 contain 4 kg/tonne and 5 kg/tonne of element B, respectively.
- Ores 1 and 2 contain 4 kg/tonne and 1 kg/tonne of element C, respectively.

It is specified that the mixture must contain at least 8 kg of element A, at most 8 kg of element B, and at least 2 kg of element C. The costs of ores 1 and 2 are 2 and 1 units/tonne respectively. Find the proportions that satisfy the conditions specified at minimum cost.

Let x_1 and x_2 be the number of tonnes of ores 1 and 2, respectively. Then we wish to find

$$\text{Min} (z = 2x_1 + x_2) \tag{5.6}$$

subject to the **constraints**

$$5x_1 + 10x_2 \geq 8,$$
$$4x_1 + 5x_2 \leq 8 \text{ and}$$
$$4x_1 + x_2 \geq 2. \tag{5.7}$$

Since the amount of material to be mined must be greater than or equal to zero, we have also that $x_1 \geq 0, x_2 \geq 0$. If we treat all of the inequalities as equalities, we may obtain the boundaries and thereby plot the feasible space of solutions in this two-dimensional problem as in Figure 5.4. What is the optimal point and solution? This will be answered below in Section 5.4.

As just stated, we shall return to the solution of this problem; here it is to be noted that there are three inequality constraints. The problem is linear since the objective function and the constraints are linear functions of the quantities x_1 and x_2.

Minimization versus maximization

One important final point that should be made in this introduction is that maximization and minimization are essentially equivalent. To change Min to Max, one merely changes the sign of the objective function. In the example above, the function (5.6) is equivalent to

$$\text{Max} \left(z' = -z = -2x_1 - x_2 \right).$$

5.3.2 Lagrange multipliers

We consider a general optimization problem, in which the functions can be non-linear. The principal application in the present work is the maximization of the entropy function, given in

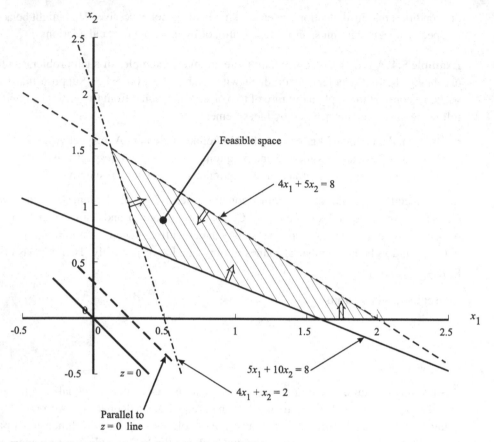

Figure 5.4   Feasible space for example in Section 5.3

the next chapter. We therefore deal with 'Max' rather than 'Min' in this subsection. As noted above, it is an easy matter to convert a problem from Min to Max or *vice versa*. The purpose here is to maximize the objective function, that is, to find

$$\text{Max} \left[z' = f\left(x_1, x_2, \ldots, x_j, \ldots, x_n\right) = f(\mathbf{x}) \right]. \tag{5.8}$$

This has to be done subject to the m constraints

$$h_1\left(x_1, x_2, \ldots, x_j, \ldots, x_n\right) = 0,$$
$$h_2\left(x_1, x_2, \ldots, x_j, \ldots, x_n\right) = 0,$$
$$\vdots$$
$$h_i\left(x_1, x_2, \ldots, x_j, \ldots, x_n\right) = 0,$$
$$\vdots$$
$$h_m\left(x_1, x_2, \ldots, x_j, \ldots, x_n\right) = 0. \tag{5.9}$$

These equations can be abbreviated to

$h_i(\mathbf{x}) = 0, i = 1, \ldots, m,$ or

$\mathbf{h}(\mathbf{x}) = \mathbf{0}.$

The boldface symbols indicate vectors.

We consider equality constraints as in (5.9) initially. Generally $m < n$; if $m = n$ then one could (at least in principle) solve for the xs. The number $(n - m)$ is termed the **degrees of freedom**; there are other kinds of degrees of freedom in probability theory; see for example Chapters 8 and 11. To solve the optimization problem, we first form the Lagrangian, or composite function, denoted L. This is the objective function minus the constraints multiplied by the 'Lagrange multipliers', λ_i:

$$L\left(x_1, x_2, \ldots, x_j, \ldots, x_n; \lambda_1, \lambda_2, \ldots, \lambda_i, \ldots, \lambda_m\right) = f(x_1, \ldots, x_n) - \sum_{i=1}^{m} \lambda_i h_i(x_1, \ldots, x_n).$$
(5.10)

There is one Lagrange multiplier for each constraint equation. If the constraint equations (5.9) are satisfied, then Equation (5.10) reduces to (5.8), and we can, under these circumstances, maximize either of these two equations. In brief we can write (5.10) as

$L(\mathbf{x}, \boldsymbol{\lambda}) = f(\mathbf{x}) - \boldsymbol{\lambda} \cdot \mathbf{h}.$

The really elegant aspect of the Lagrangian L is that we can treat the multipliers λ_i as variables, in addition to the x_j. Then we determine the stationary points of L with respect to the augmented list of variables, with the following result.

$$\frac{\partial L}{\partial x_j} = \frac{\partial f}{\partial x_j} - \sum_{i=1}^{m} \lambda_i \frac{\partial h_i}{\partial x_j} = 0,$$
(5.11)

for $j = 1, 2, \ldots, n$, and

$$\frac{\partial L}{\partial \lambda_i} = -h_i(x_1, x_2, \ldots, x_n) = 0,$$
(5.12)

for $i = 1, 2, \ldots, m$. The neat thing is that Equations (5.12) are the constraint equations, so that we have converted the original constrained problem of Equations (5.8) and (5.9) to the unconstrained one of finding the stationary point(s) of L, treating the λ_i as additional variables.

In effect, a stationary point of L at $(\mathbf{x}^*, \boldsymbol{\lambda}^*)$ corresponds to the stationary point of f at $\mathbf{x}^*$, subject to the constraints of the problem, as originally stated. In considering the nature of a stationary point using second derivatives, the original function $f(\cdot)$ should be differentiated rather than the Lagrangian $L(\cdot)$, which should not be expected to determine the character of the stationary point. See Exercise 5.5. The Lagrange multipliers are sometimes termed 'undetermined' multipliers but we shall see that they generally have meaning in physical problems and are well worth determining. In the maximization of entropy, the first multiplier gives the partition function while another gives the average energy kT, with k denoting Boltzmann's constant and T the temperature.

Example 5.5 A numerical example will assist in describing the method. This is: find the stationary point of $f(x_1, x_2) = 4x_1^2 + 5x_2^2$ subject to $h_1(x_1, x_2) = 2x_1 + 3x_2 - 28 = 0$. The solution is as follows:

$$L = 4x_1^2 + 5x_2^2 - \lambda(2x_1 + 3x_2 - 28). \tag{5.13}$$

(Here there is only one Lagrange multiplier.)

$$\frac{\partial L}{\partial x_1} = 8x_1 - 2\lambda_1 = 0,$$

$$\frac{\partial L}{\partial x_2} = 10x_2 - 3\lambda_2 = 0, \text{ and}$$

$$\frac{\partial L}{\partial \lambda} = -(2x_1 + 3x_2 - 6) = 0 \tag{5.14}$$

The first two of these equations yields $x_1 = \lambda/4$, $x_2 = 3\lambda/10$ and, using the third equation, $x_1 = 15/14$, $x_2 = 9/7$, $\lambda = 30/7$.

We can further illustrate the nature of the procedure described above by considering a problem in three dimensions:

$$\text{Max } f(x_1, x_2, x_3) \tag{5.15}$$

subject to

$$h_1(x_1, x_2, x_3) = 0 \text{ and}$$
$$h_2(x_1, x_2, x_3) = 0. \tag{5.16}$$

The constraint equations can be represented by surfaces, as in Figure 5.5. Their intersection represents the condition $h_1 = h_2 = 0$. The optimization procedure consists of moving along

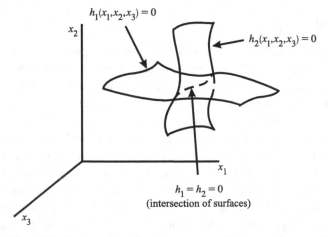

Figure 5.5 Intersection of surfaces representing constraint equations

this intersection line until a stationary value of $f(\cdot)$ is found. The solution $\{x_1^*, x_2^*, x_3^*\}$ is defined as the point at which $f(x_1, x_2, x_3)$ is a maximum. The nature of the stationary point, whether maximum, minimum or point of inflexion, must be investigated further, as discussed in the paragraph preceding Example 5.5.

We complete the discussion regarding the three-dimensional problem. At the stationary point, Equations (5.11) yield

$$\frac{\partial f}{\partial x_1} - \lambda_1 \frac{\partial h_1}{\partial x_1} - \lambda_2 \frac{\partial h_2}{\partial x_1} = 0,$$

$$\frac{\partial f}{\partial x_2} - \lambda_1 \frac{\partial h_1}{\partial x_2} - \lambda_2 \frac{\partial h_2}{\partial x_2} = 0 \text{ and}$$

$$\frac{\partial f}{\partial x_3} - \lambda_1 \frac{\partial h_1}{\partial x_3} - \lambda_2 \frac{\partial h_2}{\partial x_3} = 0, \tag{5.17}$$

all of which may be combined to give

$$\left(\frac{\partial f}{\partial x_1} - \lambda_1 \frac{\partial h_1}{\partial x_1} - \lambda_2 \frac{\partial h_2}{\partial x_1} \right) dx_1 + \left(\frac{\partial f}{\partial x_2} - \lambda_1 \frac{\partial h_1}{\partial x_2} - \lambda_2 \frac{\partial h_2}{\partial x_2} \right) dx_2$$

$$+ \left(\frac{\partial f}{\partial x_3} - \lambda_1 \frac{\partial h_1}{\partial x_3} - \lambda_2 \frac{\partial h_2}{\partial x_3} \right) dx_3 = 0. \tag{5.18}$$

Rearranging,

$$\left(\frac{\partial f}{\partial x_1} dx_1 + \frac{\partial f}{\partial x_2} dx_2 + \frac{\partial f}{\partial x_3} dx_3 \right) - \lambda_1 \left(\frac{\partial h_1}{\partial x_1} dx_1 + \frac{\partial h_1}{\partial x_2} dx_2 + \frac{\partial h_1}{\partial x_3} dx_3 \right)$$

$$- \lambda_2 \left(\frac{\partial h_2}{\partial x_1} dx_1 + \frac{\partial h_2}{\partial x_2} dx_2 + \frac{\partial h_2}{\partial x_3} dx_3 \right) = 0, \tag{5.19}$$

or

$$df - \lambda_1 dh_1 - \lambda_2 dh_2 = 0. \tag{5.20}$$

Therefore $df = dh_1 = dh_2 = 0$; the last two conditions ensure that the point remains on the constraint line.

The procedure outlined above yields stationary points of the constrained function; further analysis must be conducted to determine their character, whether maximum, minimum, or point of inflexion. Further, although the first derivatives of L with respect to the x_i are equivalent to the derivatives of f, it does not follow that the second derivatives of L and f are equivalent. These points are of concern in general, but in the case of maximizing entropy, we shall show that the stationary point of H is a global maximum (Section 6.7). As a last point, it should be noted that the technique described is in the field of the calculus of variations, and may be treated formally in that way.

The method just described will be used in Chapter 6 in maximizing the entropy function.

5.3.3   Equality and inequality constraints; slack and excess variables

As noted above, the constraints in a problem can be equalities, $=$, or inequalities, either 'greater than or equal to, $\geq$', or 'less than or equal to, $\leq$', depending on the nature of the problem. In the case of inequalities, a domain in the space under consideration can be delineated, giving the boundary and interior of the feasible region. This was shown for Example 5.4 in Figure 5.4. For a general non-linear problem, the optimal solution might be on a boundary or in the interior of the feasible space, as illustrated in Figure 5.6. In this example, we consider the maximization of $z = f(x)$, with x constrained to lie in the interval $x_a \leq x \leq x_b$. For linear problems, the solution will be on one of the edges or vertices of a convex space.

In dealing with inequality constraints, it is convenient to eliminate them by adding **slack and excess variables**. For example, if we take the inequality

$$5x_1 + 10x_2 \geq 8,$$

we can subtract an excess variable x_3 to arrive at

$$5x_1 + 10x_2 - x_3 = 8,$$

with $x_3 \geq 0$. This adds a new variable but converts the constraint to an equality. In Example 5.4 the inequalities (5.7) can be written as

$$5x_1 + 10x_2 - x_3 = 8,$$
$$4x_1 + 5x_2 + x_4 = 8 \quad \text{and}$$
$$4x_1 + x_2 - x_5 = 2, \tag{5.21}$$

where x_3 and x_5 are excess variables, and x_4 is a slack variable. Excess variables reduce the 'excess' condition of the left hand side of $\geq$ to equality, while a slack variable provides the slack in the left hand side of $\leq$ to ensure equality. On the boundaries of the feasible space of

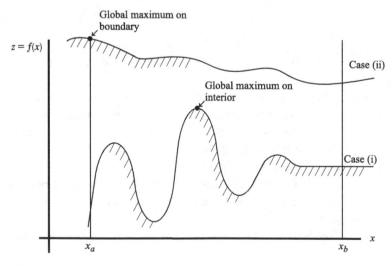

Figure 5.6   Maxima on the interior and on the boundary of feasible space

Figure 5.4, the appropriate slack and excess variables would equal zero, depending on the boundary under consideration.

We shall consider principally the linear problem, but slack and excess variables also have a role in non-linear analysis. Consider the following problem.

Example 5.6 This example is dealt with in detail in Beveridge and Schechter (1970). Find

$$\text{Min} \left(z = 2x_1 + x_1^2 - x_2^2 \right) \tag{5.22}$$

subject to

$$x_1^2 + x_2^2 - 1 \le 0. \tag{5.23}$$

The constraint confines the feasible region to the interior of the unit circle in $x_1 - x_2$ space. We can introduce a slack variable x_3 such that

$$x_1^2 + x_2^2 - 1 + x_3^2 = 0 \tag{5.24}$$

where we have used the square of the slack variable to ensure that a positive quantity is added to the left hand side of inequality (5.23). (We do not add squares in linear systems as the linearity would then be destroyed.) We can now form the Lagrangian

$$
\begin{aligned}
L &= -z - \lambda \left(x_1^2 + x_2^2 - 1 + x_3^2 \right) \\
&= -2x_1 - x_1^2 + x_2^2 - \lambda \left(x_1^2 + x_2^2 - 1 + x_3^2 \right).
\end{aligned} \tag{5.25}
$$

We differentiate this with respect to the four variables, resulting in the following equations:

$$
\begin{aligned}
1 + x_1 + \lambda x_1 &= 0 \\
x_2 - \lambda x_2 &= 0 \\
\lambda x_3 &= 0 \\
x_1^2 + x_2^2 + x_3^2 - 1 &= 0.
\end{aligned} \tag{5.26}
$$

These equations may be used to solve the problem; it is necessary to investigate the character of the stationary points. This is dealt with in Exercise 5.5.

5.4   Linear optimization

In this case, the functions $f(\cdot)$ and $h(\cdot)$ introduced in the preceding section – in Equations (5.8) and (5.9) – are linear. This is the form that we are interested in when solving games, but the methods used, of 'linear programming', have wide application in other fields as well. As an aside, linear programming is a branch of 'mathematical programming'; these terms do not refer to computer programming but rather to a program of instructions to solve the problem using mathematical methods (at the same time, computer programming does indeed assist in practice). The constraint equations can involve 'equality, $=$', 'greater than or equal to, $\ge$', or 'less than or equal to, $\le$', depending on the problem under consideration. We wish to include in our analysis the possibility that any one of these relationships is allowed. The notation $\gtreqless$ is introduced for this purpose. We pose the following problem, using minimization 'Min':

$$\text{Min} \left(z = c_1 x_1 + c_2 x_2 + \cdots + c_j x_j + \cdots + c_n x_n = \sum_{j=1}^{n} c_j x_j \right) \tag{5.27}$$

subject to

$$a_{11} x_1 + a_{12} x_2 + \cdots + a_{1j} x_j + \cdots + a_{1n} x_n \gtreqless b_1$$
$$a_{21} x_1 + a_{22} x_2 + \cdots + a_{2j} x_j + \cdots + a_{2n} x_n \gtreqless b_2$$
$$\vdots$$
$$a_{i1} x_1 + a_{i2} x_2 + \cdots + a_{ij} x_j + \cdots + a_{in} x_n \gtreqless b_i$$
$$\vdots$$
$$a_{m1} x_1 + a_{m2} x_2 + \cdots + a_{mj} x_j + \cdots + a_{mn} x_n \gtreqless b_m. \tag{5.28}$$

This may be written in matrix form:

$$\sum_{j} a_{ij} x_j = [a_{ij}] x_j \gtreqless b_i. \tag{5.29}$$

(Vector notation can also be used if desired, for example Min ($\mathbf{z} = \mathbf{c} \cdot \mathbf{x}$), and so on.)

In the jargon of mathematical programming, the c_j are the **cost coefficients**, the b_i the **stipulation**s, and the a_{ij} the **structural coefficients**. The values of these quantities are known. The expression 'linear programming' is often abbreviated to '*LP*'. The methods of calculus introduced in the preceding Section 5.3 lead to optimal solutions, but since the functions are now linear, the derivatives are constant. Setting these equal to zero does not lead to a solution. It turns out that the optimal solutions are concentrated at the boundary corresponding to the constraint equations – to the points or edges of the simplex, not the interior, as we shall see.

Example 5.7 *Graphical solution to Example 5.4.* In this example we introduced a linear optimization problem of the kind just discussed. The constraints were plotted in Figure 5.4 and formed a closed convex figure bounded by the linear constraint relationships. Referring to this figure, let us now consider the objective function

$$z = 2x_1 + x_2,$$

which we wish to minimize. This is plotted on the figure for the case where $z = 0$. If we preserve the slope and move the line in a northeast or southwest direction, we preserve the relationship between x_1 and x_2 but change the z-value. We have also plotted the objective function for $z = 0.5$ on Figure 5.4. To minimize z, the further the line moves in the southwest direction, the better.[3] The solution to the problem is to move the line to the most southwest corner of the convex region, to the intersection of the lines $5x_1 + 10x_2 = 8$ and $4x_1 + x_2 = 2$.

---

[3] This is the reverse of the situation in Figure 4.26 since we are now minimizing rather than maximizing the objective function.

This is the point (see Figure 5.4) which gives the optimal set of values. The latter are always indicated by an asterisk:

$$x_1^* = \frac{12}{35}, x_2^* = \frac{22}{35}, \text{ with } z^* = 1\frac{11}{35}. \tag{5.30}$$

The salient aspect to be observed is that the solution corresponds to a vertex of the convex region, on the boundary of the feasible space.

5.4.1   Geometric aspects

We now review some important definitions. For most purposes, we deal with Euclidean space R^n and cartesian coordinates. Consider two points $P^{(1)}$ and $P^{(2)}$ in this n-dimensional space, with coordinates $x_j^{(1)}$ and $x_j^{(2)}$ respectively, where $j = 1, 2, \ldots, n$. The **segment** joining $P^{(1)}$ and $P^{(2)}$ is the collection of points $P^{(\mu)}$ whose coordinates are given by

$$x_j^{(\mu)} = (1 - \mu) x_j^{(1)} + \mu x_j^{(2)}, \text{ for } j = 1, 2, \ldots, n, \tag{5.31}$$

and μ is a number between 0 and 1, $0 \le \mu \le 1$. The coordinates of $P^{(\mu)}$ for given μ represent a linear combination of those of $P^{(1)}$ and $P^{(2)}$, with the point $P^{(\mu)}$ 'sliding' on a straight line between $P^{(1)}$ where $\mu = 0$ and $P^{(2)}$ where $\mu = 1$. We can illustrate this definition well in two dimensions. First, we rewrite Equation (5.31) as follows:

$$x_j^{(\mu)} = x_j^{(1)} + \mu\big(x_j^{(2)} - x_j^{(1)}\big), \text{ for } j = 1, 2, \ldots, n.$$

In two dimensions, $n = 2$, the situation is as illustrated in Figure 5.7. As noted, $P^{(\mu)}$ can be at any point on the line joining $P^{(1)}$ and $P^{(2)}$, depending on the value of μ, at $P^{(1)}$ for $\mu = 0$ and at $P^{(2)}$ for $\mu = 1$.

A **convex set** is a set of points such that, if any $P^{(1)}$ and $P^{(2)}$ are members of the set, then the segment joining them is also in the set. Some two-dimensional sets are illustrated in Figure 5.8;

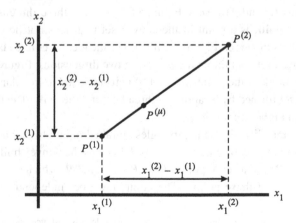

Figure 5.7   Segment between $P^{(1)}$ and $P^{(2)}$

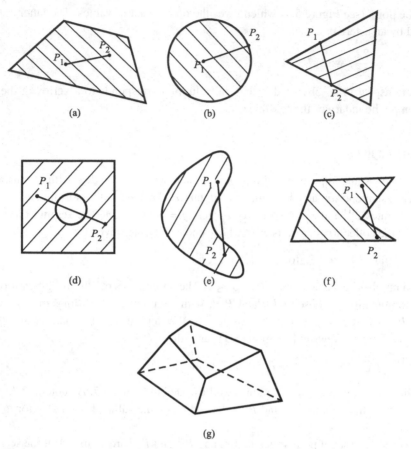

Figure 5.8 Convex and non-convex sets

(a), (b) and (c) are convex while (d), (e) and (f) are not. Figure 5.8(g) shows a three-dimensional convex set. A **vertex** or **extreme point** is a point in the convex set that does not lie on the segment joining any other two points in the set. Figure 5.9 illustrates the idea. A **convex hull** of a set of points is the smallest convex set containing the points. In two dimensions, Figure 5.10 illustrates a convex hull: one can think of the points as a set of nails attached to the plane and of the convex hull as the convex set formed by wrapping a thread around the nails. The thread becomes a plane or hyperplane in higher dimensions.

Simplexes are extensions of points, line segments, triangles, tetrahedra, into higher dimensions. For $n \geq 0$, an n-simplex, or simplex of dimension n, is defined as the convex hull in an m-dimensional Euclidean space[4] (R^m) of a set of points $P^{(0)}$, $P^{(1)}$, ..., $P^{(n)}$, which constitute the vertices. There are then $(n + 1)$ of these points. The points must be 'independent'; for

---

[4] Simplexes are sometimes defined in affine space, in which the ability to measure length, distance and angles is restricted. De Finetti emphasizes the importance of affine spaces in probability theory.

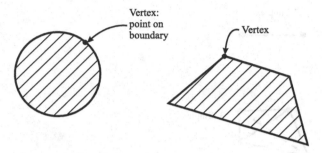

Figure 5.9   Vertices of convex sets

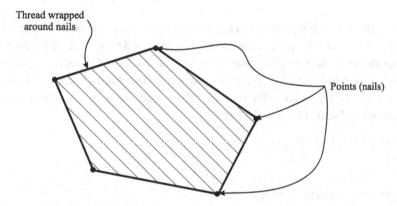

Figure 5.10   Convex hull

example three points must not be collinear. Individual and two distinct points are independent; $m + 1$ points are dependent if they are in a hyperplane, and $m + 2$ or more points are always dependent. Figure 5.11 shows the simplexes up to $n = 3$. Note that the tetrahedron contains four simplexes of dimension 2, each of which contain three simplexes of dimension 1, each of which contain two simplexes of dimension 0. Equilateral simplexes have sides of equal length. The word **edge** refers to the face of a convex hull or simplex.

5.4.2   Simplex method

We can think of the feasible space in any linear programming problem as a convex space such as that in Figure 5.4. The objective function there is a straight line. Rather than this two-dimensional figure, one can imagine the outer edges (faces) of the feasible space to consist of planes and hyperplanes rather than lines. This feasible space can be thought of as a set of simplexes meeting with common faces (a set of triangles in Figure 5.4). The objective function can generally be thought of as a hyperplane; moving it in one direction increases its value z while moving it in the other decreases z. The solution will correspond to the intersection of the z-plane with a vertex of the feasible space – the vertex of the appropriate simplex. Multiple solutions can

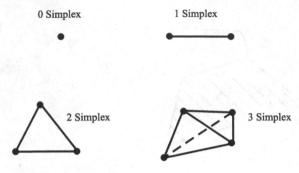

Figure 5.11 Simplexes

occur if the objective function is parallel to a face of the feasible space, and unbounded solutions can occur if part of the feasible space is not closed. We leave aside these details for the present.

The **simplex method** was devised by Dantzig in 1947. It consists of investigating one vertex of a space such as that in Figure 5.4, and then improving the solution if not optimal until an optimal solution can be recognized. We first need to convert our problem into the **standard linear programming form**. This is as follows:

$$\text{Min}\ \left(z = c_1 x_1 + c_2 x_2 + \cdots + c_j x_j + \cdots + c_n x_n\right) \tag{5.32}$$

subject to

$$a_{11} x_1 + a_{12} x_2 + \cdots + a_{1j} x_j + \cdots + a_{1n} x_n = b_1$$
$$a_{21} x_1 + a_{22} x_2 + \cdots + a_{2j} x_j + \cdots + a_{2n} x_n = b_2$$
$$\vdots$$
$$a_{i1} x_1 + a_{i2} x_2 + \cdots + a_{ij} x_j + \cdots + a_{in} x_n = b_i$$
$$\vdots$$
$$a_{m1} x_1 + a_{m2} x_2 + \cdots + a_{mj} x_j + \cdots + a_{mn} x_n = b_m, \tag{5.33}$$

with

$$x_j \geq 0,\ j = 1, 2, \ldots, n,\ \text{and}$$
$$b_i \geq 0,\ i = 1, 2, \ldots, m. \tag{5.34}$$

This form has been developed, with some variants, by convention. It is fundamental to implementing a mathematical programming method, as will become evident. To convert a problem that is not in standard form into the required form, the following are useful.

1. Any problem involving 'Max' can be converted to 'Min' by changing the sign of the objective function. Min z and Max $\left(z' = -z\right)$ are equivalent.
2. If a negative sign appears on the right hand side of a constraint equation (negative stipulation b_i), simply multiply the equation through by (-1).

3. If a variable is unrestricted in sign (non-negativity is not satisfied: in the jargon of *LP*), replace it by the difference of two variables that are. For example, if x_k is unrestricted in sign, we write

$$x_k = x_k^+ - x_k^-, \tag{5.35}$$

where x_k^+ and x_k^- are two new variables and $x_k^+, x_k^- \geq 0$. Even though both of these are non-negative, the difference of Equation (5.35) can be negative if $x_k^+ < x_k^-$. This step adds one new variable to the problem for each unrestricted variable.

4. If there are inequality constraints, we must convert these to equalities. As noted in Section 5.3.3 above, this is achieved by adding slack or excess variables. For example, the kth constraint equation

$$a_{k1}x_1 + a_{k2}x_2 + \cdots + a_{kn}x_n \leq b_k$$

becomes

$$a_{k1}x_1 + a_{k2}x_2 + \cdots + a_{kn}x_n + x_{n+1} = b_k,$$

where x_{n+1} is a slack variable, and

$$a_{k1}x_1 + a_{k2}x_2 + \cdots + a_{kn}x_n \geq h_k$$

becomes

$$a_{k1}x_1 + a_{k2}x_2 + \cdots + a_{kn}x_n - x_{n+1} = b_k,$$

where x_{n+1} is an excess variable. In both cases $x_{n+1} \geq 0$. We have used n as subscript for the number of variables in Equations (5.33), but if there were h inequality constraints in the original problem, this would increase to $n + h$.

First, we need to describe the simplex method. This consists of finding a vertex of the feasible space, for instance in Figure 5.4. This figure has been reproduced in Figure 5.12 to identify the vertices where they have been numbered. Once a vertex has been identified, the solution is improved until the objective function is minimized. Let us use Example 5.4 (also Figures 5.4 and 5.12) to explore these ideas further. In standard form, we wish to

Min $(z = 2x_1 + x_2)$

subject to Equations (5.21):

$$5x_1 + 10x_2 - x_3 = 8,$$
$$4x_1 + 5x_2 + x_4 = 8 \quad \text{and}$$
$$4x_1 + x_2 - x_5 = 2. \tag{5.21}$$

Note that the x_i are all ≥ 0, and so are the stipulations. It is natural in this type of problem for this to be the case since the physical quantities have to be positive. No doubt, this played

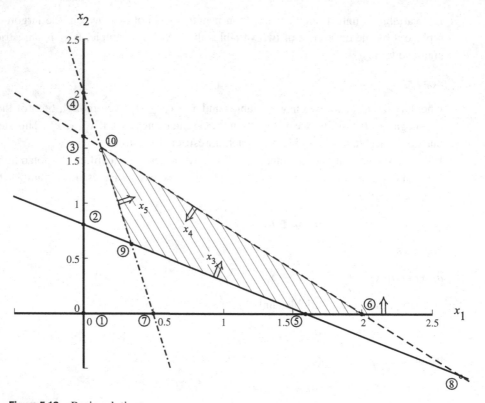

Figure 5.12   Basic solutions

a rôle in the development of the convention of non-negativity, but many problems of interest can have negative values which have to be dealt with as described above.

In this problem we have five variables and three constraint equations. We explore first the determination of the vertices of the problem space, enumerating these one by one; this becomes far too lengthy a process in real problems and hence the need for the simplex method. The points of interest are shown in Figure 5.12 – we know from the solution in Example 5.7 above that point 9 is the optimal solution. The points are determined by setting two of the five variables equal to zero and solving for the remaining three. The variables not set equal to zero are the **basic** variables.

The number of cases to be considered is

$$\frac{5!}{3!2!} = 10,$$

which are summarized in Table 5.10. In general, with n variables and m constraint equations, there will be

$$\frac{n!}{m!\,(n-m)!} \tag{5.36}$$

cases to consider. This number rapidly becomes exorbitantly large for large systems – hence

Table 5.10 *Ten vertices in Example 5.4; points refer to Figure 5.12; f=feasible; nf=not feasible; [0] indicates non-basic variable*

Point→	1	2	3	4	5	6	7	8	9	10
x_1	[0]	[0]	[0]	[0]	1.6	2	0.5	2.667	0.343	0.125
x_2	[0]	0.8	1.6	2	[0]	[0]	[0]	−0.533	0.629	1.5
x_3	−8	[0]	8	12	[0]	2	−5.5	[0]	[0]	7.625
x_4	8	4	[0]	−2	1.6	[0]	6	[0]	3.486	[0]
x_5	−2	−1.2	−0.4	[0]	4.4	6	[0]	8.133	[0]	[0]
Nature	nf	nf	nf	nf	f	f	nf	nf	f	f

the need to shorten the procedure. Let us nevertheless complete the exercise for our example, as this assists greatly in explaining the simplex method. We take two examples, points 1 and 6; the rest are summarized in Table 5.10. The shaded area only in Figure 5.12 is feasible since only there are the constraints and non-negativity satisfied.

- Point 1, column 1 in Table 5.10. With $x_1 = x_2 = 0$, we can solve the three Equations (5.21) to find the values of the basic variables: $x_3 = -8$, $x_4 = 8$, $x_5 = -2$. The condition of non-negativity is violated which is confirmed by reference to Figure 5.12.
- Point 6, column 6 in Table 5.10. Here $x_2 = x_4 = 0$. The basis is $\{x_1, x_3, x_5\}$. We substitute in Equations (5.21) to find

$$5x_1 - x_3 = 8,$$
$$4x_1 = 8 \quad \text{and}$$
$$4x_1 - x_5 = 2.$$

Then $x_1 = 2$, $x_3 = 2$ and $x_5 = 6$. The point is therefore feasible since non-negativity is satisfied. This again can be checked by reference to Figure 5.12.

As indicated the **basis** is the collection of variables not set equal to zero in each case. For point 6 the basis is, as noted, $\{x_1, x_3, x_5\}$, and for case 10 it is $\{x_1, x_2, x_3\}$. Each of the ten solutions in Table 5.10 is a basic solution. A **feasible** solution satisfies the constraint and non-negativity conditions. Points in the shaded area of Figure 5.12 are feasible. **Basic feasible** solutions satisfy the constraint and non-negativity conditions in addition to being basic; they correspond to the vertices of the feasible space. Cases 5, 6, 9 and 10 are basic and feasible (we know that 9 is optimal). It is convenient to deal with Phase II of the simplex method first – in this it is assumed that a basic feasible starting point is available. Phase I is concerned with finding this starting point.

Phase II of simplex algorithm

We need, in accordance with the preceding two sentences, to have a basic feasible starting point. Let this be point 6 above, for which $x_2 = x_4 = 0$. Here, we deduce the required equations, but Phase I – to be introduced below – will do this for us automatically. The required equations are

(5.37) below; these can be deduced from Equations (5.21), also reproduced as (i) to (iii) below for convenience. The basis is $\{x_1, x_3, x_5\}$; for ease of solution we like to have only one of these variables to appear in each equation. We therefore start with the second equation (ii) below. This is repeated, after division by 4, as the first line of Equation (5.37). It is the **canonical** equation for x_1; since $x_2 = x_4 = 0$, we can solve $x_1 = 2$.

In the second line of Equation (5.37) we have started with equation (i), multiplied by (-1), and then added equation (ii) after having multiplied this by 5/4. This eliminates x_1 (we want the basic variables to appear in one equation only), and x_3, another basic variable, appears with a coefficient of 1, as required. Noting again that $x_2 = x_4 = 0$, we can solve $x_3 = 2$. The last equation in (5.37) is deduced by taking equation (iii), subtracting equation (ii), and then multiplying through by (-1), to ensure a non-negative stipulation. This is again in canonical form, and with $x_2 = x_4 = 0$, we solve $x_5 = 6$. All of this is in agreement with Table 5.10, point 6.

$$5x_1 + 10x_2 - x_3 = 8 \qquad \text{(i)}$$

$$4x_1 + 5x_2 + x_4 = 8 \qquad \text{(ii)}$$

$$4x_1 + x_2 - x_5 = 2 \qquad \text{(iii)}$$

$$
\begin{aligned}
x_1 &+ 1.25x_2 && + 0.25x_4 && = 2 \\
&- 3.75x_2 &+ x_3 &+ 1.25x_4 && = 2 \\
&4x_2 && + x_4 &+ x_5 &= 6
\end{aligned}
\qquad \text{(5.37)}
$$

Note the structure of Equations (5.37). The basic variables appear only once, each in a single equation, with a coefficient of unity.

The general canonical form is

$$
\begin{aligned}
x_1 && + a'_{1,m+1}x_{m+1} + & \cdots & + a'_{1,n}x_n &= b'_1 \\
& x_2 & + a'_{2,m+1}x_{m+1} + & \cdots & + a'_{2,n}x_n &= b'_2 \\
& \ddots & \vdots & & \vdots & \\
&& x_m + a'_{m,m+1}x_{m+1} + & \cdots & + a'_{m,n}x_n &= b'_m.
\end{aligned}
\qquad \text{(5.38)}
$$

Here, the basis is assumed to be $\{x_1, \ldots, x_m\}$ but it could be any set of m variables from amongst the n. There are $(n - m)$ non-basic variables. The primes in Equations (5.38) refer to the fact that the original values in Equations (5.33) have been manipulated to eliminate variables and to reduce the coefficients of the basis to unity, as we did in deducing Equations (5.37). The b'_i are termed the **canonical stipulations**.

We now investigate this first basic feasible point – as given in (5.37) with $x_1 = 2, x_3 = 2, x_5 = 6, x_2 = x_4 = 0$ – for optimality. (We actually know it isn't since it corresponds to point 6 in Figure 5.12, but we will not generally know this.) To do this we write the objective function in terms of the non-basic variables:

$$
\begin{aligned}
z &= 2x_1 + x_2 = 2(2 - 1.25x_2 - 0.25x_4) + x_2 \\
&= 4 - 1.5x_2 - 0.5x_4, \text{ or}
\end{aligned}
$$

$$z - 4 = -1.5x_2 - 0.5x_4. \qquad \text{(5.39)}$$

In deducing this we have used the first of the set (5.37) to eliminate the basic variable x_1. Equation (5.39) represents the objective function in **canonical form**. The general form corresponding to Equation (5.39) is

$$z - z_0 = c'_{m+1}x_{m+1} + \cdots + c'_n x_n, \tag{5.40}$$

written in terms of non-basic variables; z_0 is a constant. The current solution is then $z = z_0$; this has to be tested for optimality.

The solution of Equation (5.39) corresponding to the assumed basis $\{x_1, x_3, x_5\}$ is

$$z - 4 = 0, \text{ or } z = 4, \text{ with } z_0 = 4 \tag{5.41}$$

since $x_2 = x_4 = 0$. Now the key point is: would the act of increasing x_2 or x_4 from zero lead to a lower value of z? We look at Equation (5.39): both the coefficients of x_2 and x_4 are negative. Increasing either will reduce z. In the simplex method, we increase only one variable at a time. We choose x_2 since it has the most negative coefficient $(= -1.5)$ as compared to (-0.5) for x_4.

How much can we increase x_2? We can determine this by inspecting Equations (5.37). We keep $x_4 = 0$ and therefore exclude it from the analysis. The first of Equations (5.37) can be written

$$x_1 + 1.25x_2 = 2 \text{ or } x_1 = -1.25x_2 + 2. \tag{5.42}$$

From this we can see that increasing x_2 will reduce x_1, and indeed when $x_2 = (2/1.25) = 8/5 = 1.6$, x_1 will have been reduced to zero. Any further reduction will result in x_1 becoming negative, thus violating the non-negativity condition for all variables x_j. We repeat the process just followed for the second and third equations of (5.37), which are

$$-3.75x_2 + x_3 = 2 \text{ or } x_3 = 3.75x_2 + 2, \text{ and} \tag{5.43}$$

$$4x_2 + x_5 = 6 \text{ or } x_5 = -4x_2 + 6. \tag{5.44}$$

In the first of these, (5.43), the fact that the coefficient $a'_{22} = -3.75$ is negative results in the situation that x_2 can be increased indefinitely without making x_3 become negative. In the second, (5.44), the increase in x_2 must be $\leq 6/4 = 1.5$, at which point x_5 becomes equal to zero.

Of the three conditions just investigated, the first states that $x_2 \leq 1.6$, the second that x_2 can be increased without limit, and the third that $x_2 \leq 1.5$. The third governs; if implemented, all three conditions of non-negativity will be satisfied. Further, we increase x_2 to the limit of 1.5; then $x_5 = 0$. Thus x_5 leaves the basis and x_2 enters it. The key in the procedure was to find the minimum ratio of b'_i to a'_{ir} provided $a'_{ir} > 0$. In the present example, this ratio was $b'_3/a'_{32} = 6/4$. The index r equals the column number of the incoming basic variable.

The rule for changing the basis is summarized as follows.

> Choose the most negative of the canonical cost coefficients, say c'_r;
> then r denotes the column of the incoming basic variable;
> then choose the smallest $\frac{b'_i}{a'_{ir}}$ provided $a'_{ir} > 0$;
> a'_{ir} is then the pivotal element for bringing x_r into the basis.

We now investigate the second vertex of the feasible space. We have to rewrite Equations (5.37) in canonical form corresponding to the new basis $\{x_1, x_2, x_3\}$. The third equation of this set is ideal for the canonical form since it contains one basic variable (x_2), and the two non-basic variables $(x_4$ and $x_5)$. We divide the equation by 4 to reduce the coefficient of x_2 to unity. Then we use this equation to eliminate x_2 from the first two. The coefficient a'_{32} is the pivotal element in this case. We repeat for convenience below Equations (5.37) and underneath them the new set of canonical equations (5.45) with the new basis $\{x_1, x_2, x_3\}$.

$$
\begin{aligned}
x_1 \quad +1.25x_2 \qquad\qquad +0.25x_4 \qquad\qquad &= 2 \\
-3.75x_2 \quad +x_3 \quad +1.25x_4 \qquad\qquad &= 2 \\
\boxed{4x_2} \qquad\qquad +x_4 \quad +x_5 &= 6
\end{aligned}
$$

$$
\begin{aligned}
x_1 \qquad\qquad -0.0625x_4 \quad -0.3125x_5 &= 0.125 \\
+x_3 \quad +2.1825x_4 \quad +0.9375x_5 &= 7.625 \\
x_2 \qquad +0.25x_4 \qquad +0.25x_5 &= 1.5.
\end{aligned} \tag{5.45}
$$

The objective function must also be written in canonical form:

$$
z - 1.75 = -0.125x_4 + 0.375x_5. \tag{5.46}
$$

This is deduced by eliminating x_2 from Equation (5.39) using the third equation containing the pivotal element in the set (5.45).

We summarize the steps so far and complete the solution in the **simplex tableau** of Table 5.11. The second solution corresponds to point 10 in Figure 5.12 and Table 5.10, with $x_1 = 0.125$, $x_2 = 7.625$ and $x_3 = 1.5$. In the objective function, x_2 has also been eliminated using the third constraint equation. There is only one negative canonical cost coefficient, $c'_4 = -0.125$. Hence $r = 4$, and inspecting the fourth column, we see that the smallest b'_i/a'_{i4} ratio is

Table 5.11 *Simplex tableau; $\boxed{\cdot}$ denotes pivotal element*

b	x_1	x_2	x_3	x_4	x_5	Basis
2	1	1.25		0.25		x_1
2		−3.75	1	1.25		x_3
6		$\boxed{4}$		1	1	x_5
$z-4$		−1.5		−0.5		
0.125	1			−0.0625	−0.3125	x_1
7.625			1	$\boxed{2.1875}$	0.9375	x_3
1.5		1		0.25	0.25	x_2
$z-1.75$				−0.125	0.375	
0.3429	1		0.0625		−0.2857	x_1
3.4857			0.4571	1	0.4286	x_4
0.6286		1	−0.1143		0.1429	x_2
$z-1.3143$			0.0571		0.4286	

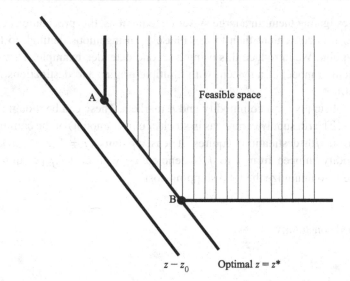

Feasible space

A

B

$z - z_0$ Optimal $z = z^*$

Figure 5.13   Multiple solutions; all points on the true joining A and B optimal

$7.625/2.1825 \simeq 3.5$ with $i = 2$. Recall that the a'_{ir} must be positive. Therefore a'_{24} is the pivotal element, with x_4 entering the basis and x_3 leaving. The same procedure is followed as before with the result given in Table 5.11 where the problem is completed. The result is

$$x_1^* = 0.3429, \ x_2^* = 0.6286, \ x_3^* = 0, \ x_4^* = 3.4857, \ x_5^* = 0, \ \text{with } z^* = 1.3143;$$

this solution is that given in Equations (5.30) corresponding to point 9 in Figure 5.12.

The method described above represents Phase II of the simplex method. We emphasize the optimality criterion.

> A basic feasible solution is minimal (i.e. minimum $z = z^*$)
> if all canonical cost coefficients are non-negative.

An *LP* can have

- a finite unique optimal solution, as in the example above. This occurs if all canonical cost coefficients are positive (no zeros).
- multiple optimal solutions, if one or more of the canonical cost coefficients is zero. One can think of the z-plane as being parallel to an edge of the feasible space, as illustrated in two dimensions in Figure 5.13. Here there are an infinity of solutions.
- an unbounded solution, or
- no solution.

Phase I of simplex algorithm

We assumed in Phase II above that a basic feasible starting point, a vertex of the convex space of feasible solutions, was available. We showed how to deduce it using an example, enumerating all the vertices and investigating them. In a large system of equations, this procedure becomes difficult or impossible. It is convenient then to use Phase I, an ingenious method, to find a basic feasible starting point. We introduce this using a second extended example. This is the **transportation problem**, a model of a system with multiple origins and destinations in the marketing of a product.

There are m centres of supply for a commodity and n markets. These can be thought of as origins and destinations. The ith supply centre has in stock s_i units (amount) of the commodity $(i = 1, 2, \ldots, m)$ and the jth destination requires at least r_j units $(j = 1, 2, \ldots, n)$. The amount of the commodity shipped from i to j is denoted x_{ij} with cost c_{ij} per unit. The purpose of the exercise is to minimize the total shipping cost

$$\sum_j \sum_i x_{ij} c_{ij} \tag{5.47}$$

subject to the prescribed constraints

$$\sum_j x_{ij} \le s_i,$$
$$\sum_i x_{ij} \ge r_j. \tag{5.48}$$

We apply this model using the following data:

Table 5.12   *Unit costs c_{ij}*

		j	
		1	2
	1	5	3
i	2	6	2
	3	1	8

The following constants are to be used in the constraint equations:

$s_1 = 5, \ s_2 = 7, \ s_3 = 11$ and

$r_1 = 12, \ r_2 = 3.$

The formulation of the problem using the data given is as follows.

$$\text{Min } (z = 5x_{11} + 3x_{12} + 6x_{21} + 2x_{22} + x_{31} + 8x_{32}) \tag{5.49}$$

subject to

$$x_{11} + x_{12} \le 5,$$
$$x_{21} + x_{22} \le 7,$$
$$x_{31} + x_{32} \le 11,$$
$$x_{11} + x_{21} + x_{31} \ge 12 \quad \text{and}$$
$$x_{12} + x_{22} + x_{32} \ge 3. \tag{5.50}$$

All x_{ij} are ≥ 0.

In standard form the constraints become

$$x_{11} + x_{12} \qquad\qquad\qquad\qquad + x_{s1} \qquad\qquad\qquad = 5,$$
$$x_{21} + x_{22} \qquad\qquad\qquad\qquad + x_{s2} \qquad\qquad = 7,$$
$$x_{31} + x_{32} \qquad\qquad\qquad + x_{s3} \qquad = 11,$$
$$x_{11} \qquad + x_{21} \qquad + x_{31} \qquad\qquad\qquad\qquad - x_{e1} \qquad = 12 \text{ and}$$
$$x_{12} \qquad + x_{22} \qquad + x_{32} \qquad\qquad\qquad\qquad - x_{e2} = 3. \tag{5.51}$$

Slack variables are denoted x_{s1}, x_{s2} and x_{s3} while excess ones are x_{e1} and x_{e2}. Inspection of these equations reveals that the slack variables provide a convenient initial basis; they automatically have a coefficient of $(+1)$, only one appears per equation, and they do not appear in the objective function. Slack variables are a gift! But what do we do about the last two equations? Excess variables have a coefficient of (-1); multiplying the equation through by (-1) does not help since this would make the stipulation negative.

Rather than attempting elimination as we did to set up the Phase II example above, we introduce the principal idea of Phase I. We simply add variables, called **artificial variables**, to the problem equations. Then the last two of the set (5.51) become

$$x_{11} \qquad + x_{21} \qquad + x_{31} \qquad - x_{e1} \qquad + x_{a1} \qquad = 12 \text{ and}$$
$$x_{12} \qquad + x_{22} \qquad + x_{32} \qquad - x_{e2} \qquad + x_{a2} = 3, \tag{5.52}$$

where x_{a1} and x_{a2} are the artificial variables, both ≥ 0. We have changed the original problem by adding these additional variables, and have enlarged the original convex space. Our purpose now is to eliminate them as soon as possible! If we do this and maintain feasibility, we hope soon to end up on a basic feasible vertex of the original problem.

Our means of eliminating x_{a1} and x_{a2} is to introduce a temporary objective function as follows:

$$w = x_{a1} + x_{a2}. \tag{5.53}$$

This can be written as

$$w = x_{a1} + x_{a2}$$
$$= 15 - x_{11} - x_{12} - x_{21} - x_{22} - x_{31} - x_{32} + x_{e1} + x_{e2}, \tag{5.54}$$

using Equations (5.52). We wish to reduce w of Equation (5.53) to zero; then $x_{a1} = x_{a2} = 0$, and we can return to the original objective function. In general the temporary objective function might be

$$w = x_{n+1} + x_{n+2} + \cdots + x_{n+\ell}, \tag{5.55}$$

for ℓ artificial variables.

We now proceed as shown in Tables 5.13(a) and (b). The temporary objective function is used in determining the incoming basic variable; it dictates the criterion for optimization until it reaches zero. During this Phase I period, the original (z) objective function is adjusted to reflect the current basis, any incoming basic variables being eliminated from the objective

Table 5.13(a)   *Solution to transportation problem; reduction of w to zero*

b	x_{11}	x_{12}	x_{21}	x_{22}	x_{31}	x_{32}	x_{s1}	x_{s2}	x_{s3}	x_{e1}	x_{e2}	x_{a1}	x_{a2}	Basis
5	1	1					1							x_{s1}
7			1	1				1						x_{s2}
11					1	1			1					x_{s3}
12	1		1		1					−1		1		x_{a1}
3		1		1		[1]					−1		1	x_{a2}
$z-0$	5	3	6	2	1	8								
$w-15$	−1	−1	−1	−1	−1	−1				1	1			
5	[1]	1					1							x_{s1}
7			1	1				1						x_{s2}
8		−1		−1	1				1		1		−1	x_{s3}
12	1		1		1					−1		1		x_{a1}
3		1		1		1					−1		1	x_{32}
$z-24$	5	−5	6	−6	1						8		−8	
$w-12$	−1		−1		−1					1			1	
5	1	1					1							x_{11}
7			1	1				1						x_{s2}
8		−1		−1	1				1		1		−1	x_{s3}
7		−1	1		[1]		−1			−1		1		x_{a1}
3		1		1		1					−1		1	x_{32}
$z-49$		−10	6	−6	1		−5				8		−8	
$w-7$		1	−1		−1		1			1			1	
5	1	1					1							x_{11}
7			1	1				1						x_{s2}
1			−1	−1			1		1	1	1	−1	−1	x_{s3}
7		−1	1		1		−1			−1		1		x_{31}
3		[1]		1		1					−1		1	x_{32}
$z-56$		−9	5	−6	0		−4			1	8	−1	−8	
$w-0$		0	0	0		0		0			0		1	1	

function. But w is to be minimized until equal to zero, the purpose of which is solely to find a basic feasible starting point of the original problem. Then we focus on z. This process of reducing w to zero endures for three cycles of the simplex operation as in Table 5.13(a). At the step when $w = 0$, one then immediately reverts to z; the temporary objective function w is no longer of any relevance; nor are the artificial variables. The completion of the problem and minimization of z is shown in Table 5.13(b). The tables (a) and (b) can be combined if desired.

Table 5.13(b)   *Transportation problem continued; determination of optimal z*

b	x_{11}	x_{12}	x_{21}	x_{22}	x_{31}	x_{32}	x_{s1}	x_{s2}	x_{s3}	x_{e1}	x_{e2}	Basis
2	1		−1		−1	1					1	x_{11}
7			1	1				1				x_{s2}
1				−1	−1		$\boxed{1}$		1	1	1	x_{s3}
10				1	1	1	1	−1		−1	−1	x_{31}
3		1	5	1		1				−1		x_{12}
$z-29$				3		9	−4			1	−1	
1	1		1			−1				−1	−1	x_{11}
7			1	1				1				x_{s2}
1				−1	−1		1		1	1	1	x_{s1}
11					1	1		1				x_{31}
3		1		$\boxed{1}$		1					−1	x_{12}
$z-25$			1	−1		9			4	5	3	
1	1		1			−1				−1	−1	x_{11}
4		−1	1			−1		1			1	x_{s2}
4		1	−1		1	1	1		1	1		x_{s1}
11					1	1		1				x_{31}
3		1		1		1					−1	x_{22}
$z-22$		1	1			10			4	5	2	

Solution. $x_{11}^* = 1$ $x_{31}^* = 11$ $x_{22}^* = 3$
          $x_{s1}^* = 4$ $x_{s2}^* = 4$ Other variables $= 0$.
          $z^* = 22$

We have introduced the simplex method by example – since it is a practical method for use. Formal theorems are available in works such as Karlin (1959) and Jones (1980), two excellent references in this area.

5.4.3   Dual linear programs

For each linear program, called the **primal** problem

$$\text{Min} \left(z = \sum_{j=1}^{n} c_j x_j \right) \tag{5.56}$$

subject to

$$\sum_{j=1}^{n} a_{ij} x_j = [a_{ij}] x_j \geq b_i, \text{ for } i = 1, 2, \ldots, m, \tag{5.57}$$
$$x_j \geq 0, \text{ for } j = 1, 2, \ldots, n;$$

there is a **dual** linear problem

$$\text{Max}\left(w = \sum_{i=1}^{m} b_i y_i\right) \tag{5.58}$$

subject to

$$\sum_{i=1}^{m} a_{ij} y_i = \left[a_{ij}\right]^T y_i = \left[a_{ji}\right] y_i \le c_j \text{ for } j = 1, 2, \ldots, n, \tag{5.59}$$
$$y_i \ge 0 \text{ for } i = 1, 2, \ldots, m.$$

In (5.59), $[\cdot]^T$ indicates the transpose of the matrix. The striking relationship between these two problems is typical of dual relationships. 'Min' in the primal becomes 'Max' in the dual. The variable x_j is replaced by the dual y_i; cost coefficients and stipulations interchange; $\left[a_{ij}\right]$ is transposed. The dual of the dual problem is the primal; Exercise 5.12.

One of the problems that stimulated interest in *LP* is the diet problem (Jones, 1980). (This is a product-mix problem, similar to that in Example 5.4 and in Exercise 5.14.) In the primal problem the x_j are units of a particular type of food. The constraints specify the minimum amounts of nutrients – for example, vitamins – that must be supplied, governed by the b_i of Equations (5.57). The a_{ij} denote the amount of each nutrient (i) contained in a unit of the foods (j). The problem is to minimize the total cost as in Equation (5.56). This might be the approach taken by a cost-conscious hospital board. The dual problem of Equations (5.58) and (5.59) represents that of a drug company that wishes to substitute synthetic mixtures of nutrients that satisfy the minimal requirements, thus containing b_i units of nutrient i. The variable y_i now represents the price to be charged for a unit of nutrient i. The drug company wishes to maximize the profit given by the function (5.58). The constraints of (5.59) indicate that the cost of any mixture of nutrients, equivalent to food j, should be less than the cost of the food itself.

Now it turns out that the solutions to the primal and dual problems coincide! This means that the hospital board and the drug company can together satisfy minimum nutritional requirements and maximum profit. An interpretation in terms of 'shadow prices' is given in Exercise 5.14. Jones (1980) and Chvátal (1980) provide good outlines of the theory of duality in linear programming.

The theory behind this begins with the following result:

$$z = \sum_{j=1}^{n} c_j x_j \ge w = \sum_{i=1}^{m} b_i y_i. \tag{5.60}$$

This is shown by considering that

$$\sum_{i=1}^{m} b_i y_i \le \sum_{i=1}^{m} y_i \left(\sum_{j=1}^{n} a_{ij} x_j\right) \tag{5.61}$$

since

$$\sum_{j=1}^{n} a_{ij} x_j \ge b_i,$$

by inequality (5.57). Now inequality (5.61) can be written as

$$\sum_{j=1}^{n} x_j \left(\sum_{i=1}^{m} a_{ij} y_i \right) \leq \sum_{j=1}^{n} c_j x_j$$

by inequality (5.59). Thus inequality (5.60) is proved.

If we manage to find a primal feasible solution $\{x_1^*, x_2^*, \ldots, x_n^*\}$ together with a dual feasible solution $\{y_1^*, y_2^*, \ldots, y_m^*\}$ such that

$$\sum_{j=1}^{n} c_j x_j^* = \sum_{i=1}^{m} b_i y_i^*, \text{ or} \tag{5.62}$$

$$\text{Min } z = \text{Max } w, \tag{5.63}$$

then the two solutions are optimal, since $w \leq z$ by inequality (5.60).

The **duality theorem** turns this around. It states that, if we find an optimal solution $\{x_1^*, x_2^*, \ldots, x_n^*\}$ of the primal problem, then the dual problem has a solution $\{y_1^*, y_2^*, \ldots, y_m^*\}$ such that Equation (5.62) is satisfied. The solution to the primal *LP* gives the solution to the dual *LP* and *vice versa*; Min z =Max w.

Proof of duality theorem

We shall base the proof on the simplex method. It is then convenient to start with the dual linear program, since it is easy to deduce a canonical form from the dual constraints of Equation (5.59):

$$\sum_{i=1}^{m} a_{ij} y_i = \left[a_{ij} \right]^T y_i \leq c_j \text{ for } j = 1, 2, \ldots, n.$$

The useful thing to observe is that the inequalities ($\leq$) in this relationship imply that slack variables y_{m+j} will be used in the simplex method; thus

$$y_{m+j} = c_j - \sum_{i=1}^{m} a_{ij} y_i, \text{ for } j = 1, 2, \ldots, n, \tag{5.64}$$

and this gives an ideal basic feasible starting point. All of the slack variables enter the first basis. The other variables are non-basic and set equal to zero. The objective function contains only non-basic variables:

$$\text{Max} \left(w = \sum_{i=1}^{m} b_i y_i \right).$$

(The motto 'slacks are a gift' presented earlier has real meaning!) To use the simplex method as described above, we convert the objective function of (5.58) from 'Max' to 'Min' :

$$\text{Min } w' = -w = -\sum_{i=1}^{m} b_i y_i = \sum_{i=1}^{m} (-b_i) y_i. \tag{5.65}$$

We now consider the situation where we have completed the simplex operation, arriving at the optimal solution w'^*:

$$w' - w'^* = \sum_{h=1}^{m+n} \left(-\bar{b}_h \right) y_h. \tag{5.66}$$

In this equation, we have written the last row of the simplex operation; h is an index that ranges over both i and j, over all $(m + n)$ of the y_h, including the slack variables (5.64). The negative sign in $(-\bar{b}_h)$ in Equation (5.66) is included so as to be consistent with Equation (5.65). Now we know that, in Equation (5.66), $(-\bar{b}_h)$ is 0 whenever y_h is a basic variable and positive or zero when it is not, since $w' = w'^*$ is optimal. The y_h are zero when non-basic as demonstrated using the simplex method above. Since all $(-\bar{b}_h)$ are positive or zero they are non-negative, and consequently the $\bar{b}_h$ are non-positive.

We know also that at optimality

$$w'^* = \sum_{i=1}^{m} (-b_i)\, y_i^* \text{ or}$$

$$w^* = \sum_{i=1}^{m} b_i y_i^*. \tag{5.67}$$

The optimal x_j in the primal problem are given by

$$x_j^* = -\bar{b}_{m+j} \text{ for } j = 1, \ldots, n. \tag{5.68}$$

We shall show that $\{x_1^*, x_2^*, \ldots, x_n^*\}$, thus defined, constitutes a primal feasible solution satisfying (5.62) and therefore optimal. First we write (5.66) as

$$w - w^* = \sum_{h=1}^{m+n} \bar{b}_h y_h;$$

then substituting for w from (5.58):

$$\sum_{i=1}^{m} b_i y_i - w^* = \sum_{h=1}^{m+n} \bar{b}_h y_h = \sum_{h=1}^{m} \bar{b}_h y_h + \sum_{h=m+1}^{m+n} \bar{b}_h y_h.$$

We now substitute for the slack variables from Equations (5.64) and for $\bar{b}_{m+j}$ from (5.68):

$$\sum_{i=1}^{m} b_i y_i - w^* = \sum_{h=1}^{m} \bar{b}_h y_h + \sum_{j=1}^{n} (-x_j^*) \left(c_j - \sum_{i=1}^{m} a_{ij} y_i \right).$$

Then we can write

$$\sum_{i=1}^{m} b_i y_i = \left(w^* - \sum_{j=1}^{n} c_j x_j^* \right) + \sum_{i=1}^{m} \left(\bar{b}_i + \sum_{j=1}^{n} a_{ij} x_j^* \right) y_i. \tag{5.69}$$

Equation (5.69) has been derived from the definition of the objective function for w in terms of the y_i in Equation (5.58), i.e. $w = \sum_{i=1}^{m} b_i y_i$, and from the definition of the slack variables, (5.64), thus eliminating any slack variables from the equation. Equation (5.69) therefore relates generally the y_i and is valid for any $\{y_1, y_2, \ldots, y_m\}$. We can choose all the y_i to be zero; then

$$w^* = \sum_{j=1}^{n} c_j x_j^*. \tag{5.70}$$

This, together with Equation (5.67), shows Equation (5.62) to be correct with the association (5.68).

If we choose now $y_i = 1$, and all other y_k to be zero $(k \neq i)$, then we find that

$$b_i = \bar{b}_i + \sum_{i=1}^{m} a_{ij} x_j^*.$$

Since the $\bar{b}_i$ are non-positive, $\bar{b}_i \leq 0$, therefore

$$\sum_{i=1}^{m} a_{ij} x_j^* \geq b_i, \tag{5.71}$$

and with $\left(x_j^* = -\bar{b}_{m+j}\right)$, Equation (5.68),

$$x_j^* \geq 0. \tag{5.72}$$

These last two equations are the primal constraints.

5.4.4   Applications of linear programming

Linear programming has been applied in various optimization problems. These include product-mix problems, for example Exercise 5.14 and the diet problem discussed above in the second paragraph of Section 5.4.3. The transportation problem introduced in Section 5.4.2 with Phase I of the simplex method is another. Shortest and longest routes through a network can be obtained using linear programming, the latter of great use in scheduling and obtaining the critical path in construction projects of various kinds. Warehouse storage and activity analysis are areas where linear methods can be used. Mathematical plasticity in structural theory can be dealt with using linear programming with elegant methodology introduced notably by Munro (see Cohn and Maier, 1979, for example). And of course there is the solution of problems in game theory, the main focus here.

5.5   Back to games: mixed strategies

Games with saddle points have solutions with pure strategies. This was shown in the example in Section 5.2 (Table 5.2). We now return to the ideas under the section 'Conflict', introduced at the beginning of the chapter. An animal might choose to fight or to retreat on a case by case basis. The idea of a mixed strategy is to play the two strategies randomly, for example to fight 50% of the time, and to retreat 50% of the time. This is a simplification of reality, since the individual decisions are made depending on the protagonist's state of mind and judgement of the strength of the antagonist. Von Neumann showed that, if mixed strategies are permitted, then equilibrium can be reached.

Each player's problem in zero-sum games can be reduced to a linear program. Further, the players' mixed strategies are dual solutions of each other's. This is a beautiful result, and adds insights into conflict and the difference between this situation and the 'game against nature' of decision theory. In the latter interpretation of a game against nature we face an impartial antagonist, but in conflict our antagonist will choose strategies to outwit and defeat us. Estimation of the probabilities of our antagonist's strategies is profoundly different where there is intelligent conflict.

Let us consider a typical zero-sum game without a saddle point, that illustrated in Table 5.4. We denote the payoff value in general as V, from B to A according to convention. The game is repeated in Table 5.14, in which we have shown that A plays the strategies σ_1 and σ_2 with probabilities p and $(1 - p)$ respectively. At the same time, B plays τ_1 and τ_2 with probabilities

Table 5.14   *Game with mixed strategies*

			B	
			τ_1	τ_2
			q	$1-q$
A	σ_1	p	3	6
	σ_2	$1-p$	5	4

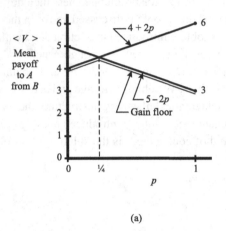 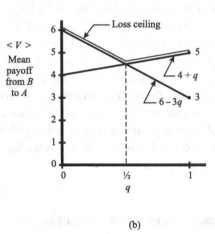

(a)   (b)

Figure 5.14   Gain floor and loss ceiling

q and $(1-q)$ respectively. We first consider the situation from A's point of view. He ponders what B will do. If she plays τ_1, the mean (expected) payoff to A is

$$\langle V \rangle = 3p + 5(1-p)$$
$$= 5 - 2p. \tag{5.73}$$

If B on the other hand plays τ_2, the mean is

$$\langle V \rangle = 6p + 4(1-p)$$
$$= 4 + 2p. \tag{5.74}$$

These are plotted in Figure 5.14(a). Player A can place a lower limit on his expected gains, shown as a **gain floor** in the figure. The greatest value of mean gain is at the intersection of the two lines, at

$$5 - 2p = 4 + 2p, \text{ or}$$
$$p = \frac{1}{4}. \tag{5.75}$$

Player A therefore adopts the strategy of playing σ_1 with probability 1/4, and σ_2 with probability 3/4, written $\{\sigma_1, 1/4; \sigma_2, 3/4\}$. This result is completely consistent with the maximin philosophy in maximizing the gain floor.

Player B undergoes a similar reasoning process and calculates the following mean losses. If A plays σ_1, her mean loss is

$$\langle V \rangle = 3q + 6(1 - q)$$
$$= 6 - 3q. \tag{5.76}$$

If A plays σ_2, this becomes

$$\langle V \rangle = 5q + 4(1 - q)$$
$$= 4 + q. \tag{5.77}$$

These are plotted in Figure 5.14(b). In this case B can construct a loss ceiling, shown in the figure. To minimize the loss, B calculates

$$6 - 3q = 4 + q, \text{ or}$$
$$q = \frac{1}{2}; \tag{5.78}$$

she adopts $\{\tau_1, 1/2; \tau_2, 1/2\}$. Again the philosophy is maximin, bearing in mind that B pays to A.

In the solution equilibrium has been reached; the value of the game is 4.5. This can be calculated from any of the four expected values above. The solution is based on mean values, averaged over many plays of the game, and constitutes a generalization of the saddle point of Equation (5.1). It brings inequality (5.2) into equality.

5.5.1 Generalization and example

We now generalize the results on mixed strategies to show that player A and B's solutions are dual linear programs. There is one point that must be introduced at the start so as to make the reasoning clear. We do this first with an example. Consider the payoff matrix (B to A):

$$\begin{bmatrix} 2 & 5 & 4 \\ 6 & 1 & 3 \\ 4 & 6 & 1 \end{bmatrix}. \tag{5.79}$$

As in the preceding section, player A evaluates the mean (expected) value of each strategy, assuming that B plays strategy τ_i. This must be evaluated using the **transpose** of matrix (5.79). In matrix (5.79), consider the case where A is considering the expected value given that B plays τ_2. The expected value is

$$(5) p_1 + (1) p_2 + (6) p_3.$$

This can be seen to be given by the second member in the vector

$$\begin{bmatrix} 2 & 6 & 4 \\ 5 & 1 & 6 \\ 4 & 3 & 1 \end{bmatrix} \begin{Bmatrix} p_1 \\ p_2 \\ p_3 \end{Bmatrix} = \begin{bmatrix} 2 & 5 & 4 \\ 6 & 1 & 3 \\ 4 & 6 & 1 \end{bmatrix}^T \begin{Bmatrix} p_1 \\ p_2 \\ p_3 \end{Bmatrix}. \tag{5.80}$$

The point of clarification is that we **transpose** the payoff matrix so as to write the theory in a manner that is compatible with our treatment of linear programming and duality above. We then define this transposed matrix as $[a_{ij}]$,

$$[a_{ij}] \equiv [\text{payoff matrix}]^T . \tag{5.81}$$

As a result,

$$[\text{payoff matrix}] = [a_{ij}]^T . \tag{5.82}$$

In the example given above,

$$[a_{ij}] = \begin{bmatrix} 2 & 6 & 4 \\ 5 & 1 & 6 \\ 4 & 3 & 1 \end{bmatrix} .$$

In general, we write a_{ij} as

$$[\text{payoff matrix}]^T = [a_{ij}] = \begin{bmatrix} a_{11} & a_{12} & \cdots & a_{1j} & \cdots & a_{1n} \\ a_{21} & a_{22} & \cdots & a_{2j} & \cdots & a_{2n} \\ \vdots & \vdots & & \vdots & & \vdots \\ a_{i1} & a_{i2} & \cdots & a_{ij} & \cdots & a_{in} \\ \vdots & \vdots & & \vdots & & \vdots \\ a_{m1} & a_{m2} & \cdots & a_{mj} & \cdots & a_{mn} \end{bmatrix} \tag{5.83}$$

and the payoff matrix is

$$[\text{payoff matrix}] = [a_{ij}]^T = [a_{ji}] = \begin{bmatrix} a_{11} & a_{21} & \cdots & a_{i1} & \cdots & a_{m1} \\ a_{12} & a_{22} & \cdots & a_{i2} & \cdots & a_{m2} \\ \vdots & \vdots & & \vdots & & \vdots \\ a_{1j} & a_{2j} & \cdots & a_{ij} & \cdots & a_{mj} \\ \vdots & \vdots & & \vdots & & \vdots \\ a_{1n} & a_{2n} & \cdots & a_{in} & \cdots & a_{mn} \end{bmatrix} . \tag{5.84}$$

This is the matrix given in a problem of game theory, by convention. For example, if A plays σ_3 and B plays τ_2, B pays a_{23} to A. We assume in the following that the values in the matrix a_{ij} are greater than or equal to zero. The reason for this is given below. To achieve this, one can add the same positive constant to all the values in a payoff matrix without altering the strategic outcome of the game.

Player A has n possible pure strategies,

$$\sigma_1, \sigma_2, \ldots, \sigma_j, \ldots, \sigma_n .$$

Similarly B has m pure strategies

$$\tau_1, \tau_2, \ldots, \tau_i, \ldots, \tau_m .$$

Player A's mixed strategies consist of

a p_1 probability of strategy σ_1,

a p_2 probability of strategy σ_2,

$\vdots$

a p_j probability of strategy σ_j,

$\vdots$

and a p_n probability of strategy σ_n.

We assume that both players are probabilistically coherent. Then for player A,

$$\sum_{j=1}^{n} p_j = 1. \tag{5.85}$$

Player B's mixed strategies are

a q_1 probability of strategy τ_1,

a q_2 probability of strategy τ_2,

$\vdots$

a q_i probability of strategy τ_i,

$\vdots$

and a q_m probability of strategy τ_m,

again with coherence

$$\sum_{i=1}^{m} q_i = 1. \tag{5.86}$$

As in the preceding section (Figure 5.14), player A evaluates the following values:

$$\langle V \rangle = a_{i1}p_1 + a_{i2}p_2 + \cdots + a_{ij}p_j + \cdots + a_{in}p_n \geq v, \text{ for } i = 1, 2, \ldots, m, \tag{5.87}$$

where V, as before, represents a payoff from matrix (5.83). In this, v is the **least** of the set of the m sums given in (5.87); it corresponds to the gain floor of Figure 5.14(a). Rather than consisting of a series of straight lines, there will now be a set of hyperplanes in $\{p_j\}$-space. For any set of probabilities $\{p_j\}$, there will always be a minimum value ($= v$) amongst these hyperplanes corresponding to the gain floor. Equations (5.87) may be written as

$$\sum_{j=1}^{n} a_{ij}p_j = [a_{ij}] p_j \geq v, \tag{5.88}$$

Player A has now weighted the columns of matrix (5.84) – the rows of matrix (5.83) – by the p_j, as required.

Now we introduce the new quantity x_j given by

$$x_j = \frac{p_j}{v}, \text{ with } j = 1, 2, \ldots, n. \tag{5.89}$$

Then the constraints (5.88) become

$$\sum_{j=1}^{n} a_{ij}x_j = [a_{ij}] x_j \geq 1 \text{ for } i = 1, 2, \ldots, m. \tag{5.90}$$

In deriving this equation, we have substituted $p_j = vx_j$ and cancelled the factor v from both sides of the result. Note that v has a fixed value for any set of values of $\{p_j\}$. We see now why the values a_{ij} were made to be ≥ 0; this is to ensure that $v > 0$, and consequently $x_j \geq 0$, required for the simplex method.

We wish to find the set of probabilities at the maximum value of v, the highest point in the gain floor. Substituting from Equation (5.89) into (5.85)

$$\sum_{j=1}^{n} x_j = \frac{1}{v};$$

(5.91)

then maximizing v implies

$$\text{Min} \sum_{j=1}^{n} x_j.$$

(5.92)

This equation, combined with the constraints (5.90), and the fact that all $x_j \geq 0$ (provided $v > 0$) shows that the determination of the optimal set $\{p_j\}$ can be achieved by the following *LP*:

$$\text{Min} \sum_{j=1}^{n} x_j \text{ subject to}$$
$$\sum_{j=1}^{n} a_{ij}x_j = [a_{ij}]x_j \geq 1 \text{ for } i = 1, 2, \ldots, m \text{ and}$$
$$x_j \geq 0, \text{ for } j = 1, 2, \ldots, n.$$

(5.93)

The striking similarity of relationships (5.93) and the pair (5.56) with (5.57) should be noted.

To ensure $v > 0$ in the above, we make all $a_{ij} \geq 0$ in the matrix $[a_{ij}]$. Then the inequalities (5.88) ensure that v is positive since it is the smallest value calculated from them. To make the entries $a_{ij} \geq 0$, one adds a constant value to each entry in the matrix $[a_{ij}]$; doing so does not alter the game strategically. Adding the same quantity to each payoff value results in all expected values increasing by the same added amount. The **relative** values of these mean values are precisely the same. One could if one wishes, after the analysis, subtract the constant from the payoff values.

We now consider player B. He wishes to find the optimal set $\{q_i\}$. The expected values for B are

$$\langle V \rangle = a_{1j}q_1 + a_{2j}q_2 + \cdots + a_{ij}q_i + \cdots + a_{mj}q_m \leq t, \text{ for } j = 1, 2, \ldots, n,$$

(5.94)

where t is the greatest of the sums – the loss ceiling of Figure 5.14. For example, using the payoff matrix (5.79), the following equations result:

$$\begin{bmatrix} 2 & 5 & 4 \\ 6 & 1 & 3 \\ 4 & 6 & 1 \end{bmatrix} \begin{Bmatrix} q_1 \\ q_2 \\ q_3 \end{Bmatrix} = [\text{payoff matrix}] \begin{Bmatrix} q_1 \\ q_2 \\ q_3 \end{Bmatrix}.$$

The payoff matrix is the **transpose** of $[a_{ij}]$. Thus the set of equations (5.94) is

$$\sum_{i=1}^{m} a_{ij}q_i = [a_{ij}]^T q_i = [a_{ji}]q_i \leq t, \text{ for } j = 1, 2, \ldots, n.$$

(5.95)

Again we consider a transformation

$$y_i = \frac{q_i}{t}, \text{ with } i = 1, 2, \ldots, m.\tag{5.96}$$

Substituting into Equation (5.95) results in

$$\sum_{i=1}^{m} a_{ij} y_i = \left[a_{ji}\right] y_i \leq 1, \text{ for } j = 1, 2, \ldots, n,$$

and together with coherence, $\sum_{i=1}^{m} q_i = 1$, gives

$$\sum_{i=1}^{m} y_i = \frac{1}{t}.$$

Since we wish to minimize t, the equations just derived constitute the following linear program.

$$\boxed{\begin{array}{c} \text{Max } \sum_{i=1}^{m} y_i \text{ subject to} \\ \sum_{i=1}^{m} a_{ij} y_i = \left[a_{ji}\right] y_i \leq 1, \text{ for } j = 1, 2, \ldots, n, \text{ and} \\ y_i \geq 0, \text{ for } i = 1, 2, \ldots, m. \end{array}}\tag{5.97}$$

This is the dual *LP* to that given in Equations (5.93). Relationships (5.97) are similar in form to the pair (5.58) and (5.59).

We arrive at the interesting result that

Player B's problem is the dual of player A's.

Example 5.8 *Match-penny.* Player A chooses heads or tails on his coin, while B makes a choice on hers. Neither player knows the other's choice. If the coins match, player B receives a payoff of unity from A. If the coins do not match, A receives a payoff of unity from B.

The payoff matrix is

		B	
		τ_1	τ_2
A	σ_1	-1	1
	σ_2	1	-1

First add unity to each entry:

		B	
		τ_1	τ_2
A	σ_1	0	2
	σ_2	2	0

Consider player B.

$$2q_2 \leq t$$
$$2q_1 \leq t.$$

Substitute $y_i = q_i/t$:

Max $(y_1 + y_2)$

subject to

$$2y_2 \leq 1$$
$$2y_1 \leq 1$$

with both $y_i \geq 0$. The solution is

$$q_1 = q_2 = \frac{1}{2};$$

the game is symmetrical so that $p_1 = p_2 = 1/2$.

Example 5.9 Find the solution to the matrix game

$$\begin{bmatrix} -2 & 1 & 0 \\ 2 & -3 & -1 \\ 0 & 2 & -3 \end{bmatrix}. \tag{5.98}$$

First, we make the entries positive by adding a positive constant, say 4, so as to ensure that the values of v and t are > 0. Then

$$\begin{bmatrix} 2 & 5 & 4 \\ 6 & 1 & 3 \\ 4 & 6 & 1 \end{bmatrix} = [a_{ji}].$$

(This was used as an example in introducing the section above.) Then

$$a_{ij} = [\text{payoff matrix}]^T = \begin{bmatrix} 2 & 6 & 4 \\ 5 & 1 & 6 \\ 4 & 3 & 1 \end{bmatrix}.$$

The mean values to be considered by A are

$$[a_{ij}] p_j = \begin{bmatrix} 2 & 6 & 4 \\ 5 & 1 & 6 \\ 4 & 3 & 1 \end{bmatrix} \begin{Bmatrix} p_1 \\ p_2 \\ p_3 \end{Bmatrix} \geq v.$$

Following the procedure given above, we then arrive at the following.

Min $(x_1 + x_2 + x_3)$ \tag{5.99}

subject to

$$
\begin{array}{llll}
2x_1 & +6x_2 & +4x_3 & \geq 1 \\
5x_1 & +x_2 & +6x_3 & \geq 1 \\
4x_1 & +3x_2 & +x_3 & \geq 1,
\end{array}
$$

with all $x_j > 0$ (since $v > 0$).

The dual LP (B's solution) is

$$\text{Max } (y_1 + y_2 + y_3) \tag{5.100}$$

subject to

$$
[a_{ji}]\, y_i =
\begin{array}{llll}
2y_1 & +5y_2 & +4y_3 & \leq 1 \\
6y_1 & +y_2 & +3y_3 & \leq 1 \\
4y_1 & +6y_2 & +y_3 & \leq 1,
\end{array}
\tag{5.101}
$$

with all $y_i > 0$.

The solution to the LPs yields the following optimal strategies:

$$\text{A}: \left\{ \sigma_1, \frac{21}{35}; \sigma_2, \frac{13}{35}; \sigma_3, \frac{1}{35} \right\} \text{ and}$$

$$\text{B}: \left\{ \tau_1, \frac{13}{35}; \tau_2, \frac{12}{35}; \tau_3, \frac{2}{7} \right\}, \tag{5.102}$$

with a value of $(124/35)$.

5.6   Discussion; design as a game against nature

The key aspect introduced by game theory is the question of conflict. This gives an entirely different complexion to the estimation of probabilities. We have considered two-person zero-sum games where the solution is based on maximin philosophy, on the assumption that one's opponent is out to win at all costs. This represents a ruthless competitor and brings out the essential aspects of the solution to the assignment of probabilities under these conditions. Many other solutions exist including those for n-person games, bimatrix games, cooperative games and bargaining models. One area of application of game theory is war, and in economics where human behaviour is important. Another very interesting recent application is to the modelling of the behaviour of genes. Once again duality appeared – now in a different but related way to that of the preceding chapter.

Various simplifications were introduced. Mean values were used when considering the value of the various strategies. This is appropriate in repeated games where the consequences are not large; see Chapter 4 on risk aversion. The analysis could be extended to consider utilities, and the utility functions could be different for the two players. In reality, the assessment of probabilities under conditions of conflict would involve a behavioural assessment of one's antagonist, and one's own mental state at the point in time. There might be no similarity in the players' behaviour assumed in the above; one player might be a lot stronger than the other. The idea of 'mental roulette wheels' in which decisions are made automatically according to a set of probabilities is clearly a simplification of reality (as are all models). Yet the important ideas of conflict are captured in the analysis. The difference from decision theory lies in the assessment of the opponent's probabilities. This reflects the attitude taken in situations of conflict. The central message to learn is that there is a different way to think about assessment of probabilities in these situations.

We have posed questions of design as problems of decision. Another way to look at this is to consider design as a game against nature (Turkstra, 1970). In this approach, one considers a set of players, a set of moves, a set of outcomes (endpoints) and a set of values (utilities) attached to each outcome. The players here are the designer and nature; the designer chooses the physical system and details (for instance structural form, dimensions,...) while nature chooses factors such as climate, loadings, good or bad workmanship,...; in the endpoints the system might function as intended or not, with associated utilities. This is the same model as decision theory would indicate: one has to suppose the impartiality of nature. There is no room for maximin (pessimist) philosophies in this idealization. Yet, when we consider that our interventions can change nature – large dams being a pertinent example – we see the need to think about how these changing conditions should be modelled in our decision theory.

5.7   Exercises

5.1 Consider any game of your choice. Ponder the set of possible strategies for the players and the normal form that arises as well as the corresponding payoff matrix.

5.2 Solve the following optimization problem, in which v represents the volume of a tank.

$$\text{Max}\,(v = x_1 x_2 x_3)$$

subject to

$$x_1 + x_2 + 2x_3 = b \text{ and}$$
$$x_1, x_2, x_3 \geq 0.$$

5.3 A closed right cylindrical pressure vessel of volume v is to be constructed so as to have minimum surface area a. Find the radius r and length ℓ that achieve this. Discuss how one might investigate the character of the stationary point found.

5.4 Solve the following optimization problem, in which H represents the entropy. There is a countable set of states x_i, $i = 0, 1, 2, \ldots$, of the random quantity X with associated probabilities p_i.

$$\text{Max}\left(H = -\sum_{i=1}^{n} p_i \ln p_i\right)$$

subject to coherence

$$\sum_{i=0}^{\infty} p_i = 1,$$

and

$$\sum_{i=0}^{\infty} p_i x_i = \langle X \rangle = \mu.$$

5.5 Sketch the feasible region and stationary points of Example 5.6. Investigate and determine the minima of Equation (5.22).

5.6 Solve by the simplex method and check the answer graphically:

$$\text{Min } z = -10x_1 + 4x_2$$

subject to $2x_1 + x_2 \leq 9$

$$x_1 - 2x_2 \leq 2$$

$$-3x_1 + 2x_2 \leq 1$$

$$x_1, x_2 \geq 0.$$

5.7 Solve by the simplex method:

$$\text{Max } z' = -3x_1 - 6x_2 + 3x_3 - 3x_4$$

subject to $x_1 - x_2 + x_3 \leq 5$

$$x_1 + x_2 + 2x_4 \leq 6$$

$$x_2 - 2x_3 + x_4 \leq 3$$

$$-x_1 + 2x_2 + x_3 \leq 2$$

$$x_1, x_4 \geq 0$$

$$x_2, x_3 \; : \; \text{unrestricted in sign.}$$

5.8 Solve by the simplex method:

$$\text{Max } z' = -2x_1 + 4x_2 - 2x_3$$

subject to $3x_1 - x_2 + x_3 \leq 4$

$$x_1 - x_2 + x_3 \leq 2$$

$$-2x_1 + x_2 - x_3 \leq 4$$

$$x_1, x_2, x_3 \geq 0.$$

5.9 Solve by the simplex method:

$$\text{Min } z = -x_1 - x_2 - 2x_3$$

subject to $4 \geq x_1 \geq 1$

$$3x_2 - 2x_3 = 6$$

$$2 \geq x_3 \geq -1$$

$$x_2 \geq 0.$$

5.10 Check the answers to Problems 5.6 to 5.9 by using computer software.

5.11 *Structural systems.*

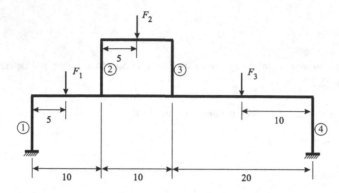

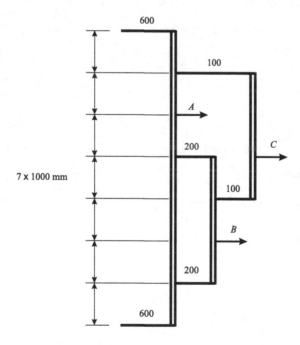

(b)

(a) An engineer is considering the structural system in Figure (a) on which the loads F_1, F_2 and F_3 represent storage loads which will be imposed for a short duration. The objective is to maximize the total storage subject to the capacity of the columns. These correspond to buckling loads of

200 tonnes for column 1, 100 tonnes for columns 2 and 3, and 150 tonnes for column 4. Treat all beams as being simply supported at their ends.

(b) The structural system in Figure (b) is comprised of horizontal tension members (for example wires) supported by vertical beams. The loads A, B and C are applied as shown in the figure, together with the capacities of the tension members. Assume that the beams can support the loads. What is the maximum load $(A + B + C)$ that can be supported?

5.12 Show that the dual of the dual problem in *LP* is the primal problem.

5.13 *Complementary slackness theorem.* There is a natural relationship between the excess or slack variables in either the primal or dual program and the decision variables in the other. Whenever an excess or slack variable is not equal to zero (at optimality), the associated decision variable in the dual program is equal to zero. When a decision variable is greater than zero, the associated excess or slack variable is zero. The solution is such that

$$\sum_{j=1}^{n} a_{ij} x_j^* = b_i \text{ or } y_i^* = 0 \text{ or both, for } i = 1, 2, \ldots, m;$$

$$\sum_{i=1}^{m} a_{ij} y_i^* = c_j \text{ or } x_j^* = 0 \text{ or both, for } j = 1, 2, \ldots, n.$$

Introduce the slack and excess variables

$$x_{n+i} = -b_i + \sum_{j=1}^{n} a_{ij} x_j, \text{ for } i = 1, 2, \ldots, m \text{ and}$$

$$y_{m+j} = c_j - \sum_{i=1}^{m} a_{ij} y_i, \text{ for } j = 1, 2, \ldots, n,$$

to show that

$$x_{n+i}^* = 0 \text{ or } y_i^* = 0 \text{ or both};$$

$$y_{m+j}^* = 0 \text{ or } x_j^* = 0 \text{ or both.}$$

$$\sum_{i=1}^{m} a_{ij} y_i^* = c_j \text{ whenever } x_j^* > 0;$$

$$y_i^* = 0 \text{ whenever } \sum_{j=1}^{n} a_{ij} x_j^* > b_i.$$

5.14 Consider the following product-mix problem. A company manufactures two products. Let x_1 and x_2 be the amounts of the products, say A and B, made. Various resources such as labour and different kinds of raw materials are needed to make the products. In the present problem, there are three resources required. The amounts available are $b_i = 5, 3, 4$ units respectively for resources labelled $i = 1, 2$ and 3. The units just mentioned would typically be person-days, tonnes of steel, cubic metres of fuel, and so on. Each unit of product A requires two units of the first resource ($i = 1$), while product B requires one unit. The corresponding values for the second resource ($i = 2$) are unity for both products while for resource 3 ($i = 3$) they are 0 and 2. The profit is 1.5 per unit of A and 1 per unit of B produced. The profit is to be maximized subject to the availability of resources.

(a) Solve using the simplex method.
(b) Find the dual of the linear program.
(c) Use the complementary slackness theorem to deduce the solution of the dual problem.
(d) Provide an interpretation of the dual problem in terms of 'shadow prices' (see for example Monahan, 2000, or search the Web).

5.15 A town produces a minimum of 200 000 tonnes of solid waste per year. There are two possible ways of disposing of the waste, first to dump the untreated waste in a landfill and second to incinerate it. Let x_1 be the mass of untreated waste and x_2 be the mass of untreated waste to be incinerated; use units of 1000 tonnes. The landfill will accept at most 160 000 tonnes of solid waste comprising untreated material, and 24 000 tonnes of ash from the incinerator. An 80% reduction of mass results during incineration, leaving a mass equal to 20% of the untreated waste. The incinerator produces 2500 units of energy per tonne of waste, which can be sold for $0.08 per unit. The table summarizes the costs associated with shipping and treatment of the waste.

	Cost per tonne
Landfill – untreated material	$100
Landfill – incinerator ash	$200
Incinerator	$60

(a) Formulate two objective functions, one that minimizes all costs (including sale of energy) and another that minimizes the mass of material dumped in the landfill. Solve the problem for both objective functions.

(b) Reformulate the problem considering other objectives in waste disposal, for example pollution due to incineration, and recycling of fibre, plastics and metals.

5.16 Two software companies – A and B – are planning to market a product. Marketing strategists have investigated four strategies open to B, and five open to A. They have prepared the following table, showing the payoffs for the various $\{\sigma, \tau\}$ pairs. The payoffs are from A's perspective, with 0 = failure of the product, 5 = satisfactory, 10 = maximum success. Use a maximin (game theory) approach to investigate whether there is a saddle point. Is there an optimal pure strategy or is a mixed strategy appropriate? Under what conditions would you opt for a different approach from the maximin?

		Company B			
		τ_1	τ_2	τ_3	τ_4
	σ_1	9	4	3	2
	σ_2	3	1	3	8
Company A	σ_3	7	6	10	7
	σ_4	4	4	5	3
	σ_5	6	3	4	4

5.17 Solve the following two-person zero-sum game using mixed strategies if necessary. Plot the expected values for each player to determine the gain floor and the loss ceiling as in Figure 5.14.

	τ_1	τ_2
σ_1	1	-2
σ_2	-1	2

5.18 Solve the following two-person zero-sum game using mixed strategies if necessary. Player A plays the rows and B plays the columns. We now denote the payoffs (from B to A, as per convention) as v. Plot the values in a two-dimensional graph as pairs $(v_i | \sigma_1)$ versus $(v_i | \sigma_2)$ for all τ_i, $i = 1, \ldots, 5$. The points show the payoffs to A for the pure strategies open to B. Describe possible mixed strategies open to B. Interpret player B's optimal strategy.

	τ_1	τ_2	τ_3	τ_4	τ_5
σ_1	1	2	2	2	4
σ_2	3	1	4	3	3

5.19 For the game of rock-paper-scissors – mentioned in the introductory section in this chapter on Conflict – construct the payoff matrix assuming payments of one unit. Is there a saddle point? Solve the game.

5.20 The following represents a payoff matrix, with A representing the row, and B the column, player. Payoffs are, according to convention, from B to A. Imagine that you are player B; the values in the table can be interpreted as (negative) utilities for you.

(a) Consider the situation where A represents a non-antagonistic opponent, as in 'design as a game against nature'. Find B's optimal strategy if it is estimated that $\mathbf{Pr}(\sigma_1) = 0.2$ and $\mathbf{Pr}(\sigma_2) = 0.8$. Plot the values of the various strategies open to B in a two-dimensional graph with axes 'value given σ_1' and 'value given σ_2'. Find the ranges of probability of σ_1 and σ_2 for which various strategies open to B are optimal, and identify those that are never optimal.

(b) Analyze the problem as a two-person zero-sum game, considering player B's game. Set up and solve the linear program.

(c) Again consider the problem as a game. Plot the expected values of player B's strategies against $\mathbf{Pr}(\sigma_1) = p$. Determine from this the optimal (mixed) strategy for A.

	τ_1	τ_2	τ_3	τ_4
σ_1	2	3	1	5
σ_2	4	1	6	0

6 Entropy

Chaos umpire sits,
And by decision more embroils the fray
By which he reigns: next him high arbiter
Chance governs all.

John Milton, *Paradise Lost,* book ii, l. 907

6.1 Introductory comments

Entropy is a concept that is basic to the second law of thermodynamics. It has been used extensively in deriving and studying probability distributions corresponding to maximum entropy. We know that it represents, in some way, disorder, and that the entropy of the universe increases. Eddington called it 'time's arrow'. Its roots are in the erratic development of the laws of thermodynamics over the last century and a half. Bridgman (1941) says of these laws that 'they smell more of their human origin' (as compared to the other laws of physics). This origin encompasses such activities as the observation by Mayer of the colour of blood of his patients in the tropics. During bleeding of these patients in Java, he noticed that the red colour of their venous blood was much brighter than that of his patients in Germany. This brightness indicated that less oxygen had been used in the processing of body fuel than in Germany. The human body, treated as a heat engine, needed less energy in the tropics since less energy was taken out by the surroundings. Mayer concluded that work and heat are equivalent. The second law is illuminated by the irreversible generation of heat during the boring of cannons using teams of horses by Count Rumford.

As a result of the varying aspects of entropy it has been described as 'a precious jewel with many facets'. But what does entropy mean? We need a single clear definition. We need also to associate it with the words 'disorder' and 'chaos'. We have used probability to reflect our feelings of uncertainty, but the relationship is not direct. We do not say that we are 'more uncertain' as probability increases (rather the opposite). Uncertainty implies the absence of certainty, and we often say that we feel 'more uncertain' about an event if the probability of it happening decreases for some reason or another. Using this sense of the word, some authors associate entropy with 'uncertainty' along with 'disorder'.

We give the following precise definition of the entropy of a quantity such as the energy level of a particle whose value corresponds to one of a set of possible events.

Entropy is the smallest expected number of YES–NO questions to determine the true event.

The determination of this expected number must be made in the most efficient manner: the approach to doing this is given in Section 6.2. The definition based on information theory, as also explained in the next two sections, derived from the work of Shannon (1948) and further advanced by Jaynes (1957, 2003), Khinchin (1957), Brillouin (1962), Tribus (1961, 1969) and others. De Finetti (1974) briefly discussed entropy in terms of binary YES–NO questions. Special mention should be made of Jaynes, who has made powerful contributions to the study of entropy using information theory (see Rosenkrantz, 1983). This approach has stimulated developments in areas other than classical thermodynamics, for example in transportation engineering and granular materials (configurational entropy). It will also be helpful in bringing a wider audience to thermodynamic thought.

6.2   Entropy and disorder

We often comment that our desk, or office, or other space, is in a state of disorder: 'my room is in chaos', and so on. Imagine that we have 16 ($= 2^4$) piles of paper on our desk, and that a missing sheet is in one of them, but we don't know which. The entropy is simply the power 4 in the expression 2^4. To explain this further, we introduce the 'entropy demon', a close relative of Maxwell's demon.[1] The entropy demon knows where your sheet of paper is, and will answer intelligently posed YES–NO (binary) questions. He insists that these be posed in such a way that the expected number of questions is the least. This then measures the **information content** of the state of disorder. Each question posed to the demon is answered by 'yes' or 'no', thus eliciting one bit of information. The term 'bit', or binary digit, comes from the binary (0–1) system, the unit of information in computing, and is our fundamental unit of disorder. The demon expects the questions to be framed so as to transfer the least expected number of bits of information, as we shall now explain. Were we to ask in order, 'is it in pile 1?', 'is it in pile 2?', ..., and so on until we obtain the answer 'yes', the number of questions ranges from 1 to 15 (if we know that the sheet is not in piles 1 to 15, it must be in pile 16). The numbers of questions and associated bits of information are all equally likely except for question 15 so that the expected number of questions is

$$\frac{1}{16}(1 + 2 + \cdots + 14) + \frac{2}{16}(15) = \frac{1}{16}(7 \times 15) + \frac{30}{16} = 8.4375.$$

---

[1] Maxwell's demon has very sharp faculties, so that, if a box is divided into two by a division with a small hole, he can open and close the hole, purposely letting the faster molecules into one side, and the slower ones into the other. He thus raises the temperature of one side, and lowers that of the other, without expenditure of work. See Brillouin (1962), Chapter 13.

The Entropy Demon

The entropy demon would be not at all pleased with this method. In a rare mood of cooperation, he advises the following. First, number the piles from 1 to 16. Then ask as follows (we assume in this demonstration that the number of the pile containing the missing number is 9).

1. You: 'Is it > 8?'   Demon: 'Yes'
2. You: 'Is it > 12?'   Demon: 'No'
3. You: 'Is it > 10?'   Demon: 'No'
4. You: 'Is it = 9?'   Demon: 'Yes'.

By this procedure of halving the possibilities each time, and asking if the 'true event' is in one or other half, one arrives at the correct pile in four questions, regardless of which pile the missing sheet is in, thus freeing the demon to pursue his other interests.

For equally likely events, the entropy is defined by the number of bits of information required to determine the true event, and thereby to remove the disorder. The entropy associated with

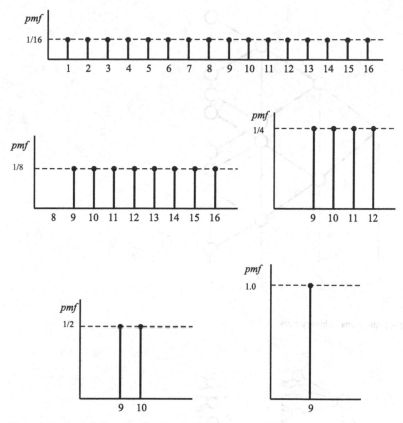

Figure 6.1 Probability distribution, initially (top), and then after the four successive questions

our original situation of the missing sheet is therefore 4 bits. The probability mass functions of our state of uncertainty at various stages in the questioning of the entropy demon are shown in Figure 6.1. The game just described, involving the entropy demon, is a variation of the parlour game of 'twenty questions', in which one tries within the limit of twenty questions to identify a person, or to guess an unknown occupation. The questions are generally of the kind that elicit one bit of information each time, and the objective is to obtain the correct answer with as few bits as possible. We can also think of drawing an object from a hat containing a certain number of distinct objects.

If there are n questions rather than four as in the example above, for events considered to be equally likely, there are 2^n possible values, and the entropy is n. The entropy H is the logarithm to the base 2 of the number of possibilities. In the analysis above, only binary questions were permitted. There is no reason why questions that permit three answers (ternary questions) should not be permitted, or in general, s-ary questions. In the latter case, the question might be 'is the missing paper in group $1, 2, 3, \ldots, s$?', where the available piles have been grouped into s piles. Binary and s-ary questions are illustrated by the trees of Figure 6.2. In the case of s-ary questions, and if the number of equally likely possibilities is s^n, the entropy is again n if

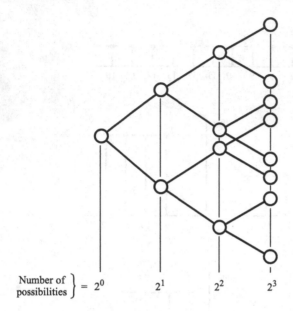

(a) Binary questions in binary trees

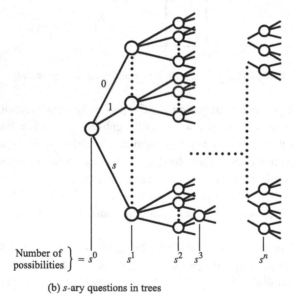

(b) s-ary questions in trees

Figure 6.2 Possible values for answers to repeated questions

the base of the logarithm is s. We can therefore write

$$H^{(s)}(s^n) = \log_s(s^n) = n, \tag{6.1}$$

with a special case being the binary one, $s = 2$, already discussed. The superscript s in $H^{(s)}$ indicates the number of possibilities in each question. We have used this as the base in measuring entropy; change of base will be dealt with in the next section.

6.3   Entropy and information: Shannon's theorem

We again consider the situation in which we had lost an important piece of paper, entropy measuring the 'missing information' required to find it. The entropy demon is usually uncooperative so that the best we can do is to conduct a search. Yet the idea of measuring disorder by means of missing information is very helpful. Entropy is often interpreted as a measure of uncertainty; this is appropriate in the sense that the more questions that we have to ask to find our missing object, the more uncertain are we.

In the previous section the entropy of s^n possibilities was considered, using the base s for our logarithmic measure: the result was n. Let us consider the same number of possibilities, but with the use of a different base, say b. First, we consider the set of s^n equally likely possibilities in Figure 6.2(b). Let us denote the entropy of s possibilities to the base b as $H^{(b)}(s)$. Of course, we suspect that the result will be $\log_b(s)$ but we must show this convincingly. An important point is that we can always decompose the s^n possibilities into a series with n nodes of s possibilities, each node of the series being identical in terms of information content (Figure 6.2). Therefore the information content of the series is

$$H^{(b)}\left(s^n\right) = n H^{(b)}(s). \tag{6.2}$$

This equation shows that the information content associated with the nodes in the tree of Figure 6.2 are additive, each contributing $H^{(b)}(s)$, and resulting in $n H^{(b)}(s)$ in total. Each path through the tree constitutes a series of encounters with identical nodes, n in total.

In the following, we continue to consider the entropy to any base b, but for simplicity we omit the superscript and use the notation $H(\cdot)$. Consider now two sets of positive integers $\{r, m\}$ and $\{s, n\}$. We wish to find the entropy of r and s (think of r and s as given and that we can play with m and n). From Equation (6.2),

$$H\left(r^m\right) = m H(r) \text{ and } H\left(s^n\right) = n H(s). \tag{6.3}$$

We can choose an m, and find an integer n such that

$$s^n \le r^m < s^{n+1}, \tag{6.4}$$

where the pairs $\left\{s^n, s^{n+1}\right\}$ cover the range $\{1, s\}, \left\{s, s^2\right\}, \left\{s^2, s^3\right\}, \ldots$, for $n = 0, 1, 2, \ldots$ It is always possible to locate r^m in one of these intervals ($r^m \ge 1$). We are going to make

m arbitrarily large in the following, while keeping r and s unchanged; therefore n will also become very large. First, we take logarithms of inequality (6.4):

$$n \ln s \leq m \ln r < (n+1) \ln s,$$

and consequently

$$\frac{n}{m} \leq \frac{\ln r}{\ln s} < \frac{n}{m} + \frac{1}{m}. \tag{6.5}$$

Natural logarithms have been used in these equations, but any other base could have been chosen and the results will be valid for any choice.

One of the basic requirements of Shannon's theory is the very reasonable one that the entropy $H(x)$ should increase monotonically with x. This requirement, combined with inequality (6.4), gives

$$H\left(s^{n}\right) \leq H(r^{m}) < H\left(s^{n+1}\right),$$

and using Equation (6.3),

$$nH(s) \leq mH(r) < (n+1)H(s).$$

It follows that

$$\frac{n}{m} \leq \frac{H(r)}{H(s)} < \frac{n}{m} + \frac{1}{m}. \tag{6.6}$$

Combining this inequality with (6.5),

$$\left| \frac{H(r)}{H(s)} - \frac{\ln r}{\ln s} \right| < \frac{1}{m}, \tag{6.7}$$

which is independent of n.

We now carry out our earlier suggestion, and make m arbitrarily large. Then Equation (6.7) implies that

$$H(t) = k \ln t, \tag{6.8}$$

with $k > 0$ if $H(t)$ is to increase with t, the number of possibilities. The derivation above was carried out for integers, and t is therefore any integer. The base of the logarithms could have been any number; we used the natural base only for convenience. We can now measure the entropy of any integer to any base.

Example 6.1 *If* $t = 7$ *and the base is 2,*

$$H = \log_2 7 = 2.81 \ldots \text{ bits.} \tag{6.9}$$

The constant k was taken as unity in (6.9) to be consistent with our previous definition of the bit. Equation (6.8) is identical to the Boltzmann definition, usually written as $S = k \log W$, notably on the monument to him in Vienna. We shall return to this definition later in this chapter.

The units if one uses natural logarithms are sometimes termed 'natural units' or 'nats'; we shall deal with thermodynamic units later. We can also interpret the value of 2.81 ... bits – Equation (6.9) – in terms of binary questions using inequality (6.4) as shown in the following example.

Example 6.2 Considering the base 2, we can write $H(7) = 0.1H\left(7^{10}\right)$, from Equation (6.2), and we note that $2^{28} < 7^{10} = \left(2^{2.81\ldots}\right)^{10} = 2^{28.1\ldots} < 2^{29}$. Therefore the number of binary questions for 7^{10} equally likely possibilities is between 28 and 29, and the entropy of 7 is bounded by a tenth of these values, i.e. it is between 2.8 and 2.9. We can approximate the entropy of 7 in this way to any desired accuracy by considering $7^{100}, 7^{1000}$, and so on.

Example 6.3 The definition (6.8) is for a random quantity with equally likely possible values. What happens if the events are not equally likely? For example, consider the events $\{a, b, c, d\}$ with probabilities $\{1/2, 1/4, 1/8, 1/8\}$ respectively. If we wish to remove the maximum amount of uncertainty at each step, we should halve the probability with each question and answer. The questions that one would ask are: is it a?; if NO, is it b?; if NO, is it c? This is illustrated as a tree in Figure 6.3, in which the branches required to complete an 'equally likely' tree, similar to those already considered, are shown in dotted lines.

We now consider a finite partition of n possible events

$$\{B_1, B_2, \ldots, B_i, \ldots, B_n\}$$

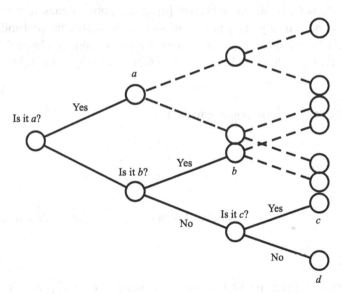

Figure 6.3   Questions for possibilities $\{a, b, c, d\}$ with probabilities $\{\frac{1}{2}, \frac{1}{4}, \frac{1}{8}, \frac{1}{8}\}$

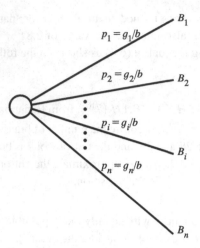

Figure 6.4 Finite partition of n events

with probabilities

$$\{p_1 = g_1/b, p_2 = g_2/b, \ldots, p_i = g_i/b, \ldots, p_n = g_n/b\}$$

and with $\sum_{i=1}^{n} g_i = b$. This is shown in the form of a tree in Figure 6.4. The p_i are therefore represented by rational numbers, and indeed correspond to the probabilities of the drawing of balls of colours $1, 2, \ldots, i, \ldots, n$ from an urn containing b balls, with g_1 of colour 1, g_2 of colour $2, \ldots, g_i$ of colour $i, \ldots,$ and g_n of colour n. We shall now convert the tree of Figure 6.4 to an equivalent case of equally likely possibilities. This is illustrated in Figure 6.5; in (a) the possibilities $\{B_1, B_2, \ldots\}$ have been split into subsets of $\{g_1, g_2, \ldots\}$ possibilities, respectively. This results in the total set of b equally likely possibilities of Figure 6.5(b), the probability of each of which is $1/b$. The situation now is different from the original one of Figure 6.4, but related. The entropy regarding the b possibilities of Figure 6.5(b) is found using (6.8):

$$H(b) = k \ln b. \tag{6.10}$$

But we wish to find the entropy corresponding to the partition

$$\{B_1, B_2, \ldots, B_n\}$$

of Figure 6.4. We denote this

$$H\{p_1, p_2, \ldots, p_n\}.$$

To find this value we note that the entropy associated with the g_i possibilities *given that B_i is true* is

$$H(g_i) = k \ln g_i; \tag{6.11}$$

imagine that we are surveying the g_i possibilities while standing at B_i in Figure 6.5(a). For identifying the true B_i, we do not care which of the g_i is true, only that one of these members of

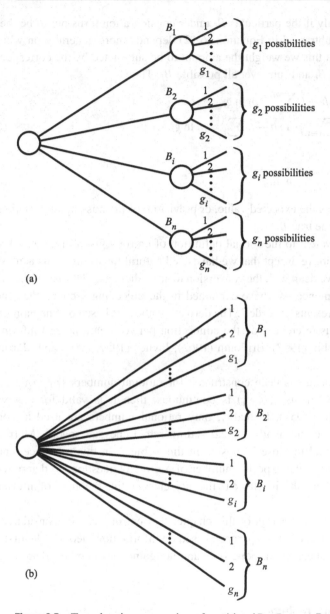

(a)

(b)

Figure 6.5 Tree showing conversion of partition $\{B_1, B_2, \ldots, B_i, \ldots, B_n\}$ into equally likely possibilities

the subset constituting B_i of Figure 6.5(a) is true. The entropy associated with B_i is therefore Equation (6.10) minus (6.11):

$$k \ln b - k \ln g_i = -k \ln \frac{g_i}{b} = -k \ln p_i. \tag{6.12}$$

But this will be correct only if the particular B_i under consideration turns out to be the true event; the amount to be subtracted in Equation (6.12) depends more generally on which of the B_i is true. To deal with this we weight the amount to be subtracted by the corresponding probability ($p_i = g_i/b$) of B_i and sum over all possible B_i. Then

$$
\begin{aligned}
H(p_1, p_2, \ldots, p_n) &= k \ln b - k \sum_{i=1}^{n} p_i \ln g_i \\
&= k \sum_{i=1}^{n} p_i \ln b - k \sum_{i=1}^{n} p_i \ln g_i.
\end{aligned}
\tag{6.13}
$$

The final result is then

$$
H(p_1, p_2, \ldots, p_n) = -k \sum_{i=1}^{n} p_i \ln p_i,
\tag{6.14}
$$

which may be interpreted as the expected value, or prevision, of the missing information that would be required to find the true B_i.

Equation (6.14) is equivalent to the formal definition of entropy given in the first section 'Introductory comments' above except that we have used natural logarithms; this is no shortcoming since we have above dealt with the conversion to any other base. The use of the natural base is a matter of convenience. We have explained in the preceding sections the sense in which Equation (6.14) represents 'disorder'. The derivation above is based on Shannon's work (1948); there are numerous references on the connection between entropy and information theory, for example Khinchin (1957), Brillouin (1962), Jaynes (1957, 2003) and Martin and England (1981).

The derivation given above has been constructed for rational numbers $\{p_1, p_2, \ldots, p_n\}$; if it is postulated that $H(p_1, p_2, \ldots, p_n)$ is continuous, then it is valid for any values of its arguments (Khinchin, 1957). In reality, only rational numbers are used in computations, and the extension of the mathematical setting into Lebesgue spaces (Martin and England, 1981) might appeal to those interested in this subject. In this reference, entropy is defined as a lower bound to the expected number of YES–NO questions to determine the true event (as given above), and this is related to the 'noiseless coding theorem' of information theory.

We stated in the introductory section of this chapter that entropy can be considered as a 'precious jewel with many facets'. The approach based on information theory is the first main facet. The second main facet can be introduced by another game played by the demon.

6.4   The second facet: demon's roulette

It is a scarcely guarded secret that the entropy demon enjoys gambling. He has devised a game to obtain the maximum-entropy solution to a problem. This turns out also to be the most likely solution, as we shall see. The germinal problem was the distribution of energy amongst 'identical' particles, each of which can be in any of n energy (or quantum) states. To model this, the demon has constructed a roulette wheel with n divisions, giving n possible outcomes.

He can set n to any finite value that he pleases. He labels these outcomes $\{E_1, E_2, \ldots, E_n\}$, which could be energy levels or any other application where the model makes sense. The demon has time on his hands and plays compulsively, N times, such that $N \gg n$. He counts the number of times that the outcomes $E_1, E_2, \ldots, E_i, \ldots, E_n$ occur, resulting in the numbers $N_1, N_2, \ldots, N_i, \ldots, N_n$, with $\sum_i N_i = N$. Probabilities are assigned to the E_i on the basis that

$$\mathbf{Pr}(E_i) \equiv p_i \equiv N_i/N. \tag{6.15}$$

The demon is not surprised that the numbers are reasonably close to each other – the resulting distribution is uniform in this case. To make further progress, the results must be constrained, he explains. The mean, for example of energy, might be constrained to be equal to a certain value. We shall see that this changes the distribution from uniform to another shape.

A standard approximation to reality in many cases is that the energy content of a set of particles in a certain closed volume is constant. We can therefore state that the mean energy, averaged over all the particles, is fixed, but that the energy of the individual particles fluctuates in some manner. We now pass this idea to our demon who fiendishly starts his game of roulette over again. As an example problem, he takes $n = 5$. The events $\{E_1, \ldots, E_i, \ldots, E_5\}$ are considered, corresponding to five equally spaced divisions of the roulette wheel. The demon assigns the values of the random quantity X, $\{x = 1, 2, 3, 4, 5\}$ to the events E_i for $i = 1, 2, \ldots, 5$, respectively. To test the idea of keeping the mean value constant, the demon agrees to specify that $\langle X \rangle = 2$. The demon proposes to carry out the entire experiment, that is, to spin the roulette wheel N times, and then to discard the result if the constraint is not satisfied. For example, if $N = 1000$, the following result would be discarded:

$$N_1 = 100, \quad N_2 = 200, \quad N_3 = 300, \quad N_4 = 400, \quad N_5 = 0, \tag{6.16}$$

since the values of p_i,

$$p_1 = 0.1, \quad p_2 = 0.2, \quad p_3 = 0.3, \quad p_4 = 0.4, \quad p_5 = 0.0, \tag{6.17}$$

would not lead to the specified mean value of 2 since

$$(0.1)(1) + (0.2)(2) + (0.3)(3) + (0.4)(4) + (0.0)(5) = 3.0 \tag{6.18}$$

On the other hand, the values of N_i given by $\{450, 300, 100, 100, 50\}$ lead to a mean value of 2.0, and therefore the imposed constraint is satisfied.

The demon poses the question: if I keep on repeating the experiment, discarding those assignments that do not agree with the constraint, what happens? The demon also agrees to increase the value N to some very large number, say $100\,000$; each experiment gives him an assignment corresponding to the set $\{N_i\}$, with associated probabilities $\mathbf{Pr}(X = i)$. The demon has been through all of this before, and is not surprised to find that certain results come up much more frequently than others: to obtain this conclusion he repeats the $100\,000$ spins of the roulette wheel many times, throwing away those results that do not conform to the imposed

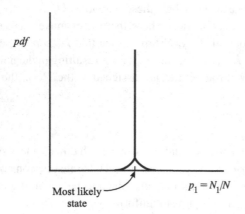

pdf

Most likely state

$p_1 = N_1/N$

Figure 6.6 Schematic illustrating sharp peak at maximum entropy solution, becoming a Dirac spike

constraints, until he has many ($\aleph$) sets of results

$$\{p_1 = N_1/N,\ p_2 = N_2/N,\ \dots,\ p_i = N_i/N,\ \dots,\ p_n = N_n/N\}\,, \tag{6.19}$$

all of which conform to the imposed constraint(s) (there could actually be several constraints). The demon constructs a graph to illustrate the point. The result is illustrated in Figure 6.6 for p_1. A certain value of $p_1 = N_1/N$ is much more likely than the others; this only becomes evident if the number of repetitions ($\aleph$) of the entire experiment yielding results that are consistent with the constraints becomes very large ($\aleph \gg N \gg n$). In the case $n = 2$, in the absence of any constraints, the assignment $N_1 = N_2 = N/2$ is the most likely; this can be confirmed by noting that there are many more paths through the 'centre' of the random walk diagram of Figure 2.19, or near the centre of Pascal's triangle (Table 2.1). A similar situation exists in the case $n > 2$, as will be made clear in the following. In a sense the demon's game represents a model of nature.

The procedure followed by the demon in fact corresponds to the idea of an 'ensemble' introduced by Gibbs. As noted, the germinal problem was the distribution of energy amongst a set of particles. A Gibbs ensemble is a set of N imagined copies of the real system under consideration. The system could be a set of particles, for instance molecules in a gas, or in the case of solids, the energy levels of an oscillator or the atomic vibrations at eigenfrequencies of the solid (normal modes of vibration). In Aston and Fritz (1959), 'elements' are elementary particles, atoms, molecules or modes of vibration (for example in crystals). A system contains many elements, and a large number of systems is an ensemble. Fundamental particles will be dealt with further in Sections 6.8 and 6.9.

We can carry out a calculation that gives a result similar to the demon's. This shows that the demon is in fact maximizing the entropy of Equation (6.14). First, we write an expression for the number of ways that the demon would obtain a particular result. We return to the case of n outcomes and N spins of the wheel, with a result $N_1, N_2, \dots$ that is in conformity with the imposed constraint(s). If the roulette wheel is perfectly balanced, as we have assumed in the foregoing, each of the possible ways w of obtaining the result is equally likely, and

$$w = \binom{N}{N_1 N_2 \cdots N_i \cdots N_n} = \frac{N!}{N_1! N_2! \cdots N_i! \cdots N_n!}. \tag{6.20}$$

For example with $n = 2$ and $N = 6$, and no constraints, there are $2^6 = 64$ possible outcomes with likelihoods of $\{1, 6, 15, 20, 15, 6, 1\}$ for the outcomes $N_1 = \{0, 1, 2, 3, 4, 5, 6\}$ respectively (Pascal's triangle). If the experiment (consisting of six trials) is repeated many times, the result $N_1 = 3$ occurs (on average) in the proportion $(20/64)$ of the time, where the value 20 is obtained from Pascal's triangle. For the maximum entropy solution, we wish to increase $\aleph$; we could for example repeat the six trials a million times. But we also need to increase N. When N becomes very large there is a solution that is overwhelmingly likely. We shall demonstrate this in the following.

Returning to the general case, we wish to maximize Equation (6.20); this is a convenient relative measure of the proportion of times a particular result will occur within the total number of repetitions $\aleph$ of the entire experiment, all of which yield results conforming to the constraints. It is convenient in the present case to maximize $(\ln w)$ rather than w, since they are monotonically related. Now

$$\ln w = \ln N! - \sum_{i=1}^{n} \ln N_i!. \tag{6.21}$$

Next, we replace N_i with $N \cdot p_i$ where p_i is the probability assignment to the ith event E_i. Thus

$$\ln w = \ln N! - \sum_{i=1}^{n} \ln (Np_i)!. \tag{6.22}$$

We now use the Stirling approximation

$$N! \rightarrow \sqrt{2\pi N}\, N^N \exp(-N) \tag{6.23}$$

as $N \rightarrow \infty$. The proof of this formula is given in Feller (1968). Consequently,

$$\ln N! \rightarrow \left(N + \frac{1}{2}\right) \ln N - N + \frac{1}{2}(\ln 2\pi) \tag{6.24}$$

as $N \rightarrow \infty$. For large N, this tends to

$$(N \ln N - N). \tag{6.25}$$

Substituting this into Equation (6.22), and noting that $N - \sum_i (Np_i) = 0$,

$$\ln w \rightarrow N \ln N - \sum_{i=1}^{n} [Np_i \ln Np_i]$$
$$= N \ln N - N \ln N \sum_{i=1}^{n} p_i - N \sum_{i=1}^{n} p_i \ln p_i.$$

Therefore

$$\ln w \rightarrow -N \sum_{i=1}^{n} p_i \ln p_i; \tag{6.26}$$

apart from the coefficient N, this expression is identical to Equation (6.14). In fact, N is eventually replaced by k; this has no effect on the deliberations that follow since it is the summation in the expression (6.26) that we wish to maximize. Later on, we will deal with thermodynamic units.

The spirit of the mathematical rendering of the entropy demon's game leading to the expression (6.26) was to find the probability distribution that occurs with greatest likelihood. The coefficient N in this expression represents the number of spins of the roulette wheel in one realization of Equation (6.19), and the entire expression (6.26) represents the natural logarithm of the number of ways that a particular assignment $\{N_1, N_2, \ldots\}$ would be achieved. This is then a convenient measure (to a logarithmic scale) of the proportion of times a particular result will occur if the whole procedure (spinning N times) were repeated many times ($\aleph$), all of which yield results conforming to the constraints.

We wish then to choose the assignment that occurs with by far the greatest probability in the demon's game, and therefore we choose that distribution which maximizes

$$H = -\sum_{i=1}^{n} p_i \ln p_i, \tag{6.27}$$

in which we have omitted the factor N in expression (6.26). But the number of ways, w, is given by

$$w = \exp(NH), \tag{6.28}$$

and this becomes very large for large N and maximum H. See also Equation (6.85).

6.5   Maximizing entropy; Maxwell–Boltzmann distribution; Schrödinger's nutshell

Methodology and introductory example

We saw in the preceding section that the probability distribution that maximises entropy is by far the most likely distribution to occur in a random game such as that played by the demon, subject to the constraints of the problem. This suggests the idea of using mathematics to find this distribution, resulting in the maximum-entropy method. The basic idea in the maximum-entropy method then is to maximize (6.27),

$$H = -\sum_{i=1}^{n} p_i \ln p_i,$$

subject to any constraints of the problem. This yields a probability distribution which has two main interpretations. In the first, the maximum-entropy distribution corresponds to that for which the maximum expected number of questions have to be asked to obtain the true event, as in Section 6.3 and Equation (6.14). There is a minimum amount of information embodied in the probability distribution, other than what is contained in the constraints. It is therefore 'maximally non-committal with regard to missing information', to use Jaynes' phrase. It is also the 'flattest' probability distribution consistent with the constraints. In the second interpretation, we obtain the probability distribution that is most likely in the demon's roulette game, overwhelmingly, as in Section 6.4 and Equations (6.26) and (6.27).

In accordance with the prescription just given, we want to maximize Equation (6.27) subject to any constraints in the problem for which we wish to derive a distribution of probability. This is best achieved by the use of Lagrange multipliers, which provide a useful and elegant way to proceed. The relevant theory was outlined in Section 5.3.2. It is helpful to introduce the procedure using an example; see also Jaynes (2003).

Example 6.4 An interesting and important problem to discuss so as to illustrate the development of the theory is the germinal one noted before, in which the outcomes represent energy levels X. We wish to find the maximum-entropy distribution, subject to the condition that the mean value of X is fixed. To illustrate the method, let us assume that the random quantity has three possible outcomes. $X = 0, 1, 2$ for $i = 1, 2, 3$ respectively. Let $\mathbf{Pr}(X = 0) = p_1$, $\mathbf{Pr}(X = 1) = p_2$ and $\mathbf{Pr}(X = 2) = p_3$. The problem to be solved can be summarized as follows.

$$\text{Max}\left(H = -\sum_{i=1}^{3} p_i \ln p_i\right) \tag{6.29}$$

subject to

$$\sum_{i=1}^{3} p_i = 1 \tag{6.30}$$

and

$$\sum_{i=1}^{3} x p_i = \langle X \rangle = \mu,$$

which becomes

$$0 \cdot p_1 + 1 \cdot p_2 + 2 \cdot p_3 = \langle X_i \rangle = \mu$$

or

$$p_2 + 2 p_3 = \mu. \tag{6.31}$$

In the last three equations, μ stands for the mean value.

The solution requires us to maximize the entropy, Equation (6.29), subject to the constraints of (6.30) and (6.31). The Lagrangian is

$$L\left(p_1, p_2, p_3; \lambda_0', \lambda_1\right) = -\sum_{i=1}^{3} p_i \ln p_i - \lambda_0'\left(\sum_{i=1}^{3} p_i - 1\right) - \lambda_1(p_2 + 2p_3 - \mu); \tag{6.32}$$

the reason for the use of the notation $\left\{\lambda_0', \lambda_1\right\}$ will become clear in the following. Differentiating the Lagrangian:

$$\frac{\partial L}{\partial p_1} = (-\ln p_1 - 1) - \lambda_0' = 0$$

$$\frac{\partial L}{\partial p_2} = (-\ln p_2 - 1) - \lambda_0' - \lambda_1 = 0$$

$$\frac{\partial L}{\partial p_3} = (-\ln p_3 - 1) - \lambda_0' - 2\lambda_1 = 0$$

$$\frac{\partial L}{\partial \lambda_0'} = -(p_1 + p_2 + p_3 - 1) = 0$$

$$\frac{\partial L}{\partial \lambda_1} = -(p_2 + 2p_3 - \mu) = 0. \tag{6.33}$$

In differentiating $H = -\sum_i p_i \ln p_i$ with respect to p_i, we always obtain the term $(-\ln p_i - 1)$, as indicated above. It is most convenient to write

$$-\lambda_0 = -\lambda_0' - 1, \tag{6.34}$$

in which we have borrowed (-1) from the derivative of H; this will always be done in the following. From the first three derivatives of the Lagrangian, we can write

$$p_1 = \exp(-\lambda_0), \quad p_2 = \exp(-\lambda_0 - \lambda_1), \quad p_3 = \exp(-\lambda_0 - 2\lambda_1), \tag{6.35}$$

and where we have used the first three derivatives of the Lagrangian L. The first of the two remaining conditions, which is the constraint giving the normalization of probabilities, reduces to

$$\exp(-\lambda_0)\left[1 + \exp(-\lambda_1) + \exp(-2\lambda_1)\right] = 1$$

or

$$\exp \lambda_0 = 1 + \exp(-\lambda_1) + \exp(-2\lambda_1); \tag{6.36}$$

this is termed the **partition function** Z. This function and its important uses are described in more detail below. In the present case,

$$Z(\lambda_1) = \exp \lambda_0 = \sum_j \exp(-\lambda_1 x_j), \quad j = 0, 1, 2.$$

The second constraint yields the following equation:

$$\exp(-\lambda_0)\left[\exp(-\lambda_1) + 2\exp(-2\lambda_1)\right] = \mu. \tag{6.37}$$

We can solve for the probabilities using Equations (6.35), (6.36) and (6.37). We assume first that $\mu = 1$. Then Equations (6.35) and (6.36) can be combined to eliminate $\exp(-\lambda_0)$, giving $\exp(-2\lambda_1) = 1$, or $\lambda_1 = 0$. Hence $p_1 = p_2 = p_3 = 1/3$, as expected since the mean is at the centre of the three values. For the case where $\mu = 0.5$, and following the same procedure, it is found that

$$1.5 \exp(-2\lambda_1) + 0.5 \exp(-\lambda_1) - 0.5 = 0,$$

which may be solved as a quadratic equation to yield $p_1 = 0.616$, $p_2 = 0.267$, and $p_3 = 0.116$ (see Figure 6.7).

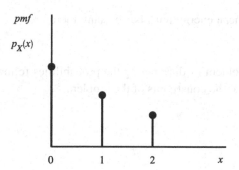

Figure 6.7   Maximum entropy solution to problem with three possibilities

Canonical aspects and the nutshell

The derivation in Example 6.4 can be generalized to the result that is termed the 'canonical ensemble'. Do not be put off by this rather forbidding expression. Of course the demon really dislikes the canonical aspect! We are trying to obtain the distribution of energies amongst systems that have a fixed number of particles. We can interpret for this example the system as a set of particles, or a macroscopic sample which the demon interrogates to find the energy level of a particle in the sample. The energy is denoted as E and the possible energy states are $\{e_0, e_1, e_2, \ldots, e_i, \ldots\}$, a countably infinite set (we are now starting with e_0 rather than e_1 as it is convenient to do so; no following arguments are affected by the choice). The probability that any system is in a state corresponding to energy e_i is denoted p_i. Imagine that we have a set of N mental replicas of this system, and that the systems are placed in a heat bath at temperature T, thermally insulated so that the total energy content is constant. This could be thought of as a 'supersystem' comprising the N replicas, with a convenient thermodynamic interpretation.

In Section 3.6.3, when discussing stochastic processes, it was pointed out that consideration of an 'ensemble' of outcomes was not necessary in modelling a random quantity in time; one has only to think of the random quantity. The same applies in the present case. The mental replicas of the preceding paragraph are similarly convenient ways of thinking about and representing the situation to assist in developing a thought-experiment. A 'mechanistic' world of this kind can be constructed following inductive reasoning and is useful in modelling and carrying out simulations. But we could have merely stated that we are uncertain about the energy level of particle i and proceeded directly to Equations (6.40) to (6.42) below.

To continue with the thought-experiment: each replica corresponds to a spin of the demon's roulette game, which is repeated N times. This leads to an assignment of probabilities

$$p_0 = \frac{N_0}{N}, \, p_1 = \frac{N_1}{N}, \, p_2 = \frac{N_2}{N}, \cdots, \frac{N_i}{N}, \cdots, \text{ and so on.} \tag{6.38}$$

As before, he then repeats the whole experiment (N trials) $\aleph$ times to find the most likely distribution. Probabilities are to be assigned subject to the only constraint that the total energy of the supersystem should be constant – since it is thermally insulated. If we average this energy over the systems, we find that the mean energy must be constant; therefore

$$\sum_{i=0}^{\infty} p_i e_i = \langle E \rangle = \mu, \tag{6.39}$$

where μ denotes the mean energy. The problem of distributing the probabilities follows the demon's method; we play roulette subject to the constraints of the problem.

Mathematically, this is equivalent to

$$\text{Max} \left(H = -\sum_{i=0}^{\infty} p_i \ln p_i \right) \tag{6.40}$$

subject to

$$\sum_{i=0}^{\infty} p_i = 1 \tag{6.41}$$

and

$$\sum_{i=0}^{\infty} p_i e_i = \langle E \rangle = \mu. \tag{6.42}$$

We then form the Lagrangian

$$L = H - \lambda_0' \left(\sum_{i=0}^{\infty} p_i - 1 \right) - \lambda_1 \left(\sum_{i=0}^{\infty} p_i e_i - \mu \right), \tag{6.43}$$

and differentiate:

$$\frac{\partial L}{\partial p_i} = -\ln p_i - \lambda_0 - \lambda_1 e_i = 0,$$

where $\lambda_0 = \lambda_0' + 1$, as before in Equation (6.34). Hence

$$p_i = \exp(-\lambda_0) \exp(-\lambda_1 e_i). \tag{6.44}$$

Now the probabilities are coherent – Equation (6.41) – and therefore

$$\exp \lambda_0 = \sum_{i=0}^{\infty} \exp(-\lambda_1 e_i) \tag{6.45}$$

which defines the **partition function**

$$Z \equiv \exp \lambda_0, \tag{6.46}$$

as before. The normalization of probabilities appears in the Lagrangian as the first constraint multiplied by the multiplier λ_0'; the second constraint fixes the mean. We write (6.44) as

$$p_i = \frac{\exp(-\lambda_1 e_i)}{Z(\lambda_1)}. \tag{6.47}$$

The partition function Z is termed the 'Zustandssumme' in German, or 'sum-over-states', which has led to the use of the symbol Z; thus, from Equation (6.45),

$$Z(\lambda_1) = \sum_i \exp(-\lambda_1 e_i). \tag{6.48}$$

A term of this kind appears as a normalizing constant in all maximum-entropy distributions. (It arises from the condition of coherence $\sum_i p_i = 1$.) The following equations may be deduced with relative ease; the first fixes the mean of the distribution.

$$\mu = \langle E \rangle = -\frac{\partial \ln Z}{\partial \lambda_1} \tag{6.49}$$

$$p_i = -\frac{1}{\lambda_1} \frac{\partial \ln Z}{\partial e_i}. \tag{6.50}$$

Schrödinger (1952) suggested that the second of these equations, (6.50), contains the whole of thermodynamics 'in a nutshell'. Indeed, it specifies the entire probability distribution of energy.

If we use Equation (6.14), i.e. $H = -k \sum p_i \ln p_i$, and if there are w events which we judge to be equiprobable ($p_i = 1/w$ for each i), the expression for entropy is

$$H = k \ln w, \tag{6.51}$$

which is Boltzmann's formula. The constant k is Boltzmann's constant, equal to 1.38×10^{-23} Joules per degree K. We can write (6.47) as

$$p_i = \frac{\exp(-\lambda_1 e_i)}{\sum_i \exp(-\lambda_1 e_i)}. \tag{6.52}$$

For equally spaced small intervals de with e_i in the ith interval, $i = 0, 1, 2, \ldots,$

$$p_i = f_E(e)\,de = \frac{\exp(-\lambda_1 e)\,de}{\int_{e=0}^{\infty} \exp(-\lambda_1 e)\,de},$$

where $f_E(e)$ is now a density (see also Example 6.5 below). Carrying out the integration,

$$f_E(e) = \lambda_1 \exp(-\lambda_1 e), \tag{6.53}$$

which may also be written as

$$f_E(e) = \frac{\exp(-e/kT)}{kT}, \tag{6.54}$$

where $\lambda_1 = 1/kT$. The term kT is the average energy of the oscillators (Eisberg and Resnick, 1974). Thus (e/kT) is dimensionless, as is appropriate.

6.6   Maximum-entropy probability distributions

We have just maximized entropy subject to the single constraint that the mean is fixed, resulting in the exponential distribution (6.52). The maximization of entropy can be carried out for the case where there is a **set** of constraints. We consider, as before, a random quantity associated with the set of numbers $\{0, 1, 2, 3, \ldots, i, \ldots\}$. We denote this quantity as X_i with $\mathbf{Pr}(X_i = x_i) = p_i$. The set of constraints is associated with the functions $g_j(x_i)$. The problem is then posed as follows:

$$\text{Max} \left(H = -\sum_i p_i \ln p_i \right)$$

subject to

$$\sum_i p_i = 1$$

and the following specified values of the mean v_j:

$$\sum_i g_1(x_i) p_i = \langle g_1(X) \rangle = v_1,$$

$$\sum_i g_2(x_i) p_i = \langle g_2(X) \rangle = v_2,$$

$$\vdots$$

$$\sum_i g_j(x_i) p_i = \langle g_j(X) \rangle = v_j,$$

$$\vdots$$

$$\sum_i g_m(x_i) p_i = \langle g_m(X) \rangle = v_m. \tag{6.55}$$

Then the Lagrangian can be written as

$$L = -\sum_i p_i \ln p_i - \lambda_0' \left(\sum_i p_i - 1 \right) - \sum_{j=1}^m \lambda_j \left[\sum_i g_j(x_i) p_i - v_j \right], \tag{6.56}$$

where the constraints (6.55) have been condensed into a sum of terms, these having first been multiplied by the appropriate Lagrange multiplier. Following the same procedure as before, differentiating the Lagrangian and equating to zero, it is found that

$$p_i = \exp \left[-\lambda_0 - \sum_{j=1}^m \lambda_j g_j(x_i) \right]. \tag{6.57}$$

From the condition of coherence ($\sum p_i = 1$)

$$\sum_i \exp \left[-\lambda_0 - \sum_{j=1}^m \lambda_j g_j(x_i) \right] = 1,$$

which gives the partition function

$$Z(\lambda_1, \lambda_2, \ldots, \lambda_m) = \exp \lambda_0 = \sum_i \exp \left[-\sum_{j=1}^m \lambda_j g_j(x_i) \right]. \tag{6.58}$$

Hence

$$p_i = \frac{\exp \left[-\sum_{j=1}^m \lambda_j g_j(x_i) \right]}{Z(\lambda_1, \lambda_2, \ldots, \lambda_m)}. \tag{6.59}$$

Also

$$\lambda_0 = \ln Z. \tag{6.60}$$

There are several important applications of this distribution. First, in statistical mechanics, we may derive the Gibbs' grand canonical ensemble. This is an even more forbidding title, and again the demon takes exception to the canonical aspect. (Although he likes mathematics in its most authoritative form.) The methodology outlined above follows a relatively straightforward recipe; the ingredients are entropy, the constraints, the formation of the Lagrangian, and the differentiation of the latter to obtain the desired distribution. The partition function then provides a summary of the whole result.

The following relationships are interesting and useful.

$$\langle g_j(X)\rangle = v_j = -\frac{\partial \ln Z}{\partial \lambda_j}. \tag{6.61}$$

$$H_{\max} \equiv S = \lambda_0 + \lambda_1 \langle g_1(X)\rangle + \cdots + \lambda_m \langle g_m(X)\rangle$$
$$= \lambda_0 + \lambda_1 v_1 + \cdots + \lambda_m v_m. \tag{6.62}$$

This defines the maximum entropy $H_{\max} \equiv S$, which is a function of $\lambda_0 = \ln Z$ and the expected values only. The variance of the distribution of $g_j(X)$ is given by

$$\sigma_{G_j}^2 = \frac{\partial^2 \lambda_0}{\partial \lambda_j^2}, \tag{6.63}$$

and the covariance by

$$\sigma_{G_j,G_k} = \frac{\partial^2 \lambda_0}{\partial \lambda_j \lambda_k}. \tag{6.64}$$

See Exercise 6.4 for more details on these and related relationships.

The exponential and normal distributions in their continuous form can also be seen as part of the present theory, as illustrated in the following two examples.

Example 6.5 *Exponential distribution.* Consider Equation (6.59) in the following form:

$$p_i = \frac{\exp\left[-\sum_{j=1}^{m} \lambda_j g_j(x_i)\right]}{\sum_k \exp\left[-\sum_{j=1}^{m} \lambda_j g_j(x_k)\right]}. \tag{6.65}$$

We consider first the case where $j = m = 1$, and $g_1(x_i) = x_i$, and obtain essentially Equation (6.47) again, but with X as the random quantity rather than E:

$$p_i = \frac{\exp(-\lambda_1 x_i)}{\sum_k \exp(\lambda_1 x_k)}.$$

Now divide the x-axis into equally spaced intervals Δx such that x_i is contained in the ith interval; $i = 0, 1, 2, \ldots, i, \ldots$ Then

$$p_i = \frac{\left[\exp(-\lambda_1 x_i)\right] \cdot \Delta x}{\sum_k \left[\exp(-\lambda_1 x_k)\right] \cdot \Delta x}.$$

In the limit as $\Delta x \to 0$,

$$p_i = f_X(x)\,dx = \frac{\exp(-\lambda_1 x)\,dx}{\int_{x=0}^{\infty} \exp(-\lambda_1 x)\,dx}$$

or, carrying out the integration,

$$f_X(x) = \lambda_1 \exp(-\lambda_1 x),\tag{6.66}$$

which is the familiar exponential distribution with mean and standard deviation equal to $1/\lambda_1$; this was derived before as (6.53). If the constraint equation is written as $\sum p_i x_i = \mu$, then $1/\lambda_1 \equiv \mu$.

Example 6.6 *Normal distribution.* We consider now the case where $g_1(x_i)$ is again equal to x_i, with $\sum p_i x_i = \mu$, and the second constraint is

$$\sum p_i (x_i - \mu) = \sigma^2.\tag{6.67}$$

Equation (6.67) suggests that the variance (or equivalently the standard deviation) is fixed. The constraint may be written as

$$\sum p_i x_i^2 = \sigma^2 + \mu^2,\tag{6.68}$$

so that we take $g_2(x_i) = x_i^2$. Using Equation (6.65) and dividing the x-axis into equally spaced intervals $i = \ldots, -2, -1, 0, 1, 2, \ldots$, with x_i contained in the ith interval, and following the same procedure as before, it is found that

$$p_i = f_X(x)\,dx = \frac{\exp\left(-\lambda_1 x - \lambda_2 x^2\right)dx}{\int_{-\infty}^{\infty} \exp\left(-\lambda_1 x - \lambda_2 x^2\right)dx}.\tag{6.69}$$

This can be written in the familiar form

$$f_X(x) = \frac{1}{\sqrt{2\pi}\sigma}\exp\left[-\frac{1}{2}\left(\frac{x-\mu}{\sigma}\right)^2\right],\tag{6.70}$$

in which λ_1 and λ_2 have been solved for in terms of $\mu = \int x f_X(x)\,dx$ and $\sigma^2 = \int (x - \mu)^2 f_X(x)\,dx$. The distribution is dealt with in Sections 3.3.3 and 7.5. In the latter case, the central limit theorem is used to derive the distribution as the 'model of sums'. The maximum-entropy derivation above is the simplest way to deduce this distribution from a theoretical standpoint, but all derivations give special insights.

In the case that all first and second moments are specified in the case of several random quantities $\{X, Y, Z, \ldots\}$ the maximum-entropy distribution is the standard multidimensional normal distribution with the given moments. See Exercise 6.7.

Transition from discrete to continuous distributions   Jaynes in his 1963 Brandeis lectures (Jaynes 1963b; Rosenkrantz, 1983) discussed the question of the transition from discrete to continuous probability distributions. In the expression for entropy derived in the preceding,

$$H = -\sum p_i \ln p_i,\tag{6.71}$$

we can take

$$p_i = \mathbf{Pr}\,(x_i \leq X < x_{i+1}) = f_X\,(x)\,\mathrm{d}x,$$

where $f_X\,(x)$ is the density corresponding to discrete probabilities p_i and p_{i+1} at points x_i and x_{i+1} respectively. The difference $(x_i - x_{i+1})$ is equal to $\mathrm{d}x$. The equivalence of p_i and $f_X\,(x)\,\mathrm{d}x$ seems natural in the probabilistic weighting p_i of $(\ln p_i)$ in (6.71). But in the term $(\ln p_i)$, Jaynes argues for the inclusion of a function $m\,(x)$ measuring the density of points in the transition from the discrete to the continuous case (in units of proportion of points per unit length). If there are h points in total over the range of x, then

$$\frac{1}{\mathrm{d}x} = hm\,(x)$$

since there is one point added in the interval x to $x + \mathrm{d}x$, or

$$\mathrm{d}x = \frac{1}{hm\,(x)}.$$

Then

$$p_i \rightarrow \frac{f_X\,(x)}{hm\,(x)} \tag{6.72}$$

in the transition from discrete to continuous. The final result is

$$H = -\int f_X\,(x)\ln\left[\frac{f_X\,(x)}{m\,(x)}\right]\mathrm{d}x \tag{6.73}$$

from which an infinite term corresponding to $(\ln h)$ has been subtracted, leaving the terms of interest. This strategy hinges on the use of the function $m\,(x)$. It also deals with problems of invariance in the continuous form of the expression for entropy. In the transition to the continuous distributions, Equations (6.66) and (6.70) above, we have taken $m\,(x)$ as a constant.

6.6.1   Grand canonical ensemble

We now outline the grand canonical ensemble and the grand partition function. These are derived for the case where there is uncertainty regarding the number of particles in the system, in addition to that regarding the energy level or quantum state. We can think of our system as a 'leaky box', a device used by Tribus (1961) to explain the present situation. Other references such as Gopal (1974), Eisberg and Resnick (1974) and Kapur and Kesavan (1992) might be consulted. Previously, we considered that the number of particles in our system was fixed; this corresponded to the number n for which N trials $(N \gg n)$ by the demon were conducted. Fundamental particles have a finite probability of penetrating a wall; photons are absorbed and emitted by the walls in a blackbody, with higher energy particles being converted into a number of lower energy ones and *vice versa*, so that the number of particles is not fixed.

In Section 6.5, each trial corresponded to the demon 'interrogating' the system to find the energy level of a particle in a spin of the roulette wheel. In the grand canonical ensemble, the systems (leaky boxes) are once again placed in a 'supersystem' containing N (leaky) boxes such that the number of particles in this supersystem is fixed. The implication of this is that the mean number of particles is fixed, but that the number varies from one realization to another, modelled as a random quantity in the demon's game. The demon would be faced with having to sample a different number of particles from L (random) with particular values $\ell = 0, 1, 2, \ldots$, each time he wishes to sample an energy level. For each of these combinations, he again plays N times, selecting only those results that conform to the constraints. For ℓ particles, let there be k_ℓ energy levels corresponding to the quantum states. Let i denote all of the combinations of ℓ and k_ℓ. This can reasonably be modelled as being countably infinite. The whole experiment is repeated $\aleph$ times. The number of ways of obtaining the result is counted over all combinations of ℓ and k_ℓ, represented by i. We consider here only one type of particle; the analysis is easily extended to more than one (see Tribus, 1961).

The problem reduces to that introduced above, that of finding

$$\text{Max} \left(H = - \sum_i p_i \ln p_i \right)$$

subject to

$$\sum_i p_i = 1$$

with the following equations representing the constraints corresponding to Equations (6.55):

$$\sum_i p_i e_i = \langle E \rangle \tag{6.74}$$

$$\sum_i p_i \ell_i = \langle L \rangle. \tag{6.75}$$

The functions $g_1(x_i)$ and $g_2(x_i)$ in Equations (6.55) become e_i and ℓ_i with $x_i = i$. Following Equations (6.58) and (6.59) above, the solution is

$$p_i = \frac{\exp(-\lambda_1 e_i - \lambda_2 \ell_i)}{Z(\lambda_1, \lambda_2)}, \tag{6.76}$$

where

$$Z(\lambda_1, \lambda_2) = \exp \lambda_0 = \sum_i \exp(-\lambda_1 e_i - \lambda_2 \ell_i), \tag{6.77}$$

and where there are two Lagrange multipliers, with $m = 2$ in Equation (6.55) above. The results (6.76) and (6.77) form the basis of Bose–Einstein and Fermi–Dirac statistics, to be discussed further in Sections 6.8 and 6.9 below. Equation (6.77) is usually written in the form

$$Z(\mu, T) = \sum_i \exp[(\mu \ell_i - e_i)/kT], \tag{6.78}$$

where μ is the chemical potential. In the function Z of Equations (6.76) and (6.78) one can think of i as ranging over all the values of ℓ, and for each ℓ, ranging over all the ℓ-particle quantum states of the system.

Again, the use of 'mental replicas' in the reasoning above provides a mechanistic framework within which we can ask the demon to perform experiments provided the mood of the moment takes him in that direction, or for us to perform Monte Carlo experiments of our own. But the fundamental idea is merely to maximize the entropy function subject to specified constraints.

6.7   Nature of the maximum: interpretations

We have mentioned that the entropy maximum is an absolute maximum, and that we do not have to worry about saddle points and the like. We shall give a proof of this statement, based on the presentation of Tribus (1969). First, we use the main result of the first part of the preceding section, with $H_{\max} \equiv S$, (6.62), and contrast this with another result that also satisfies the imposed constraints, but does not maximize entropy. We term these results A and B respectively. The probabilities assigned to the random quantity X_i are denoted p_i and q_i respectively. Both assignments are coherent.

We may now calculate

$$S - T = - \sum p_i \ln p_i + \sum q_i \ln q_i.$$

First, we add and subtract $\sum q_i \ln p_i$, giving

$$S - T = - \sum (p_i - q_i) \ln p_i + \sum q_i \ln \left(\frac{q_i}{p_i} \right). \tag{6.79}$$

But we know that

$$\ln p_i = - \ln Z - \sum_j \lambda_j g_j (x_i) \tag{6.80}$$

from (6.59), where $\ln Z \equiv \lambda_0$. Substituting for $\ln p_i$ in the first term on the right hand side of Equation (6.79),

$$S - T = \lambda_0 \sum_i (p_i - q_i) + \left[\sum_i p_i \sum_j \lambda_j g_j (x_i) - \sum_i q_i \sum_j \lambda_j g_j (x_i) \right] + \sum_i q_i \ln \left(\frac{q_i}{p_i} \right).$$

Now the first term in this equation disappears as a result of coherence of both systems (Table 6.1). The terms in the square brackets add to zero since the first one is

$$\sum_i p_i \sum_j \lambda_j g_j (x_i) = \sum_j \lambda_j \sum_i p_i g_j (x_i) = \sum_j \lambda_j \nu_j,$$

with the same result for the second since they both satisfy the constraint equations (Table 6.1). As a consequence

$$S - T = \sum_i q_i \ln \left(\frac{q_i}{p_i} \right). \tag{6.81}$$

Table 6.1 *Comparison of systems, A satisfying*
$S \equiv \mathrm{Max}(H = -\sum p_i \ln p_i)$, *and B with different*
probabilities q_i

A	B
$S = -\sum p_i \ln p_i$	$T = -\sum q_i \ln q_i$
$\sum p_i = 1$	$\sum q_i = 1$
$\sum_i g_j(x_i)\, p_i = v_j$	$\sum_i g_j(x_i)\, q_i = v_j$
$p_i = \frac{1}{Z} \exp\left[-\sum_j \lambda_j g_j(x_i)\right]$	The q_i differ from case A

To prove that this represents a positive quantity with a minimum at $q_i = p_i$, we make the following change to Equation (6.81):

$$S - T = \sum_i \Delta S_i = \sum_i \left[q_i \ln\left(\frac{q_i}{p_i}\right) - q_i + p_i \right], \tag{6.82}$$

where the notation ΔS_i has been introduced, and the addition and subtraction of probabilities q_i and p_i make no difference to the quantity $S - T$, since both q_i and p_i normalize to unity. Taking each term ΔS_i at a time we find

$$\frac{\partial \Delta S_i}{\partial q_i} = \ln q_i - \ln p_i \tag{6.83}$$

and

$$\frac{\partial^2 \Delta S_i}{\partial^2 q_i} = \frac{1}{q_i}. \tag{6.84}$$

The first derivative shows that there is a point of inflexion of each ΔS_i at $q_i = p_i$, and the second, being always positive, shows that they reach a minimum at this point. The expression $\Delta S_i = [q_i \ln(q_i/p_i) - q_i + p_i]$ is zero at $q_i = p_i$. Therefore the ΔS_i and their sum, $S - T$, are always positive, except when $q_i = p_i$. The maximum entropy value is then an unconditional maximum. Tribus (1969) ascribes the elegant last part of the proof to Gibbs.

We can now use Equation (6.26) to state that

$$\frac{w}{w'} = \exp\left[N\left(S - T\right)\right], \tag{6.85}$$

where w' is the number of ways in which the solution corresponding to T can be realized, rather than S, for which there are w ways. This can be made as large as we like, so that for large enough N, the maximum-entropy solution has a likelihood corresponding to a probability spike of unit mass. (See also Figure 6.6 above.)

Bayesian thought experiment   We could also consider the maximization of entropy as a 'thought experiment' in which Bayesian updating is used, conditional on the thought experiment, with vague prior probabilities. The experiment is aimed at comparing various solutions such as the sets corresponding to p_i and q_i above. We can compare the 'A' and 'B' solutions using

$$\mathbf{Pr}\,(\text{solution}|\text{results of thought experiment}) \propto \mathbf{Pr}\,(\text{results}|\text{solution method})$$

$$\propto w, \text{ or } w', \qquad (6.86)$$

for solutions A and B respectively. The result is in accordance with (6.20) and the vague prior information being absorbed since it is essentially a constant and dominated by the likelihood function. The maximum-entropy solution is much more likely than any other, consistent with (6.85).

6.8 Blackbodies and Bose–Einstein distribution

6.8.1 Blackbodies

The idea of the blackbody – seemingly vague – is indeed precise enough to have enabled Max Planck to discover the quantization of energy, first communicated in 1900. This is an elegant piece of theory, yet there were serendipitous aspects to his discovery, as we shall describe. Segré (1980, 1984), in two historical works, gives an interesting account of the development of the theory, tracing the twists and turns of the experimental and theoretical reasoning. He termed it an 'encompassing problem'. It is a common experience on a bright day that a room with an uncovered window will appear dark from the outside. The room acts as a trap for radiation which enters through the window. We know that black objects absorb radiation, and appear black for this reason. Examples are the room just mentioned, or the pupil of the eye. Conversely, white objects reflect light at all frequencies, and green objects absorb light radiation at all frequencies except those corresponding to green, which are reflected.

All bodies emit thermal radiation as a result of their temperature and absorb radiation from their surroundings. The frequency of the radiation increases with temperature and hot bodies – including black ones – become visible at high temperatures even without any light reflected from them. The porcelain maker Wedgwood observed and noted in 1792 that all bodies when heated become red at the same temperature. This can be confirmed in a rough way by observation of a campfire, grill, stones, and other objects in a fire turning red at the same temperature. Kirchhoff considered the power emitted per unit surface area (emissivity e) for an interval of frequency, as well as the energy absorbed per unit time; these are in units of power per unit frequency per unit area. The surface is at temperature T and emits energy in the frequency interval v to $v + dv$ at the rate $e\,(v, T)\,dv$ per unit area and per unit time. The quantity e depends upon the nature of the surface. The fraction of power falling on the surface that is absorbed by it is a, where a is a number between 0 and 1.

The two opposing surfaces shown in Figure 6.8(a) are black and white. These are enclosed in a box such that there is no contact between the materials, and they can exchange heat only by radiation. The interior of the box is maintained at the same temperature T. For a blackbody

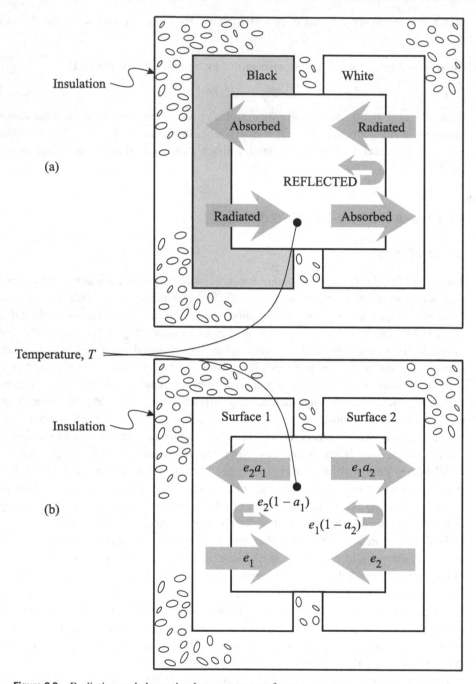

Figure 6.8 Radiation and absorption between two surfaces

$a = 1$, since all radiation is absorbed and none is reflected. The white surface reflects some radiation, but also absorbs and radiates the remainder. The surfaces are at constant temperature and therefore there must be no net absorption of radiation by either. The flows in the two directions in Figure 6.8(a) must then be the same, and this demonstrates that blackbodies must be the best radiators of energy. Good absorbers are therefore good radiators.

We now consider the situation in Figure 6.8(b), where two surfaces 1 and 2 are shown. These can be black or white, or any other colour or texture. The radiation from the surfaces is denoted e_1 and e_2 respectively. The absorption is $e_2 a_1$ and $e_1 a_2$ respectively, with corresponding reflections of $e_2 (1 - a_1)$ and $e_1 (1 - a_2)$. For constant temperature of the left hand surface,

$$e_1 + e_2 (1 - a_1) = e_2 a_1 \tag{6.87}$$

or

$$e_1 + e_2 = 2 e_2 a_1. \tag{6.88}$$

Similarly, for the right hand surface,

$$e_1 + e_2 = 2 e_1 a_2. \tag{6.89}$$

Therefore

$$e_1 a_2 = e_2 a_1 \tag{6.90}$$

or

$$\frac{e_1 (v, T)}{a_1 (v, T)} = \frac{e_2 (v, T)}{a_2 (v, T)} = u (v, T), \tag{6.91}$$

in which the function u is independent of the surface, and therefore a universal function, giving the spectral density. The result is Kirchhoff's law of heat exchange, showing that the ratio of emissivity to absorption coefficient is independent of the nature of the body and a function of frequency and temperature only.

The use of a cavity with the interior walls maintained at constant temperature and which communicates with the exterior through a small aperture provides a very good experimental approximation to a blackbody. This strategy results in radiation that is independent of the material. For a blackbody $a = 1$, since all radiation is absorbed and none is reflected. Black radiation has the characteristic of perfect randomness, with propagation in all directions and a definite spectral distribution of energy. The apparatus employed in the early blackbody experiments is shown in Figure 6.9, together with some of the results. It should be noted that at high temperatures, blackbodies will not appear black; stars and planets are modelled as blackbodies. In this way, the sun could be modelled as a blackbody at 6000 K.

6.8.2   Partition function, probabilities and solution

The derivation of the Bose–Einstein distribution provides the solution to the problem. The methodology is embodied in Section 6.6, and the grand partition function. The particles in the

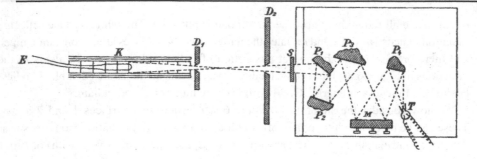

(a) Apparatus

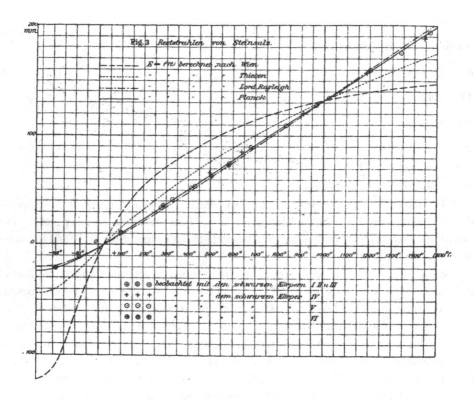

(b) Emission curves at constant frequency and variable temperature

Figure 6.9   Measurements of blackbody radiation. Based on H. Rubens and F. Karlbaum, *Annalen der Physik* **4**, 649 (1901)

present case are bosons; more accurately, we shall eventually deal with a special case of this, photons. We repeat the main result previously obtained, Equation (6.77):

$$Z(\lambda_1, \lambda_2) = \exp \lambda_0 = \sum_i \exp(-\lambda_1 e_i - \lambda_2 \ell_i), \tag{6.77}$$

with

$$p_i = -\frac{1}{\lambda_1} \frac{\partial \ln Z(\lambda_1, \lambda_2)}{\partial e_i} = \frac{\exp(-\lambda_1 e_i - \lambda_2 \ell_i)}{Z(\lambda_1, \lambda_2)}. \tag{6.92}$$

The usual form for Equation (6.77) was given in Equation (6.78):

$$Z(\mu, T) = \sum_i \exp[(\mu \ell_i - e_i)/kT]. \tag{6.78}$$

As we noted above, Schrödinger suggested that the results can be embodied in the logarithm of the partition function, containing everything in a nutshell. In the present case, $(\ln Z)$ is known as the grand potential, derivatives of which give the probabilities of various energy levels and the equations of state. Here we outline the theory to give the solution to the radiation problem.

In the case of bosons, the energy of the system is the sum of the 'private' energies of the particles. Each particle is therefore treated as being independent of the others. For ℓ particles with energy ϵ, the energy level in the expression for Z is

$$e = \ell\epsilon \tag{6.93}$$

for the quantum state considered. Equation (6.78) for Z can be shown to reduce to

$$Z = \prod_{i=1}^{\infty} \{1 - \exp[(\mu - \epsilon_i)/kT]\}^{-1} = \prod_{i=1}^{\infty} \xi_i, \tag{6.94}$$

where ϵ_i is the particle energy and

$$\xi_i = \{1 - \exp[(\mu - \epsilon_i)/kT]\}^{-1}. \tag{6.95}$$

Equation (6.94) can be shown directly, and this is the subject of Exercise 6.8.

We shall here demonstrate its character using a related method (Tribus, 1961). This first requires us to recall that the summation in Equations (6.77) and (6.78) consists of i covering all the values of ℓ, and for each ℓ, all the ℓ-particle quantum states of the system. There is therefore a double summation implied. Now there is a neat and simple way to express the partition function. First note that

$$\xi_i = \frac{1}{1 - \exp[(\mu - \epsilon_i)/kT]} = \{1 - \exp[(\mu - \epsilon_i)/kT]\}^{-1}$$
$$= 1 + \exp[(\mu - \epsilon_i)/kT] + \exp[(2\mu - 2\epsilon_i)/kT] + \exp[(3\mu - 3\epsilon_i)/kT] + \cdots \tag{6.96}$$

We shall demonstrate the idea by considering two states ϵ_1 and ϵ_2. Consider the product of the following two terms taken from (6.94), using (6.96).

$$
\begin{aligned}
\xi_1\xi_2 &= \frac{1}{1 - \exp[(\mu - \epsilon_1)/kT]} \cdot \frac{1}{1 - \exp[(\mu - \epsilon_2)/kT]} \\
&= \{1 + \exp[(\mu - \epsilon_1)/kT] + \exp[(2\mu - 2\epsilon_1)/kT] \\
&\quad + \exp[(3\mu - 3\epsilon_1)/kT] + \cdots\}\{1 + \exp[(\mu - \epsilon_2)/kT] \\
&\quad + \exp[(2\mu - 2\epsilon_2)/kT] + \exp[(3\mu - 3\epsilon_2)/kT] + \cdots\} \\
&= 1 \\
&\quad + \exp[(\mu - \epsilon_1)/kT] + \exp[(\mu - \epsilon_2)/kT] \\
&\quad + \exp[(2\mu - 2\epsilon_1)/kT] + \exp[(2\mu - \epsilon_1 - \epsilon_2)/kT] + \exp[(2\mu - 2\epsilon_2)/kT] \\
&\quad + \exp[(3\mu - 3\epsilon_1)/kT] + \exp[(3\mu - 2\epsilon_1 - \epsilon_2)/kT] \\
&\quad + \exp[(3\mu - \epsilon_1 - 2\epsilon_2)/kT] + \exp[(3\mu - 3\epsilon_1)/kT] + \cdots
\end{aligned}
\tag{6.97}
$$

The series includes a term for no particles and consequently zero energy, represented by the value of unity at the beginning. Then two terms represent the likelihood of a single particle being in the two energy states ϵ_1 and ϵ_2 respectively. The next three terms (third line of (6.97) above) represent the likelihood of two particles being in the first state, one in the first and one in the second, or both in the second state, respectively. The states can be thought of as boxes in which the (indistinguishable) particles are arranged. This is illustrated in Figure 6.10. The arrangements of three particles, and the corresponding terms in the fourth and fifth lines of expression (6.97) above are made clear in this figure. (Reference is also made to the occupancy problems and the introduction to Bose–Einstein statistics given in Section 2.6.3 and Table 2.3.) The terms of expression (6.97) represent all of the possible arrangements of any number of particles between two states.

It can be checked that if all terms of the kind

$$
\xi_i = \frac{1}{1 - \exp\left[(\mu - \epsilon_i)/kT\right]}
$$

are multiplied into the products of kind in expression (6.97) above, the expansion of this equation in a manner similar to (6.97) results in terms in which every possible number ℓ of particles are arranged in all possible i quantum states. This results in

$$
Z = \prod_{i=1}^{\infty} \{1 - \exp\left[(\mu - \epsilon_i)/kT\right]\}^{-1},
\tag{6.98}
$$

which is the first equality of (6.94). If we use (6.95), then

$$
Z = \prod_i \xi_i.
\tag{6.99}
$$

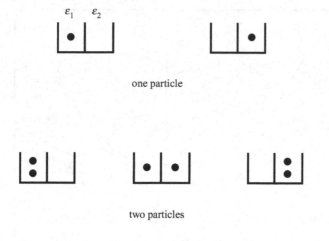

Figure 6.10 Particles in boxes (occupancy)

For completeness, we provide a précis of the solution to the blackbody problem. The chemical potential μ for a photon gas is zero. We consider therefore the partition function

$$Z = \prod_i \left[1 - \exp\left(-h\nu_i/kT\right)\right]^{-1},\tag{6.100}$$

where we have made the substitution for the quantum of energy, $\epsilon_i = h\nu_i$, in which ν is frequency and h is Planck's constant. Then

$$\ln Z = -\sum_i \ln\left[1 - \exp\left(-h\nu_i/kT\right)\right].\tag{6.101}$$

The summation in Equation (6.101) can be replaced by an integral:

$$\ln Z = -\frac{8\pi V}{c^3}\int_0^\infty \nu^2 \ln\left[1 - \exp\left(-h\nu/kT\right)\right]\mathrm{d}\nu$$

where c is the speed of light and V is the volume of the blackbody enclosure. Now the internal energy

$$\langle E\rangle = kT^2\left(\frac{\partial \ln Z}{\partial T}\right)_V.\tag{6.102}$$

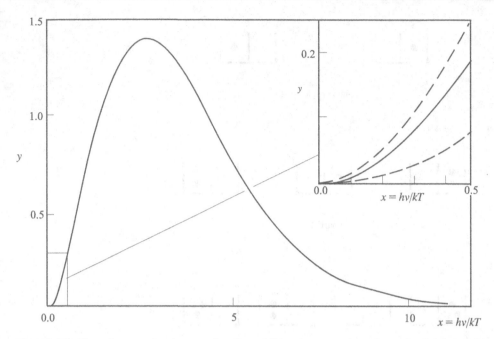

Figure 6.11   Plot of energy density equation. Dotted lines show equations due to Rayleigh and Wien. See also Segré (1980)

The spectral density $u(v, T)$ is defined by

$$\langle E \rangle = V \int_0^\infty u(v, T)\,dv \tag{6.103}$$

and hence

$$u(v, T) = \frac{8\pi v^3}{c^3} \frac{h}{\exp(hv/kT) - 1}. \tag{6.104}$$

This is Planck's celebrated equation for the energy density as a function of frequency, which was found to agree with the experimental results.

The serendipitous nature of the discovery is famous; Planck originally divided the energy into small finite quantities for computational reasons, and expected that in the limit of infinitesmal intervals the calculations would be correct. It turned out that the relationship

$$\epsilon = hv \tag{6.105}$$

gave the best agreement, and he had then to develop a physical interpretation, which led to the quantization of energy, Equation (6.105) representing a quantum of energy.

Figure 6.11 gives a graph of the result (see also Figure 6.9) using

$$y = \frac{x^3}{\exp x - 1},$$

where

$$x = \frac{h\nu}{kT}.$$

The material given above on the blackbody problem represents an outline, with the intention of showing that it is quite accessible to a person with the introduction to entropy based on probability given in this chapter.

6.9   Other facets; demon's urns

We can feel, at this point, some satisfaction. We have derived, on different grounds, a simple formula for entropy:

$$H = -\sum p_i \ln p_i,$$

and have applied it effectively to a number of physical situations. The demon detests this mood of self-satisfaction. He has come up with alternative explanations of the results, and different sets of games. He suggests that we investigate the Bose–Einstein case a little further. Consider, he says, Equation (6.94) above:

$$Z = \prod_{i=1}^{\infty} \{1 - \exp[(\mu - \epsilon_i)/kT]\}^{-1}. \tag{6.106}$$

Hence

$$\ln Z = -\sum_i \{1 - \exp[(\mu - \epsilon_i)/kT]\}. \tag{6.107}$$

It can be shown that the mean number of particles for the partition function of (6.94) is given by

$$\langle L \rangle = \sum_i \frac{1}{\{\exp[(\epsilon_i - \mu)/kT]\} - 1},$$

and for state i by

$$\langle L_i \rangle = \frac{1}{\{\exp[(\epsilon_i - \mu)/kT]\} - 1}. \tag{6.108}$$

The derivation is dealt with in Exercise 6.11. The demon adds the question of degeneracy. If one can group several single-particle states g_i in a group corresponding to the energy ϵ_i, or in a small interval around ϵ_i, there is degeneracy g_i for state i. There might be states with different behaviour that can be so grouped. Then, using Equation (6.108),

$$\langle L_i \rangle = \frac{g_i}{[\exp(\epsilon_i - \mu)/kT] - 1}. \tag{6.109}$$

Ah, says the demon, but I can get this result with a new game using an urn rather than a roulette wheel! Furthermore, you are yourself rather fond of urn models! Recall that, in Pólya's urn of Section 2.6.5, containing b balls of which we sample n, the probability of drawing $\{x_1, x_2, \ldots, x_i, \ldots, x_k\}$ balls[2] comprising $\{g_1, g_2, \ldots, g_i, \ldots g_k\}$ of the various colours $\{1, 2, \ldots, i, \ldots, k\}$, is

$$\frac{1}{\binom{b+n-1}{n}} \prod_{i=1}^{k} \binom{g_i + x_i - 1}{x_i}. \tag{2.62}$$

Let us consider a particular term in the product such as

$$\binom{g_i + x_i - 1}{x_i}; \tag{2.63}$$

this represents the number of ways that one can draw x_i balls from g_i of colour i, but it also represents the number of ways that one can distribute x_i indistinguishable objects (particles) amongst g_i cells (states in energy level i). Reference is made to Section 2.6.5 and Equations

---

[2] The symbols x_i used for the remainder of this section are consistent with the urn models used in Chapter 2 and elsewhere, and not with the usage earlier in this chapter.

(2.62) and (2.63). Then the product

$$w_{BE} = \prod_{i=1}^{k} \binom{g_i + x_i - 1}{x_i} \tag{6.110}$$

represents the number of ways that

$$n = x_1 + x_2 + \cdots + x_i + \cdots + x_k \tag{6.111}$$

particles can be distributed amongst i energy levels with degeneracy

$$b = g_1 + g_2 + \cdots + g_i + \cdots + g_k. \tag{6.112}$$

Now, says the demon, I sample $n = \sum x_i$ balls from an urn containing $b = \sum g_i$ balls of colour i $(i = 1, \ldots, k)$ **with double replacement**, in accordance with Pólya's urn model. As in the roulette game, I repeat the game repeatedly until I find the most probable distribution. Mathematically, this means that I maximize Equation (6.110); it must be obvious now that the subscript 'BE' in this equation refers to 'Bose–Einstein'. We can just as well maximize the logarithm of Equation (6.110), that is, maximize

$$\ln w_{BE} = \sum_i \ln (g_i + x_i - 1)! - \sum_i \ln (x_i)! - \sum_i \ln (g_i - 1)!. \tag{6.113}$$

We know from Section 6.4 above that Stirling's approximation reduces to $\ln N! \to N \ln N - N$. Using this in the expression for $\ln w_{BE}$, we have

$$\ln w_{BE} = \sum_i \left[(g_i + x_i - 1) \ln (g_i + x_i - 1) - x_i \ln x_i - (g_i - 1) \ln (g_i - 1) \right]. \tag{6.114}$$

We can write the Lagrangian as

$$L = \ln w_{BE} - \lambda_0' \left(\sum_i x_i - n \right) - \lambda_1 \left(\sum_i x_i \epsilon_i - n \langle \epsilon \rangle \right), \tag{6.115}$$

where we have included the appropriate constraint equations. Note that these are merely variations of the constraints that we considered before:

$$\sum p_i - 1 = 0 \to \sum x_i - n = 0 \tag{6.116}$$
$$\sum p_i \epsilon_i - \langle \epsilon \rangle = 0 \to \sum x_i \epsilon_i - n \langle \epsilon \rangle = 0. \tag{6.117}$$

We can throw out as before those results that do not conform to the constraints, says the demon. The 'fixing' of the mean number of particles arises from the sampling itself. For given number of balls n, the numbers g_i are fixed. Note also that we have assumed $x_i, g_i \gg 1$ in making the Stirling approximation, and therefore $g_i + x_i - 1 \simeq g_i + x_i$. Differentiating the Lagrangian with respect to x_i, and using this excellent approximation, we have

$$x_i = \frac{g_i}{\left[\exp (\lambda_0 + \lambda_1 \epsilon_i) \right] - 1}. \tag{6.118}$$

Table 6.2　*Urn equivalents to thermodynamic models*

Thermodynamics	Urns
Particles, ℓ_i	Balls drawn from urn, x_i
Energy level, ϵ_i	Colour i
Degeneracy g_i for level i	Balls g_i of colour i
Indistinguishable, symmetrical, Bose–Einstein	Double replacement: $\prod_{i=1}^{k} \binom{g_i + x_i - 1}{x_i}$
Indistinguishable, antisymmetrical, Fermi–Dirac	Without replacement: $\prod_{i=1}^{k} \binom{g_i}{x_i}$
Classical, Maxwell–Boltzmann, distinguishable	Single replacement: $n! \prod_{i=1}^{k} \frac{g_i^{n_i}}{n_i!}$
Classical, Maxwell–Boltzmann, indistinguishable	Single replacement: $\prod_{i=1}^{k} \frac{g_i^{n_i}}{n_i!}$

We compare this to Equation (6.109) above, with appropriate substitution of the constants $\lambda_1 = 1/kT$ and $\lambda_0 = -\mu/kT$:

$$x_i = \frac{g_i}{\left[\exp\left(\epsilon_i - \mu\right)/kT\right] - 1}. \tag{6.119}$$

The equations are essentially the same, taking into account the use of the symbol x_i in place of $\langle L_i \rangle$ when using the urn model. The energy level for specifying $\langle L_i \rangle$ can be chosen to be appropriately small (Tolman, 1979) so we use ℓ_i in Table 6.2 and below.

Furthermore, says the demon, we can construct other probabilistic distributions by considering similar models. Consider Table 6.2. In general there is a correspondence between

- ℓ_i particles in energy level ϵ_i with degeneracy g_i and
- x_i balls of colour i drawn from an urn containing g_i balls of colour i,

where i ranges from 1 to k.

6.10　Applications outside thermodynamics

There have been many applications of the entropy concept outside the traditional ones in thermodynamics, and we will consider some of these in this section. We return in this section to the basic expression for entropy,

$$H = -\sum p_i \ln p_i.$$

Jessop (1990) in problems of transport and Kapur and Kesavan (1992) in a variety of areas

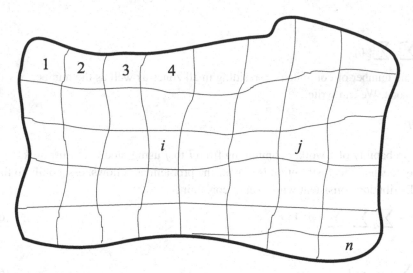

Figure 6.12   Geographical zone

give overviews of the use of the principle. Jowitt (1979) used the principle to derive the Gumbel distribution (see Chapter 9 for detail of this distribution) as a way of interpreting *pdf*s of maximum annual flood discharges, together with an interesting analysis of related parameter estimation. He also used the principle in a rainfall runoff model. This is summarized in Jowitt (1999) together with an application; the catchment storage elements and the catchment water were described in probabilistic terms using in both cases the principle of maximum entropy.

A further application of the entropy concept lies in the area of granular media. This was developed by Jowitt and Munro (1975), and was concerned with the use of mean porosity as a constraint in determining the microstucture distribution. It was found that the microstructure was constrained to a number of feasible packing arrangements. An analysis incorporating these constraints gives a theoretical interpretation of the critical void ratio.

Example 6.7 *Detailed illustration in transport: origin to destination work trips.* This application lies in the subject of transportation planning (Cohen, 1961; Wilson, 1970). We consider a region, that could be a city or an area including several urban centres. We consider origin-to-destination work trips within the region. The region is subdivided into n discrete zones $\{1, 2, 3, \ldots, i, \ldots, j, \ldots, n\}$; see Figure 6.12; see also Section 3.4. We wish, for planning purposes, to estimate the number of persons t_{ij}^k commuting from i to j, using mode of transportation k (bus, car, rail transit and so on). The total number of trips

$$t = \sum_i \sum_j \sum_k t_{ij}^k. \tag{6.120}$$

We know the number of commuters r_i residing in all zones as well as the number of workers s_j in each zone. We can write

$$p_{ij}^k = t_{ij}^k/t;$$

this is the probability of a worker commuting from i to j using mode of transport k.

One way to study this distribution is to pose the problem as follows, essentially to find the 'flattest' distribution consistent with a set of constraints.

$$\text{Max } H = -\sum_i \sum_j \sum_k p_{ij}^k \ln p_{ij}^k \tag{6.121}$$

subject to

$$\sum_i \sum_j \sum_k p_{ij}^k = 1 \tag{6.122}$$

and

$$\sum_j \sum_k p_{ij}^k = r_i/t, \tag{6.123}$$

$$\sum_i \sum_k p_{ij}^k = s_j/t. \tag{6.124}$$

One further constraint is added; it is supposed that the cost c_{ij}^k for travel from i to j using k is known, as well as the total spent on travel c. (These could represent some kind of generalized cost, or impedance.) Then

$$\sum_i \sum_j \sum_k p_{ij}^k c_{ij}^k = c/t = \bar{c}. \tag{6.125}$$

We follow the usual procedure of forming the Lagrangian

$$L = H - \lambda_0' \left(\sum_i \sum_j \sum_k p_{ij}^k - 1 \right) - \sum_i \lambda_i^{(1)} \left(\sum_j \sum_k p_{ij}^k - r_i/t \right)$$
$$- \sum_j \lambda_j^{(2)} \left(\sum_i \sum_k p_{ij}^k - s_j/t \right) - \lambda^{(3)} \left(\sum_i \sum_j \sum_k p_{ij}^k c_{ij}^k - \bar{c} \right). \tag{6.126}$$

In this equation, there are $(2n - 2)$ Lagrange multipliers. Then differentiating and solving, it is found that

$$p_{ij}^k = \exp\left(-\lambda_0 - \lambda_i^{(1)} - \lambda_j^{(2)} - \lambda^{(3)} c_{ij}^k \right). \tag{6.127}$$

Substitution of the constraints leads to a version of the so-called gravity model, in which the 'attraction' between origin i and destination j decreases with cost c_{ij} (and, by implication, distance) for like modes of transport. In Newton's gravity model, the force of attraction between two bodies is governed by a distance deterrence function proportional to $1/(\text{distance})^2$. We shall not enter into a detailed discussion of the assumptions in the model presented in this section, but note that the constraints can be modified to give a variety of results.

6.10.1 Prior probabilities

One of the perceived difficulties in inference – the subject of Chapter 8 – is the assignment of prior probabilities to be used with Bayes' theorem to determine posterior probabilities related to parameter estimation. There is a desire often to obtain a firm indisputable prior probability distribution. Entropy methods offer a possible avenue in this direction. But it is rare that one's prior knowledge corresponds to firm knowledge of moments. More usually it is a question of background information and knowledge, often not easily put into a maximum-entropy formulation. An exception is the case when no moments are known, leading to a uniform distribution. This is the case of prior ignorance.

6.11 Concluding remarks

The objective of the chapter was to introduce an important line of probabilistic thinking, using a single unified definition of entropy from which all distributions of interest can be derived. The definition was given in the introduction to this chapter:

Entropy is the smallest expected number of YES–NO questions to determine the true event.

Many applications and interpretations rich in meaning have been seen to flow from this definition. These were termed 'facets' of the jewel, entropy. The interpretation of some thermodynamic results in terms of urn models (Section 6.9) can also be viewed as examples of the use of the same fundamental definition.

The maximum-entropy method is based on the assumption that certain parameters giving expected values (mean, variance, . . .) are fixed but that in any particular realization the random quantity is free to take on any value provided that the expected values are maintained 'on average'. This led to derivations of common probabilistic models such as the exponential and normal distributions. These are derived elsewhere in the book on a different basis, for example the normal distribution is derived in the next chapter on the basis of the central limit theorem. Insights are to be gained by seeing the different interpretations of a particular distribution.

The fixing of moments works particularly well in the case of thermodynamic quantities. The continual fluctuations of energy of particles with the total (or mean) energy remaining constant is a vivid and satisfying idealization. But there have been many instances where this idealization of a particular situation has to be reconsidered in the light of reality, and in particular of measurements. The best approach is that a discrepancy between theory and experience indicates the possibility of new constraints or a misapprehension regarding those already included. In this way, insights into the underlying mechanism may be obtained.

One has to take special care to avoid treating the concept as a sinecure, in the use of entropy to situations outside thermodynamics. It is tempting to say 'the mean is known and nothing else'. Problems in the world usually contain pieces of information that are not of this 'clean' nature. In fact the situation is often murky, with vague knowledge of the mean and other statistical parameters. Further, there is not the vivid kind of interpretation described in the previous paragraph. Maximum-entropy methods are beautiful because of their elegance and depth of thought, but they do not provide a refuge for solution of the 'murky' problems of the world. An example is the use of maximum-entropy prior probability distributions as a matter of dogma; this is simply not convincing.

6.12 Exercises

The chapter deals with a probabilistic definition and treatment of entropy. Applications in various areas have been introduced. The problems below can be supplemented with detailed analysis of informa- tion theory, probability theory, physics and thermodynamics as well as other applications outside of thermodynamics. The maximum-entropy derivations provide additional insights into distributions de- rived in other ways elsewhere in the book.

6.1 Rare events, those events with a low probability p, are said to cause greater surprise when they occur and therefore contain greater information. The functions 'surprisal' and 'information' s_u are defined as

$$s_u \propto \ln \frac{1}{p} = -\ln p;$$ (6.128)

see Tribus (1969) and Lathi (1998). Discuss this concept in terms of events such as

- aircraft crashes
- suicidal aircraft crashes (Twin Towers)
- frequent letters in the English language, e.g. t, e, a as against rare ones x, q, z (use of code words)
- events that cause surprise generally.

Discuss entropy as expected information

$$\langle S_u \rangle = \sum p_i (s_u)_i$$ (6.129)

and the relevance to source encoding (Lathi, 1998).

6.2 Consider an event E with probability p. Plot the entropy associated with this event (and its complement) against p. What would the maximum-entropy assignment of probability be, in the absence of constraints other than coherence?

Explain and contrast why the maximum-entropy solution becomes very likely, with a 'sharp' distribution while the plot of H against p is quite smooth.

6.3 Solve Example 6.4, obtaining the probabilities in terms of the mean μ. Plot the three p_i against μ, the latter ranging from 0 to 2.

6.4 Complete the proof of Equations (6.61) to (6.64). Also show that

$$\lambda_j = \frac{\partial S}{\partial \nu_j}, \tag{6.130}$$

where S represents $H_{\max}$. Form the matrix of second order partial derivatives of λ_0 with respect to $\lambda_1, \ldots, \lambda_j, \ldots, \lambda_m$ (the Hessian matrix). Show that this is the variance–covariance matrix. Since this is positive definite, what conclusions can be reached?

6.5 Show that the maximum-entropy distribution subject to no constraints other than coherence is a uniform distribution.

6.6 Find maximum-entropy solutions for the gamma and beta distributions.

6.7 Show that the maximum-entropy distribution in the case where the first and second moments (means, variances and covariances) of a set of random quantities $X_1, X_2, \ldots, X_i, \ldots, X_n$ are fixed is the multidimensional normal distribution. Note that here we use the subscript i to denote individual random quantities and not particular values of a single random quantity as in (6.55).

The treatment of 'statistical physics' in the various text books resembles somewhat the Tower of Babel, as would be expected for a subject rapidly developed. There is an opportunity to write it all clearly and succinctly.

6.8 Show the equivalence (6.94) directly starting with Equations (6.77) and (6.78). Confirm in examples that the occupancy corresponds to the information given in Section 2.6.3 and Table 2.3.

6.9 Again let $Z = \prod_i \xi_i$. Obtain the function ξ_i in (6.99) for the Fermi–Dirac gas; the result is

$$\xi_i = 1 + \exp\left[(\mu - \epsilon_i)/kT\right],$$

since not more than one particle can go to any energy state.

6.10 Derive the Bose–Einstein distribution using Jaynes' widget method. It concerns a salesperson who is in charge of a factory producing differently painted widgets. The question is which colour should be produced each day. Constraints are concerned with the average number of widgets ordered each day and the average size of individual orders for the different colours. Details can be found in Jaynes (1963a).

6.11 Show that the mean number of particles for the partition function of (6.94) is given by

$$\langle L \rangle = kT \frac{\partial \ln Z}{\partial \mu}, \tag{6.131}$$

and in state i

$$\langle L_i \rangle = kT \frac{\partial \xi_i}{\partial \mu}, \tag{6.132}$$

and that this results in

$$\langle L \rangle = \sum_i \frac{\exp[(\mu - \epsilon_i)/kT]}{1 - \exp[(\mu - \epsilon_i)/kT]}$$

$$= \sum_i \frac{1}{\{\exp[(\epsilon_i - \mu)/kT]\} - 1} \tag{6.133}$$

and

$$\langle L_i \rangle = \frac{1}{\{\exp[(\epsilon_i - \mu)/kT]\} - 1}. \tag{6.134}$$

Use the result $\xi_i = \{1 - \exp[(\mu - \epsilon_i)/kT]\}^{-1}$.

6.12 Derive the Fermi–Dirac distribution using an urn model.

6.13 Derive the Maxwell–Boltzmann distribution using an urn model.

6.14 Review one or more areas of interest of application of the entropy expression outside thermodynamics.

6.15 Derive from first principles the expression for the rate of a spontaneous reaction using the theory of activation energy,

$$\nu \exp\left(-\frac{\epsilon_A}{kT}\right),$$

where ϵ_A is the energy barrier and ν the number of runs per second at the barrier.

7 Characteristic functions, transformed and limiting distributions

Like the crest of a peacock, so is mathematics at the head of all knowledge.

Old Indian saying

In mathematics, you don't understand things. You just get used to them.

John von Neumann

7.1 Functions of random quantities

Often we need to manipulate distributions and to consider functions of random quantities. We dealt with an important example in Chapter 4, where utility U was a function of one or more attributes, say X_j. For optimal decision-making, we needed to calculate the expected utility $\langle U \rangle$, that is, the first moment of U about the origin. The probability assignment had originally been made over the attributes, and we used the methods of Chapter 3 to obtain the expected value $\langle U \rangle$ of a **function** of these. But we might wish to derive the entire probability distribution of U in detail, given the distribution over the attributes X_j, for instance in obtaining sensitivity to assumptions.

There are many other cases where we consider functions of random quantities. The voltage V across a resistor may be random with known probability density, and we might wish to find the probability density associated with the power W:

$$W = \frac{V^2}{r},$$

where r is the resistance. A further example is the calculation of the force F acting on a tubular member in an offshore structure associated with a current velocity of magnitude V. This could be estimated using Morison's equation

$$F = kV^2,$$

in which k is a constant. The problem to be addressed is: knowing the probability distribution of V, how does one find the probability distribution of F? Knowing the moments of V, how does one find the corresponding moments of F?

These questions will initially be addressed. Flowing from these ideas, characteristic functions and limit theorems provide further solutions, fundamental insights and background to probability theory.

7.2   Transformation of distributions

A most useful and general idealization has been introduced in Chapter 4, which will help us to conceptualize the present situation. This is, simply, to treat the probability distribution as representing 'masses' of probability and to attach them to the curve giving the relationship between the new random quantity, say Y, and the old one, X. Figure 7.1(a) represents a typical relationship. It doesn't matter whether the distributions are discrete or continuous; in the first case the probability is treated as a set of 'point masses' on the curve, while in the second, the mass is 'smeared' out as a distributed mass. This way of thinking about the problem will assist throughout the following analyses. We showed in Chapter 4 that it is convenient to think of the probability masses as attached to the curve or relationship between U and the values of X, and to take moments using U or X as 'distance' (see Figure 4.1).

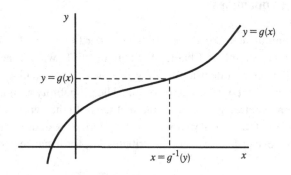

(a) Monotonic relationship between x and y

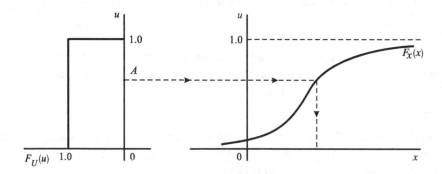

(b) Probability integral transform

Figure 7.1   Transformation of random quantities

7.2.1 Single random quantities

We consider the case shown in Figure 7.1(a) where the quantity Y is a one-to-one increasing function of the quantity X,

$$Y = g(X). \tag{7.1}$$

This function could relate a set of points $\{x_i\}$ to $\{y_i\}$, at which probability masses are attached in the discrete case, or relate a continuous set of $\{x\}$ values to the corresponding $\{y\}$ in the case of continuous quantities. We assume that we can invert the relationship to obtain

$$X = g^{-1}(Y). \tag{7.2}$$

Then we can see from Figure 7.1(a) that if $Y \leq y$ then

$$\left\{X = g^{-1}(Y)\right\} \leq \left\{x = g^{-1}(y)\right\}. \tag{7.3}$$

Therefore, thinking of the probability as masses on the curve,

$$\mathbf{Pr}(Y \leq y) = \mathbf{Pr}\left[X = g^{-1}(Y) \leq x = g^{-1}(y)\right].$$

In terms of cumulative distribution functions,

$$F_Y(y) = F_X\left[g^{-1}(y)\right]. \tag{7.4}$$

In general we can map regions or intervals of X to the corresponding regions of Y, given the one-to-one monotonic relationship between the two. Equation (7.4) is valid for both discrete and continuous random quantities.

Example 7.1 Given the standard logistic distribution

$$F_X(x) = \frac{1}{1 + \exp(-x)} = \frac{\exp x}{1 + \exp x}, \text{ for } -\infty < x < \infty, \tag{7.5}$$

find the distribution function of $Y = \exp X$. The distribution (7.5) has been used for models of population growth (Gumbel, 1958). The transformation maps the random quantity X into the space $(0, \infty)$. We have that the inverse relationship is

$$X = \ln Y,$$

and using Equation (7.4),

$$F_Y(y) = \frac{y}{1 + y} \text{ for } y \geq 0.$$

Example 7.2 *Probability Integral Transform*; Figure 7.1(b). Given a continuous random quantity X with *cdf* $F_X(x)$, we consider the random quantity U defined by

$$U = F_X(X). \tag{7.6}$$

In other words, U is defined by the cumulative distribution function itself, so that $g \equiv F_X$ in Equation (7.1). What is the distribution of U? In fact, it is uniform on $[0, 1]$. Note that $0 \leq F_X(x) \leq 1$, so that U is confined to this range. The inverse relationship corresponding to Equation (7.6) is

$$X = F_X^{-1}(U).$$

Also, by Equation (7.4),

$$F_U(u) = F_X\left[F_X^{-1}(u)\right] = u \text{ for } 0 \leq u \leq 1. \tag{7.7}$$

The distribution of U is therefore uniform on $[0, 1]$. It is useful also to consider a continuous function F_X that is strictly increasing (no flat regions); the reasoning can be extended to conclude that, if U has a continuous uniform distribution on $[0, 1]$, the random quantity X defined by

$$F_X(X) \equiv U$$

has a *cdf* $F_X(x)$. U defines the probability integral transform.

The results in this example are very useful. In Monte Carlo simulation, we often wish to obtain a sample from a given distribution $F_X(x)$. It is relatively easy to sample from a uniform distribution. With the method just outlined, we can sample from a uniform distribution, and use the relationship between $u = F_X(x)$ and x to find the value of x. This is shown in Figure 7.1(b). An example of application is given in Exercise 7.2 and in Chapter 11.

For the case where the relationship (7.1) is negative, for example $Y = -3X$, the approach has to be modified slightly. Then, for this example (Figure 7.2)

$$F_Y(y) = \mathbf{Pr}(Y \leq y) = \mathbf{Pr}\left(X \geq x = -\frac{y}{3}\right)$$
$$= 1 - F_X\left(x = -\frac{y}{3}\right) \tag{7.8}$$

for a continuous distribution. In the case of a discrete distribution, the probability mass (if any) at $X = x$ must be added back into this equation since it has been subtracted out in the term $F_X(x)$. In this case, Equation (7.8) becomes

$$F_Y(y) = \mathbf{Pr}(Y \leq y) = \mathbf{Pr}\left(X \geq x = -\frac{y}{3}\right)$$
$$= 1 - F_X\left(x = -\frac{y}{3}\right) + p_X\left(x = -\frac{y}{3}\right). \tag{7.9}$$

If the relationship between Y and X is not monotonic, one can map from one domain to another in such a way as to obtain the required result. For example, in Figure 7.3, one might obtain the *cdf* of Y in the case of a continuous random quantity as follows.

$$\mathbf{Pr}(Y \leq y) = F_Y(y) = F_X(x_1) + F_X(x_3) - F_X(x_2),$$

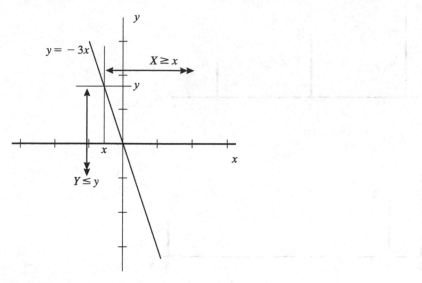

Figure 7.2 y as a negative function of x

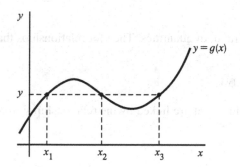

Figure 7.3   Non-monotonic relationship between y and x

where $y = g(x_1) = g(x_2) = g(x_3)$. For the discrete case, this has to be modified so as to include any probability masses at these three points. One can again imagine that the probability masses, discrete or continuous, are placed on the curve $y = g(x)$.

Returning to relationships that are one-to-one, the masses of probability for discrete random quantities will simply be transferred to the corresponding y-value, as in

$$p_Y(y) = p_X\left[g^{-1}(y)\right].\tag{7.10}$$

Example 7.3 An honest coin is tossed twice. The probabilities of zero, one or two heads are as shown in Figure 7.4(a). Dollar rewards are given, equal to the number of heads raised to the power three. The probability distribution of the dollar rewards is given in Figure 7.4(b).

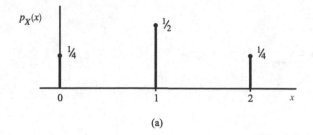

(a)

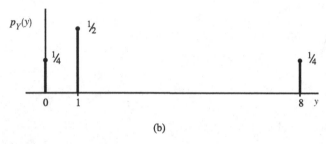

(b)

Figure 7.4   Example where $y = x^3$

We consider now the case of continuous random quantities. Then for relationships that are one-to-one, continuous and differentiable,

$$\mathbf{Pr}\,(y \le Y \le y + \mathrm{d}y) = \mathbf{Pr}\,(x \le X \le x + \mathrm{d}x),$$

as shown in Figure 7.5, with two sets of values that are linked by the relationship $y = g(x)$. Thus

$$f_Y(y)\,\mathrm{d}y = f_X(x)\,\mathrm{d}x,$$

since this is the probability mass on the short segment $\mathrm{d}s$ (Figure 7.5). Consequently

$$f_Y(y) = f_X(x)\,\frac{\mathrm{d}x}{\mathrm{d}y}$$

$$= f_X\left[g^{-1}(y)\right]\frac{\mathrm{d}g^{-1}(y)}{\mathrm{d}y}.$$

In this, it has been assumed that the function $g(x)$ is increasing, and that it can be inverted. For a decreasing function (right hand illustration of Figure 7.5),

$$f_Y(y)\,\mathrm{d}y = -f_X(x)\,\mathrm{d}x$$

and the result is the same as before, except for the presence of the minus sign. But the derivative $\mathrm{d}x/\mathrm{d}y$ is also negative in this case, so that we obtain generally, for monotonic increasing or

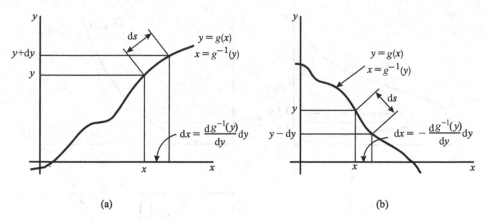

Figure 7.5 Increasing and decreasing functions

decreasing functions,

$$f_Y(y) = f_X\left[g^{-1}(y)\right]\left|\frac{dg^{-1}(y)}{dy}\right|_+,$$

(7.11)

where $||_+$ indicates the absolute value.

Example 7.4 The depth X of a beam of rectangular cross section is normally distributed with a mean d and a standard deviation σ_d. The probability distribution of the second moment of area can be found as follows. First,

$$f_X(x) = \frac{1}{\sqrt{2\pi}\sigma_d}\exp\left[-\frac{1}{2}\left(\frac{x-d}{\sigma_d}\right)^2\right].$$

Let

$$Y = \frac{bX^3}{12} = g(X)$$

be the second moment of area, where $b =$ width of beam, taken as a constant. Then

$$g^{-1}(Y) = \left(\frac{12Y}{b}\right)^{\frac{1}{3}}$$

and

$$\frac{dg^{-1}(y)}{dy} = \left(\frac{12}{b}\right)^{\frac{1}{3}}\frac{1}{3y^{\frac{2}{3}}}.$$

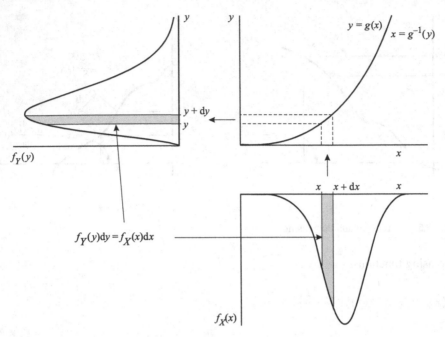

Figure 7.6 Density $f_X(x)$ for X transformed to the density $f_Y(y)$ of Y

Hence

$$f_Y(y) = \left(\frac{12}{b}\right)^{\frac{1}{3}} \frac{1}{\sqrt{2\pi}\sigma_d 3 y^{\frac{2}{3}}} \exp\left\{-\frac{1}{2}\left[\frac{\left(\frac{12y}{b}\right)^{\frac{1}{3}} - d}{\sigma_d}\right]^2\right\}.$$

This is plotted in Figure 7.6, which shows the rearrangement of the probability mass.

7.2.2 Standardized distributions using location and scale parameters

For many continuous distributions, it is convenient to make a linear transformation of the random quantity by changing the location and scale, with the result that the distribution remains the same, except for the new location and scale. We have seen this already for the normal distribution (Section 3.3.3) in developing the standard normal distribution using

$$Z = \frac{X - \mu}{\sigma},$$

where μ is the mean and σ the standard deviation. These are location and scale parameters, respectively.

In the exponential distribution

$$f_X(x) = \frac{1}{\beta} \exp\left(-\frac{x - \alpha}{\beta}\right) \text{ for } x \geq \alpha, \tag{7.12}$$

α represents a shift of the probability mass of the distribution (α is not the mean) and the standard deviation is β. These two are the location and scale parameters in this case. For the Gumbel distribution,

$$f_Z(z) = \frac{1}{b} \exp\left[-\exp\left(-\frac{z - a}{b}\right) - \frac{z - a}{b}\right], \tag{7.13}$$

where a is the mode and b a measure of dispersion.

In each case we can produce a **standardized distribution** by subtracting a location parameter and dividing by a scale parameter. Denoting these in general by the symbols c and d, we can make the substitution

$$Y = g(X) = \frac{X - c}{d} \tag{7.14}$$

so that

$$X = g^{-1}(Y) = c + dY, \tag{7.15}$$

$$\frac{dg^{-1}(y)}{dy} = d, \tag{7.16}$$

$$f_Y(y) = d \cdot f_X(c + dy), \tag{7.17}$$

and

$$f_X(x) = \frac{1}{d} f_Y\left(\frac{x - c}{d}\right). \tag{7.18}$$

In this, Y is the standardized random quantity.

Example 7.5 For the three cases discussed above (normal, exponential and Gumbel distributions), the standardized distributions are

$$f_Y(y) = \frac{1}{\sqrt{2\pi}} \exp\left(-\frac{y^2}{2}\right) \text{ for } -\infty < y < \infty, \tag{7.19}$$

$$f_Y(y) = \exp(-y) \text{ for } 0 \leq y < \infty, \text{ and} \tag{7.20}$$

$$f_Y(y) = \exp\left[-\exp(-y) - y\right] \text{ for } -\infty < y < \infty. \tag{7.21}$$

The density of the standardized distribution is always divided by d to get the distribution in terms of the non-standardized parameter. This ensures that the final probability 'mass' will be measured in units of probability, without physical dimension. The three forms just given are precisely the same regardless of the initial values of c and d. All distributions can be subjected to a linear transformation, but a standardized distribution does not necessarily result. The lognormal distribution, for example, can only be standardized in terms of the logarithm of the random quantity.

In some instances, the standardized distribution becomes a family of distributions that are a function of a parameter or parameters, as in the following example.

Example 7.6 The gamma distribution

$$f_X(x) = \frac{(x-a)^{\alpha-1}\exp[-(x-a)/\beta]}{\beta^\alpha\Gamma(\alpha)}, \quad \text{for } x \geq a \text{ with } \alpha > 0, \ \beta > 0 \tag{7.22}$$

becomes, upon making the transformation $y = (x-a)/\beta$,

$$f_Y(y) = \frac{y^{\alpha-1}\exp(-y)}{\Gamma(\alpha)}, \quad \text{for } y \geq 0 \text{ with } \alpha > 0, \beta > 0. \tag{7.23}$$

The constant α is the shape parameter of the standardized random quantity Y. Another example is the Weibull distribution (Exercise 7.3).

Interesting discussions of theoretical aspects in terms of stable distributions, a subclass of those that are infinitely divisible (see also Section 7.4.4 below) are given in de Finetti (1974) and Feller (1971).

7.2.3   Two random quantities

We consider now the transformation of two random quantities X_1 and X_2 with cdf $F_{X_1,X_2}(x_1, x_2)$ into Y_1 and Y_2 using the one-to-one mapping

$$Y_1 = g_1(X_1, X_2) \text{ and} \tag{7.24}$$
$$Y_2 = g_2(X_1, X_2). \tag{7.25}$$

The use of cumulative distributions in one-to-one increasing relationships is similar to that given for single random quantities above in Equation (7.4). In this case, provided the relationships can be inverted as

$$X_1 = g_1^{-1}(Y_1, Y_2) \text{ and} \tag{7.26}$$
$$X_2 = g_2^{-1}(Y_1, Y_2), \tag{7.27}$$

then

$$F_{Y_1,Y_2}(y_1, y_2) = F_{X_1,X_2}\left[g_1^{-1}(y_1, y_2), g_2^{-1}(y_1, y_2)\right]. \tag{7.28}$$

For the discrete case, given a one-to-one relationship not necessarily increasing, the probability masses at the X coordinates transfer to the corresponding image points in Y coordinates,

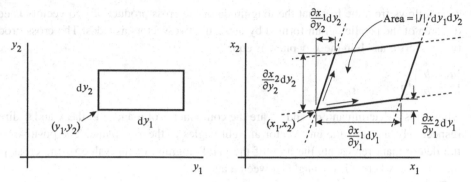

Figure 7.7 Mapping of (x_1, x_2) to (y_1, y_2) using inverse relationship

as in the one-dimensional case above of Equation (7.10). Thus

$$p_{Y_1,Y_2}(y_1, y_2) = p_{X_1,X_2}\left[g_1^{-1}(y_1, y_2), g_2^{-1}(y_1, y_2)\right]. \tag{7.29}$$

For continuous random quantities, we again consider one-to-one relationships of the kind specified in Equations (7.24) and (7.25), invertible, and which have continuous partial derivatives. Consider now the inverse mapping of Figure 7.7. This shows an element $dy_1 dy_2$ in $y_1\ y_2$ space that has been transformed into the element $dx_1 dx_2$ in x_1–x_2 space. Our objective is once again to redistribute the probability masses; the mass that is placed on $dx_1 dx_2$ must now be placed on $dy_1 dy_2$. In terms of the densities, written as f_{X_1,X_2} and f_{Y_1,Y_2}, we have

$$f_{X_1,X_2}(x_1, x_2)\, dx_1 dx_2 = f_{Y_1,Y_2}(y_1, y_2)\, dy_1 dy_2. \tag{7.30}$$

We need to determine the relationship between the areas $dx_1 dx_2$ and $dy_1 dy_2$. Now

$$dx_1 = \frac{\partial x_1}{\partial y_1} dy_1 + \frac{\partial x_1}{\partial y_2} dy_2 \text{ and}$$
$$dx_2 = \frac{\partial x_2}{\partial y_1} dy_1 + \frac{\partial x_2}{\partial y_2} dy_2, \tag{7.31}$$

where $x_1 = g_1^{-1}(y_1, y_2)$ and $x_2 = g_2^{-1}(y_1, y_2)$ are the inverse relationships. The coefficients in the relationships for dx_1 and dx_2, just given, form the Jacobian matrix:

$$\begin{bmatrix} \dfrac{\partial x_1}{\partial y_1} & \dfrac{\partial x_1}{\partial y_2} \\[2ex] \dfrac{\partial x_2}{\partial y_1} & \dfrac{\partial x_2}{\partial y_2} \end{bmatrix}. \tag{7.32}$$

We can see that the two columns in this matrix are the coefficients just mentioned, and these can be represented as vectors that are indicated by arrows in Figure 7.7. The area of the parallelogram of Figure 7.7 is given by the positive determinant $|J|$ of the Jacobian matrix.

This follows from the fact that the magnitude of the cross product of two vectors is equal to the area of the parallelogram formed by using the two vectors as sides. The cross product of two vectors $\mathbf{a}$ and $\mathbf{b}$ in the x–y plane is equal to

$$\begin{vmatrix} a_x & b_x \\ a_y & b_y \end{vmatrix} \mathbf{k},$$

where $||$ is a determinant, a_x and b_y are the components of $\mathbf{a}$ and $\mathbf{b}$ in the x and y directions respectively, and $\mathbf{k}$ is the unit vector at right angles to the x–y plane. The positive value of the determinant represents the area of the parallelogram, or the value of the cross product $|ab \sin \theta|_+$, where θ is the angle between $\mathbf{a}$ and $\mathbf{b}$.

Thus

$$dx_1 dx_2 = \begin{vmatrix} \dfrac{\partial x_1}{\partial y_1} & \dfrac{\partial x_1}{\partial y_2} \\ \dfrac{\partial x_2}{\partial y_1} & \dfrac{\partial x_2}{\partial y_2} \end{vmatrix}_+ dy_1 dy_2 = |J|_+ dy_1 dy_2 \tag{7.33}$$

(where the $+$ sign indicates the positive value of the determinant), and from Equation (7.21),

$$f_{Y_1,Y_2}(y_1, y_2) \, dy_1 dy_2 = f_{X_1,X_2}(x_1, x_2) \, dx_1 dx_2$$
$$= f_{X_1,X_2} \left[g_1^{-1}(y_1, y_2), g_2^{-1}(y_1, y_2) \right] |J|_+ \, dy_1 dy_2$$

or

$$f_{Y_1,Y_2}(y_1, y_2) = f_{X_1,X_2} \left[g_1^{-1}(y_1, y_2), g_2^{-1}(y_1, y_2) \right] |J|_+. \tag{7.34}$$

The determinant had to be specified as being positive for the same reason as in the case of single random quantities; the relationships between X and Y could involve negative gradients and thus produce negative areas. We assign probability mass for positive increments of random quantities only.

Example 7.7 The transformation

$$y_1 = x_1 - x_2$$
$$y_2 = x_2$$

is useful, as will be seen. The inverse relationships are

$$x_1 = y_1 + y_2$$
$$x_2 = y_2.$$

The positive value of the determinant of the Jacobian is

$$|J|_+ = \begin{vmatrix} 1 & 1 \\ 0 & 1 \end{vmatrix} = 1.$$

From this we see that

$$f_{Y_1,Y_2}(y_1, y_2) = f_{X_1,X_2}(y_1 + y_2, y_2). \tag{7.35}$$

We can use this result in reliability theory, to be further considered in Chapter 10. Let $X_1 = R$, the capacity (or resistance in the terminology of structural engineering) of a component, for example an electrical circuit, a transportation network or a structural component. If the capacity is exceeded by demand, the system fails. Let $X_2 = Y_2 = Z$, the demand, for example peak voltage, traffic, or loading in the three examples cited. Then we write $Y_1 = M = R - Z$, the 'margin of safety', the difference between capacity and demand. If $M \leq 0$, we have failure of the system. From Equation (7.35)

$$f_{M,Z}(m, z) = f_{R,Z}(m + z, z). \tag{7.36}$$

We can evaluate probabilities related to various values of M by obtaining the marginal distribution

$$f_M(m) = \int_{-\infty}^{\infty} f_{R,Z}(m + z, z)\, dz. \tag{7.37}$$

If R and Z are independent of each other,

$$f_M(m) = \int_{-\infty}^{\infty} f_R(m + z) f_Z(z)\, dz. \tag{7.38}$$

The function f_M is termed a convolution of f_R and f_Z. (See Section 7.3 below.)

To illustrate, we use exponential functions:

$$f_R(r) = \lambda_1 \exp(-\lambda_1 r) \text{ and} \tag{7.39}$$
$$f_Z(z) = \lambda_2 \exp(-\lambda_2 z), \tag{7.40}$$

for $r, z \geq 0$, and 0 otherwise. Then, from Equation (7.37),

$$f_M(m) = \lambda_1 \lambda_2 \int_{-m}^{\infty} \exp[-\lambda_1(m + z) - \lambda_2 z]\, dz \text{ for } M \leq 0 \text{ and}$$
$$= \lambda_1 \lambda_2 \int_{0}^{\infty} \exp[-\lambda_1(m + z) - \lambda_2 z]\, dz \text{ for } M \geq 0.$$

The limits for these integrals are illustrated in Figure 7.8, and the following gives the results:

$$f_M(m) = \frac{\lambda_1 \lambda_2}{\lambda_1 + \lambda_2} \exp(\lambda_2 m) \text{ for } M \leq 0 \text{ and} \tag{7.41}$$
$$f_M(m) = \frac{\lambda_1 \lambda_2}{\lambda_1 + \lambda_2} \exp(-\lambda_1 m) \text{ for } M \geq 0. \tag{7.42}$$

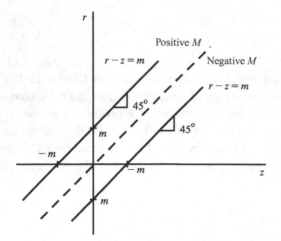

Figure 7.8 Integration limits in reliability problems; broken line indicates $m = 0$

Failure corresponds to the event $\{M < 0\}$ with probability

$$p_f = \int_{-\infty}^{0} \frac{\lambda_1 \lambda_2}{\lambda_1 + \lambda_2} \exp(\lambda_2 m)\, dm = \frac{\lambda_1}{\lambda_1 + \lambda_2}. \tag{7.43}$$

Models of choice. Another interpretation of the analysis of this example lies in models of choice. If R and Z represent two generalized costs for the choices A and B, the randomness expressing the uncertainty with regard to the perception of the next customer, then the events $\{M < 0\}$ and $\{M > 0\}$ might represent the choice of B and A respectively.

Example 7.8 *Rayleigh distribution from Gaussian process.* Voltage, sound, and the surface elevation of the sea can be modelled using a Gaussian process. The Rayleigh distribution can be obtained from this process. We consider

$$X(t) = A(t) \cos\left[\omega t + \Xi(t)\right], \tag{7.44}$$

where ω is a constant; the amplitude $A(t)$ and the phase $\Xi(t)$ are random quantities such that $0 \le A < \infty$ and $0 \le \Xi \le 2\pi$. On the assumption that these two random quantities vary slowly with time, so that $\dot{A}$ and $\dot{\Xi}$ are negligible with respect to $\dot{X}$, we obtain

$$\dot{X}(t) \simeq -\omega A(t) \sin\left[\omega t + \Xi(t)\right]. \tag{7.45}$$

If the joint distribution of X and $\dot{X}$ is such that they are independent and normal, $N\left(0, \sigma^2\right)$ and $N\left(0, \omega^2\sigma^2\right)$ respectively,

$$f_{X,\dot{X}}(x, \dot{x}) = \frac{1}{2\pi\omega\sigma^2} \exp\left[-\frac{1}{2\sigma^2}\left(x^2 + \frac{\dot{x}^2}{\omega^2}\right)\right]. \tag{7.46}$$

Then we can deduce the joint distribution of $\{A, \Xi\}$ from this equation using Equation (7.34).

Consequently,

$$f_{A,\Xi}(a,\xi) = \frac{1}{2\pi\omega\sigma^2}\exp\left\{-\frac{1}{2\sigma^2}\left[\left(a\cos\overline{\omega t+\xi}\right)^2 + \frac{\left(-\omega a\sin\overline{\omega t+\xi}\right)^2}{\omega^2}\right]\right\}$$

$$\times\left|\begin{matrix}\cos(\omega t+\xi) & -a\sin(\omega t+\xi)\\ -\omega\sin(\omega t+\xi) & -\omega a\cos(\omega t+\xi)\end{matrix}\right|_+$$

$$= \frac{a}{2\pi\sigma^2}\exp\left(-\frac{a^2}{2\sigma^2}\right), \tag{7.47}$$

where we have written $a(t)$ and $\xi(t)$ simply as a and ξ. This equation shows that Ξ is uniformly distributed from 0 to 2π. The distribution of A is then

$$f_A(a) = \int_0^{2\pi} f_{A,\Xi}(a,\xi)\,d\xi$$

$$= \int_0^{2\pi}\frac{a}{2\pi\sigma^2}\exp\left(-\frac{a^2}{2\sigma^2}\right)d\xi$$

$$= \frac{a}{\sigma^2}\exp\left(-\frac{a^2}{2\sigma^2}\right)\ \text{for } a \geq 0. \tag{7.48}$$

This is the Rayleigh distribution, discussed further in Chapter 9, Section 9.4.3.

7.2.4   Several random quantities

The need arises to consider several random quantities $X_1, X_2, \ldots, X_i, \ldots, X_n$ which transform into another set $Y_1, Y_2, \ldots, Y_i, \ldots, Y_n$. We shall use the vector notation illustrated by

$$\mathbf{X} = (X_1, X_2, \ldots, X_i, \ldots, X_n).$$

We also write

$$p_{\mathbf{X}}(\mathbf{x}) = p_{X_1, X_2, \ldots, X_i, \ldots, X_n}(x_1, x_2, \ldots, x_i, \ldots, x_n)$$

and

$$f_{\mathbf{X}}(\mathbf{x}) = f_{X_1, X_2, \ldots, X_i, \ldots, X_n}(x_1, x_2, \ldots, x_i, \ldots, x_n)$$

as the multidimensional probability distributions for the discrete and continuous cases, respectively. We wish to transform these to the distributions

$$p_{\mathbf{Y}}(\mathbf{y}) = p_{Y_1, Y_2, \ldots, Y_i, \ldots, Y_n}(y_1, y_2, \ldots, y_i, \ldots, y_n)$$

and

$$f_{\mathbf{Y}}(\mathbf{y}) = f_{Y_1, Y_2, \ldots, Y_i, \ldots, Y_n}(y_1, y_2, \ldots, y_i, \ldots, y_n).$$

The transformation using cumulative distribution functions and of discrete random quantities follows the procedures outlined above; see, for example, Equations (7.28), (7.29) and (7.9). For example, the equation

$$p_Y(\mathbf{y}) = p_X\left[\mathbf{g}^{-1}(\mathbf{y})\right]$$

assigns probability masses from the $\mathbf{X}$-space to the $\mathbf{Y}$-space, in a similar way to that in Equation (7.29), provided that the relationships

$$\mathbf{g}(\mathbf{x}) = [g_1(x_1, \ldots, x_n), \ldots, g_n(x_1, \ldots, x_n)] \tag{7.49}$$

and their inverses

$$\mathbf{g}^{-1}(\mathbf{y}) = \left[g_1^{-1}(x_1, \ldots, x_n), \ldots, g_n^{-1}(x_1, \ldots, x_n)\right] \tag{7.50}$$

are one-to-one.

We shall now outline the generalization of the transformation of densities using Jacobians. The results for the two-dimensional case generalize readily. Once again, we consider that the probability mass is distributed over elementary volumes. For example, in the three-dimensional case a small cube in the y_1–y_2–y_3 space transforms into a small parallelepiped in the x_1–x_2–x_3 space, in a manner entirely analogous to the transformation depicted in Figure 7.7. The volume of the parallelepiped formed by three vectors in a Euclidean space is given by the determinant formed by the components in three-dimensional cartesian coordinates. It is often introduced as the mixed triple product of the three vectors. This product is also the moment of a force about an axis in mechanics; see for instance Beer and Johnston (1981).

The components of our vectors are the values in the Jacobian; in three dimensions, the determinant of the Jacobian matrix (7.33) becomes

$$\begin{vmatrix} \dfrac{\partial x_1}{\partial y_1} & \dfrac{\partial x_1}{\partial y_2} & \dfrac{\partial x_1}{\partial y_3} \\[2ex] \dfrac{\partial x_2}{\partial y_1} & \dfrac{\partial x_2}{\partial y_2} & \dfrac{\partial x_2}{\partial y_3} \\[2ex] \dfrac{\partial x_3}{\partial y_1} & \dfrac{\partial x_3}{\partial y_2} & \dfrac{\partial x_3}{\partial y_3} \end{vmatrix}.$$

Then we have the transformation of volumes

$$dx_1 dx_2 dx_3 = \begin{vmatrix} \dfrac{\partial x_1}{\partial y_1} & \dfrac{\partial x_1}{\partial y_2} & \dfrac{\partial x_1}{\partial y_3} \\[2ex] \dfrac{\partial x_2}{\partial y_1} & \dfrac{\partial x_2}{\partial y_2} & \dfrac{\partial x_2}{\partial y_3} \\[2ex] \dfrac{\partial x_3}{\partial y_1} & \dfrac{\partial x_3}{\partial y_2} & \dfrac{\partial x_3}{\partial y_3} \end{vmatrix}_+ dy_1 dy_2 dy_3 = |J|_+ \, dy_1 dy_2 dy_3, \tag{7.51}$$

and we find, analogously to Equation (7.34),

$$f_{Y_1, Y_2, Y_3}(y_1, y_2, y_3)$$
$$= f_{X_1, X_2, X_3}\left[g_1^{-1}(y_1, y_2, y_3), g_2^{-1}(y_1, y_2, y_3), g_3^{-1}(y_1, y_2, y_3)\right] |J|_+ , \tag{7.52}$$

where $|J|_+$ is now given in Equation (7.51).

In n dimensions,

$$f_{\mathbf{Y}}(\mathbf{y}) = f_{\mathbf{X}}\left[\mathbf{g}^{-1}(\mathbf{y})\right] |J|_+ , \tag{7.53}$$

where

$$|J|_+ = \begin{vmatrix} \dfrac{\partial x_1}{\partial y_1} & \dfrac{\partial x_1}{\partial y_2} & \cdots & \dfrac{\partial x_1}{\partial y_j} & \cdots & \dfrac{\partial x_1}{\partial y_n} \\[2mm] \dfrac{\partial x_2}{\partial y_1} & \dfrac{\partial x_2}{\partial y_2} & \cdots & \dfrac{\partial x_2}{\partial y_j} & & \dfrac{\partial x_2}{\partial y_n} \\[2mm] \vdots & \vdots & & \vdots & & \vdots \\[2mm] \dfrac{\partial x_i}{\partial y_1} & \dfrac{\partial x_i}{\partial y_2} & \cdots & \dfrac{\partial x_i}{\partial y_j} & \cdots & \dfrac{\partial x_i}{\partial y_n} \\[2mm] \vdots & \vdots & & \vdots & & \vdots \\[2mm] \dfrac{\partial x_n}{\partial y_1} & \dfrac{\partial x_n}{\partial y_2} & \cdots & \dfrac{\partial x_n}{\partial y_j} & \cdots & \dfrac{\partial x_n}{\partial y_n} \end{vmatrix}_+ \tag{7.54}$$

and the notation $\mathbf{g}^{-1}(\mathbf{y})$ is given in Equations (7.49) and (7.50) (which are valid for discrete or continuous quantities).

There is an elegant generalization of the visually perceived parallelepiped in three-dimensional space to that in a linear vector space of n-dimensions (Greub, 1975). If one has n linearly independent vectors, they define an n-dimensional parallelepiped, and the volume is defined analogously to the three-dimensional case. Spaces in which norms provide the abstraction of distance and inner products the abstraction of scalar products and consequently angles provide extensions of the applicability of the theory.

Example 7.9 *Multidimensional normal distribution.* For convenience we define here the vector **X** as a column vector

$$\mathbf{X} = (X_1, X_2, \dots, X_i, \dots, X_n)^T . \tag{7.55}$$

We continue to use the notation

$$f_{\mathbf{X}}(\mathbf{x}) = f_{X_1, X_2, \dots, X_i, \dots, X_n}(x_1, x_2, \dots, x_i, \dots, x_n) . \tag{7.56}$$

We start the example with the standardized multidimensional normal distribution with means of zero and standard deviations of unity,

$$f_{\mathbf{X}}(\mathbf{x}) = \frac{1}{(2\pi)^{n/2}} \exp\left(-\frac{1}{2}\sum_{i=1}^{n} x_i^2\right)$$

$$= \frac{1}{(2\pi)^{n/2}} \exp\left(-\frac{1}{2}\mathbf{X}^T \cdot \mathbf{X}\right). \tag{7.57}$$

Thus

$$\langle \mathbf{X} \rangle = \mathbf{0}, \tag{7.58}$$

where $\mathbf{0}$ is a column vector of zeros and the covariance matrix is

$$[D] = [I], \tag{7.59}$$

the unit matrix. The Xs are therefore independent of each other with zero means and standard deviations of unity. We noted in Section 3.5.7 that if covariances are zero the random quantities are not necessarily independent in general. The present case of normal distributions constitutes an exception to this rule.

Now we transform $\mathbf{X}$ into $\mathbf{Y}$, where

$$\mathbf{Y} = (Y_1, Y_2, \ldots, Y_i, \ldots, Y_n)^T, \tag{7.60}$$

using the linear transformation

$$\mathbf{Y} = [A]\mathbf{X} + \boldsymbol{\mu}, \tag{7.61}$$

where

$$\boldsymbol{\mu} = (\mu_1, \mu_2, \ldots, \mu_n)^T \tag{7.62}$$

and $[A]$ is the matrix

$$[A] = [a_{ij}] = \begin{bmatrix} a_{11} & a_{12} & \cdots & a_{1j} & \cdots & a_{1n} \\ a_{21} & a_{22} & \cdots & a_{2j} & \cdots & a_{2n} \\ \vdots & \vdots & & \vdots & & \vdots \\ a_{i1} & a_{i2} & \cdots & a_{ij} & \cdots & a_{in} \\ \vdots & \vdots & & \vdots & & \vdots \\ a_{n1} & a_{n2} & \cdots & a_{nj} & \cdots & a_{nn} \end{bmatrix}. \tag{7.63}$$

The inverse relationships (7.50)

$$\mathbf{g}^{-1}(\mathbf{y}) = \left[g_1^{-1}(x_1, \ldots, x_n), \ldots, g_n^{-1}(x_1, \ldots, x_n)\right]$$

are deduced from (7.61).

$$\mathbf{X} = [A]^{-1}(\mathbf{Y} - \boldsymbol{\mu}) \tag{7.64}$$

which appears as a column vector of linear relationships.

We wish to obtain

$$f_{\mathbf{Y}}(\mathbf{y}) = f_{Y_1,\ldots,Y_i,\ldots,Y_n}(y_1,\ldots,y_i,\ldots,y_n) \tag{7.65}$$

by using Equation (7.53),

$$f_{\mathbf{Y}}(\mathbf{y}) = f_{\mathbf{X}}\left[\mathbf{g}^{-1}(\mathbf{y})\right] |J|_+ .$$

Now

$$|J|_+ = \begin{vmatrix} \dfrac{\partial x_1}{\partial y_1} & \cdots & \dfrac{\partial x_1}{\partial y_j} & \cdots & \dfrac{\partial x_1}{\partial y_n} \\ \vdots & & \vdots & & \vdots \\ \dfrac{\partial x_i}{\partial y_1} & \cdots & \dfrac{\partial x_i}{\partial y_j} & \cdots & \dfrac{\partial x_i}{\partial y_n} \\ \vdots & & \vdots & & \vdots \\ \dfrac{\partial x_n}{\partial y_1} & \cdots & \dfrac{\partial x_n}{\partial y_j} & \cdots & \dfrac{\partial x_n}{\partial y_n} \end{vmatrix}_+ = \frac{1}{|A|_+} = |A|_+^{-1} . \tag{7.66}$$

One gets

$$f_{\mathbf{Y}}(\mathbf{y}) = \frac{1}{(2\pi)^{n/2}} |A|_+^{-1} \exp\left[-\frac{1}{2}(\mathbf{y} - \boldsymbol{\mu})^T \left([A]^{-1}\right)^T [A]^{-1} (\mathbf{y} - \boldsymbol{\mu})\right],$$

where we have used the fact that $([C][D])^T = [D]^T [C]^T$.

Now let

$$[\Sigma] = [A][A]^T ; \tag{7.67}$$

Σ is then symmetric. Consequently

$$f_{\mathbf{Y}}(\mathbf{y}) = \frac{1}{(2\pi)^{n/2}} |\Sigma|^{-1/2} \exp\left[-\frac{1}{2}(\mathbf{y} - \boldsymbol{\mu})^T [\Sigma]^{-1} (\mathbf{y} - \boldsymbol{\mu})\right]. \tag{7.68}$$

The matrix $[\Sigma]$ is the covariance matrix

$$[\Sigma] = \begin{bmatrix} \sigma_1^2 & \rho_{12}\sigma_1\sigma_2 & \cdots & \rho_{1j}\sigma_1\sigma_j & \cdots & \rho_{1n}\sigma_1\sigma_n \\ \rho_{21}\sigma_2\sigma_1 & \sigma_2^2 & \cdots & \rho_{2j}\sigma_2\sigma_j & \cdots & \rho_{2n}\sigma_2\sigma_n \\ \vdots & \vdots & & \vdots & & \vdots \\ \rho_{i1}\sigma_i\sigma_1 & \rho_{i2}\sigma_i\sigma_2 & \cdots & \rho_{ij}\sigma_i\sigma_j & \cdots & \rho_{in}\sigma_i\sigma_n \\ \vdots & \vdots & & \vdots & & \vdots \\ \rho_{n1}\sigma_n\sigma_1 & \rho_{n2}\sigma_n\sigma_2 & \cdots & \rho_{nj}\sigma_n\sigma_j & \cdots & \sigma_n^2, \end{bmatrix}, \tag{7.69}$$

where the σs are the standard deviations and the ρs the correlation coefficients. See also Exercise 6.4.

In this example, we started with independent random quantities and ended with dependent ones. In structural reliability theory the reverse is sometimes needed (Melchers, 1999).

7.3 Convolutions

We are often concerned with a linear combination of two random quantities; one case was given in Example 7.7. Here we consider

$$Y = X_1 + X_2. \tag{7.70}$$

This might be the total time for transmission of two packets of information, 1 and 2, or the total waiting time for two modes of transport, or the sum of two individual loadings. Considering first the discrete case,

$$p_Y(y) = \sum_{y=x_1+x_2} p_{X_1,X_2}(x_1, x_2),$$

where the summation is taken, as indicated, over all combinations of x_1 and x_2 that constitute a particular value y. For each fixed y we can write x_1 (say) as $(y - x_2)$, so that this expression can be written as

$$p_Y(y) = \sum_{x_2=-\infty}^{\infty} p_{X_1,X_2}(y - x_2, x_2). \tag{7.71}$$

(This could have been written in terms of x_1 by substituting $(y - x_1)$ for x_2.) In the case of independent random quantities X_1 and X_2,

$$p_Y(y) = \sum_{x_2=-\infty}^{\infty} p_{X_1}(y - x_2) p_{X_2}(x_2) \text{ or}$$

$$p_Y(y) = \sum_{x_1=-\infty}^{\infty} p_{X_1}(x_1) p_{X_2}(y - x_1). \tag{7.72}$$

These equations are termed **convolutions** of p_{X_1} and p_{X_2}, and can be written succinctly as

$$p_Y = p_{X_1} * p_{X_2}. \tag{7.73}$$

For continuous random quantities, we consider again the sum in Equation (7.70). The method used in Example 7.7 gives the result that follows, with a change of sign of x_2 (since we considered there $y_1 = x_1 - x_2$ rather than $x_1 + x_2$). For variety, we give a somewhat different derivation. Referring to Figure 7.9, we integrate over the appropriate region $(x_1 + x_2 \leq y)$ to find

$$F_Y(y) = \mathbf{Pr}(X_1 + X_2 \leq y) = \int_{x_2=-\infty}^{\infty} \int_{x_1=-\infty}^{y-x_2} f_{X_1,X_2}(x_1, x_2) \, dx_1 dx_2. \tag{7.74}$$

From this we obtain by differentiation with respect to y that

$$f_Y(y) = \int_{-\infty}^{\infty} f_{X_1,X_2}(y - x_2, x_2) \, dx_2. \tag{7.75}$$

One can also derive this result directly from Figure 7.9.

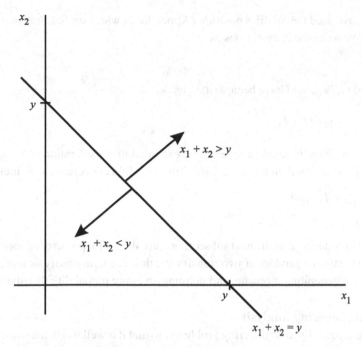

Figure 7.9 Integration limits for convolutions

Again independence provides the further result that

$$f_Y(y) = \int_{-\infty}^{\infty} f_{X_1}(y - x_2) f_{X_2}(x_2)\, dx_2. \tag{7.76}$$

The related result

$$f_Y(y) = \int_{-\infty}^{\infty} f_{X_1}(x_1) f_{X_2}(y - x_1)\, dx_1 \tag{7.77}$$

may be found in a similar way to the above, for instance by reversing the order of integration in Equation (7.74). Equations (7.76) and (7.77) are further examples of convolutions, written in a similar manner to the discrete version given in expression (7.73):

$$f_Y = f_{X_1} * f_{X_2}. \tag{7.78}$$

We can also express the convolutions in terms of *cdf*s. In this case

$$\begin{aligned}
F_Y(y) &= \int_{-\infty}^{\infty} F_{X_2}(y - x_1)\, dF_{X_1}(x_1) \\
&= \int_{-\infty}^{\infty} F_{X_2}(y - x_1) f_{X_1}(x_1)\, dx_1.
\end{aligned} \tag{7.79}$$

In the first line, we have used the Stieltjes notation (Appendix 2) which unifies the continuous and discrete cases. We write the convolutions as

$$F_Y = F_{X_1} * F_{X_2};$$ (7.80)

note that the integral (7.79) could have been written as

$$F_Y(y) = \int_{-\infty}^{\infty} F_{X_1}(y - x_2) \, dF_{X_2}(x_2).$$ (7.81)

The two-fold convolutions discussed above can be extended to consideration of the sum of several (say n) independent, random quantities, resulting in n-fold convolutions, written

$$f_{X_1} * f_{X_2} * \cdots * f_{X_{n-1}} * f_{X_n} \text{ and}$$ (7.82)

$$F_{X_1} * F_{X_2} * \cdots * F_{X_{n-1}} * F_{X_n}.$$ (7.83)

Transforms, to be introduced after the next subsection, have the useful property of converting a convolution integral into the product of two transforms, thus rendering analysis more easy. They are useful also in generating moments and defining an entire probability distribution.

Rôle of convolutions in engineering analysis

Convolutions appear frequently in engineering problems, so that it is well worth the while spent on this subject. The use in probability theory has been introduced in the preceding paragraphs, and we use them to calculate failure probabilities in Chapter 10. But there are analogous uses in the various areas of circuit theory, mechanics, hydrology and mortality of persons or equipment, to name a few. These analyses are all based on linear superposition in time.

In the case of equipment mortality, let n_0 be the number of pieces of new equipment in service at time 0, the initial point of analysis. This decreases to

$$n_0 g(t)$$

at time t, such that $g(0) = 1$, with $g(t)$ declining from unity eventually to zero. Now let the replacement rate be $f(t)$ per unit time at time t. Then the replacements during the time interval from time τ to $\tau + d\tau$ are $f(\tau) \, d\tau$, and at time t – after an elapsed time $(t - \tau)$ – the surviving replacements are

$$f(\tau) g(t - \tau) \, d\tau.$$

At time t, the total population is

$$n_0 g(t) + \int_0^t f(\tau) g(t - \tau) \, d\tau,$$ (7.84)

where it has been assumed as before that the n_0 pieces of equipment are new.

This can be extended to consideration of human populations by considering a mixture of ages at the start of the analysis, and $f(t)$ becomes the birth rate. The analysis is based on constancy in time of the various rates and does not account for events such as plagues. There

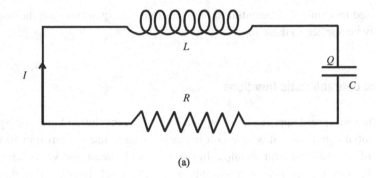

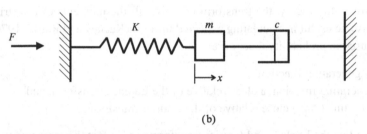

Figure 7.10   Electrical and mechanical systems

are ways to deal with such factors, beyond the present introduction. The second (convolution) term of Equation (7.84) could be used also to model the flow in a river, say, as a result of a storm of intensity $f(t)$ in centimetres per hour (for example). In this case, $g(t)$ is the 'unit hydrograph', that is, the response of the system to a unit input at time 0. This again assumes linearity of response of the system.

Figure 7.10 shows electrical and mechanical systems for which the equations are identical. Springs and capacitors act as storage units, dashpots and resistances cause dissipation and consequent entropy increase, while the mass and the inductance provide inertial components to the equations. Two examples of the use of convolutions in these systems will be given. First, a linear spring–mass system subject to a forcing function $f(t)$ constitutes an oscillator with one degree of freedom, with the response in terms of displacement x given by

$$x(t) = \frac{1}{\omega} \int_0^t f(\tau) \sin[\omega(t - \tau)] \, d\tau, \qquad (7.85)$$

where ω is the natural frequency of the oscillator. Second, the response of a spring–dashpot or capacitor–resistor system subjected to a force or potential follows equations similar in form to (7.84).

The main effect of the transforms of the following section is to convert the convolution to the product of two transformed quantities, thus providing a means of solution. The equations can often be expressed in terms of differential rather than integral equations, and the transforms can conveniently be applied to these too.

7.4   Transforms and characteristic functions

As a result of the widespread appearance of equations of the convolution kind – or equivalent forms as differential equations – it is common in engineering to use various transforms. The transforms are of several kinds, for example in the electrical circuit and viscoelastic theory just discussed, Laplace transforms are commonly used (Churchill, 1972), and in data analysis, signal processing and spectral analysis, Fourier transforms are central. The transforms convert equations involving convolutions into simpler multiplicative forms, but there are other advantages. In probability theory, the transform can give all the moments of a distribution. Transforms can also be useful in calculating expected utilities (Keeney and Raiffa, 1976). The main transforms used in probability theory are:

- the probability generating function,
- the moment generating function, a close relative of the Laplace transform, and
- the characteristic function, a close relative of the Fourier transform.

It is also of interest that the Laplace and Fourier transforms are in fact the same transform but with a change of variable; see de Finetti (1974). We introduce the three main transforms just noted, for completeness.

7.4.1   Probability generating functions (*pgfs*)

We consider initially a discrete random quantity X, which takes values $x = 0, 1, 2, \ldots$ The definition that follows is therefore confined to random quantities that take the possible values of 0 and the natural numbers. Typical examples are the binomial, Poisson and geometric distributions. The probability generating function $g_X(t)$ is defined as the expected value

$$
\begin{aligned}
g_X(t) = \left\langle t^X \right\rangle &= \sum_{x=0}^{\infty} t^x p_X(x) \\
&= p_X(0) + t p_X(1) + t^2 p_X(2) + t^3 p_X(3) + \cdots
\end{aligned} \tag{7.86}
$$

One can see from the last line of this definition why $g_X(t)$ is termed a probability generating function. If one can evaluate the series, the probability $p_X(n)$ is the coefficient of t^n. The *pgf*s arise quite naturally as power series.

Example 7.10 Consider the geometric distribution of Equation (3.7)

$$
\mathbf{Pr}(X = x) = (1 - p)^{x-1} p,
$$
$$
x = 1, 2, \ldots
$$

Then

$$g_X(t) = p \sum_{y=0}^{\infty} t^y (1-p)^y$$
$$= p \sum_{y=0}^{\infty} [t(1-p)]^y$$
$$= \frac{p}{1-t(1-p)}, \tag{7.87}$$

in which $y = x - 1$. We see from these equations that

$$g_X(t) = p + t[p(1-p)] + t^2[p(1-p)^2] + t^3[p(1-p)^3] + \cdots \tag{7.88}$$

The coefficients of t^n give the probabilities of $y = 0, 1, 2, \ldots$, or equivalently of $x = 1, 2, 3, \ldots$

Example 7.11 For the Poisson distribution of Equation (3.13),

$$p_X(x) = \frac{v^x \exp(-v)}{x!}, \text{ for } x = 0, 1, 2, \ldots,$$

the *pgf*, from which the reader might wish to obtain probabilities, is

$$g_X(t) = \exp(-v) \sum_{x=0}^{\infty} \frac{(vt)^x}{x!}$$
$$= \exp(-v) \exp(vt) = \exp - [v(1-t)]. \tag{7.89}$$

The series (7.86) certainly converges for $-1 \le t \le 1$. Further discussion of this subject is given, for example, in Lloyd (1980), who also deals with the case where the values of the random quantity do not correspond to the integers.

Convolutions

We discussed above in Section 7.3 the sum of independent random quantities X_1 and X_2, and the resulting expressions were written as

$$p_{X_1} * p_{X_2}.$$

The early motivation for the development of transform techniques arose out of the fact that, for two quantities X_1 and X_2,

$$t^{X_1+X_2} = t^{X_1} t^{X_2}, \tag{7.90}$$

for any number t. Because of the result expressed in Equation (7.90), the present method is intimately connected with the use of exponents. We have previously seen that for two independent random quantities X and Y, $\langle XY \rangle = \langle X \rangle \langle Y \rangle$, as shown in Equation (3.79). In the present case,

$$\langle t^{X_1+X_2} \rangle = \langle t^{X_1} t^{X_2} \rangle = \sum \sum t^{X_1} t^{X_2} p_{X_1}(x_1) p_{X_2}(x_2)$$
$$= \sum t^{X_1} p_{X_1}(x_1) \sum t^{X_2} p_{X_2}(x_2)$$
$$= \langle t^{X_1} \rangle \langle t^{X_2} \rangle . \tag{7.91}$$

We can therefore write

$$g_{X_1+X_2}(t) = \langle t^{X_1+X_2} \rangle = g_{X_1}(t) g_{X_2}(t); \tag{7.92}$$

the *pgf* of the sum of two independent random quantities is the product of the individual *pgf*s. This can be extended to the sum of three or more independent random quantities: the *pgf* of the sum of the independent random quantities is the product of the individual *pgf*s.

Example 7.12 The *pgf* of the sum of two independent Poisson random quantities X_1 and X_2 with expected values v_1 and v_2, respectively, is

$$g_{X_1+X_2}(t) = \{\exp - [v_1(1-t)]\} \cdot \{\exp - [v_2(1-t)]\}$$
$$= \exp - [(v_1 + v_2)(1-t)] , \tag{7.93}$$

by Equation (7.92); see also Example 7.11. The resulting expression (7.93) is in fact the transform of a Poisson random quantity with expected value $(v_1 + v_2)$. We see from this result that the sum of two Poisson random quantities is another Poisson random quantity with expected value $(v_1 + v_2)$.

Moments

Probability generating functions can be used to calculate means, variances, and other moments. The derivative of $g_X(t)$ with respect to t is

$$g'_X(t) = p_X(1) + 2t p_X(2) + 3t^2 p_X(3) + \cdots$$
$$= \sum_{x=1}^{\infty} x p_X(x) t^{x-1},$$

and therefore

$$g'_X(t=1) = \langle X \rangle . \tag{7.94}$$

Example 7.13 Applying this to the Poisson distribution using Equation (7.89), for example, we find that $\langle X \rangle = v$.

The second derivative

$$g''_X(t) = 2p_X(2) + 3 \cdot 2t p_X(3) + 4 \cdot 3t^2 p_X(4) + \cdots$$
$$= \sum_{x=2}^{\infty} x(x-1) p_X(x) t^{x-2},$$

and therefore

$$g_X''(1) = \langle X(X-1) \rangle = \langle X^2 \rangle - \langle X \rangle. \tag{7.95}$$

We can use this expression to calculate the variance, which from Equation (3.87) is given by

$$\sigma_X^2 = \langle X^2 \rangle - \langle X \rangle^2. $$

Example 7.14 For the Poisson distribution, using Equation (7.89),

$$g_X''(1) = \nu^2;$$

from Equation (7.95)

$$\langle X^2 \rangle = \nu^2 + \nu$$

and therefore

$$\sigma_X^2 = \nu.$$

7.4.2 Moment generating and characteristic functions (*mgf*s and *cf*s)

Both transforms to be described here will be seen to generate moments – rather than probabilities, as in the case of *pgf*s. We shall use the usually accepted terms, moment generating and characteristic functions (*mgf*s and *cf*s) although both generate moments. To extend the ideas of the previous section to continuous random quantities, and to those that take values other than integers, it is convenient to use the following functions, replacing t of the previous section with

$$t = \exp s, \text{ and} \tag{7.96}$$

$$u = \exp(iu) \tag{7.97}$$

in the case of *mgf*s and *cf*s respectively, where $i = \sqrt{-1}$. Then t^X in Equation (7.86) becomes

$$\exp(sX) \text{ and} \tag{7.98}$$

$$\exp(iuX), \text{ respectively.} \tag{7.99}$$

The two transforms are now defined.

The **moment generating function** (*mgf*) is given by

$$m_X(s) = \langle \exp(sX) \rangle = \begin{cases} \sum_{\text{all } x} \exp(sx)\, p_X(x) & \text{for discrete } X, \\ \int_{\text{all } x} \exp(sx)\, f_X(x)\, dx & \text{for continuous } X. \end{cases} \tag{7.100}$$

We see the essential role of the exponent, now formally as the power to the natural base 'e'.

Equations (7.100) may, as elsewhere, be combined into one using the Stieltjes form

$$m_X(s) = \int_{\text{all } x} \exp(sx) \, dF_X(x). \tag{7.101}$$

We define the **characteristic function** (*cf*) as

$$\phi_X(u) = \langle \exp(iuX) \rangle = \begin{cases} \sum_{\text{all } x} \exp(iux) \, p_X(x) & \text{for discrete } X, \\ \int_{\text{all } x} \exp(iux) \, f_X(x) \, dx & \text{for continuous } X. \end{cases} \tag{7.102}$$

These definitions may, as before, be combined in the Stieltjes integral:

$$\phi_X(t) = \int_{\text{all } x} \exp(iux) \, dF_X(x). \tag{7.103}$$

The *cf* may also be written as

$$\begin{aligned} \phi_X(u) &= \langle \cos ux + i \sin ux \rangle \\ &= \int_{\text{all } x} (\cos ux) \, f_X(x) \, dx + i \int_{\text{all } x} (\sin ux) \, f_X(x) \, dx, \end{aligned} \tag{7.104}$$

where we have used the integral as an example; the same equation may be rewritten with summations.

The *mgf* does not necessarily converge. It may converge for only an interval around the origin, especially when there is a lot of probability mass in the tails of the distribution. On the other hand, the *cf* exists for all real values of u. There is a one-to-one correspondence between an *mgf* (when it exists) or a *cf* and the corresponding probability distribution. The *cf* is more general, and the one-to-one correspondence applies to all proper distributions. The important result is that knowledge of the *cf* is sufficient to determine the probability distribution. See de Finetti (1974, vol. 1), Fraser (1976) and Lloyd (1980) for details. The inverse relation to Equation (7.102) in the continuous case is

$$f_X(x) = \frac{1}{2\pi} \int_{-\infty}^{\infty} \exp(-iux) \phi_X(u) \, du. \tag{7.105}$$

Example 7.15 For the binomial distribution of Equation (3.3),

$$p_X(x) = \binom{n}{x} p^x (1-p)^{n-x},$$

the *mgf* is

$$\begin{aligned} m_X(s) &= \sum_{x=0}^{n} \exp(sx) \binom{n}{x} p^x q^{n-x} = \sum_{x=0}^{n} \binom{n}{x} \left[p \exp(s) \right]^x q^{n-x} \\ &= \left[p \exp(s) + q \right]^n, \end{aligned} \tag{7.106}$$

in which $q = 1 - p$.

In a similar way it can be shown that the *cf* is

$$\phi_X(u) = \left[p\exp(iu) + q\right]^n.$$

Example 7.16 For the Poisson distribution of Equation (3.13),

$$p_X(x) = \frac{v^x \exp(-v)}{x!} \text{ for } x = 0, 1, 2, \ldots,$$

$$\phi_X(u) = \exp(-v)\sum_{x=0}^{\infty}\frac{\left[v\exp(iu)\right]^x}{x!} = \exp(-v)\exp\left[v\exp(iu)\right] \tag{7.107}$$
$$= \exp\left\{-v\left[1 - \exp(iu)\right]\right\}.$$

The *mgf* is similarly obtained; we dealt with the *pgf* in Equation (7.89).

Example 7.17 For the standard normal distribution, $f_Z(z) = \left(1/\sqrt{2\pi}\right)\exp(-z^2/2)$,

$$m_Z(s) = \frac{1}{\sqrt{2\pi}}\int_{-\infty}^{\infty}\exp(sz)\exp\left(-\frac{z^2}{2}\right)dz$$
$$= \left(\exp\frac{s^2}{2}\right)\frac{1}{\sqrt{2\pi}}\int_{-\infty}^{\infty}\exp\left[-\frac{(z-s)^2}{2}\right]dz$$
$$= \exp\left(\frac{s^2}{2}\right). \tag{7.108}$$

The *cf* may be shown to be given by

$$\phi_Z(u) = \exp\left(-\frac{u^2}{2}\right). \tag{7.109}$$

The proof for the *cf* requires contour integration in the complex plane; see Exercise 7.12.

Laplace and Fourier transforms

The *mgf* and the *cf* are closely related to the Laplace and Fourier transforms, respectively. The Laplace transform is defined as the integral, usually from 0 to ∞, of the same argument as above, except that the term $\exp(sX)$ is replaced by $\exp(-sX)$. In a similar way, $\phi_X(-u)$ is the Fourier transform of $f_X(x)$.

7.4.3   Linear transformations and convolutions

The *mgf* and *cf* of

$$Y = aX + b, \tag{7.110}$$

where a and b are constants, are, respectively,

$$m_Y(s) = \langle \exp(sY) \rangle = \langle \exp[s(aX + b)] \rangle = \exp(sb)\, m_X(as) \text{ and} \tag{7.111}$$

$$\phi_Y(u) = \langle \exp(iuY) \rangle = \langle \exp[iu(aX + b)] \rangle = \exp(iub)\, \phi_X(au). \tag{7.112}$$

Example 7.18 Transformation of the standard normal random quantity Z to the $N(\mu, \sigma^2)$ distribution by the linear transformation $Y = \mu + \sigma Z$. Using the *cf*, for example,

$$\begin{aligned} \phi_Y(u) &= \exp(iu\mu)\, \phi_Z(\sigma u) \\ &= \exp\left(iu\mu - \sigma^2 u^2/2\right), \end{aligned} \tag{7.113}$$

using Equation (7.109). Similarly the *mgf* is

$$m_Y(s) = \exp\left(s\mu + \sigma^2 s^2/2\right). \tag{7.114}$$

The sum of independent random quantities X_1 and X_2 constitutes a convolution, as discussed above in Section 7.3. Writing $Y = X_1 + X_2$, the distribution of Y is a convolution of those of X_1 and X_2, written in summary form as

$$p_Y = p_{X_1} * p_{X_2} \text{ and}$$
$$f_Y = f_{X_1} * f_{X_2}$$

for discrete and continuous distributions, respectively. We use again the multiplicative property of the exponential function shown in Equation (7.90), that is, $t^{X_1 + X_2} = t^{X_1} t^{X_2}$:

$$\langle \exp(sY) \rangle = \langle \exp[s(X_1 + X_2)] \rangle = \langle \exp(sX_1) \rangle \cdot \langle \exp(sX_2) \rangle \text{ and}$$

$$\langle \exp(iuY) \rangle = \langle \exp[iu(X_1 + X_2)] \rangle = \langle \exp(iuX_1) \rangle \cdot \langle \exp(iuX_2) \rangle.$$

We can conclude that the *mgf* and *cf* of Y are

$$m_Y(s) = m_{X_1}(s) \cdot m_{X_2}(s) \text{ and} \tag{7.115}$$

$$\phi_Y(u) = \phi_{X_1}(u) \cdot \phi_{X_2}(u), \tag{7.116}$$

respectively. This can be extended to the sum of several, say n independent random quantities. If

$$Y = \sum_{i=1}^{n} X_i, \tag{7.117}$$

where the X_i are independent, then

$$\begin{aligned} m_Y(s) &= \prod_{i=1}^{n} m_{X_i}(s) \\ &= m_{X_1}(s) \cdot m_{X_2}(s) \cdot \ldots \cdot m_{X_i}(s) \cdot \ldots \cdot m_{X_n}(s) \text{ and} \end{aligned} \tag{7.118}$$

$$\begin{aligned} \phi_Y(u) &= \prod_{i=1}^{n} \phi_{X_i}(u) \\ &= \phi_{X_1}(u) \cdot \phi_{X_2}(u) \cdot \ldots \cdot \phi_{X_i}(u) \cdot \ldots \cdot \phi_{X_n}(u). \end{aligned} \tag{7.119}$$

Addition of independent random quantities corresponds to multiplication of *mgf*s and *cf*s. This very useful property will be exploited in several applications, not least in the Central Limit Theorem.

For the *iid* (independent and identically distributed) case, using *cf*s as an example,

$$\phi_Y(u) = \prod_{i=1}^{n} \phi_{X_i}(u) = \phi_X^n(u). \tag{7.120}$$

Example 7.19 The sum of two independent random quantities X_1 and X_2, following the normal distribution and with parameters (μ_1, σ_1^2) and (μ_2, σ_2^2) respectively, will have the following *cf*, using Equations (7.113) and (7.116).

$$\begin{aligned}
\phi_Y(u) &= \phi_{X_1}(u) \cdot \phi_{X_2}(u) \\
&= \exp\left(iu\mu_1 - \sigma_1^2 u^2/2\right) \cdot \exp\left(iu\mu_2 - \sigma_2^2 u^2/2\right) \\
&= \exp\left[iu(\mu_1 + \mu_2) - \left(\sigma_1^2 + \sigma_2^2\right) u^2/2\right].
\end{aligned} \tag{7.121}$$

From this we see that the sum of the two independent normal random quantities is another with a normal distribution having mean $(\mu_1 + \mu_2)$ and variance $\left(\sigma_1^2 + \sigma_2^2\right)$. Recall that there is a one-to-one correspondence between a *cf* and the corresponding probability distribution.

Example 7.20 The convolution just discussed can be applied to a sum Y of n independent normal random quantities $X_1, X_2, X_3, \ldots, X_i, \ldots X_n$. The parameters are denoted as (μ_i, σ_i^2) for the ith random quantity. One can apply the convolution of Example 7.19 initially to the first two of the random quantities, then to this result, and then to the third, and so on. Using *mgf*s, based on Equation (7.114),

$$\begin{aligned}
m_Y(s) &= m_{X_1}(s) \cdot m_{X_2}(s) \cdot m_{X_3}(s) \cdot \ldots \cdot m_{X_i}(s) \cdot \ldots \cdot m_{X_n}(s), \\
&= \prod_{i=1}^{n} \exp\left(s\mu_i + \sigma_i^2 s^2/2\right) \\
&= \exp\left[s \sum_{i=1}^{n} \mu_i + \left(\sum_{i=1}^{n} \sigma_i^2\right) s^2/2\right].
\end{aligned} \tag{7.122}$$

This is the *mgf* of a normal random quantity with parameters $\left(\sum_{i=1}^{n} \mu_i, \sum_{i=1}^{n} \sigma_i^2\right)$. Addition of normal random quantities results in another normal random quantity with these parameters.

Example 7.21 *Gamma densities.* The gamma distribution

$$f_X(x) = \frac{x^{\alpha-1} \exp(-x/\beta)}{\beta^\alpha \Gamma(\alpha)} \tag{7.123}$$

has a *cf* as follows:

$$\phi_X(u) = (1 - iu\beta)^{-\alpha}. \tag{7.124}$$

(See Exercise 7.18.) The sum Y of two random quantities with gamma distributions, X_1 (parameters α_1, β) and X_2 (parameters α_2, β), that are independent results in another gamma density,

$$
\phi_Y(u) = \phi_{X_1}(u) \cdot \phi_{X_2}(u)
$$
$$
= (1 - iu\beta)^{-\alpha_1} \cdot (1 - iu\beta)^{-\alpha_2}
$$
$$
= (1 - iu\beta)^{-(\alpha_1 + \alpha_2)}. \tag{7.125}
$$

This is a gamma density with parameters $(\alpha_1 + \alpha_2, \beta)$. The parameter β is the scale parameter while $(\alpha_1 + \alpha_2)$ is the shape parameter resulting from the original two shape parameters α_1 and α_2. It is also said that the family of gamma distributions is 'closed under convolution', expressed by

$$
f^{\alpha_1, \beta} * f^{\alpha_2, \beta} = f^{\alpha_1 + \alpha_2, \beta}, \tag{7.126}
$$

where the meanings of the symbols are clear from the discussion above.

7.4.4   Infinite divisibility

The gamma and normal densities are examples of **infinitely divisible** distributions. The distribution F_Y is denoted as infinitely divisible if it can be written as the sum of n independent random quantities with a common distribution F_X, for every n. Thus

$$
F_Y = F_X * F_X * \cdots * F_X * F_X, \text{ with } n \text{ terms in the convolution.} \tag{7.127}
$$

The distribution function F_Y corresponds to the distribution of the sum of n independent and identically distributed (*iid*) random quantities, resulting in the n-fold convolution of the common distribution F_X. In terms of *cfs*,

$$
\phi_Y = \phi_X^n. \tag{7.128}
$$

Continuous stochastic processes with independent increments require infinite divisibility to decompose into the increments, as mentioned briefly at the end of Section 3.6.1.

Example 7.22 The exponential distribution

$$
f_Y(x) = \frac{1}{\beta} \exp(-y/\beta)
$$

has a *cf* as follows:

$$
\phi_Y(u) = (1 - iu\beta)^{-1}.
$$

This can be written as

$$
\phi_Y(u) = \left[(1 - iu\beta)^{-\frac{1}{n}}\right]^n, \tag{7.129}
$$

and $(1 - iu\beta)^{-\frac{1}{n}}$ is the *cf* of a random quantity with a gamma distribution (parameters $\frac{1}{n}, \beta$). Therefore an exponentially distributed random quantity is the sum of n *iid* random quantities with gamma distributions as indicated, for every n.

For more details on infinitely divisible distributions, see de Finetti (1974) and Feller (1971).

7.4.5   Generation of moments

Both *mgfs* and *cfs* generate moments. As a reminder, the *k*th moment of a probability distribution about the origin was defined in Equations (3.82) and (3.83) as

$$\mu_X^{(k)} = \langle X^k \rangle.$$

We make the following expansions. First, for the *mgf*,

$$
\begin{aligned}
m_X(s) &= \langle \exp(sX) \rangle \\
&= \left\langle 1 + sX + \frac{1}{2!}(sX)^2 + \frac{1}{3!}(sX)^3 + \cdots + \frac{1}{k!}(sX)^k + \cdots \right\rangle \\
&= 1 + s\mu_X^{(1)} + \frac{1}{2!}s^2\mu_X^{(2)} + \frac{1}{3!}s^3\mu_X^{(3)} + \cdots + \frac{1}{k!}s^k\mu_X^{(k)} + \cdots
\end{aligned}
\tag{7.130}
$$

The corresponding series for the *cf* is as follows:

$$
\begin{aligned}
\phi_X(u) &= \langle \exp(iuX) \rangle \\
&= \left\langle 1 + iuX - \frac{1}{2!}(uX)^2 - \frac{1}{3!}i(uX)^3 + \cdots + \frac{1}{k!}(iuX)^k + \cdots \right\rangle \\
&= 1 + iu\mu_X^{(1)} - \frac{1}{2!}u^2\mu_X^{(2)} - \frac{1}{3!}iu^3\mu_X^{(3)} + \cdots + \frac{1}{k!}(iu)^k\mu_X^{(k)} + \cdots
\end{aligned}
\tag{7.131}
$$

One sees in both series a full set of moments for the probability distribution: the *mgfs* and *cfs* generate a series containing all the moments of the distribution of X. We now see why the expression 'moment generating function' is used; in the usual usage, the term is reserved for the *mgf*, as has been done in the above.

From Equations (7.130) and (7.131), it follows by differentiation and by setting s and u equal to zero that

$$
\left. \frac{d^r m_X(s)}{ds^r} \right|_{s=0} = \mu_X^{(r)} \text{ and}
\tag{7.132}
$$

$$
\left. \frac{d^r \phi_X(u)}{du^r} \right|_{u=0} = i^r \mu_X^{(r)}.
\tag{7.133}
$$

Central moments $m_X^{(k)} = \langle (X - \mu_X)^k \rangle$ of a random quantity X, as defined in Equations (3.84) and (3.85), can also be generated. Taking *cfs* as an example, one can write

$$
\begin{aligned}
\left[\exp(-iu\mu_X) \right] \phi_X(u) &= \left[\exp(-iu\mu_X) \right] \langle \exp(iuX) \rangle = \langle \exp[iu(X-\mu_X)] \rangle \\
&= 1 + ium_X^{(1)} - \frac{1}{2!}u^2 m_X^{(2)} - \frac{1}{3!}iu^3 m_X^{(3)} + \cdots + \frac{1}{k!}(iu)^k m_X^{(k)} + \cdots
\end{aligned}
\tag{7.134}
$$

Example 7.23 A random quantity Y with a normal distribution (μ, σ^2) has a *cf* as follows; see Equation (7.113):

$$\phi_Y (u) = \exp\left(iu\mu - \sigma^2 u^2/2\right).$$

This can be written as

$$\left[\exp\left(-iu\mu_X\right)\right]\phi_X (u) = \exp\left(-\sigma^2 u^2/2\right)$$

$$= 1 - \frac{1}{2}u^2\sigma^2 + \frac{1}{2!}u^4 \left(\frac{1}{2}\sigma^2\right)^2 + \cdots \tag{7.135}$$

From this we deduce that the variance $m_X^{(2)}$ (second central moment) is σ^2, and that the fourth central moment, $m_X^{(4)}$, the coefficient of $u^4/4!$, is $3\sigma^4$. This can be used to deduce the coefficient of kurtosis given in Equations (3.89) and (3.90),

$$\gamma_2' = \left[\frac{m_X^{(4)}}{\left(m_X^{(2)}\right)^2}\right] - 3 = 0 \tag{7.136}$$

in the present instance. The normal distribution is the standard, termed mesokurtic $(\gamma_2' = 0)$. Distributions with positive values are leptokurtic (more peaked); those with negative values are platykurtic (flat, or less peaked). See Figure 3.20.

7.4.6 Fourier transforms and spectral analysis

Laplace and Fourier transforms are related by a change of variable. Usage and convenience has resulted in one or other of them being used in different applications. The transforms – as noted above – are used to solve a variety of problems, many of them outside of probability theory. Tables of transform pairs (G and g below) are available for these applications. The question of power spectra is generally dealt with using the Fourier transform. The Fourier transform $G(\omega)$ of the function $g(t)$ is defined as follows:

$$G(\omega) = \int_{-\infty}^{\infty} g(t) \exp(-i\omega t)\, dt. \tag{7.137}$$

The inverse is

$$g(t) = \frac{1}{2\pi} \int_{-\infty}^{\infty} G(\omega) \exp(i\omega t)\, d\omega. \tag{7.138}$$

Note opposite signs in the exponent, as compared to characteristic functions in (7.102) and (7.105), but these are essentially the same transforms. The transform $\phi(-\omega)$ from (7.102) gives the Fourier transform of (7.137).

The **symmetry** property relates the pairs of transforms and inverses; if

$$g(t) \iff G(\omega) \tag{7.139}$$

then

$$G(t) \iff 2\pi g(-\omega). \tag{7.140}$$

(There are equivalent forms that avoid the 2π factor which mars the symmetry.) In this subsection, the application which we wish to introduce is that of stationary random processes. The **power spectrum** or **power spectral density** (*psd*) $s(\omega)$ is the Fourier transform of the autocorrelation defined in (3.128) and (3.131); here we write $\varrho_{X,X}(\tau)$ as $\varrho(\tau)$.

$$s(\omega) = \int_{-\infty}^{\infty} \varrho(\tau) \exp(-i\omega\tau) \, d\tau, \tag{7.141}$$

with the Fourier inversion formula yielding

$$\varrho(\tau) = \frac{1}{2\pi} \int_{-\infty}^{\infty} s(\omega) \exp(i\omega\tau) \, d\omega. \tag{7.142}$$

For real processes, both the autocorrelation $\varrho(\tau)$ and the power spectrum $s(\omega)$ are real and even functions. Note that the average power $\langle X^2 \rangle$ of the process is given by

$$\varrho(0) = \langle X(t) X(t) \rangle = \langle X^2 \rangle. \tag{7.143}$$

Transforms of the kind discussed here relate generally a process in time to its frequency content. In viscoelastic theory, for example, one is interested in the spectrum of retardation times in the compliance. In time series, there is generally a mixture of frequency components with the *psd* giving the relative power at the various frequencies. An interesting interpretation is given in de Finetti (1974, Vol. 1).

Physically we might apply a band pass filter with sharp cutoff giving the time history in a frequency range from ω to $\omega + \Delta\omega$, for an observation time T. Then a realization of the *psd* is given by the average of the squared output of the filter:

$$\lim_{T \to \infty} \left[\frac{1}{T} \int_0^T x^2(t, \omega, \Delta\omega) \, dt \right]. \tag{7.144}$$

Example 7.24 Find the autocorrelation function and the spectral density for the process of Example 3.38,

$$X(t) = a \cos(\omega_1 t + \Xi), \tag{7.145}$$

in which a and ω_1 are constants and Ξ is random with a uniform distribution in $[-\pi, \pi]$. From Example 3.38,

$$\varrho(\tau) = \frac{a^2}{2} \cos(\omega_1 \tau) \tag{7.146}$$

since $\langle X(t) \rangle = 0$. Note that since ϱ is a function of τ only, this is a weakly stationary process.

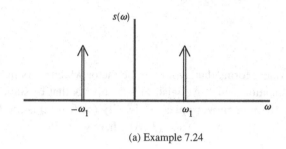

(a) Example 7.24

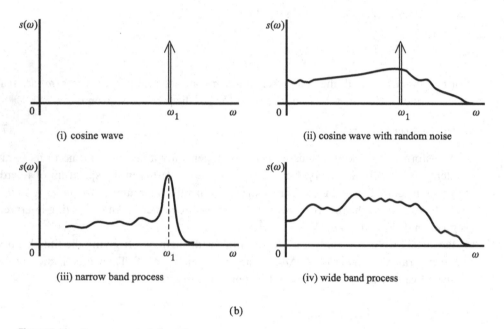

(b)

Figure 7.11   Power spectral densities

Using (7.146),

$$s(\omega) = \frac{\pi a^2}{2} \left[\delta(\omega + \omega_1) + \delta(\omega - \omega_1) \right],$$ (7.147)

where $\delta()$ is the Dirac delta function (see Appendix 2). This is plotted in Figure 7.11(a). The Fourier transforms are standard values which the reader might wish to check.

Example 7.25 Figure 7.11(b) shows power spectral densities for several processes which relate to Example 3.39 and Figure 3.24. These are given for positive values of ω. A process

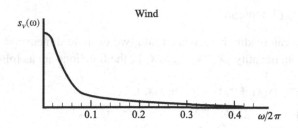

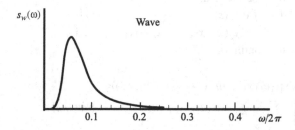

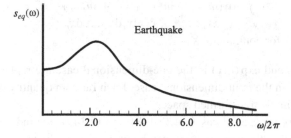

Figure 7.12   Power spectral densities for load processes

with a 'peaked' autocorrelogram declining rapidly to zero will indicate a wide band process, with white noise having an autocorrelogram corresponding to a Dirac spike at zero time displacement ($\tau = 0$).

Example 7.26 Figure 7.12 shows power spectral densities for positive ω for several processes leading to loads on engineered structures (after Wen, 1990).

Example 7.27 The power spectral density for a blackbody was given in Equation (6.104):

$$u(v, T) = \frac{8\pi v^3}{c^3} \frac{h}{\exp(hv/kT) - 1}.$$

7.4.7   Distributions in more than one dimension

The definitions for *mgfs* and *cfs* can readily be extended into two or more dimensions. Considering an n-dimensional random quantity $\{X_1, X_2, \ldots, X_n\}$, the functions are as follows:

$$m_{X_1, X_2, \ldots, X_n}(s_1, s_2, \ldots, s_n) = \langle \exp(s_1 x_1 + s_2 x_2 + \cdots + s_n x_n) \rangle$$

$$= \begin{cases} \sum \cdots \sum \exp(s_1 x_1 + s_2 x_2 + \cdots + s_n x_n) \cdot \\ p_{X_1, X_2, \ldots, X_n}(x_1, x_2, \ldots, x_n) \\ \text{for discrete } X \text{ and} \\ \int \cdots \int \exp(s_1 x_1 + s_2 x_2 + \cdots + s_n x_n) \cdot \\ f_{X_1, X_2, \ldots, X_n}(x_1, x_2, \ldots, x_n) \, dx_1 \, dx_2 \ldots dx_n \\ \text{for continuous } X. \end{cases} \tag{7.148}$$

$$\phi_{X_1, X_2, \ldots, X_n}(u_1, u_2, \ldots, u_n) = \langle \exp(iu_1 x_1 + iu_2 x_2 + \cdots + iu_n x_n) \rangle$$

$$= \begin{cases} \sum \cdots \sum \exp(iu_1 x_1 + iu_2 x_2 + \cdots + iu_n x_n) \cdot \\ p_{X_1, X_2, \ldots, X_n}(x_1, x_2, \ldots, x_n) \\ \text{for discrete } X, \text{ and} \\ \int \cdots \int \exp(iu_1 x_1 + iu_2 x_2 + \cdots + iu_n x_n) \cdot \\ f_{X_1, X_2, \ldots, X_n}(x_1, x_2, \ldots, x_n) dx_1 dx_2 \ldots dx_n \\ \text{for continuous } X. \end{cases} \tag{7.149}$$

It is seen that the terms $\exp(sx)$ and $\exp(iux)$ in the one-dimensional case are replaced by $\exp\left(\sum s_j x_j\right)$ and $\exp\left(\sum iu_j x_j\right)$ in the multidimensional case. Each random quantity X_j has a companion quantity (s_j or u_j) in the transformed space.

Equations (7.118) and (7.119) are special cases of (7.148) and (7.149); for independent random quantities $\{X_1, X_2, \ldots, X_n\}$, the *pmfs* and *pdfs* can be written as products:

$$p_{X_1, X_2, \ldots, X_n}(x_1, x_2, \ldots, x_n) = p_{X_1}(x_1) \cdot p_{X_2}(x_2) \cdot \ldots \cdot p_{X_n}(x_n) \tag{7.150}$$

and

$$f_{X_1, X_2, \ldots, X_n}(x_1, x_2, \ldots, x_n) = f_{X_1}(x_1) \cdot f_{X_2}(x_2) \cdot \ldots \cdot f_{X_n}(x_n). \tag{7.151}$$

For their transforms, we find the product form previously obtained:

$$m_Y(s) = \prod_{i=1}^{n} m_{X_i}(s) \text{ and}$$
$$\phi_Y(u) = \prod_{i=1}^{n} \phi_{X_i}(u),$$

which are found by substituting Equations (7.150) and (7.151) into Equations (7.148) and (7.149), and then separating the summations or integrals into multiplicative terms, each involving only one random quantity and its companion transform parameter.

The joint moment generating or characteristic functions for two random quantities X_1 and X_2 are a special case of Equations (7.148) and (7.149). Taking in this case *cfs* as an example,

$$\phi_{X_1,X_2}(u_1, u_2) = \langle \exp(iu_1x_1 + iu_2x_2) \rangle$$

$$= \begin{cases} \sum\sum \exp(iu_1x_1 + iu_2x_2)\, p_{X_1,X_2}(x_1, x_2) \\ \text{for discrete } X, \text{ and} \\ \iint \exp(iu_1x_1 + iu_2x_2)\, f_{X_1,X_2}(x_1, x_2)\, dx_1 dx_2 \\ \text{for continuous } X. \end{cases} \tag{7.152}$$

The joint moment $\langle X_1^k X_2^\ell \rangle$ of order $k + \ell$ can be obtained as before by expanding the exponential terms as series; thus

$$\langle X_1^k X_2^\ell \rangle = \frac{1}{i^{k+\ell}} \left. \frac{\partial^{k+\ell}\phi_{X_1,X_2}(u_1, u_2)}{\partial x_1^k \partial x_2^\ell} \right|_{\substack{x_1=0 \\ x_2=0}}. \tag{7.153}$$

Special cases of this equation may be obtained as

$$\langle X_1 \rangle = \frac{1}{i} \left. \frac{\partial\phi_{X_1,X_2}(u_1, u_2)}{\partial u_1} \right|_{\substack{u_1=0 \\ u_2=0}}, \qquad \langle X_2 \rangle = \frac{1}{i} \left. \frac{\partial\phi_{X_1,X_2}(u_1, u_2)}{\partial u_2} \right|_{\substack{u_1=0 \\ u_2=0}},$$

$$\left.\langle X_1^2 \rangle = \frac{1}{i^2} \left. \frac{\partial^2\phi_{X_1,X_2}(u_1, u_2)}{\partial u_1^2} \right|_{\substack{u_1=0 \\ u_2=0}}, \qquad \langle X_2^2 \rangle = \frac{1}{i^2} \left. \frac{\partial^2\phi_{X_1,X_2}(u_1, u_2)}{\partial u_2^2} \right|_{\substack{u_1=0 \\ u_2=0}} \text{ and }\right\} \tag{7.154}$$

$$\langle X_1 X_2 \rangle = \frac{1}{i^2} \left. \frac{\partial\phi_{X_1,X_2}(u_1, u_2)}{\partial u_1 \partial u_2} \right|_{\substack{u_1=0 \\ u_2=0}}.$$

These ideas may be extended into as many dimensions as we please.

Example 7.28 The random quantities X_1 and X_2 are modelled by the two-dimensional normal probability distribution, given by the following joint probability density function. The means are zero and the standard deviation is unity with a correlation coefficient ρ:

$$f_{X_1,X_2}(x_1, x_2) = \frac{1}{2\pi\sqrt{1-\rho^2}} \exp\left[-\frac{1}{2\sqrt{1-\rho^2}} \left(x_1^2 - 2\rho x_1 x_2 + x_2^2 \right) \right] \tag{7.155}$$

for $-\infty < x_1, x_2 < \infty$.

The joint characteristic function is, by definition, using Equation (7.152),

$$\phi_{X_1,X_2}(u_1, u_2) = \iint \exp(iu_1x_1 + iu_2x_2)$$

$$\times \frac{1}{2\pi\sqrt{1-\rho^2}} \exp\left[-\frac{1}{2\sqrt{1-\rho^2}} \left(x_1^2 - 2\rho x_1 x_2 + x_2^2 \right) \right] dx_1 dx_2$$

$$= \exp\left[-\frac{1}{2} \left(u_1^2 + 2\rho u_1 u_2 + u_2^2 \right) \right]. \tag{7.156}$$

Equations (7.154) yield, for example,

$$\langle X_1 X_2 \rangle = \rho, \tag{7.157}$$

the correlation coefficient.

7.5   Central Limit Theorem – the model of sums

It can be stated quite generally that if one adds together or averages many random quantities, one obtains a random quantity that approaches the normal distribution. This certainly holds for independent and identically distributed (*iid*) random quantities with finite variances. It has also been proved for less demanding conditions, for instance where the random quantities are independent but not identically distributed, provided each one of these contributes only a small amount to the sum. The theorem will also apply if there is some dependence between the random quantities, provided that the correlation is very small except for that between one quantity and a limited number of others. In concise terms, the distribution of one random quantity must not dominate the final result.

These statements introduce the Central Limit Theorem consciously in a rather loose fashion. But the theorem is most useful in making judgements regarding the distribution of a quantity that can be seen to be the sum or average of other random quantities which are largely independent of each other, in such a way that no one contributor dominates the result. We often average a set of readings to obtain estimates of the mean value. Examples where processes are additive are packets of information that are the sum of many components, or a deformation which is the sum of many individual deformations (for example the macroscopic creep experienced in a specimen may be the sum of a large number of individual thermally activated movements). A measured length may, in a survey, be the sum of many individual lengths.

All of these processes result in sums or averages and the Central Limit Theorem suggests that the final distributions should be close to the normal distribution. In one historic example, the use of the word 'population' arose from measurement of the height, weight and so on of human and other populations. The resulting distributions have been found to fit the normal distribution. Perhaps many independent genetic factors add together to result in the normality. The theorem can be characterized as the 'model of sums'. We shall not prove it in its entire generality, but will introduce it by using an example, then give the transition from binomial to normal distributions, and subsequently give a proof based on characteristic functions.

7.5.1   Illustration

The introductory example is concerned with *iid* random quantities X_i that are uniformly distributed on the interval $[0, 1]$, that is

$$f_{X_i}(x_i) = 1 \text{ for } 0 \le x \le 1$$
$$= 0 \text{ otherwise.} \tag{7.158}$$

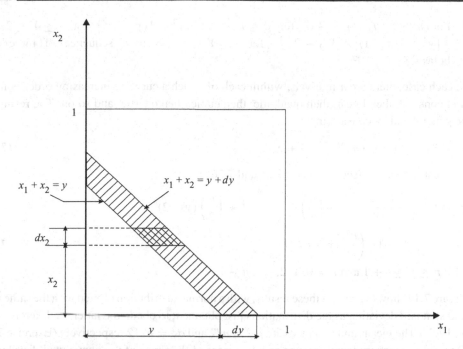

Figure 7.13   Integration over unit square

We consider now the sum of several of these. Let us start with two; consider

$$Y = X_1 + X_2,$$ (7.159)

where the X_is are modelled using Equation (7.158). Since X_1 and X_2 are independent, $Y = X_1 + X_2$ is uniformly distributed over the unit square with a corner at the origin, as illustrated in Figure 7.13. Lines corresponding to $y = x_1 + x_2$ and $y + dy = x_1 + x_2$ are shown. The area between these lines of constant Y increases linearly from $y = 0$ to $y = 1$ and then decreases linearly until $y = 2$. The probability density of Y is proportional to the area noted and therefore forms a triangular distribution over $0 \leq y \leq 2$, with a peak at $y = 1$. A more formal proof can be obtained by using Equations (7.76) and (7.77); for example, using the first of these, for $y \leq 1$ (see Figure 7.13):

$$f_Y(y) = \int_{-\infty}^{\infty} f_{X_1}(y - x_2) f_{X_2}(x_2) \, dx_2 = \int_0^y dx_2 = y.$$

One obtains the same result as before.

One can now convolve X_3 with the triangular distribution just found. Proceeding in this way, it can be found that $f_Y(y)$ can be represented by the following.

1. For $n = 1$, $f_Y(y) = 1$ for $0 \leq Y \leq 1$, the starting point.
2. For $n = 2$, $f_Y(y) = y$ for $0 \leq Y \leq 1$, and $= 2 - y$ for $1 \leq Y \leq 2$, the triangular distribution, a sequence of two straight lines.

3. For $n = 3$, $f_Y(y) = \frac{1}{2}y^2$ for $0 \le Y \le 1$, $\frac{1}{2}\left[y^2 - 3(y-1)^2\right]$ for $1 \le Y \le 2$, and $\frac{1}{2}\left[y^2 - 3(y-1)^2 + 3(y-2)^2\right]$, for $2 \le Y \le 3$, giving a sequence of three parabolas.

In each case, there are n intervals, within each of which a curve of increasing order is found, first constant, then linear, then quadratic, then cubic, then quartic, and so on. The result may be generalized by considering

$$Y = X_1 + X_2 + \cdots + X_i + \cdots + X_n. \tag{7.160}$$

The resulting *pdf* is (see Exercises 7.25 and 7.26)

$$f_Y(y) = \frac{1}{(n-1)!}[y^{n-1} - \binom{n}{1}(y-1)^{n-1} + \binom{n}{2}(y-2)^{n-1} - $$
$$\cdots (-1)^r \binom{n}{r}(y-r)^{n-1}] \tag{7.161}$$

for $r \le y \le r+1$ and $r = 0, 1, 2, \ldots, n-1$.

Figure 7.14 shows several of these results, with a normal distribution plotted with the same mean and standard deviation as the distributions just obtained; to obtain the latter see Exercises 7.17 and 7.25. The means and variances are $\mu_Y = n/2$ and $\sigma_Y^2 = n/12$ respectively (Exercise 7.26). One can see the rapid convergence as n increases of the shape of the convoluted distributions to that of the normal distribution. But it takes quite large n to ensure that the tails of the distributions are reasonably similar.

Both the mean and standard deviation of Y tend to infinity as $n \to \infty$. The standard deviation is proportional to $\sqrt{n}$, as shown in Exercise 7.26. Had we considered averages $\bar{Y} = \sum X_i/n$ instead of sums, the mean would remain constant and the standard deviation would decrease, in proportion to $1/\sqrt{n}$. To summarize, the convoluted distribution becomes wider and more spread out (sums), or shrinks, becoming narrower and more concentrated (averages). If we were to consider the random quantity $\sum X_i/\sqrt{n}$, then we might expect the distribution to be stable in lateral extent. This strategy will be considered in the following sections.

7.5.2   Normal approximation to the binomial

In the special case where the X_i are Bernoulli random quantities that are independent, Y in Equation (7.160),

$$Y = X_1 + X_2 + \cdots + X_i + \cdots + X_n,$$

represents the binomial random quantity with mean np and variance $\sigma^2 = npq$, where $q = \tilde{p}$. Although discrete, the binomial distribution can be approximated well for large n by the normal distribution, provided the probability masses in the binomial distribution are approximated by

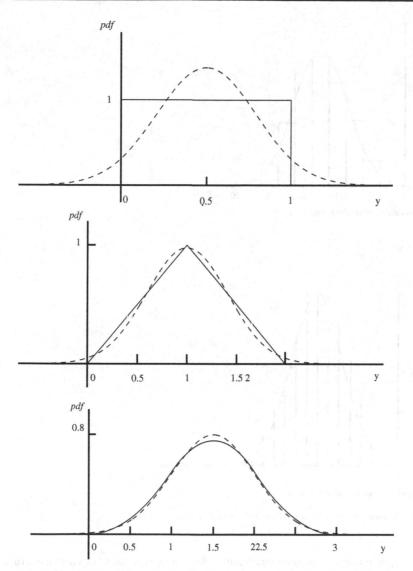

Figure 7.14   Addition of uniformly distributed random quantities (dotted line is normal distribution)

areas, for example by small rectangles – trapezoids are another possibility. Or the area under the normal curve can be calculated from a value $i - 1/2$ to $i + 1/2$ when approximating the probability at the integral value i. This is used in Exercise 7.27. Figure 7.15(a) shows the binomial random quantity for $p = 1/2$ and $n = 10$. A good approximation is obtained even for this relatively small value of n; for $p = 1/4$, Figure 7.15(b) shows that for a larger value

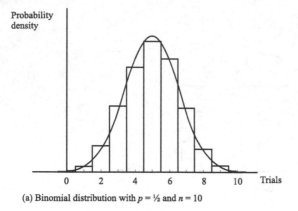

(a) Binomial distribution with $p = \frac{1}{2}$ and $n = 10$

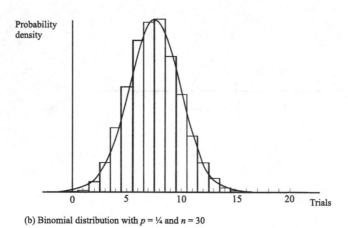

(b) Binomial distribution with $p = \frac{1}{4}$ and $n = 30$

Figure 7.15   Normal approximation to binomial distribution

of n a good approximation can also be obtained. This approximation provides a useful tool for calculation; see Exercise 7.27.

Transition from binomial to normal distributions

We now consider more precisely the transition from the binomial to the normal distributions, based on de Finetti (1974). As noted at the end of Section 7.5.1, the mean and variance of the sum Y of n *iid* random quantities increases with n. The average of n binomial random quantities is the frequency Y/n, the distribution of which becomes more and more peaked as

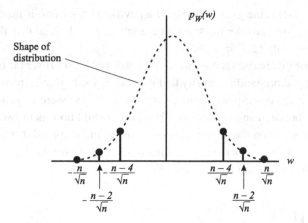

(a) Schematic illustration of probability distribution of W

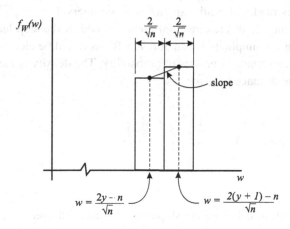

(b) Approximation of a discrete probability "spike" with a rectangle

Figure 7.16   Normal and binomial distributions

n increases. A better behaved quantity is

$$W = \frac{2Y - n}{\sqrt{n}}, \tag{7.162}$$

which has a mean of $\sqrt{n}(2p - 1)$ and a standard deviation equal to $2\sqrt{pq}$. Figure 7.16(a) shows the binomial probabilities distributed in this way. The usual binomial quantity Y (number of successes) takes values $0, 1, 2, \ldots, n$ which are mapped onto W. One can also observe that the numerator of Equation (7.162) is the gain in n trials if a payment of ± 1 unit is made for a success or failure, respectively. We consider the special case where $p = 1/2$, so that W has a mean of zero and a variance of unity $(2\sqrt{pq} = 1)$.

We now spread out the probability masses at the various values of $W = w(y)$ over the interval $\pm(1/\sqrt{n})$, resulting in a stepped probability density that is continuous. Figure 7.16(b) shows two successive spread-out masses corresponding to y and $y + 1$. If one were to simulate a point in this continuous distribution using random numbers, one would do this in two steps. First, one would simulate a point w on the distribution of W using the binomial distribution, and then a point in the interval $w \pm (1/\sqrt{n})$ using a uniform distribution for the second step. The first step involves the use of Equation (7.10), giving

$$p_W(w) = \mathbf{Pr}(Y = y) = \binom{n}{y}\left(\frac{1}{2}\right)^n,$$

where y is given by Equation (7.162); $y = (w\sqrt{n} + n)/2$.

To denote the continuous random quantity spread over the interval $\pm(1/\sqrt{n})$, we could change the letter W denoting the discrete random quantity used in the step leading to the continuous distribution; but for simplicity we shall retain W, as it will be clear when we are using densities, as against concentrated masses of probability. The density is the probability mass 'smeared out' over the distance $(2/\sqrt{n})$:

$$f_W(w) = \mathbf{Pr}(Y = y) \div \left(\frac{2}{\sqrt{n}}\right) = \binom{n}{y}\left(\frac{1}{2}\right)^n \div \left(\frac{2}{\sqrt{n}}\right)$$
$$= \frac{1}{2}\sqrt{n}\binom{n}{y}/2^n. \tag{7.163}$$

As we increase n we wish to calculate the slope to determine whether we approach the normal density. The slope between the midpoints $w(y)$ and $w(y + 1)$ of the stepped density – Figure 7.16(b) – is

$$f'_W(w) = \left\{[\mathbf{Pr}(Y = y + 1) - \mathbf{Pr}(Y = y)] \div \left(\frac{2}{\sqrt{n}}\right)\right\} \div \frac{2}{\sqrt{n}}$$
$$= \frac{1}{4}n[\mathbf{Pr}(Y = y + 1) - \mathbf{Pr}(Y = y)]. \tag{7.164}$$

Now we know that

$$\frac{\mathbf{Pr}(Y = y + 1)}{\mathbf{Pr}(Y = y)} = \frac{n - x}{x + 1} \cdot \frac{p}{\bar{p}}. \text{ (See Exercise 7.22.)}$$

We have that $p = \tilde{p} = 1/2$; substituting into Equation (7.164),

$$f'_W(w) = \frac{1}{4}n \Pr(Y = y)\left(\frac{n-y}{y+1} - 1\right)$$
$$= \frac{1}{4}n \Pr(Y = y)\left(\frac{n-2y-1}{y+1}\right). \tag{7.165}$$

Substituting for $\Pr(Y = y) = f_W(w) \cdot 2/\sqrt{n}$ from Equation (7.163), and $y = (w\sqrt{n} + n)/2$ from Equation (7.162),

$$f'_W(w) = \frac{1}{2}\sqrt{n}\, f_W(w)\left[\frac{-w\sqrt{n} - 1}{(w\sqrt{n} + n)/2 + 1}\right]$$
$$= -f_W(w)\left(\frac{wn + \sqrt{n}}{w\sqrt{n} + n + 2}\right)$$
$$= -w\, f_W(w)\left[\frac{1 + 1/(w\sqrt{n})}{1 + w/\sqrt{n} + 2/n}\right]. \tag{7.166}$$

We consider first non-zero values of w. In this case, Equation (7.166) can be written as

$$\frac{f'_W(w)}{f_W(w)} = -w + \text{ terms that tend to zero as } n \to \infty. \tag{7.167}$$

Therefore, in this limit,

$$\frac{f'_W(w)}{f_W(w)} = \frac{d\ln f_W(w)}{dw} = -w. \tag{7.168}$$

In the case where $w = 0$, we can write Equation (7.166) as

$$f'_W(w) = -f_W(w)\left[\frac{w + 1/\sqrt{n}}{1 + w/\sqrt{n} + 2/n}\right] \to 0. \tag{7.169}$$

This merely shows that the slope at $w = 0$ is zero, as would be expected for a distribution symmetrical about the origin.

Equation (7.168) yields

$$\ln f_W(w) = -\frac{w^2}{2} + \text{constant} \tag{7.170}$$

or

$$f_W(w) = K \exp\left(-\frac{w^2}{2}\right), \tag{7.171}$$

where $K = 1/\sqrt{2\pi}$ for coherence. Equation (7.171) is the standard normal distribution – recall that we arranged matters so that W had a mean of zero and a standard deviation of unity.

Similar results can be obtained if the distribution is not standardized – a normal distribution is still arrived at, but with mean and standard deviation corresponding to the case under consideration.

It is remarkable how the slopes across the stepped densities approach those of the normal density, matching this curve exactly in the limit of $n = \infty$. The derivation above is that of a limiting distribution – seen also, for example in the derivation of the Poisson distribution from the binomial (Section 3.2.6). In these instances a sequence of distributions F_n or f_n are considered; note that here the subscripts n do not refer to the random quantity but to the value of n (deterministic number) in the sequence. The transition from the sequence to the limit is often written as

$$F_n \to F, \text{ or} \tag{7.172}$$
$$f_n \to f. \tag{7.173}$$

7.5.3   Proof of the Central Limit Theorem

We consider now a more general situation, again using Equation (7.160), where the distribution of the X_i is not specified. Yet we retain the assumption that they are *iid* with common mean μ and finite variance σ^2. Then

$$Y = \sum_{i=1}^{n} X_i$$

with mean $\mu_Y = n\mu$ and variance $\sigma_Y^2 = n\sigma^2$. We consider the standardized random quantity

$$W = \frac{Y - \mu_Y}{\sigma_Y} = \frac{Y - n\mu}{\sqrt{n}\sigma}. \tag{7.174}$$

This quantity has a mean of zero and a standard deviation $\sigma_W = 1$. The distribution will not flatten or shrink, and W behaves in this respect similarly to Equation (7.162). Equation (7.174) can be written as

$$W = \frac{\sum_{i=1}^{n} (X_i - \mu)}{\sqrt{n}\sigma} = \frac{\sum_{i=1}^{n} V_i}{\sqrt{n}\sigma} = \sum_{i=1}^{n} \frac{V_i}{\sqrt{n}\sigma} = \sum_{i=1}^{n} T_i, \tag{7.175}$$

where $V_i = X_i - \mu$ has a mean of zero and a standard deviation of σ. Note that $T_i = V_i/\sqrt{n}\sigma$ must also have a mean equal to zero but a standard deviation of $1/\sqrt{n}$, resulting in the standard deviation of unity for W.

The random quantity of Equation (7.175) has the characteristic function

$$\phi_W (u) = \phi_T^n (u) = \phi_V^n \left(\frac{u}{\sqrt{n}\sigma} \right). \tag{7.176}$$

where we have used Equations (7.120) and (7.112). Now

$$\phi_V \left(\frac{u}{\sqrt{n}\sigma} \right) = \int_{-\infty}^{\infty} \exp \left(\frac{iuv}{\sqrt{n}\sigma} \right) f_V(v) \, dv$$

$$= \int_{-\infty}^{\infty} \left[1 + \frac{iuv}{\sqrt{n}\sigma} + \frac{1}{2} \left(\frac{iuv}{\sqrt{n}\sigma} \right)^2 + \cdots \right] f_V(v) \, dv$$

$$= 1 + \frac{iu}{\sqrt{n}\sigma} \langle V \rangle + \frac{1}{2} \left(\frac{iu}{\sqrt{n}\sigma} \right)^2 \langle V^2 \rangle + \cdots \tag{7.177}$$

We know that $\langle V \rangle = 0$; consequently $\langle V^2 \rangle = \sigma^2$ so that Equation (7.177) becomes

$$\phi_V \left(\frac{u}{\sqrt{n}\sigma} \right) = 1 - \frac{1}{2} \frac{u^2}{n} + \text{ terms involving } n^{-\frac{3}{2}} \text{ and smaller.} \tag{7.178}$$

Using Equation (7.176),

$$\phi_W(u) = \left(1 - \frac{1}{2} \frac{u^2}{n} + \cdots \right)^n$$

$$\rightarrow \exp \left(-\frac{u^2}{2} \right) \text{ as } n \rightarrow \infty. \tag{7.179}$$

But this is the *cf* of the standard normal distribution; see Equation (7.109). This proves the Central Limit Theorem for *iid* random quantities X_i. We normalized the random quantity W so as to have a mean of zero and a standard deviation of one; we can reverse this by using Equation (7.174) resulting in a very 'flat' distribution for Y, or a very 'sharp' one for Y/n, as discussed earlier. The key point in the derivation is the behaviour of $\phi_W(u)$ as n increases.

7.5.4   Second characteristic function and cumulants

This derivation can also be achieved by considering the **second characteristic function** which is merely the logarithm of the characteristic function:

$$\psi_X(u) = \ln \phi_X(u). \tag{7.180}$$

Then

$$\phi_X(u) = \exp \psi_X(u). \tag{7.181}$$

Also the **cumulants** $\varkappa_s$ for $s = 1, 2, 3, \ldots$ of the second characteristic function are defined by the power series

$$\psi_X(u) = \sum_{s=1}^{\infty} \varkappa_s \frac{(iu)^s}{s!} \tag{7.182}$$

with

$$\varkappa_s = i^s \frac{d^s \psi_X(u = 0)}{du^s}. \tag{7.183}$$

The cumulants $\varkappa_1$ and $\varkappa_2$ are the mean and variance of X; the reader may wish to check.

The following expansion results:

$$\psi_X(u) = \varkappa_1(iu) + \frac{1}{2}\varkappa_2(iu)^2 + \frac{1}{3!}\varkappa_3(iu)^3 + \cdots + \frac{1}{s!}\varkappa_s(iu)^s + \cdots \tag{7.184}$$

Here it is convenient to consider small u, near the origin, and in the case where the mean is zero, as in the standardized distributions above, a parabolic approximation suffices:

$$\psi_X(u) \simeq -\frac{1}{2}\sigma_X^2 u^2. \tag{7.185}$$

Equation (7.176) becomes

$$\psi_W(u) = n\psi_T(u) = n\psi_V\left(\frac{u}{\sqrt{n}\sigma}\right) \tag{7.186}$$

$$= -n \cdot \frac{1}{2}\frac{\sigma^2 u^2}{n\sigma^2} = -\frac{u^2}{2},$$

or

$$\psi_W(u) = \ln\phi_W(u) = -\frac{u^2}{2}. \tag{7.187}$$

This is the same result as that in Equation (7.179).

7.5.5   Non-identical independent random quantities

At the beginning of this section (7.5) we noted that the Central Limit Theorem also applies in more general circumstances, for example where the distributions are not identical, or where there is some dependence; generally it is required that one of the X_i does not dominate the result. An interesting review of extensions of the Central Limit Theorem is given by de Finetti (1974); we shall give a brief introduction to this. If we consider distributions of the X_i that are similar except for the variances, denoted σ_i, which vary from one quantity to another, then, instead of Equation (7.187) we have

$$-\frac{u^2}{2}\left[1 + \sum_{i=1}^n \frac{\sigma_i^2}{\sigma_T^2}\varepsilon\left(\frac{\sigma_i u}{\sigma_T}\right)\right], \tag{7.188}$$

where $\sigma_T^2 = \sum_{i=1}^n \sigma_i^2$. Now we have the same expression as before, except for the summation term, which can be thought of as a 'correction' to the previous result. If this term tends to zero, then the Central Limit Theorem remains valid. This can be shown to be true if the contribution of each term is small compared to the total of the preceding ones; also, the total variance σ_T^2 must diverge. A myriad of further extensions has been obtained, but for the purposes of the present work, the flavour of these given here is deemed sufficient.

7.6   Chebyshev inequality

We consider a random quantity X with finite mean μ and finite variance σ^2, and any distribution that satisfies these conditions, continuous or discrete. The inequality is that

$$\mathbf{Pr}\,(\mu - c\sigma < X < \mu + c\sigma) \geq 1 - \frac{1}{c^2}, \tag{7.189}$$

in which a positive number $c \geq 2$ is needed for useful results. The method for obtaining this inequality is shown in Figure 7.17. This shows a continuous distribution, but the idea is equally applicable to discrete distributions. The method will be described in two steps.

Step 1 In the upper illustration of the figure, think of the variance as the second moment of area of the probability distribution about the centroid (mean) g (one can also think of the moment of inertia for unit distributed mass). If we remove the central area (or probability mass), between the limits $\pm c\sigma$, then the second moment is reduced.

Step 2 If we take the probability in the remaining areas (p_1 and p_2 in Figure 7.17) and concentrate them as point masses on the edge of the central area, then the second moment is reduced even further.

Mathematically, these steps amount to the following.

Step 1 Since $\sigma^2 = \langle (X - \mu)^2 \rangle$,

$$\sigma^2 = \int_{\text{all } x} (x - \mu)^2 \mathrm{d}F_X\,(x)$$

$$= \int_{x \leq \mu - c\sigma} (x - \mu)^2 \mathrm{d}F_X\,(x) + \int_{\mu - c\sigma < x < \mu + c\sigma} (x - \mu)^2 \mathrm{d}F_X\,(x)$$

$$+ \int_{x \geq \mu - c\sigma} (x - \mu)^2 \mathrm{d}F_X\,(x)$$

$$\geq \int_{x \leq \mu - c\sigma} (x - \mu)^2 \mathrm{d}F_X\,(x) + \int_{x \geq \mu - c\sigma} (x - \mu)^2 \mathrm{d}F_X\,(x)\ .$$

Step 2 Now, for those regions over which these integrals are taken, provided $c > 0$,

$$(x - \mu)^2 \geq (c\sigma)^2,$$

so that

$$\sigma^2 \geq c^2 \sigma^2 \left[\int_{x \leq \mu - c\sigma} \mathrm{d}F_X\,(x) + \int_{x \geq \mu - c\sigma} \mathrm{d}F_X\,(x) \right]. \tag{7.190}$$

Therefore

$$\frac{1}{c^2} \geq \mathbf{Pr}\,[(X \leq \mu - c\sigma) \vee (X \geq \mu + c\sigma)] = p_1 + p_2$$

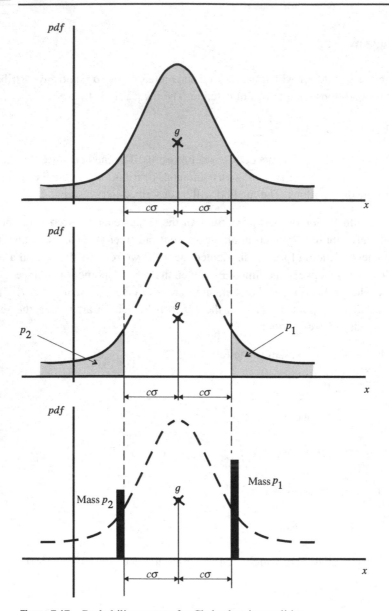

Figure 7.17 Probability masses for Chebyshev inequalities

or

$$\mathbf{Pr}(\mu - c\sigma < X < \mu + c\sigma) \geq 1 - \frac{1}{c^2}.$$

This may also be written as

$$\mathbf{Pr}(|X - \mu| \geq c\sigma) \leq \frac{1}{c^2}. \tag{7.191}$$

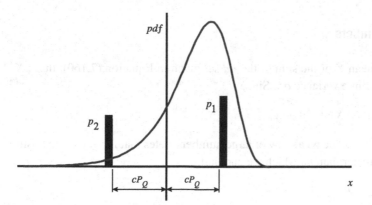

Figure 7.18   Quadratic prevision example

The Chebyshev bound is rather crude if c is small (e.g. 2, 3 or 4) but is very useful in limit theorems. The method for deducing the bound is also valid for higher moments than the second. Also, we need not have commenced the analysis at the centroid. We could have started at any point, for example the origin.

Example 7.29 The 'quadratic prevision' $P_Q(X)$ has been introduced by de Finetti (1974), where

$$P_Q(X) = \sqrt{P(X^2)} = \sqrt{\langle X^2 \rangle}.$$

Show that

$$\frac{1}{c^2} \geq \mathbf{Pr}\left[|X| \geq cP_Q(X)\right]. \tag{7.192}$$

This is an example where we commence the analysis at the origin rather than the centroid (Figure 7.18). We repeat steps 1 and 2. Thus

$$P_Q^2(X) \geq c^2 P_Q^2(X)\left[\int_{x \leq cP_Q} dF_X(x) + \int_{x \geq cP_Q} dF_X(x)\right]$$
$$= c^2 P_Q^2(X)\left[\mathbf{Pr}\,|X| \geq cP_Q(X)\right],$$

corresponding to Equation (7.190). This leads directly to the required result.

The results of this section may be summarized in terms of standardized quantities as

$$\mathbf{Pr}\left(\frac{|X - \mu|}{\sigma} \geq c\right) \leq \frac{1}{c^2} \text{ and} \tag{7.193}$$

$$\mathbf{Pr}\left(\frac{|X|}{P_Q} \geq c\right) \leq \frac{1}{c^2}. \tag{7.194}$$

The inequality will be applied in the following section.

7.7   Laws of large numbers

We consider the mean $\bar{Y}$ of the sum of the *iid* set given in Equation (7.160); these X_i have a finite mean μ and finite variance σ^2. Since

$$Y = X_1 + X_2 + \cdots + X_i + \cdots + X_n,$$

we consider $\bar{Y} = Y/n$. The **weak** law of large numbers states that for any given positive ε and θ, then for all n greater than n_0 which we can find,

$$\mathbf{Pr}\left(\left|\bar{Y} - \mu\right| > \varepsilon\right) < \theta. \tag{7.195}$$

We know that $\langle \bar{Y} \rangle = \mu$, and $\sigma_{\bar{Y}}^2 = \sigma^2/n$. Applying the Chebyshev inequality of Equation (7.191),

$$\mathbf{Pr}\left(\left|\bar{Y} - \mu\right| \geq \frac{c\sigma}{\sqrt{n}}\right) \leq \frac{1}{c^2}. \tag{7.196}$$

If we associate θ with $1/c^2$, we can choose n_0 using $\varepsilon = c\sigma/\sqrt{n}$ so that inequality (7.195) is satisfied. A further consequence of inequality (7.195) is that

$$\mathbf{Pr}\left(\left|\bar{Y} - \mu\right| > \varepsilon\right) \to 0 \tag{7.197}$$

as $n \to \infty$.

The result just obtained pertains to a particular $n > n_0$ for which we can satisfy Equation (7.195). If we now consider several cases together, say for $n = n_0, n_0 + 1, \ldots, n_0 + n_1$ (where n_1 is arbitrary), then the probability of at least one occurrence of the event $\left(\left|\bar{Y} - \mu\right| > \varepsilon\right)$ grows as n_1 increases, and Equation (7.195) is no guarantee that this will not become large.

The situation just discussed is in fact covered by the **strong** law of large numbers, which ensures that for any given positive numbers ε and θ, no matter how small, there is a number n_0 such that

$$\mathbf{Pr}\left(\max_{n_0 \leq n \leq n_0 + n_1} \left|\bar{Y} - \mu\right| > \varepsilon\right) < \theta \tag{7.198}$$

for arbitrary n_1. This is the version emphasized by de Finetti (1974). It is also sometimes stated that $\left(\bar{Y} - \mu\right) \to 0$ with probability one. There are also more general proofs of the weak and strong laws, for example for non-identical random quantities, or in the case that only the mean exists.

Probability and frequency; simulation

The theory just presented casts some light on the connection between probability and frequency. In the case where the X_i are Bernoulli (0:1) random quantities, $\bar{Y}$ is the frequency. In one view of inference – which is not advocated here – the parameter of a distribution is a 'fixed but unknown' quantity. In the Bernoulli process, this would be the probability of success in the next trial, used as a parameter in the binomial distribution. The value μ might be interpreted as the supposed probability; for example in the case of coin-tossing, $\bar{Y}$ would be frequency of heads and the supposed probability of heads might be $1/2$.

The laws of large numbers would then suggest that the frequency $\bar{Y}$ will be very close to the 'supposed probability', with a high probability approaching unity. (See also Section 10.3.4 on probability and frequency.) The results above might be used to justify an interpretation of probability that is based only on frequency. As a definition of probability, this would be inappropriate, since we have used the concept of probability in deriving the laws above, and it is illogical then to use the theorem to justify the definition. It is much better to assess matters using the proper focus of our subjective judgement. In Chapter 8, a full analysis of inference and analysis of the parameters of a distribution is given in a manner consistent with the approach of this book.

With regard to bets 'in the long run', let us assume that we have a coin that we judge to be perfectly 'honest' with an exactly equal probability of heads and tails, and furthermore that we judge that independence will apply in a long series of tosses, that we judge the probability is $1/2$ regardless of the results of the preceding tosses. In betting, say on heads in coin-tossing, 'in the long run' variations tend to cancel each other out. Yet there is no guarantee that this will happen. If one is losing because of a series of tails, there is in fact no reason to suppose that a run of heads will subsequently occur to 'cancel out' the adverse series of tails. Indeed, at this stage – after the run of tails – the expected number of heads in a long series in the future is again 50%. One might place more faith in cycles in the stock market, for example, if there are interest rate changes aimed at retarding or stimulating growth. Here at least there is a mechanism that might have some favourable consequences.

Simulation

One area where the laws of large numbers do provide significant assistance is in simulation using Monte Carlo methods (Chapter 11). In this method, one attempts to generate as closely as possible a set of *iid* random quantities for computational purposes. Then the law of large numbers can be used to estimate deviations from probabilistic estimates.

7.8   Concluding remarks

The manipulation of probabilities is an essential activity in probabilistic analysis. We learn many interesting things from the theory presented in this chapter. The transformation of random quantities with a known distribution function is frequently required in engineering analysis. Convolutions are regularly encountered. Transforms provide a powerful means of analysis. The use of convolutions and transforms shows a way to analyse sums of random quantities, with the result that we find, for example, that the sum of independent normal distributions results in another normal distribution. The sum of arbitrary independent random quantities results in an asymptotic tendency to the normal distribution in the Central Limit Theorem. We can deduce from this that the lognormal distribution results from the product of random quantities. The theoretical underpinnings are an essential part of probability theory.

7.9   Exercises

There are many excellent books on the mathematics of probability. These contain numerous exercises which may be used to supplement the following.

7.1 Confirm the use of (7.10) in an example of your choice. For example, one might transform the number of successes X in the hypergeometric distribution by

$$g(X) = X^k,$$

where k is a positive constant.

7.2 Transform the uniform distribution on $[0, 1]$, ($f_X(x) = 1$ for $0 \leq x \leq 1$, $= 0$ otherwise) using the function

$$g(X) = \ln\left(\frac{1}{1 - X}\right).$$

How may this be used with the probability integral transform in Monte Carlo simulation?

7.3 Obtain the standardized form of the Weibull distribution in terms of the parameter k:

$$f_X(x) = \frac{k}{\alpha}\left(\frac{x - \xi}{\alpha}\right)^{k-1} \exp\left[-\left(\frac{x - \xi}{\alpha}\right)^k\right] \text{ for } x \geq \xi.$$

7.4 Find the standardized Rayleigh distribution in terms of μ:

$$f_X(x) = \frac{2(x - a)}{\mu} \exp\left[-(x - a)^2/\mu\right].$$

7.5 Find the location and scale parameters for the standardization of the uniform distribution

$$f_X(x) = \frac{1}{b - a} \text{ for } a \leq x \leq b$$

into the distribution

$$f_Y(y) = 1 \text{ for } 0 \leq y \leq 1$$

7.6 If X is normally distributed, consider the transformation

$$X = \ln Y.$$

Then Y has a **lognormal** distribution.

(a) Write down the distribution of Y.
(b) Write down the mean and variance of the distribution.

(c) Show that $1/Y$ also has a lognormal distribution. Write down the mean and variance of this distribution.

(d) If Y_1 and Y_2 are two independent random quantities with lognormal distributions, show that $Y_P = Y_1 Y_2$ also has a lognormal distribution. Determine the parameters of Y_P in terms of the parameters of Y_1 and Y_2.

(e) Discuss the lognormal distribution as a 'model of products' in the light of the Central Limit Theorem as the 'model of sums'.

7.7 *Exponential growth.* Let the size Z of a population be given by

$$Z = c \exp(Xt)$$

for given time t; c is a constant. If X is random with the density

$$f_X(x) = d(1-x)^2 \text{ for } 0 < x < 1,$$

and 0 otherwise, find $f_Z(z)$.

7.8 Given that $f_{X_1, X_2}(x_1, x_2) = 1$ for $0 < x_1 < 1$ and $0 < x_2 < 1$, find $f_{Y_1, Y_2}(y_1, y_2)$ for the transformation

$$\begin{Bmatrix} y_1 \\ y_2 \end{Bmatrix} = \begin{bmatrix} 1 & -1 \\ 2 & 1 \end{bmatrix} \begin{Bmatrix} x_1 \\ x_2 \end{Bmatrix} + \begin{Bmatrix} 2 \\ 1 \end{Bmatrix}. \tag{7.199}$$

Why is this transformation termed affine?

7.9 Consider $Y = X_1 + X_2$. Use Equations (7.76) and (7.77) to obtain $f_Y(y)$ if X_1 and X_2 are *iid* with exponential distributions $\exp(-x)$ for $x \geq 0$, $= 0$ otherwise.

7.10 Consider two *iid* random quantities X_1 and X_2 that are uniformly distributed on $(0, 1)$. Use the transformations

$$y_1 = \sqrt{-2 \ln x_1} \cos(2\pi x_2) \text{ and} \tag{7.200}$$

$$y_2 = \sqrt{-2 \ln x_1} \sin(2\pi x_2) \tag{7.201}$$

to show that Y_1 and Y_2 are independent random quantities with standard normal distributions. This algorithm can be used to generate normal random quantities from the uniform distribution (Chapter 11).

7.11 Two independent random quantities Z_1 and Z_2 are modelled using the standard normal distribution. The quantities R and Θ are polar coordinates related to Z_1 and Z_2 as follows for $0 \leq R < \infty$ and $0 \leq \Theta \leq 2\pi$:

$$R = \left(Z_1^2 + Z_2^2\right)^{1/2} \text{ and} \tag{7.202}$$

$$\Theta = \arctan\left(\frac{Z_2}{Z_1}\right). \tag{7.203}$$

Determine $f_R(r)$ and $f_\Theta(\theta)$.

7.12　Derive Equation (7.109).

7.13　*Gamma and beta functions.* It is a good idea to become familiar with the gamma function

$$\Gamma(s) = \int_0^\infty u^{s-1} \exp(-u) \, du. \tag{7.204}$$

Show that

$$\Gamma(s) = (s-1) \Gamma(s-1) \tag{7.205}$$

and that for positive integers s, $\Gamma(s) = (s-1)!$. Also show that $\Gamma(1/2) = \sqrt{\pi}$, and that the beta function

$$B(r, m-r) = \int_0^1 a^{r-1}(1-a)^{m-r-1} \, da \tag{7.206}$$

used to normalize the beta distribution is given by

$$B(s, t-s) = \frac{\Gamma(s)\Gamma(t-s)}{\Gamma(t)}. \tag{7.207}$$

7.14　*Human error.* In a study of the effect of human error in design, Nessim used two independent gamma distributions with the same scale factor (β in Equation (7.210) below) to model the distributions of undetected and detected errors (M and N respectively) (see Nessim and Jordaan, 1985). Show that the probability of detecting an error,

$$P = \frac{N}{M+N}, \tag{7.208}$$

is given by the beta distribution.

Note that these distributions of error rates are consistent with the idea of gamma distributions modelling the uncertainty in rates and the errors themselves occurring as a Poisson process. In this context the gamma distributions are conjugate prior distributions for the rates of the Poisson process. This situation is modelled in Chapter 8. There are other applications using this idea; for example in arctic shipping, the two processes could correspond to successfully detected and avoided ice features on the one hand, and collisions with ice, on the other; see Jordaan *et al.* (1987).

7.15　For the binomial distribution of Equation (3.3),

$$p_X(x) = \binom{n}{x} p^x (1-p)^{n-x},$$

(a)　find the *pgf*;
(b)　use the result to show that the sum of two independent binomial random quantities X_1 and X_2, both with parameter p, but with n_1 and n_2 trials respectively, is another binomial distribution.

7.16 *Use of transforms for expected utility.* Keeney and Raiffa (1976) note that transforms can be used for expected utility. Use (7.100), (7.102) or (7.137) to obtain the expected utility if the utility function is given by (4.18):

$$u(x) = -\exp(-cx).$$

Obtain the equations of expected utility for various distributions such as the binomial and normal.

7.17 Show that the *cf* for a random quantity with a uniform distribution on [0, 1] is

$$\phi_X(u) = \frac{\exp\left(\frac{iu}{2}\right)\sin\left(\frac{u}{2}\right)}{\frac{u}{2}} = \frac{\exp(iu) - 1}{iu}. \tag{7.209}$$

7.18 Deduce from first principles the *cf* of the gamma density

$$f_X(x) = \frac{x^{\alpha-1}\exp(-x/\beta)}{\beta^\alpha \Gamma(\alpha)}, x \geq 0, \alpha, \beta > 0, \tag{7.210}$$

to give Equation (7.124),

$$\phi_X(u) = (1 - iu\beta)^{-\alpha}.$$

7.19 Find the *cf* of the bilateral exponential distribution:

$$f_X(x) = \exp(-|x|) \text{ for } -\infty < x < \infty. \tag{7.211}$$

7.20 Find the characteristic function of the Cauchy distribution (introduced in Exercise 3.14),

$$f_X(x) = \frac{a}{\pi(x^2 + a^2)}, \text{ for } -\infty < x < \infty,$$

using the inversion formula. Show that the distribution is infinitely divisible and stable.

7.21 *Ball bearing problem.* The diameter X of a spherical ball bearing taken from a lot is normally distributed with a mean of 10.0 mm and a standard deviation of 0.5 mm. The ball is supposed to fit into a steel tube taken from another lot. The tube is circular in cross section, and the diameter Y is also normally distributed but with a mean of 11.3 mm and a standard deviation of 1.2 mm. Use the result of Example 7.19 to find the probability that a randomly chosen ball will fit into a randomly chosen tube. Five randomly chosen balls are paired off with five randomly chosen tubes. What is the probability that all five balls will fit inside their tubes?

7.22 Show that, for the binomial random quantity X,

$$\frac{\mathbf{Pr}(X = x + 1)}{\mathbf{Pr}(X = x)} = \frac{n - x}{x + 1} \cdot \frac{p}{\tilde{p}}.$$

7.23 It is a useful exercise to review transform pairs in an application. The use of Fourier transforms is well known to electrical engineers; Laplace transforms are used in the same discipline and in mechanics for viscoelastic theory and dynamics.

7.24 Prove the relation between (7.139) and (7.140):

$$g(t) \iff G(\omega)$$

then

$$G(t) \iff 2\pi g(-\omega).$$

Give an illustration of its use.

7.25 Deduce Equation (7.161) using convolution formulae and mathematical induction.

7.26 Using the result of Exercise 7.17 to show that the means and standard deviations of the distributions of Equation (7.161) are given by

$$\mu_Y = \frac{n}{2} \text{ and } \sigma_Y^2 = \frac{n}{12} \tag{7.212}$$

respectively.

7.27 *Continuity correction.* In approximating discrete distributions by continuous ones, as noted in Section 7.5.2, probability masses in the discrete distribution are approximated by areas, for example by small rectangles or trapezoids. The areas can be better approximated using the 'continuity correction'. In practice, this merely means approximating the probability of a particular value i in the discrete distribution by the area under the continuous curve from $i - 1/2$ to $i + 1/2$.

The arrival of loaded carriers at a work station in a conveyor system occurs with probability $p = 0.3$, the remaining occurrences being unloaded (empty) carriers. The arrival of 20 carriers is being studied using the binomial distribution. Let X be the number of loaded carriers. Use the normal approximation to the binomial distribution to calculate

(a) $\mathbf{Pr}(X = 5)$ and
(b) $\mathbf{Pr}(2 \leq X \leq 10)$.

Compare the results with the values from the binomial distribution.

7.28 For the multidimensional normal distribution of Example 7.9, determine the characteristic functions of the forms given in Equations (7.57) and (7.68).

7.29 *Sums of random numbers of random quantities.* Very often we are interested in the sum

$$Y = X_1 + X_2 + \cdots + X_i + \cdots + X_N,$$

where the X_i and N are random. Assume that the X_i are *iid*.

(a) Give examples of practical interest of such a situation.
(b) Find an expression for the *mgf* of Y in terms of the *mgf* of X and the probability distribution of N.

(c) A mathematical model of the travel of tourists is to be set up for a group of islands. The number of tour groups per day, N, visiting a particular island is random with a Poisson distribution and a mean equal to μ. The size X of a tour group is also Poisson-distributed with a mean equal to v. Find the *mgf* of Y, the number of visitors per day at the specified island. Find the mean and variance of Y.

7.30 There are 35 components in an assembly. Each component has a mean life of 5 years with a standard deviation of 2 years. Their lifetimes may be considered to be independent in the application considered. Find the probability that the mean life of all 35 components is greater than 6 years.

7.31 The strong law of large numbers cannot be proved using the Chebyshev inequality. Find the proof.

7.32 Research the following concepts: convergence in the quadratic mean; weak convergence (convergence in probability); strong convergence (almost sure convergence). Which implies the others and why?

8 Exchangeability and inference

There is a history in all men's lives,
Figuring the nature of the times deceas'd
The which observed, a man may prophesy,
With a near aim, of the chance of things
As yet not come to life, which in their seeds
And weak beginnings lie entreasured.

William Shakespeare, *King Henry IV, Part II*

8.1 Introduction

We have described various probability distributions, on the assumption that the parameters
are known. We have dealt with classical assignments of probability, in which events were
judged as being equally likely, and their close relation, the maximum-entropy distributions of
Chapter 6. We have considered the Central Limit Theorem – the model of sums – and often this
can be used to justify the use of the normal distribution. We might have an excellent candidate
distribution for a particular application. The next step in our analysis is to estimate the model
parameters, for example the mean μ and variance σ^2 for a normal distribution. We often wish to
use measured data to estimate these parameters. The subject of the present chapter, 'inference',
refers to the use of data to estimate the parameter(s) of a probability distribution. This is key
to the estimation of the probability of a future event or quantity of a similar kind.[1] The fitting
of distributions to data is outlined in Chapter 11.

References of interest for this chapter are de Finetti (1937, 1974), Raiffa and Schlaifer (1961),
Lindley (1965, Part 2), Heath and Sudderth (1976), O'Hagan (1994) and Bernardo and Smith
(1994). The book by Raiffa and Schlaifer remains a tremendous store of information – albeit
without an index. Lindley pioneered the use of Bayesian methods in inference, particularly
in the context of classical methods. The other books are more recent and contain up-to-date
information on developments in the succeeding decades.

---

[1] In some cases, we might estimate the probability of a past event – there is no real difference in approach, if the
outcome is unknown. For example, we might be uncertain as to who wrote a particular poetic work – William
Shakespeare or the Earl of Oxford. An engineering example might be to estimate the strength of a member that has
already failed.

8.1.1   Frame of reference

In accordance with the preceding comments, there will always be consideration of a **model** of some kind, for the problems considered in this chapter.

Example 8.1 An example that encapsulates the key point is a binomial experiment, say the tossing of a coin. The parameter of interest in successive tosses of the coin is the probability p of heads in any of the trials, Equation (3.3):

$$p_X(x) = \binom{n}{x} p^x (1-p)^{n-x}.$$

Let us assume that the coin is suspected to be biased; how do we estimate p? We might toss the coin, say, ten times and if we find three heads we might be tempted to estimate the probability as 3/10. On the other hand, if we knew from previous experience that the coin was unbiased ($p = 1/2$) and obtained the same result (three heads in ten tosses) we would be unlikely to change our probability assignment: we would put the result down to 'chance'. These two extremes emphasize the role of prior information regarding the coin and our subjective judgement regarding this information.

Example 8.2 Consider again Example 2.7, the case of two boxes which are indistinguishable in appearance, one containing a two-headed coin, and the other an 'honest' coin. The model here is the binomial one in which the probability of heads is either 1/2 (honest coin) or certainty (two-headed coin). The coins were labelled $\tilde{E}$ and E respectively. This is a special case of Example 8.1, in which $p =$ either 1/2 or 1.

Example 8.3 We wish to weigh an object and to estimate the uncertainty regarding the measurement. Savage used the example of a potato that we wish to weigh (Savage *et al.*, 1962). We have some initial information on the weight of the object. We have a weighing device that has been calibrated using known weights and repeated weighing of these. As a result we know that the readings **given a known weight** follow the normal distribution of Equation (3.25) with standard deviation σ. This then represents the model of interest. We wish to weigh the new object (potato, for example) and to estimate the uncertainty associated with this measurement. In order to do this, we need to study the **parameters of the model**; this might be the standard deviation σ obtained from the calibration just mentioned.

Inference is concerned with the estimation of parameters, for example p or σ in the examples above, together with associated uncertainties. An example of this uncertainty would be that associated with small data sets, for example the ten tosses of the coin in Example 8.1 above. To deal with these problems correctly, Bayes' theorem is used. This requires an estimate of prior information to be combined with new information (generally measured data). As in all probabilistic reasoning, there are subjective elements, here in the estimation of prior knowledge and in the treatment of data. There is a desire sometimes to have some sort of 'security blanket', for example a prior distribution that can be universally accepted. Sometimes this has been proposed as the maximum-entropy distribution consistent with the state of knowledge. Unfortunately our knowledge is almost always not of the kind that leads to such assignments. The use of our prior knowledge based on subjective judgement should not be ducked; nor should our subjective assessment of data. Matthews (1998) points to the many problems that have arisen in scientific inference regarding everyday problems in health and other areas through the attempt to avoid this subjective element.

A few words on the use of the term 'population', often used in inference. We often sample the attributes of a real population. Bernstein (1996) gives an account of the history of this measurement. He describes the statistics collected by John Graunt on births and deaths in London in the seventeenth century, and the studies of population statistics by the Belgian Quetelet in the first half of the nineteenth century, the measurement of heights of adults and the size of parent peas and their offspring by Galton later in the nineteenth century, together with the accompanying growth of methodology, especially the use of the normal distribution. These examples involve real human or other living populations, and this terminology has been carried through to applications generally, for example a 'population' of steel specimens of which we wish to measure the strength. 'Population parameter' is an expression that is sometimes used, referring to the value that would be obtained if we sampled the entire 'population'. We shall simply use the term 'parameter' in the following.

8.1.2   Seeming simplicity: inference dominated by data

A question that might be posed is 'why not use the ratio of successes to trials as the parameter in the binomial process, or estimate the mean and variance of a distribution from sufficient data?' The answer is that this is what one in effect does, given sufficient data. But there are two extremes of knowledge, as noted above and previously in Section 2.3.2. The first corresponds to cases where the probability is well defined by past experience and by extensive data. A new data point will have little influence on the result. The other extreme is where we have little previous knowledge. Consider an experimental device, for instance a rocket or space probe (as also discussed below) for which there is little previous experience and only a small data set. The estimation of the parameter will inevitably contain uncertainties related to the small data set. The Bayesian method given in this chapter includes the case where there are lots of data, and inference is then relatively straightforward, as a special case. But the Bayesian method gives a rigorous and proper basis for the calculation, and uncertainty in the parameter is faced squarely.

8.1.3   Inference and decisions

Decisions based on probabilistic inference can affect large numbers of people; for example, estimates of fish population parameters affect fishing policy, and whole communities can suffer

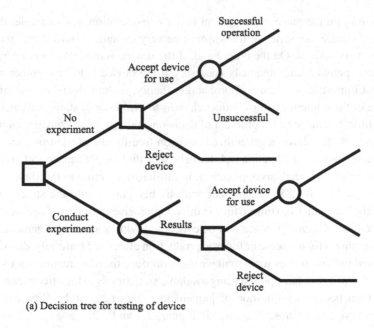

(a) Decision tree for testing of device

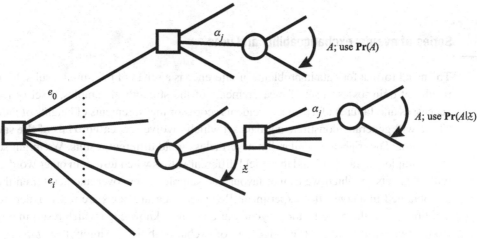

(b) General choice of experiment for parameter A

Figure 8.1   Experimentation followed by main decision

if poor decisions are made (not least the one in Newfoundland, where the cod fishery was overfished almost to extinction). It is important that **all problems of inference** be visualized as problems of decision. Figure 8.1(a) shows a typical decision tree for the following problem. We have a device or piece of machinery, and we have to decide whether to conduct a series of tests on similar pieces of equipment. The consequences of the various courses of action vary, depending on the piece of equipment under consideration. For example, if the equipment is a space shuttle, the series of trials would be very expensive indeed, and the consequences of failure very serious. On the other hand, if the device is a can opener, a series of trials would not be expensive, and obviously necessary if the device is to be commercially successful. It is most important, in the use of probability theory, to draw decision trees and to visualize the nature of the inference being conducted, so as to avoid sterile statistical calculations that might have little relevance to the problem of decision-making under consideration.

Figure 8.1(b) shows a generalized decision tree for the present problem of inference. The parameter under consideration is A. It can be seen that the present subject fits in very well indeed with the decision-analysis approach. It is similar in structure to the Bayesian trees in Figures 3.19(b), 4.24 and 4.28. The present analysis fits exactly into the structure of 'preposterior' analysis. We shall, from time to time in this chapter, give examples of decision trees for problems of inference. Often, inference will be concerned with scientific judgement. For example, is a new treatment for a disease or for crop production effective? Naturally, decisions are intimately involved with such questions (shall we mass-produce the new treatment for sale?), but we often wish also to give a fair opinion using available statistics as to the effectiveness of the treatment. This then becomes a question of judgement regarding a probability, and the decision tree becomes similar to those illustrated in Figures 2.1 and 2.2.

8.2   Series of events; exchangeability and urns

The usual format for data in problems of inference is a series of measured results, for example results of coin tosses, a set of measurements of the strength of samples, a set of records of successes and failures of a device, accident records or measurements of heights of individuals. These will generally constitute our data set which involves **repeated trials of the same phenomenon**. The trials are similar, but nevertheless yield different results. We might select our steel samples 'at random' and they yield different values when tested. In other words, we deal with data sets in which we cannot favour one sample or test over another. Often the results are obtained in a controlled experiment. Repeated measurements are made under 'identical' conditions, but there are factors beyond our control or knowledge which result in differences from one trial to another. The judgement of exchangeability is simply that '*the order of the results is of no significance*'.

Let us return to the problem discussed with regard to Figure 8.1(a). The series of events might be simple coin-tossing, or testing of a set of steel samples or can openers. We contrast these to tests on a new aircraft or a space shuttle or probe.[2] The shuttle is a case where there are serious consequences – seven people lost their lives in the Challenger accident and seven more on the Columbia. In such cases previous trials are seldom the same as later ones; design

---

[2] With regard to the space shuttle Challenger, see Feynman (1986) and Dalal *et al.* (1989). Feynman states that estimates of the probability of failure ranged from 1 in 100 to 1 in 100 000.

improvements will inevitably result from past experience so that the trials are not of the same phenomenon. We introduce this to show that in some cases, exchangeability is not appropriate: the order of the results can be important.

But order will not be important in the case of coin tossing or testing of a set of seemingly identical steel samples and in many practical problems. This is the traditional case that we shall analyse, that of 'repeated trials of the same phenomenon'. These consist of n trials of the *same* device with no design changes along the way. The possible results of the trials are embodied in the series

$$\left\{ E_1 \text{ or } \tilde{E}_1, \ E_2 \text{ or } \tilde{E}_2, \dots, E_i \text{ or } \tilde{E}_i, \dots, E_n \text{ or } \tilde{E}_n \right\}, \tag{8.1}$$

consisting altogether of 2^n possible events, or constituents, forming a partition. Equation (8.1) refers to the situation before the results are obtained, when the E_i are random. In any particular realization, there will be n results, each corresponding to one or other of the pairs of possible results noted in (8.1). Let us denote this set of results (data) as D. There is no longer any randomness, since the results are a data chain. As an example, for $n = 10$, the following is a typical string of results, using the usual notation $E_i = 1$ (success) and $\tilde{E}_i = 0$ (failure):

$$D = \{0, 1, 0, 0, 1, 1, 1, 1, 0, 1\}. \tag{8.2}$$

In this set of values, there are six successes ($\sum_{i=1}^{10} E_i$) and four failures. (Recall that the allocation 'success and failure' is arbitrary.) We might estimate from the results that the probability of success on the next trial is 0.60, based on the data D, or we might rely on extensive past experience with a similar device to estimate that (say) only 30% of the devices will succeed. But there is no seeming logic to the situation: what are the rules of the game?

The rule has already been given in Section 2.4. Bayes' theorem tells us all we have to know:

$$\mathbf{Pr}\,(E_{n+1}|D) = \frac{\mathbf{Pr}\,(DE_{n+1})}{\mathbf{Pr}\,(D)} = \frac{\mathbf{Pr}\,(D|E_{n+1})\,\mathbf{Pr}\,(E_{n+1})}{\mathbf{Pr}\,(D)}, \tag{8.3}$$

where E_{n+1} is the next event for which we wish to estimate the probability. We have noted that the new information D is now a set of data, and that there is no randomness associated with it. The terms $\mathbf{Pr}(D|E_{n+1})$ and $\mathbf{Pr}(D)$ seem to involve probabilities of non-random data! And we go further to criticize the classical method for dealing with 'random data' when it is not random. In fact, there is no difficulty here; we are considering how likely D is, given the event of interest E_{n+1}, compared to the likelihood of the same data given the other possible result $\tilde{E}_{n+1}$.

To see this, note that the term after the conditional sign, E_{n+1}, is in fact the quantity of interest in $\mathbf{Pr}(D|E_{n+1})$; in this regard, we saw in Section 2.4, Equations (2.30) and (2.31), that

$$\mathbf{Pr}\,(D) = \mathbf{Pr}\,(D|E_{n+1})\,\mathbf{Pr}(E_{n+1}) + \mathbf{Pr}(D|\tilde{E}_{n+1})\mathbf{Pr}(\tilde{E}_{n+1}). \tag{8.4}$$

Then the right hand side of (8.3),

$$\frac{\mathbf{Pr}\,(D|E_{n+1})\,\mathbf{Pr}\,(E_{n+1})}{\mathbf{Pr}\,(D)} = \frac{\mathbf{Pr}\,(D|E_{n+1})\,\mathbf{Pr}\,(E_{n+1})}{\mathbf{Pr}\,(D|E_{n+1})\,\mathbf{Pr}(E_{n+1}) + \mathbf{Pr}(D|\tilde{E}_{n+1})\mathbf{Pr}(\tilde{E}_{n+1})} \tag{8.5}$$

describes how likely the data is given the event of interest, E_{n+1}, compared to the other alternatives (here $\tilde{E}_{n+1}$ only). There will be more on this aspect of inference in the following sections.

8.2.1   Exchangeable sequences

The cornerstone of most probabilistic inference of the classical kind is the concept of **exchangeability**, resulting in an edifice that is often poorly explored. It was delightfully and effectively explored by de Finetti, resulting in clarification of the fundamental ideas.[3] The concept may be summarized in the following judgement, referring again to the series of trials of the 'same phenomenon', as discussed above:

The order of the results is of no significance.

In the example above, string (8.2), if we judge the results to be exchangeable, we judge that the order of the 0s and 1s is of no significance. Therefore, all that is important is the fact that there were six successes and four failures. These are termed **sufficient statistics,** in the sense that these numbers should be sufficient for us to make our inference (we shall soon show how!). But if we make design changes to our device during the testing period, the results will **not** be exchangeable.

There is an extremely simple interpretation of a finite set of exchangeable random events or quantities (let us assume that there are b of them). Since we consider that the order of obtaining them is completely unimportant, we might just as well throw the results into a hat and draw them at random. Of course we do not use hats in probability theory; the urn is our preferred tool. Therefore we have b balls in an urn. Since we are dealing with events, which take the values 0 and 1, these can be represented by pink and red balls respectively, as before in Section 2.5.1. Often the assumption of independence is made at this point but we shall show that this is not the appropriate assumption. What we have is exchangeability, not independence.

---

[3] There is much to be gained from reading the early paper by de Finetti (1937).

The drawing of successive balls from an urn without replacement (Section 2.5.1) does not lead to independent trials, since the composition of the urn is changing with each draw – yet any order of results is considered as likely as any other. In other words they are exchangeable. The more important point is that we are using the data to change our probability assessments (this is what inference is all about), and any statement that the probability remains the same after knowledge of the results is obtained (independence) is incorrect.

Let us assume that there are g red balls in the urn. The string of results (8.2) above could correspond to $n = 10$ drawings from the urn, as illustrated in Figure 8.2(a). A particular example is where $b = n = 10$, and $g = 6$, in which case the number of drawings n and the number of balls b coincide. More generally below, we will make $b > n$. Raiffa's urns are also identified in the figure.

The parameter p in Example 8.1, the weight of the object and standard deviation σ in Example 8.3 are examples of parameters that we wish to estimate. We now introduce the key step in our reasoning. We must treat the parameter – represented for now by the urn composition – as being random. We model this situation by making the number of red balls random, represented by the symbol G, and taking possible values $0, 1, 2, \ldots, b$. There are then $(b + 1)$ possible values of G. (We shall soon take the analogy from this discrete set to continuous sets of values.) Now we know that sampling from an urn with b balls corresponds to all of the possible paths through the 'random walk' diagrams of Chapter 2 (for example, Figures 2.9, 2.10, 2.18 and 2.19).

*The **composition** of the urn is our analogue for our model or 'population' parameter which we wish to analyse.*

Example 8.4 Figure 8.2(a) illustrates the possible values of G for the case where $b = 10$. Raiffa's urns form two of this set. The possible values of G are the 11 values $\{0, 1, 2, \ldots, 10\}$. These correspond to the vertical axis in Figure 8.2(a), illustrated using arbitrarily a gain of $+1$ for the drawing of a red ball or -1 for the drawing of a pink. The final set of possible gains is $\{-10, -8, \ldots, -2, 0, 2, \ldots, 8, 10\}$. Recall Equation (2.48), Chapter 2, which states that

$$\omega_{r,n} = \binom{n}{r} p_{r,n},$$

as illustrated in Figure 8.2(b); p in this equation represents the probability corresponding to a particular path with r successes, and ω to all paths.

The key aspect of the present demonstration is that the urn composition represents the parameter (or parameters) regarding which we wish to make inferences. Let the uncertainty regarding the composition of the urn be denoted as follows:

$$\mathbf{Pr}(G = g) = q_g, \quad \text{with } \sum_{g=0}^{b} q_g = 1.$$

Suppose that we now sample n balls from the urn $(n < b)$, and that r balls in the sample are

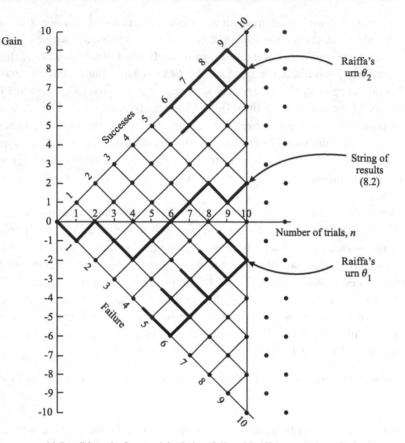

(a) Possible paths for ten trials. String (8.2) and Raiffa's urns are emphasized

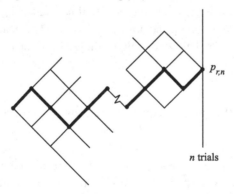

(b) Paths to endpoint $\{r,n\}$

Figure 8.2   Exchangeable sequences

red. Since all orders of results – paths in Figure 8.2(b) – are exchangeable, their probabilities are equal. For the particular result $R = r$, we denote the probability as $p_{r,n}$. Since the paths are exchangeable, we might as well consider that all the 1s occur first, and then the 0s:

$$p_{r,n} = \mathbf{Pr}(E_1 = 1, E_2 = 1, \ldots, E_r = 1, E_{r+1} = 0, \ldots, E_n = 0). \tag{8.6}$$

For any particular path, including the one just introduced, and given the composition $G = g$,

$$p_{r,n} = \frac{(g)_r (b - g)_{n-r}}{(b)_n}. \tag{8.7}$$

This corresponds to a single path in the hypergeometric distribution; see Section 3.2.5 and Equation (3.10). Then, for a random composition,

$$p_{r,n} = \sum_{g=0}^{b} \frac{(g)_r (b - g)_{n-r}}{(b)_n} q_g, \tag{8.8}$$

where we have weighted the probabilities of Equation (8.7) by the corresponding q_g. For any path rather than a single path, we write, as before, the probability as $\omega_{r,n}$. Then

$$\omega_{r,n} = \binom{n}{r} p_{r,n} = \binom{n}{r} \sum_{g=0}^{b} \frac{(g)_r (b - g)_{n-r}}{(b)_n} q_g. \tag{8.9}$$

We are going to let the number, b, of members of our exchangeable set tend to infinity. Although our sample size n will be finite, we are trying to make inferences about a 'population' that is very large. In effect, we are trying to find a rule for a 'stable' population, and we interpret this as a very long string of results, of which we may have only a relatively small sample. We can write Equation (8.8) as follows:

$$p_{r,n} = \int_0^1 \frac{(\theta b)_r \, [(1 - \theta) b]_{n-r}}{(b)_n} \, \mathrm{d}F_\Theta(\theta), \tag{8.10}$$

where $\theta = (g/b)$, and $F_\Theta(\theta)$ is a distribution function with jumps $\mathrm{d}F_\Theta(\theta) = q_g$ at the corresponding values of (g/b) with $0 \le g/b \le 1$. The representation of Equation (8.10) is essentially that of the Stieltjes integral (Section 3.3.5 and Appendix 2); one can think merely of an integral that includes discontinuous jumps that correspond to a summation. We now permit the number of balls b to increase indefinitely while keeping n and r constant. We then obtain that

$$p_{r,n} = \int_0^1 \theta^r (1 - \theta)^{n-r} \, \mathrm{d}F_\Theta(\theta). \tag{8.11}$$

From Equation (8.9),

$$\omega_{r,n} = \binom{n}{r} \int_0^1 \theta^r (1 - \theta)^{n-r} \, \mathrm{d}F_\Theta(\theta). \tag{8.12}$$

These are extremely important results. Equation (8.12) is essentially the result obtained by Bayes, but he used a rather special distribution function F_Θ; this will be discussed in the following. Indeed, Equations (8.11) and (8.12) have the binomial form $\left[\theta^r \left(1-\theta\right)^{n-r}\right]$ as part of the result. This can be used regarding inferences for the binomial parameter (A, as discussed in the sequel) but represents a much more general situation, as given in the sections below, particularly 8.4 dealing with random quantities. The symbol Θ emphasizes this generality. If Θ is continuous, we can write Equation (8.11) as

$$p_{r,n} = \int_0^1 \theta^r \left(1-\theta\right)^{n-r} f_\Theta(\theta)\mathrm{d}\theta, \tag{8.13}$$

where $f_\Theta(\theta)$ is the probability density of Θ.

The judgement of exchangeability

A few words on the judgement of exchangeability. We know from personal experience that certain trials will not admit of this judgement. For example, our first few serves in a tennis match might have more 0s (failures) than 1s; this may change as the number of trials increases. In the case of the space shuttle mentioned above, design improvements between trials could affect our judgement dramatically. Or the decision not to launch the shuttle in low temperatures (at which the O-ring seals in the booster rockets were prone to failure) would affect the exchangeability of the series of launches. A subset of the data might nevertheless be judged exchangeable. In repeated tests on the same machine, a phenomenon such as crack growth or fatigue might lead to failure, and the order of results could be rather significant; this would not apply if we test different 'identical' devices. Fatigue, human or mechanical, using the same person or machine, would result in a judgement in which one might not accept exchangeability. In studies of safety of advanced devices such as aircraft, one must be careful to note when technological changes have resulted in improvements in performance as regards safety.

The judgement is basically a question of probabilistic – and therefore subjective – opinion in which we assume that there is no systematic variation of a relevant property with the progression in the number of the test: $1, 2, 3, \ldots, n$. It should be recognized that exchangeability is the basis of almost all common forms of inference, as outlined below. In dealing with samples for testing it is generally reasonable. Altogether, it is a question of common sense.

8.2.2   Inference for exchangeable events

The quantity Θ represents the parameter space regarding which we wish to make inferences – the entire subject of inference revolves around this space. For the following, a single parameter will be considered. We have some initial opinion regarding Θ, say expressed in terms of the probability density $f_\Theta^{(i)}(\theta)$, where 'i' denotes 'initial'. We wish to use the result of new information D, given by the string of data represented by (8.1), a typical example being (8.2) to arrive at a revised and updated density $f_\Theta^{(u)}(\theta)$ using Bayes' theorem:

$$f_\Theta^{(u)}(\theta) \propto \mathbf{Pr}\left(D|\theta\right) f_\Theta^{(i)}(\theta). \tag{8.14}$$

Referring to Equation (8.11) or (8.13), we see that, given $\Theta = \theta$, the likelihood

$$\ell(\theta) \equiv \mathbf{Pr}\left(D|\theta\right) \tag{8.15}$$

is given by

$$\ell(\theta) = \theta^r (1 - \theta)^{n-r}. \tag{8.16}$$

The result amounts to 'independence given the model parameter $\Theta = \theta$'. The form of Equations (8.16), (8.11) and (8.13) is binomial in character; for this distribution,

$$p_{r,n|\Theta=\theta} = \ell(\theta) = \theta^r (1 - \theta)^{n-r} \tag{8.17}$$

(as always, for a particular path). We see from this equation how the 'sufficient statistics' $\{r, n\}$ are used. These two values only are used from the data; other details are not important.

We shall now deal with binomial processes, but it is important to bear in mind (and will be shown) that the analysis above can be used generally for other processes; Θ stands generally for 'hypothesis space' or 'parameter space'.

8.3   Inference using Bayes' theorem; example: binomial distribution; Bayes' postulate

We shall use the symbol A to represent the binomial parameter. We assume that the probability of x successes in n trials is given by the binomial distribution of Equation (3.3):

$$\mathbf{Pr}(X = x | A = a, n) = p_{X|A}(x, a) = \binom{n}{x} a^x (1 - a)^{n-x}, \tag{8.18}$$

in which we have denoted the parameter as A, taking the particular value a. It is in the spirit of the Bayesian method that the quantity A is treated as being random; the contrast to the classical method of making inferences is given in Sections 8.6 and 8.7 below. We wish to model our uncertainty about A.

The relative importance of the data and previous experience (including previous data) is guided by Bayes' theorem and the two terms on the right hand side of Equation (8.14). On the assumption of exchangeability, the final probability of x successes in n trials is given by

$$\mathbf{Pr}(X = x | n) = \binom{n}{x} \int_0^1 a^x (1 - a)^{n-x} \, dF_A(a). \tag{8.19}$$

In this equation, $F_A(a)$ represents our current state of knowledge regarding the parameter of the binomial distribution. Let us assume that A is well represented by a density. Then

$$\mathbf{Pr}(X = x | n) = \binom{n}{x} \int_0^1 a^x (1 - a)^{n-x} f_A(a) da. \tag{8.20}$$

We suppose in the present example that our initial assessment of the uncertainty in A can adequately be represented by the beta distribution:

$$f_A^{(i)}(a) = \frac{a^{r_0-1} (1 - a)^{n_0-r_0-1}}{B(r_0, n_0 - r_0)}, \quad 0 \leq a \leq 1. \tag{8.21}$$

The parameters $\{r_0, n_0\}$ characterize our prior opinion. In order that the probability density normalizes,

$$B(r_0, n_0 - r_0) = \int_0^1 a^{r_0-1} (1-a)^{n_0-r_0-1} \, da. \tag{8.22}$$

Here the beta function is given by

$$B(s, t - s) = \frac{\Gamma(s)\Gamma(t-s)}{\Gamma(t)}, \tag{8.23}$$

where the gamma function

$$\Gamma(s) = \int_0^\infty u^{s-1} \exp(-u) \, du. \tag{8.24}$$

Note that for a positive integer s, $\Gamma(s) = (s-1)!$. See Exercise 7.13 regarding gamma and beta functions.

We assume now that we have conducted n_1 tests with r successes, and that we deem the order of obtaining the successes and failures not to be important. Applying Bayes' theorem as in Equation (8.14) as well as the principle of exchangeability,

$$f_A^{(u)}(a) \, da \propto a^r (1-a)^{n_1-r} \cdot a^{r_0-1} (1-a)^{n_0-r_0-1} \, da$$
$$= a^{r+r_0-1} (1-a)^{n_1+n_0-r-r_0-1} \, da,$$

and since the denominator in the following is the integral (sum) of the numerator over all values of the argument,

$$f_A^{(u)}(a) = \frac{a^{r+r_0-1} (1-a)^{n_1+n_0-r-r_0-1}}{B(r+r_0, n_1+n_0-r-r_0)}. \tag{8.25}$$

We see that the revised probability distribution is of the same form as the initial one, a beta density, with parameters changed from $\{r_0, n_0\}$ to $\{r + r_0, n_1 + n_0\}$. We can write

$$r' = r + r_0 \text{ and} \tag{8.26}$$
$$n' = n_1 + n_0. \tag{8.27}$$

Then (8.25) becomes

$$f_A^{(u)}(a) = \frac{a^{r'-1} (1-a)^{n'-r'-1}}{B(r', n'-r')}. \tag{8.28}$$

Such distributions, which preserve their form during the application of Bayes' theorem, are termed **conjugate** families of distributions (more on this in Section 8.5 below).

8.3.1   Applications of the binomial distribution; decisions

First, it is useful to explore Equation (8.28).

Example 8.5 Figure 8.3 illustrates Equation (8.25), using the notation $r' = r + r_0$, $n' = n_1 + n_0$, and $s' = n' - r'$. Then

$$\langle A \rangle = \frac{r'}{n'}, \tag{8.29}$$

and each graph represents a situation with $r' =$ number of successes, and $s' =$ number of failures. Bayes' postulate, to be introduced below, corresponds to one success and one failure, $r' = s' = 1$; this occupies the central position in the figure. The upper left quadrant shows values that correspond to information that is less informative than this postulate. As one proceeds down and to the right, as n' increases, the distributions become more peaked, with the probability mass becoming more concentrated near the mean value, eventually in the limit, a Dirac spike. This shows the transition to the results based on data using Equation (8.29) with

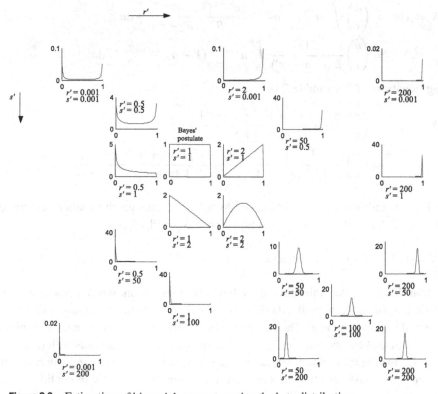

Figure 8.3   Estimation of binomial parameter using the beta distribution

probability of success in the next trial given by

$$\frac{\text{past successes}}{\text{past trials}}.$$

The uncertainty associated with the estimation of the parameter is clearly shown and modelled in the distributions of Figure 8.3. Classical estimation does not face this aspect in a clear manner; at the same time the uncertainty can be seen to be considerable. There is a richness to the Bayesian approach that is not available in classical estimation.

Equations (8.19) and (8.20) give the probability of x future successes in n trials:

$$\mathbf{Pr}(X = x|n) = \binom{n}{x} \int_0^1 a^x (1-a)^{n-x} f_A(a) da. \tag{8.30}$$

This equation, with $f_A(a)$ from (8.28) – on the basis of the conjugate analysis – will give probabilities taking into account uncertainty in the parameter A.

Example 8.6 Evaluate (8.20) using (8.28). We have

$$\mathbf{Pr}(X = x|n) = \binom{n}{x} \int_0^1 a^x (1-a)^{n-x} \cdot \frac{a^{r'-1} (1-a)^{n'-r'-1}}{B(r', n'-r')} da$$

$$= \binom{n}{x} \frac{1}{B(r', n'-r')} \int_0^1 a^{x+r'-1} (1-a)^{n+n'-x-r'-1} da. \tag{8.31}$$

Using Equations (7.206) and (8.23),

$$\mathbf{Pr}(X = x|n) = \binom{n}{x} \frac{B(x+r', n+n'-x-r')}{B(r', n'-r')}$$

$$= \binom{n}{x} \frac{\Gamma(n')}{\Gamma(r')\Gamma(n'-r')} \cdot \frac{\Gamma(x+r')\Gamma(n+n'-x-r')}{\Gamma(n+n')}. \tag{8.32}$$

This is the beta-binomial distribution; see Appendix 1. This result should be contrasted with the binomial distribution embodied in Equation (3.3) or (8.18),

$$\mathbf{Pr}(X = x|A = a, n) = p_{X|A}(x, a) = \binom{n}{x} a^x (1-a)^{n-x}.$$

Equation (8.32) is a generalization of the binomial distribution, where parameter uncertainty is modelled using the beta distribution. The standard binomial distribution of (3.3) can be recaptured for large r' and n'. Then the posterior distribution of the parameter becomes a Dirac spike of unit probability mass on $a = r'/n'$ (Exercise 8.2). This accords with (8.29) for large data sets, showing the classical estimator to correspond to a large data set with certainty of the parameter. The classical situation is rescued from calamity by the use of confidence intervals; see below.

Decisions

In decision analysis, we need expected values to rank the various actions. These can often be obtained from the moments. For example, if the decision hinges on the utility

$$U = k \cdot A,$$

where k is a constant, then

$$\langle U \rangle \propto \langle A \rangle .$$

Second moments and higher, $\langle A^2 \rangle, \langle A^3 \rangle, \ldots$, as well as transforms (see Exercise 7.16) can also be used in expected-utility analysis. Raiffa and Schlaifer (1961) deal with the case of linear loss – essentially the linear case just discussed.

Example 8.7 The mean value of A is given in (8.29):

$$\langle A \rangle = \frac{r'}{n'} \text{ or}$$

$$\langle A \rangle = \frac{r + r_0}{n_1 + n_0} . \tag{8.33}$$

In the present interpretation, the prior distribution of Equation (8.21) is seen, by substituting $r = n_1 = 0$ in Equation (8.33), to be equivalent to having had r_0 successes in n_0 trials, with the mean value of A represented by (r_0/n_0).

Example 8.8 If $r_0 = n_0 = 0$, then the mean value of A is given by

$$\langle A \rangle = \frac{r}{n_1},$$

the frequency interpretation; note that this is a mean value. As we shall see in the following Example 8.9, a somewhat different frequency is obtained if one starts with a uniform prior distribution, and then adds r successes in n_1 trials.

Example 8.9 Assume that our initial information is consistent with the beta distribution, and in particular to the values $r_0 = 1, n_0 = 2$. Then our initial (prior) distribution is

$$f_A^{(i)}(a) = 1, \ 0 \leq a \leq 1, \tag{8.34}$$

a uniform prior distribution. This is in fact Bayes' postulate, which will be discussed further below. Suppose that we are investigating the testing strategy for a device, in which there will be a string of results such as in the expressions (8.1) or (8.2) above. Then the revised distribution for A is given by (8.25) with $r_0 = 1, n_0 = 2$:

$$f_A^{(u)}(a) = \frac{a^r (1 - a)^{n_1 - r}}{B(r + 1, n_1 - r + 1)} . \tag{8.35}$$

The mean of Equation (8.35), by reference to Equation (8.33), is

$$\langle A \rangle = \frac{r+1}{n_1 + 2}. \tag{8.36}$$

This amounts to the use of Equation (8.33) with $r_0 = 1$ and $n_0 = 2$, i.e. the use of the observed frequency, (r/n_1), modified by the addition of two equivalent observations, one success and one failure. The prior corresponding to $r_0 = 0$ and $n_0 = 0$, which gives the observed frequency as the mean is discussed in Section 8.6. Equation (8.36) is Laplace's 'rule of succession'. Further interpretations are given at the end of the present section, in terms of Bayes' original analysis, and in terms of Pólya's urn model.

Example 8.10 In the case of the data string (8.2), using (8.35),

$$f_A^{(u)}(a) = \frac{a^6 (1-a)^4}{(6!4!/11!)} = 2310 \, a^6 (1-a)^4 \, ; \tag{8.37}$$

the mean value of this distribution is

$$\langle A \rangle = \frac{7}{12}. \tag{8.38}$$

Example 8.11 *A run of successes.* We now consider further variations on the problem of testing a device. Let us suppose that the device is costly, for instance a space probe, as discussed above – but there are many less dramatic examples such as a new engine, the design of which we have just completed, requiring the testing of several prototypes. We have so far conducted no tests but we wish to investigate the decision as to how many tests to conduct to achieve a certain level of confidence in the device, as modelled in Figure 8.4(a). We can interpret the probability of success in the next trial as the reliability of the device (the complement of the probability of failure: see Chapter 10). This, in turn, is represented by the parameter A, which is random, and the uncertainty in A should be taken into account. We now analyse the case where we have a series of n_1 successes, and no failures, so that $r = n_1$. We judge that $r_0 = 1$, $n_0 = 2$ represents a reasonable prior, so that (8.35) represents our opinion. Substituting for r,

$$f_A^{(u)}(a) = (n_1 + 1) a^{n_1}. \tag{8.39}$$

This function is shown in Figure 8.4(b).

A possible 'estimator', in other words a single value to represent the reliability, could be given by the mean

$$\langle A \rangle = \int_0^1 (n_1 + 1) a^{n_1} \cdot a \, da$$
$$= \frac{n_1 + 1}{n_1 + 2}; \tag{8.40}$$

but for small n_1 there is a lot of uncertainty in this estimate. We might instead calculate the

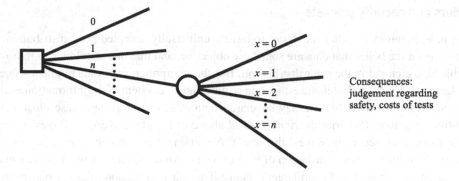

(a) Choice of number of trial tests for space probe

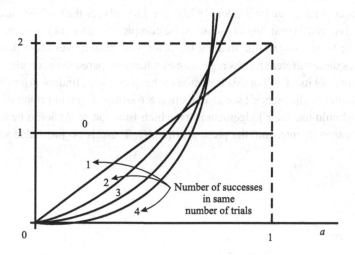

(b) Behaviour of parameter A

Figure 8.4 Tests on space probe

probability that a certain value of $A = a_1$ is exceeded:

$$\int_{a_1}^{1} (n_1 + 1) a^{n_1} da = 1 - a_1^{n_1+1}. \tag{8.41}$$

This is discussed in Lipson and Sheth (1973) in terms of the 'success-run' theorem (essentially the result just obtained).

Decision-making at this point should be subject to the fundamental methodology introduced throughout this book; decision trees should be drawn, consequences evaluated, for example of success or failure as represented by A and $(1 - A)$, combined if required with the result (8.41).

8.3.2   Priors and security blankets

We have mentioned earlier the desire to have a universally accepted prior distribution. This arises from the belief that data are somehow objective, and that one should try to remove the subjective element in the prior distribution. But the interpretation of data requires subjective judgement too. All probabilistic estimation requires an assessment of the information available to the person involved (which may be shared with others). The present case illustrates this attitude to priors. The prior distribution used above in Equation (8.34) is 'Bayes' postulate' (not to be confused with 'Bayes' theorem'). A certain degree of controversy surrounds the use of the uniform prior distribution of Θ over $[0, 1]$. An interpretation that has been made is based on the 'principle of insufficient reason'; if we have no reason to prefer one value of θ than any other, then Θ should be taken as uniformly distributed.

This has been criticized by several well known authors, including Fisher and Pearson; an interesting account is given by Stigler (1982). The grounds for the criticism are that, if we are 'completely ignorant' regarding Θ, we must be completely ignorant regarding any monotone function of θ; Stigler shows that treating the prior distribution over these functions as being uniform gives quite different results regarding Θ when compared to the results using a uniform distribution over Θ itself. All of this is related to the question of finding prior distributions that express 'complete indifference'; see also Section 8.6 below. From our point of view, a sincere subjectivist should use their judgement as to which function of θ should be uniform, so that there is no reason to enter into the present difficulty. There is no justifiable security blanket.

Thomas Bayes

8.3.3   Historical note: Bayes' postulate

To continue the discussion of 'Bayes' postulate', Stigler argues that a careful reading of Bayes' paper shows his reasoning to have been sound. Bayes proposed an argument along the following lines. A ball is rolled across a flat and levelled table; we can take the table as being square and of unit dimension (a kind of pool or billiard table). The final resting point of the ball is uniformly distributed in the unit square. Let us measure the distance in one of the two directions from one of the sides. This distance is a realization of the random quantity $A(= a$, say), the parameter in the binomial distribution. This is an experimental simulation of the uniform distribution, applied here to the parameter A. A **second** ball is now rolled n times across the table, and if the final position in any trial is within the distance a specified by the position of the first ball (say to the left of the first ball) then the trial is deemed a success. The whole procedure is repeated many times, the procedure representing precisely an analogue of the reasoning above, leading to Equation (8.35). In fact, had Bayes possessed a modern computer, he might well have conducted a Monte Carlo simulation of this situation (see Chapter 11 and Exercise 11.21).

Bayes added a 'scholium' to the argument. We give a brief summary here; full detail can be found in Stigler's paper. In effect, Bayes showed that, if one started with his 'uniform prior', and no other information or result, then the probability of x successes in n trials is $1/(n + 1)$, for $x = 0, 1, 2, \ldots, n$. (The result is for the case where the successes are achieved by any path; see Exercise 8.3.) For given n, this is a discrete uniform distribution; the probabilities of 0 or 1 success in one trial are both equal to $1/2$; the probabilities of 0, 1 or 2 successes in two trials are all $1/3$; and so on. Bayes then inferred that if our state of mind is such that we would assign this uniform discrete distribution, then the uniform distribution $f_A^{(i)}(a) = 1, 0 \leq A \leq 1$ is appropriate. This is much more cogent than the assertion that the principle of insufficient reason implies this distribution *per se*. Bayes' reasoning also cleverly connects observables (X) with unobservables (Θ). A further interpretation is given in terms of Pólya's urn model at the end of the present section.

Stigler also discusses the possible motives for the poor interpretation of Bayes' reasoning by eminent researchers. He concludes that much of the criticism arose from the desire to enlist Bayes as a supporter of their probabilistic point of view.

Pólya interpretation

A further interpretation of the uniform prior distribution is Pólya's urn. Consider this model in which there are initially one red and one pink ball. A ball is drawn, and this is replaced together with another of the same colour (double replacement; see also Section 2.6.5). The procedure is continued for n drawings. Given that there are r red balls drawn amongst these, we know that the composition is then $r + 1$ red balls in the total of $n + 2$. Then the probability of success (red ball) on the next draw is

$$\frac{r + 1}{n + 2},$$

which is the same as the probability obtained from Equation (8.36).

Example 8.12 If there have been eight drawings, of which six are red, the probability of red on the next draw is $7/10$.

Example 8.13 Given the following situation (de Finetti, 1974, vol. 2): there have been 100 drawings from the urn, and we know only the result of eight drawings: numbers 1, 3, 8, 19, 52, 53, 92, 100. These balls are red except for two (1 and 92), which are pink; what is the probability (1) that the ball on the second drawing is red, (2) that the ball on the 95th drawing is red, and (3) given that the drawings continue to 1000 in total without any further results becoming available, that the 1000th ball is red. The results of the 100 drawings are

$$\{0, E_2, 1, E_4, \ldots, E_7, 1, E_9, \ldots, E_{18}, 1, E_{20}, \ldots, E_{51}, 1, 1, E_{54}, \ldots, E_{91}, 0, E_{93}, \ldots, E_{99}, 1\}.$$

The data are exchangeable. The answers are: (1) not $1/3$ (this would be correct if we had only the result of the first drawing), but $7/10$; (2) and (3), also $7/10$. The lesson of this example is that probability is a function of information.

Another interpretation of Bayes' scholium now is apparent (see also Blom *et al.*, 1994). Consider the drawings from Pólya's urn. Let X denote the number of red balls in n trials. If we obtain a sequence of reds, say n reds in n trials, then

$$\Pr(X = n) = \frac{1}{2} \cdot \frac{2}{3} \cdot \frac{3}{4} \cdot \ldots \cdot \frac{n}{n+1} = \frac{1}{n+1}. \tag{8.42}$$

Also, since the series is exchangeable, then de Finetti's result of Equations (8.13) and (8.20) apply; denoting the parameter again as A, then

$$\Pr(X = r) = \omega_{r,n} = \binom{n}{r} \int_0^1 a^r (1 - a)^{n-r} f_A(a) da; \tag{8.43}$$

but $r = n$, so that

$$\Pr(X = n) = \omega_{n,n} = \int_0^1 a^n f_A(a) P da. \tag{8.44}$$

Equating (8.42) and (8.44),

$$\int_0^1 a^n f_A(a) da = \frac{1}{n+1}. \tag{8.45}$$

But this equation represents the moments ($n = 1, 2, \ldots$) about the origin of the distribution $[f_A(a)]$ in question. Further, the values $[1/(n + 1)]$ are the moments of the (continuous) uniform distribution on $[0, 1]$. This is the only distribution with these moments, and therefore

$$f_A(a) \equiv 1, \ 0 \le a \le 1.$$

This is Bayes' postulate. Thus Pólya's urn is equivalent to Bayes' postulate. Substitution in Equation (8.43) gives the answer to the question posed with regard to Bayes' scholium (Exercise 8.3 below), in which it is found that

$$\omega_{x,n} = \frac{1}{n+1}, \tag{8.46}$$

for $x = 1, 2, \ldots, n$. Thus X has a discrete uniform distribution, as in Bayes' scholium.

8.4   Exchangeable random quantities

Up to the present we have considered sequences of the kind

$$\{0, 1, 0, 0, 1, 1, 1, 1, 0, 1, E_{11} \text{ or } \tilde{E}_{11}, E_{12} \text{ or } \tilde{E}_{12}, \ldots\},$$ (8.47)

where we have added uncertain (unknown) events to the string of 8.2. If, instead of events, we are considering **quantities** X_i, then we might have the sequence

$$\{x_1, x_2, \ldots, x_{n_1}, X_{n_1+1}, X_{n_1+2}, \ldots, X_{n_1+n}\}.$$ (8.48)

The n_1 lower case letters $\{x_1, \ldots, x_{n_1}\}$ represent measurements, and there is nothing random about these. They might consist of readings of the weight of Savage's potato, or recordings of ice thickness in an arctic field expedition, and they assist us in making inferences. The upper case quantities X_i refer to random quantities, for example future values, or unknown values from the past. In the present idealization, **both** the data set and the random quantities are exchangeable. But they have **different rôles** as we have seen above; data are used to improve, change and update our probability distribution of the parameter, represented by Θ. The remaining random quantities must then be based on this updated distribution, but we cannot improve our probabilistic situation without further information.

Let us consider a sequence of n exchangeable random quantities

$$\{X_1, X_2, \ldots, X_i, \ldots, X_n\}.$$ (8.49)

Since the order is judged to be of no significance in making probabilistic statements, the $n!$ permutations of (8.49) have the same n-dimensional probability distribution. Further, we may choose any order by permuting the subscripts, for example

$$X_{i_1}, X_{i_2}, X_{i_3}, \ldots, X_{i_n},$$

in which the subscripts $i_1, \ldots, i_n$ have been chosen in any manner whatsoever. For example, we could put the original set $1, 2, \ldots, n$ into a hat, and choose them at random, with the first number drawn being i_1, and so on. De Finetti (1974) states that:

... every condition concerning n of the X_h has the same probability, no matter how the X_h are chosen or labelled.

We define the events E_i as follows:

$$E_t = X_i \leq x.$$ (8.50)

Since the X_i are exchangeable, so are the E_i. We now make the following association:

$$X_i \leq x \rightarrow E_i, \quad \text{'success', red ball in urn,}$$
$$X_i > x \rightarrow \tilde{E}_i, \quad \text{'failure', pink ball in urn.}$$ (8.51)

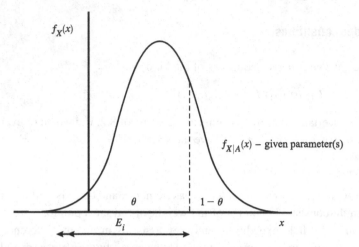

Figure 8.5 Distribution of X, given parameters. The event E_i corresponds to $X_i \leq x$.

Therefore, conditional on the value x, we have the same situation as before: a series of exchangeable successes and failures in drawings from an urn.

In problems of inference we always consider a probabilistic model. The situation is illustrated in Figure 8.5. An example is the normal distribution in weighing an object. The quantities X_i represent unknown results of (say, future) weighings. We have a direct analogy to the case of events through the associations (8.51) and consequently

$$\mathbf{Pr}(X_i \leq x) = F_X(x) = \mathbf{Pr}(E_i). \tag{8.52}$$

If our distribution parameters are known, we can calculate this probability directly as $F_X(x)$ from the distribution $f_X(x)$ with known parameters. This value will under these circumstances be fixed and known, corresponding to an urn of known composition.

But we wish to include uncertainty regarding the parameter value in our analysis, so we proceed as we did in Sections 8.2 and 8.3 above, and treat the parameter as being random, corresponding to an urn with unknown composition. The symbol Θ now stands for the parameter of the distribution. Hence

$$\mathbf{Pr}(X_i \leq x | \theta) = F_{X|\Theta=\theta}(x | \theta). \tag{8.53}$$

Events were dealt with in Sections 8.2 and 8.3 above, and $\Theta = \theta$ there stood for the probability of the event E_i. This probability is now represented by $F_{X|\Theta=\theta}(x|\theta)$. For convenience we repeat Equations (8.11) and (8.12) but using the density $\mathrm{d}F_\Theta(\theta) = f_\Theta(\theta)\mathrm{d}\theta$:

$$p_{r,n} = \int_0^1 \theta^r (1-\theta)^{n-r} f_\Theta(\theta)\mathrm{d}\theta, \tag{8.54}$$

and

$$\omega_{r,n} = \binom{n}{r} \int_0^1 \theta^r (1-\theta)^{n-r} f_\Theta(\theta)\mathrm{d}\theta. \tag{8.55}$$

We can therefore write the two preceding equations for $p_{r,n}$ and $\omega_{r,n}$ replacing θ by $F_{X|\Theta=\theta}(x|\theta)$:

$$p_{r,n} = \int_0^1 F_{X|\Theta}^r(x|\theta) \left[1 - F_{X|\Theta}(x|\theta)\right]^{n-r} f_\Theta(\theta) d\theta, \tag{8.56}$$

and

$$\omega_{r,n} = \binom{n}{r} \int_0^1 F_{X|\Theta}^r(x|\theta) \left[1 - F_{X|\Theta}(x|\theta)\right]^{n-r} f_\Theta(\theta) d\theta, \tag{8.57}$$

respectively.

We can think of Θ as corresponding to the parameter space such that $\mathbf{Pr}(\Theta \leq \theta) = F_\Theta(\theta)$. This has been modelled by the contents of an urn. If there is more than one parameter involved in the distribution of X, then $(\Theta \leq \theta)$ represents the appropriate part of the parameter space, which contains several dimensions corresponding to the parameters required; the examples of the next section should make this clear. We know from the above that if the random quantities of interest

$$\{X_1, \ldots, X_n\}$$

are exchangeable, then they have a common distribution function $F_{X|\Theta}(x|\theta)$ for all X_i. Further, from the structure of Equations (8.56) and (8.57), which are valid for any choice of $X = x$, we can see that the X_i are independent *given* the parameter $\Theta = \theta$. In raising $F_{X|\Theta}(x|\theta)$ to the power r in Equations (8.56) and (8.57), or raising its complement to the power $(n - r)$, independence is clearly implied. And this is true for any $X = x$ and for any $\Theta = \theta$. Therefore, given $\Theta = \theta$, the X_i are independent and identically distributed (*iid*).

8.4.1 Inference for exchangeable quantities

Let the parameter of interest be denoted A – this is intended as an example of a single-valued parameter space. (A is really no different from Θ except that we reserve Θ for the general idea of parameter or hypothesis space.) Therefore A could be the mean of the distribution for the weight of our object or for a steel strength that is being measured. From the discussion in the preceding paragraph, *given* the parameter $A = a$, the X_i are independent and identically distributed (*iid*) random quantities. In applying Bayes' theorem, it is convenient to think of intervals in the neighbourhood of the value of continuous random quantities considered, since a continuous distribution gives a density. Let the data be

$$D = \{x_1, x_2, \ldots, x_n\}. \tag{8.58}$$

The appropriate form of Bayes' theorem then becomes

$$f_A^{(u)}(a) da = \frac{\left[f_{X_1,\ldots,X_n}(x_1, \ldots, x_n|A = a) dx_1 \cdots dx_n\right] \cdot f_A^{(i)}(a) da}{\int \left[f_{X_1,\ldots,X_n}(x_1, \ldots, x_n|A = a) dx_1 \cdots dx_n\right] \cdot f_A^{(i)}(a) da}, \tag{8.59}$$

where the term in square brackets of the numerator is the **likelihood** $\ell \equiv \mathbf{Pr}(D|A = a)$, for the intervals stated. Now it always happens that the intervals 'da' cancel from the left and right hand side of the Bayesian equation; the one on the denominator gets absorbed into the integral. Further, the intervals '$dx_1 \cdots dx_n$' cancel from the numerator and the denominator. So we can rewrite the equation as

$$f_A^{(u)}(a) = \frac{\left[f_{X_1,\ldots,X_n}(x_1,\ldots,x_n|A=a)\right] \cdot f_A^{(i)}(a)}{\int \left[f_{X_1,\ldots,X_n}(x_1,\ldots,x_n|A=a)\right] \cdot f_A^{(i)}(a)\,da}, \tag{8.60}$$

dealing only with densities. In this case, the likelihood is

$$\ell = f_{X_1,\ldots,X_n}(x_1,\ldots,x_n|A=a), \tag{8.61}$$

omitting the intervals dx_i, which are always omitted from Equation (8.60) in any event since they cancel.

Since the numerator of the right hand side of Equation (8.60) deals with *iid* quantities, we can write it as

$$f_{X_1,\ldots,X_n}(x_1,\ldots,x_n|A=a) = f_{X|A=a}(x_1,a) \cdot f_{X|A=a}(x_2,a) \cdot \ldots \cdot f_{X|A=a}(x_n,a),$$

$$= \prod_{i=1}^{n} f_{X|A=a}(x_i,a). \tag{8.62}$$

Therefore

$$f_A^{(u)}(a) = \frac{\prod_{i=1}^{n} f_{X|A=a}(x_i,a) \cdot f_A^{(i)}(a)}{\int \prod_{i=1}^{n} f_{X|A=a}(x_i,a) \cdot f_A^{(i)}(a)\,da}, \tag{8.63}$$

and

$$\ell = \prod_{i=1}^{n} f_{X|A=a}(x_i,a). \tag{8.64}$$

Examples of the application of this equation are given in the next section.

As a final point of passing interest in this section, we note that in Equation (8.50), we considered the events $(E_i = X_i \leq x)$. We could equally well have considered the event $F_i = (x \leq X_i \leq x + dx)$.

8.5   Examples; conjugate families of distributions

We have already encountered an example of a conjugate family in making inferences regarding the binomial distribution in Section 8.3. In the example given, the uncertainty regarding the binomial parameter A was modelled using the beta distribution. We found that the revised distribution, after updating with Bayes' theorem, was also of the beta family. Given this situation, the term for the initial (prior) distribution that has been adopted in the literature is the **conjugate prior** distribution, and the families are termed **conjugate families**; the name and technique are due to Raiffa and Schlaifer (1961).

One should express one's prior beliefs by means of any distribution that expresses one's opinion correctly – it does not have to correspond to any family. The conjugate families represent a convenient way to introduce the updating of beliefs about parameters, and, in the cases presented, have an interesting relationship to classical methods which are based entirely on the data. But any distribution can be used, not necessarily of the conjugate family. In general, numerical computer-based techniques offer a flexible approach in cases where there is no convenient mathematical 'family'.

8.5.1   Unknown mean of normal distribution

We shall commence with an example from the theory of errors, for example an object weighed on a known and calibrated measuring device. Other than weighing, there are many other measurements in physics and astronomy which have been analysed in this way; a further example is measuring distance between two points in a geodesic survey. For the present analysis it is assumed that the accuracy of the device has been ascertained from previous measurements; this is interpreted as the variance – or standard deviation – of the distribution being known. Readings have been observed to follow a normal distribution $N\left(\mu, \sigma^2\right)$.

We now undertake new measurements, for example of an unknown weight. As a result, we have a number of new readings, represented by $D = \{x_1, x_2, \ldots, x_n\}$. The number n of data points may be small. Consequently the mean A is considered uncertain, and the data are to be used to make inferences about A. The readings are therefore modelled as $N(A, \sigma^2)$, in which A is random and σ^2 is the known variance. Thus

$$f_{X|A=a}(x, a) = \frac{1}{\sqrt{2\pi}\sigma} \exp\left[-\frac{1}{2}\left(\frac{x-a}{\sigma}\right)^2\right]$$

$$= K_1 \exp\left[-\frac{1}{2}(x-a)^2/\sigma^2\right], \tag{8.65}$$

in which the constant $K_1 = 1/\left(\sqrt{2\pi}\sigma\right)$. We may now use Equation (8.63) to write

$$f_A^{(u)}(a) = \frac{\prod_{i=1}^n f_{X|A=a}(x_i, a) \cdot f_A^{(i)}(a)}{N_1},$$

where we have written the denominator, which is nothing but a normalizing factor, as N_1. The product term of the numerator is the likelihood ℓ, as defined in Equation (8.64) for densities. This term, in the case of the distribution (8.65), can be written as

$$\ell(a) = K_1^n \exp\left[-\frac{1}{2\sigma^2} \sum_{i=1}^n (x_i - a)^2\right]. \tag{8.66}$$

Now

$$\sum_{i=1}^n (x_i - a)^2 = \sum_{i=1}^n x_i^2 - 2a \sum_{i=1}^n x_i + na^2;$$

noting that $\sum_{i=1}^n x_i = n\bar{x}$, where $\bar{x}$ is the mean of the sample $\{x_1, \ldots, x_n\}$, we may write that

$$\sum_{i=1}^n (x_i - a)^2 = \sum_{i=1}^n (x_i - \bar{x})^2 + n(\bar{x} - a)^2. \tag{8.67}$$

The two terms involving $n\bar{x}^2$ from the first (summation) term of Equation (8.67), and the one from the second term cancel (the reader might wish to check; these were included to complete the square). Now we see in this equation that the random quantity $A = a$ appears only in the second term so that the first term can be included in the constant (this will not be the case when we deal with 'unknown mean and variance' below). Then

$$\ell(a) = K \exp\left[-\frac{n}{2\sigma^2}(\bar{x} - a)^2\right],$$ (8.68)

where $K = K_1^n \exp\left[-(1/2\sigma^2)\sum_{i=1}^n (x_i - \bar{x})^2\right]$. The term does not involve a, so that (as noted) it contributes only to the constant.

Consequently,

$$
\begin{aligned}
f_A^{(u)}(a) &= \frac{K \exp\left[-\frac{n}{2\sigma^2}(\bar{x} - a)^2\right] \cdot f_A^{(i)}(a)}{N_1} \\
&= \frac{\exp\left[-\frac{n}{2\sigma^2}(\bar{x} - a)^2\right] \cdot f_A^{(i)}(a)}{N_2},
\end{aligned}
$$ (8.69)

where we have absorbed all the constants into $N_2 = N_1/K$. It is emphasized that the density $f_A(a)$, initial or updated, is a function only of a; everything else contributes merely to the normalizing constant that ensures coherence. This fact is extremely useful in the updating exercises of this chapter, and it is most helpful to exploit it. One concentrates on those terms that are functions of the random quantity under consideration, and all the other multiplicative terms become absorbed into the normalizing constant.

A few points should be made at this juncture. First, often a looser definition of likelihood ℓ is used, simply by the omission of the factor K in Equation (8.68); thus

$$\ell(a) = \exp\left[-\frac{n}{2\sigma^2}(\bar{x} - a)^2\right]$$ (8.70)

is often used. It is also seen that the likelihood function has the form of a normal distribution, with a standard deviation $\sigma/\sqrt{n}$, in other words σ reduced by the factor $\sqrt{n}$. It is useful to introduce the terminology emphasized by de Finetti. For the distribution $N(\mu, \sigma^2)$, $(1/\sigma)$ is termed the 'precision', and $(1/\sigma^2) = \sigma^{-2}$ is termed the 'weight' which we shall denote as w. The precision in the likelihood of Equation (8.70) is therefore increased from $(1/\sigma)$ by a factor $\sqrt{n}$, and the weight by a factor of n:

$$w = nw_1,$$ (8.71)

where $w_1 = $ weight of one data point $(1/\sigma^2)$ and $w = $ weight of data set.

Now we have set the stage for the use of a conjugate prior distribution. For the present example, this is another normal distribution. We therefore assume that our initial information is reasonably consistent with a normal density; in other words

$$f_A^{(i)}(a) = \frac{1}{\sqrt{2\pi}\,\sigma_0} \exp\left[-\frac{1}{2}\left(\frac{a-\mu_0}{\sigma_0}\right)^2\right] \tag{8.72}$$

$$= \frac{w_0^{1/2}}{\sqrt{2\pi}\,\sigma_0} \exp\left[-\frac{w_0}{2}(a-\mu_0)^2\right]. \tag{8.73}$$

In this equation, the distribution $N(\mu_0, \sigma_0^2)$ represents our prior belief regarding the parameter μ represented by A. We use the notation $\sigma_0^{-2} = w_0$. Substituting Equation (8.72) into (8.69), we find

$$f_A^{(u)}(a) = \frac{\exp\left[-\frac{n}{2\sigma^2}(\bar{x}-a)^2\right] \cdot \exp\left[-\frac{1}{2\sigma_0^2}(a-\mu_0)^2\right]}{N_2 \cdot \sqrt{2\pi}\,\sigma_0}. \tag{8.74}$$

Again, we may absorb all constants into a single normalizing value; what matters is the function of a. We combine the two exponential functions containing a as follows:

$$f_A^{(u)}(a) \propto \exp\left(-\frac{1}{2}\left\{[(n\sigma^{-2})(\bar{x}-a)^2] + [(\sigma_0^{-2})(a-\mu_0)^2]\right\}\right)$$

$$\propto \exp[-\frac{1}{2}\left(n\sigma^{-2} + \sigma_0^{-2}\right) \times$$

$$\left(a^2 - 2a\frac{\bar{x}n\sigma^{-2} + \mu_0\sigma_0^{-2}}{n\sigma^{-2} + \sigma_0^{-2}} + \text{terms not involving } a\right)]. \tag{8.75}$$

The last terms 'not involving a' are again constant, and we may adjust (8.75) to complete the square:

$$f_A^{(u)}(a) \propto \exp\left[-\frac{1}{2}\left(n\sigma^{-2} + \sigma_0^{-2}\right)\left(a - \frac{\bar{x}n\sigma^{-2} + \mu_0\sigma_0^{-2}}{n\sigma^{-2} + \sigma_0^{-2}}\right)^2\right].$$

We may conclude by writing the following (with a version where $(1/\sigma^2)$ is written as the weight w):

$$f_A^{(u)}(a) = \frac{1}{\sqrt{2\pi}\,\sigma_f} \exp\left[-\frac{1}{2}\left(\frac{a-\mu_f}{\sigma_f}\right)^2\right]$$

$$= \frac{w_f^{1/2}}{\sqrt{2\pi}} \exp\left[-\frac{w_f}{2}(a-\mu_f)^2\right], \tag{8.76}$$

where

$$\mu_f = \frac{\bar{x}n\sigma^{-2} + \mu_0\sigma_0^{-2}}{n\sigma^{-2} + \sigma_0^{-2}},$$

$$= \frac{\bar{x}nw_1 + \mu_0 w_0}{nw_1 + w_0}, \tag{8.77}$$

and

$$\sigma_f^{-2} = n\sigma^{-2} + \sigma_0^{-2},$$
$$w_f = nw_1 + w_0. \tag{8.78}$$

In these equations, w_0 is the weight of the prior distribution, $w = nw_1$ is the weight of the data, in accordance with (8.71) and w_f is the final weight.

As indicated earlier, the normal distribution represents the conjugate family for the parameter A. The mean of the updated distribution is the weighted sum of the prior mean and the mean of the likelihood, the weights being that of the prior distribution $(1/\sigma_0^2)$ and that of the likelihood $(n/\sigma^2 = n\sigma^{-2})$. The posterior weight is the sum of the weights from the prior distribution and the likelihood. On the question of notation, had our prior information corresponded to a set of n_0 readings, each with weight $(1/\sigma_0^2)$, then the terms σ_0^{-2} in Equations (8.77) and (8.78) above would be replaced by $(n_0\sigma_0^{-2})$ and w_0 with $n_0 w_0$.

Example 8.14 *Bridge elevation 1.* The elevation of a newly constructed bridge deck is most important so as to give adequate clearance to a railway line below. A set of measurements with a particular measuring system showed the mean to be normally distributed with an average of 20.70 m with a standard deviation of 0.150 m. Take these values as representing the prior knowledge.

Six new measurements of the elevation have been made using a new (improved) measuring device. The results are as follows (in metres). The device used is known to have a standard deviation of 0.050.

    20. 51, 20. 53, 20. 47, 20. 61, 20. 51, 20. 47,
    Mean: 20. 52, standard deviation: 0.0516.

The answer may be deduced as follows.

    Prior mean = 20.70.
    Prior weight = $1/(0.150)^2 = 44.4$.
    Data (likelihood) mean = 20.52.
    Data (likelihood) weight = $n\sigma^{-2} = 6/(0.05)^2 = 2400$.
    Posterior mean = $[(20.52)(2400) + (20.70)(44.4)]/(2400 + 44.4) = 20.52$.
    Posterior weight = 2444.4.
    Posterior standard deviation = $0.0202 \simeq 0.05/\sqrt{6}$.

Figure 8.6(a) shows the relevant distributions for this example.

Example 8.15 *Bridge elevation 2.* In this case, we have the following prior information: $\mu_0 = 20.525$ m; $\sigma_0 = 0.0221$ m. The same data as in Example 8.14 are available, comprising the six measurements above. Then we obtain the following (using units of metres).

    Prior mean = 20.525.
    Prior weight = $1/(0.0221)^2 = 2047$.
    Data (likelihood) mean = 20.52.

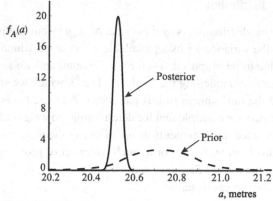

(a) First prior distribution

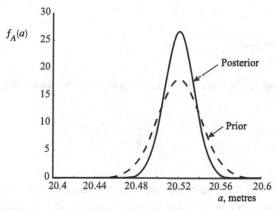

(b) Second prior distribution

Figure 8.6   Change of opinion regarding bridge elevation

Data (likelihood) weight $= n\sigma^{-2} = 6/(0.05)^2 = 2400$, as before.
Posterior mean $= [(20.52)(2400) + (20.525)(2047)]/(2400 + 2047) = 20.522$.
Posterior weight $= 4807$.
Posterior standard deviation $= 0.0150$.
Figure 8.6(b) shows the relevant distributions for this example.

The two examples just completed show the influence of different prior distributions. In the first case the prior distribution is relatively diffuse with a low weight whereas in the second case the weight is similar to that of the data.

Future values

Rather than modelling the mean value, we might be interested in a future value. For example, we might know the mean height of waves during a storm but might wish to address the uncertainty in the next wave height. See Exercise 8.6.

8.5.2   Unknown variance of normal distribution

We now consider the case where the distribution is again normal $N(\mu, \sigma^2)$, but where the mean is known and we wish to study the variance σ^2 using available information which might not be sufficient for a definitive value to be proposed. Then σ^2 is random, and we again follow Raiffa and Schlaifer, and de Finetti, in modelling the 'weight' $(1/\sigma^2)$, which we shall denote as W, the upper case indicating the randomness of this parameter. An example is where we are calibrating our measuring device, for example that for determining the weight of an object. This we might do by making repeated measurements using an object the weight of which is accurately known. Recall that this is an analogue for a much wider set of problems; also we discussed calibration and Bayes' theorem in Section 2.4.3.

Applying Bayes' theorem in the present instance gives

$$f_W^{(u)}(w) = \frac{\prod_{i=1}^{n} f_{X|W=w}(x_i, w) \cdot f_W^{(i)}(w)}{N_1}, \tag{8.79}$$

where

$$f_{X|W=w}(x, w) = \frac{w^{\frac{n}{2}}}{\sqrt{2\pi}} \exp\left[-\frac{w}{2} \sum_{i=1}^{n} (x_i - \mu)^2\right] \tag{8.80}$$

and N_1 is a normalizing constant, as before. The likelihood is therefore as follows (omitting constants):

$$\begin{aligned}
\ell(w) &= w^{\frac{n}{2}} \exp\left[-\frac{w}{2} \sum_{i=1}^{n} (x_i - \mu)^2\right] \\
&= w^{\frac{n}{2}} \exp\left[-\frac{w}{2} s_d^2\right],
\end{aligned} \tag{8.81}$$

where $s_d^2 = \sum_{i=1}^{n}(x_i - \mu)^2$. Now, ℓ is of the same form as the gamma distribution; we shall use this for the prior distribution of W, anticipating it as the appropriate conjugate family. In particular we shall use the chi-square distribution, which is a particular form of the gamma density since this is commonly used in inferences regarding the variance.

The standard form of the gamma distribution is

$$f_X(x) = \frac{x^{\alpha-1} \exp\left(-\frac{x}{\beta}\right)}{\beta^\alpha \Gamma(\alpha)}. \tag{8.82}$$

where α and β are positive constants, and $\Gamma(\cdot)$ is the gamma function. In studying variance, it is usual to write this as a chi-square distribution and we shall follow this convention. Formally, for a random quantity $Y = 2X/\beta$, the chi-square distribution with ν degrees of freedom is

$$f_Y(y) = \frac{y^{\frac{\nu}{2}-1} \exp\left(-\frac{y}{2}\right)}{2^{\frac{\nu}{2}} \Gamma\left(\frac{\nu}{2}\right)}, \tag{8.83}$$

where α in (8.82) has been written as $\nu/2$ and $\beta = 2$. The chi-square distribution can also be seen as a special case of the gamma distribution with shape parameter $\nu/2$ and a scale parameter of 2.

The prior distribution is written in terms of

$$W' = \frac{W}{w_0},$$ (8.84)

where w_0 is a normalizing constant (this allows us surreptitiously to change the scale parameter while keeping $\beta = 2$ as noted above). Using the form (8.83), we write

$$f_{W'}^{(i)}(w') = \frac{(w')^{\frac{v_0}{2}-1} \exp\left(-\frac{w'}{2}\right)}{2^{\frac{v_0}{2}} \Gamma\left(\frac{v_0}{2}\right)},$$ (8.85)

where $(W' = W/w_0)$ is chi-square, and where v_0 and w_0 express our prior beliefs regarding the parameter W. The ratio

$$\frac{W}{W'} = w_0$$ (8.86)

is used to scale the parameter W; note that

$$\left\langle \frac{W}{w_0} \right\rangle = v_0$$ (8.87)

so that

$$\langle W \rangle = v_0 w_0.$$ (8.88)

Also the standard deviation is $2\sqrt{v_0}$, and the mode (for $v_0 > 2$) is at $(w/w_0) = 2[(v_0/2) - 1]$. See also Exercise 8.8 (relates gamma and chi-square).

Substituting Equation (8.85) and the likelihood function into Equation (8.79) yields a distribution of the same chi-square form as Equation (8.85):

$$f_{W'}^{(u)}(w') = \frac{(w')^{\frac{v_f}{2}-1} \exp\left(-\frac{w'}{2}\right)}{2^{\frac{v_f}{2}} \Gamma\left(\frac{v_f}{2}\right)},$$ (8.89)

but with

$$W' = \frac{W}{w_f},$$ (8.90)

where

$$v_f = v_0 + n$$ (8.91)

and

$$\frac{1}{w_f} = \frac{1}{w_0} + s_d^2.$$ (8.92)

The chi-square distribution of Equation (8.89) has a mean

$$\left\langle \frac{W}{w_f} \right\rangle = v_f,$$

$$\langle W \rangle = v_f w_f,$$ (8.93)

a standard deviation of $2\sqrt{v_f}$, and a mode (for $v_f > 2$) at $(w/w_f) = 2[(v_f/2) - 1]$. These are of assistance in obtaining a feeling for one's degree of belief.

There are various versions of the distribution of Equation (8.89). Lindley (1965) considers the variance σ^2 as the random quantity; prior knowledge is expressed via v_0, as above, and σ_0^2; the chi-square random quantities are $(v_0\sigma_0^2/\sigma^2)$ and $(v_f\sigma_f^2/\sigma^2)$, where

$$v_0\sigma_0^2 = \frac{1}{w_0}, \text{ with}$$

$$v_f\sigma_f^2 = \frac{1}{w_f} = v_0\sigma_0^2 + s_d^2. \tag{8.94}$$

This is consistent with our formulation and is quite attractive since we can write $s_d^2 = vs^2$, where s^2 is the sample variance. Then

$$v_f\sigma_f^2 = v_0\sigma_0^2 + vs^2. \tag{8.95}$$

Note that σ_0^2 and σ_f^2 just discussed have a somewhat different meaning from that in Section 8.5.1 since we are now modelling the variance as random, and these parameters now specify the distribution of W'.

Example 8.16 Referring to Example 8.14 above, we assume now that the measurements are taken to a known height in order to estimate the variance. The measurements are as follows (in metres).

> 20. 51, 20. 53, 20. 47, 20. 61, 20. 51, 20. 47,
> Mean: 20. 52, standard deviation: 0.0516.
> Prior information on $(W = 1/\sigma^2)$: $w_0 = 250$; $v_0 = 4$; $\langle W \rangle = 1000 = v_0w_0$, $\sigma_0^2 = 0.001$.
> Prior standard deviation of $W' = 2\sqrt{v_0} = 4.0$.
> The updated parameters are found to be $w_f = 61.07$; $v_f = 10$, $\sigma_f^2 = 1/610.7 = 0.00164$;

The prior and posterior distributions of W are plotted in Figure 8.7.

Note: to obtain the distribution of W from that of W', see Exercise 8.8. For confidence intervals, see Exercise 8.20.

8.5.3   Unknown mean and variance of normal distribution

It is often the case that neither the mean nor the variance are known before taking the measurement, yet we might know from past experience that the distribution is normal (and tests can be conducted to verify this using the data; see Chapter 11). This is probably the most common form of inference – we mentioned in the introduction to this chapter the measurements that had been made historically, for example the heights (and other attributes) of human beings.

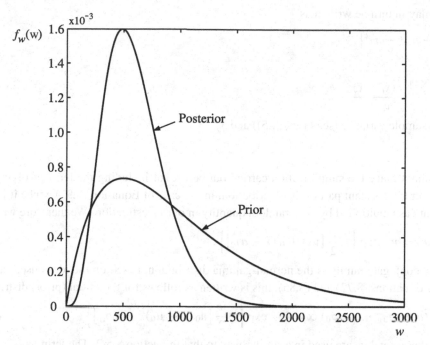

Figure 8.7 Inference regarding variance

Many processes are naturally additive, and the Central Limit Theorem will result in a tendency to normality. We often study the mean of a set of results, and if the sample is large enough, no matter what the underlying distribution, the mean will be normally distributed to a sufficient degree of accuracy for the problem at hand.

The distribution of the random quantity X is given by

$$f_{X|A=a,W=w}(x, a, w) \propto w^{\frac{1}{2}} \exp\left[-\frac{w}{2}(x - a)^2\right], \tag{8.96}$$

where W is a random quantity representing the 'weight' $(1/\sigma^2)$ and A is the mean, also random. Given the usual data set $D = \{x_1, x_2, \ldots, x_n\}$, the likelihood is developed along the same lines as for Equations (8.66) and (8.81) but recognising both A and W as being random:

$$\ell(a, w) = w^{\frac{n}{2}} \exp\left[-\frac{w}{2}\sum_{i=1}^{n}(x_i - a)^2\right]. \tag{8.97}$$

Now the summation term may be written, as before – see Equation (8.67) – in the following way:

$$\sum_{i=1}^{n}(x_i - a)^2 = \sum_{i=1}^{n}(x_i - \bar{x})^2 + n(\bar{x} - a)^2.$$

This may in turn be written as

$$\left[vs^2 + n\left(\bar{x} - a\right)^2 \right],$$ (8.98)

where

$$s^2 = \frac{\sum_{i=1}^{n}\left(x_i - \bar{x}\right)^2}{v}$$ (8.99)

is the sample variance (see also (11.5)) and

$$v = n - 1.$$ (8.100)

We cannot make the simplification carried out before, of including the term involving s^2 of (8.98) in the constant part of the distribution, in the case of Equation (8.97), since it is in the present case multiplied by w, a random quantity in our investigation. We therefore write

$$\ell\left(a, w\right) = w^{\frac{n}{2}} \exp\left\{-\frac{w}{2}\left[vs^2 + n\left(\bar{x} - a\right)^2\right]\right\}.$$ (8.101)

The conjugate family is the normal-gamma distribution, based on the ideas used in developing Equations (8.72) and (8.85); this is written as follows for the initial (prior) distribution:

$$f_{A,W}^{(i)}\left(a, w \mid n_0, \mu_0, v_0, \sigma_0\right) \propto w^{\frac{v_0}{2} - \frac{1}{2}} \exp\left\{-\frac{w}{2}\left[n_0\left(a - \mu_0\right)^2 + v_0\sigma_0^2\right]\right\}.$$ (8.102)

Note that v_0 and σ_0 are used in a similar way to that in Section 8.5.2. The term

$$\exp\left[-\frac{w}{2}v_0\sigma_0^2\right]$$

is separable (as a multiplier) from the exponential term in Equation (8.102); it is attractive to write $w_0 = 1/(v_0\sigma_0^2)$, as in Section 8.5.2. Combining the likelihood with this prior distribution, the following is obtained:

$$f_{A,W}^{(u)}\left(a, w \mid n_f, \mu_f, v_f, \sigma_f\right) \propto w^{\frac{v_f}{2} - \frac{1}{2}} \exp\left\{-\frac{w}{2}\left[n_f\left(a - \mu_f\right)^2 + v_f\sigma_f^2\right]\right\},$$ (8.103)

where

$$n_f = n_0 + n,$$ (8.104)

$$\mu_f = \frac{n_0\mu_0 + n\bar{x}}{n_0 + n},$$ (8.105)

$$v_f = v_0 + n = v_0 + v + 1 = n_0 + n - 1 = n_f - 1$$ (8.106)

and

$$v_f\sigma_f^2 = v_0\sigma_0^2 + vs^2 + \frac{n_0 n\left(\bar{x} - \mu_0\right)^2}{n_0 + n}.$$ (8.107)

There are interesting mechanical interpretations of these equations. Think of two masses, n_0 and n, situated with their centres of mass at μ_0 and $\bar{x}$ from the origin, respectively. Each mass is distributed around these centres, with second moments of $(\nu_0 \sigma_0^2)$ and (νs^2) respectively (sums of deviations squared; recall that $\nu s^2 = s_d^2$). The results can be viewed in an interesting way, such that Equation (8.105) represents the distance from the origin to the centre of mass of the combined masses n_0 and n, and that Equation (8.107) represents the second moment of the distributed masses about the centre. See Exercise 8.10. We should keep in mind the constants involved in the proportionalities; see Exercise 8.11.

We need to obtain the marginal distributions of A and W. We do this by integrating out the random quantity not required. In the case where we wish to evaluate the distribution of A,

$$f_A^{(u)}(a) = \int_0^\infty f_{A,W}^{(u)}\left(a, w | n_f, \mu_f, \nu_f, \sigma_f\right) dw. \tag{8.108}$$

This can be evaluated by substituting

$$z = \frac{w}{2}\left[n_f\left(a - \mu_f\right)^2 + \nu_f \sigma_f^2\right] \tag{8.109}$$

in expression (8.103). Then Equation (8.108) becomes

$$f_A^{(u)}(a) \propto \left[n_f\left(a - \mu_f\right)^2 + \nu_f \sigma_f^2\right]^{-\frac{1}{2}(\nu_f + 1)} \int_0^\infty z^{\frac{\nu_f}{2} - \frac{1}{2}} \exp\left(-z\right) dz. \tag{8.110}$$

Now the integral in this expression is simply a gamma function equal to $\Gamma(\frac{\nu_f}{2} + \frac{1}{2})$. (Recall the definition $\Gamma(s) = \int_0^\infty u^{s-1} \exp\left(-u\right) du$.) Then

$$f_A^{(u)}(a) \propto \left(1 + \frac{t^2}{\nu_f}\right)^{-\frac{1}{2}(\nu_f + 1)}, \tag{8.111}$$

where

$$t \equiv \frac{a - \mu_f}{\sigma_f / \sqrt{n_f}}. \tag{8.112}$$

The expression (8.111) is the student's t-distribution with ν_f degrees of freedom. Note that when $\nu_f \sigma_f = \nu s^2$ with $\nu_f = \nu$, $n_f = n$ and $\mu_f = \bar{x}$, then

$$t = \frac{a - \bar{x}}{s / \sqrt{n}}.$$

This is the familiar data-based value of t. More on this in the next section (8.6).

For inferences regarding W we wish to evaluate

$$f_W^{(u)}(w) = \int_{-\infty}^\infty f_{A,W}^{(u)}\left(a, w | n_f, \mu_f, \nu_f, \sigma_f\right) da.$$

Now the quantity $A = a$ appears only in the exponential term of expression (8.103), and we have

$$f_W^{(u)}(w) \propto w^{\frac{\nu_f}{2} - \frac{1}{2}} \exp\left(-\frac{w}{2}\nu_f \sigma_f^2\right) \int_{-\infty}^\infty \exp\left\{-\frac{w}{2}\left[n_f\left(a - \mu_f\right)^2\right]\right\} da. \tag{8.113}$$

The integral has the same form as that for the area under the normal distribution; recall that

$$\int_{-\infty}^{\infty} \exp\left[-\frac{1}{2}\left(\frac{x-a}{\sigma}\right)^2\right] dx = \sqrt{2\pi}\sigma.$$

Consequently the integral in expression (8.113) is proportional to $w^{-\frac{1}{2}}$, and substituting this result into (8.113), we have

$$f_W^{(u)}(w) \propto w^{\frac{\nu_f}{2}-1} \exp\left(-\frac{w}{2}\nu_f\sigma_f^2\right). \tag{8.114}$$

The random quantity $(w\nu_f\sigma_f^2)$ therefore has a chi-square distribution with ν_f degrees of freedom. We can again write $w_f = 1/(\nu_f\sigma_f^2)$; then

$$W' = \frac{W}{w_f}$$

as in (8.90) is the chi-square random quantity with

$$f_{W'}^{(u)}(w') = \frac{(w')^{\frac{\nu_f}{2}-1} \exp\left(-\frac{w'}{2}\right)}{2^{\frac{\nu_f}{2}} \Gamma\left(\frac{\nu_f}{2}\right)}, \tag{8.115}$$

as in (8.89) but with the values of ν_f, σ_f, and w_f calculated as in the present section.

Note that Equation (8.102) is of the same form as (8.103) so that the same form of marginal distributions apply for the prior distribution $f_{A,W}^{(i)}$ as for $f_{A,W}^{(u)}$ with appropriate change of parameters. The expression for $f_W^{(i)}(w)$ is the same as (8.114) and (8.115), with ν_0, σ_0 and w_0 replacing ν_f, σ_f and w_f.

Example 8.17 We address again the bridge elevation of Examples 8.14 and 8.15. The mean and variance are unknown. The same data as in Example 8.14 are available, comprising the six measurements given there. But we shall base the analysis on an entirely different set of prior information. This is as follows.

$\mu_0 = 20.40$ m; $n_0 = 4$; then $\nu_0 = 3$; $\sigma_0 = 0.060$ m.

$$w_0 = \frac{1}{3(0.060)^2} = 92.6.$$

From the data, $\bar{x} = 20.52$ m; $s = 0.0516$ m; $n = 6$.

Then we obtain the following (based on units of metres throughout).

$n_f = 10$; $\nu = 5$; $\nu_f = 9$,

$$\mu_f = \frac{4(20.4) + 6(20.52)}{10} = 20.472 \text{ and}$$

$$9\sigma_f^2 = 3(0.060)^2 + 5(0.0516)^2 + \frac{4 \cdot 6(20.52 - 20.40)^2}{4+6}, \text{ or}$$

$$\sigma_f^2 = 0.0065192; \sigma_f = 0.08074; \frac{1}{\sigma_f^2} = 153.4; w_f = \frac{1}{\nu_f\sigma_f^2} = 17.044.$$

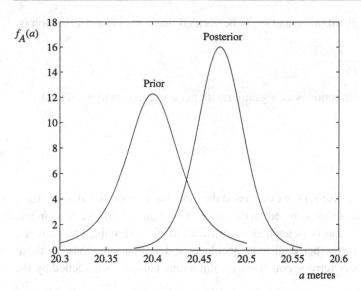

Figure 8.8 Example problem with unknown mean and variance

The updated marginal distributions are a t-distribution for the mean and a chi-square distribution for the weight W. Figure 8.8 shows the prior and posterior distributions of the parameter A, the mean elevation.

8.5.4   Poisson distribution

As a final example of the use of conjugate prior distributions, we discuss a second distribution that is not normal (we have already dealt with the binomial). We consider the distribution of Equation (3.13):

$$\Pr(X = x|v) = \frac{\exp(-v)v^x}{x!}. \tag{8.116}$$

In the Bayesian analysis, we treat the parameter v, the expected number of events in the specified period of time, as being random. We denote this as A with distribution $f_A(a)$. Consider a sample

$$\{r_1, r_2, \ldots, r_i, \ldots, r_n\}, \tag{8.117}$$

where the values represent, for example, values of the total number of vehicles passing a point in the specified period of time, with n time periods in total. The likelihood is

$$\ell(r_1, r_2, \ldots, r_i, \ldots, r_n|A = a) = \frac{[\exp(-na)]\, a^{\sum_{i=1}^{n} r_i}}{\prod_{i=1}^{n} r_i!} \propto [\exp(-na)]\, a^{n\bar{r}}, \tag{8.118}$$

where $\bar{r} = \sum r_i/n$, the mean of the r_is. The number n corresponds to the number of periods sampled, each corresponding to the time specified in defining v. The conjugate prior distribution is again the gamma distribution, and it can be verified that if the prior distribution is

$$f_A^{(i)}(a) = \frac{a^{\alpha_i - 1} \exp(-a/\beta_i)}{\beta_i^{\alpha_i} \Gamma(\alpha_i)}, \tag{8.119}$$

then the posterior distribution is also gamma with parameters α_f and β_f, where

$$\alpha_f = \alpha_i + n\bar{r} \text{ and}$$
$$\frac{1}{\beta_f} = \frac{1}{\beta_i} + n. \tag{8.120}$$

Example 8.18 In Exercise 7.14, we discussed the modelling of errors in design. The errors are modelled as a Poisson process, based on the work of Nessim (1983); see Nessim and Jordaan (1985). The same idea has been applied to collisions in arctic shipping (Jordaan *et al.*, 1987) and could be applied to other collisions, accidents and similar situations. In these models, the number of events N (errors, collisions, ...) in a time period t is modelled by the Poisson distribution

$$\mathbf{Pr}(N = n | \Lambda = \lambda) = \frac{\left[\exp(-\lambda t)\right](\lambda t)^n}{n!} \tag{8.121}$$

based on Equation (8.116). The parameter ν is modelled in (8.121) as a random rate (Λ) multiplied by the time interval t. The conjugate family for Λ is the gamma density, here given by

$$f_\Lambda(\lambda) = \frac{\left[\exp(-\lambda t_d)\right](\lambda t_d)^{n_d - 1}}{\Gamma(n_d)} t_d. \tag{8.122}$$

The distribution (8.122) is of the same form as (8.82), with $\alpha = n_d$ and $\beta^{-1} = t_d$. One can think of data as consisting of the observation of n_d events in the time interval t_d. See also Example 8.19. Updating of (8.122) with new data consisting of n_{d_1} events in time t_{d_1} results in a distribution of the same form as (8.122) but with new parameters $n_d + n_{d_1}$ and $t_d + t_{d_1}$. One can integrate (8.121) over the possible values of Λ using (8.122) to find

$$p_N(n) = \frac{\Gamma(n + n_d)}{\Gamma(n + 1)\Gamma(n_d)} \cdot \frac{t^n t_d^{n_d}}{(t + t_d)^{n + n_d}}, n = 0, 1, 2, \ldots \tag{8.123}$$

Figure 8.9 shows the density $f_\Lambda(\lambda)$ with $\langle \Lambda \rangle = 0.50$, and the event probabilities $p_N(n)$ from (8.123) for the various distributions of Λ. In Exercise 7.14, the probability of detecting an error is modelled by the beta distribution, based on distributions of the kind in (8.122).

In this example, the rate Λ is modelled as being constant in time (but random). The model of Equations (8.116) to (8.120) uses the expected value as the parameter. In this case the rate does not have to be constant since one uses the expected value as theparameter of interest;

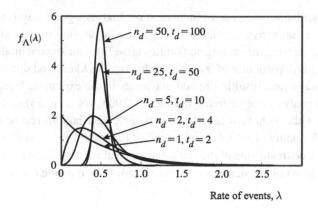

(a) Different degrees of uncertainty regarding the rate of event occurrence, $\langle \Lambda \rangle = 0.5$

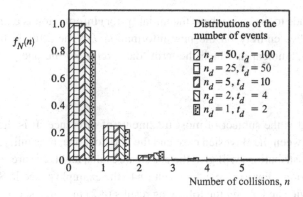

(b) The effect of uncertainty regarding the rate of events

Figure 8.9   Poisson process including uncertainty regarding rate

recall Equation (3.14):

$$v = \int_0^t \lambda(\tau)\,d\tau.$$

A data-related interpretation of (8.120) is that $1/\beta_i$ corresponds to the initial number of observations ($\beta \to \infty$ denoting a small number) and α_i to the initial number of events observed. See Example 8.19.

8.5.5   Summarizing comments

We have dealt with the use of conjugate prior distributions, but emphasize that this has been a matter of convenience in introducing Bayesian methods. One's prior beliefs can be expressed by means of any coherent distribution. The use of numerical methods is a good way forward, and Monte Carlo sampling is an attractive alternative if one wishes to develop models that are not confined to any particular family. The conjugate families arise from the exponential form of the distributions considered, in particular of $f_{X|\Theta}$; in each case, the likelihood consists of a multiplication of exponential terms, resulting in additive terms in the exponent. Details of the exponential families and analysis can be found in Lindley (1965, vol. 2, p. 55) and in de Finetti (1974, vol. 2, p. 242). Other details of families of conjugate distributions can be found in books such as Raiffa and Schlaifer (1961), Lindley (1965), O'Hagan (1994) and Bernardo and Smith (1994). These include treatments of the multidimensional normal distribution.

There is a strong need for generally available numerical schemes for updating distributions using Bayesian methods.

8.6   Basis for classical methods; reference prior distributions expressing 'indifference'

Classical estimation corresponds to the case where the initial (prior) information is extremely vague such that the result relies entirely on the new information D. The attitude may be summarized in the following comment in which the term 'data' refers to the one particular data set that is to be analysed.

Let the set of data speak for itself.

This is an important case, and is the subject of most treatments of inference. It is therefore necessary to make the link between the Bayesian case and the classical one, thus unifying the theory. A certain set of prior distributions will give the classical results, and these are referred to as 'reference' prior distributions. For example, let us return to the example posed in Section 8.3 above; we are testing a device and obtain the following results (8.2) in which success $= 1$, and $0 =$ failure:

$$D = \{0, 1, 0, 0, 1, 1, 1, 1, 0, 1\}.$$

There are six successes in ten trials; if we relied entirely on the data, we would estimate the probability of success in the next trial as the frequency $6/10$, and this will be referred to as the classical estimate. This is fine under conditions which we shall make clear in the following. We have also made clear in the preceding what to do if we have prior information, for example in the form of other relevant tests. We also should be able to ascertain what uncertainty is associated with our estimate, and how to deal with it. We know in the example just given that the uncertainty is more than in the case where there are 1000 tests with, say 606 successes, the probability estimate being 606/1000. But how much more, and how is this included in decision-making? These questions can be answered effectively within the present Bayesian format.

We shall deal with the connection to the classical case first in several sections, then introduce and discuss various classical methods, some of which are closer to the spirit of the present

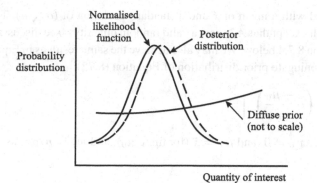

Figure 8.10   Diffuse prior distribution; data dominates the inference (schematic)

analysis than others, and then present a set of examples of classical analysis in Section 8.7.6. As a result, we will be able to address problems using methods that coincide in terms of results with the classical ones, but that have a Bayesian structure, and then we can deal convincingly with the questions raised in the preceding paragraph as well.

8.6.1   Unknown mean of normal distribution

In the case of 'unknown mean' discussed above, where the precision of the testing device is known, the classical estimate of the mean μ of the distribution would be, simply, the sample mean, $\bar{x}$, based on the data $D = \{x_1, x_2, \ldots, x_n\}$:

$$\mu \simeq \bar{x}. \tag{8.124}$$

This is a good example for a first illustration of the approach. Our generic example was the object of unknown weight, which we are about to weigh. (We have noted that there are many other applications in physics, engineering and astronomy.) The weighing device has a known standard deviation of, say, 1 g. Over a wide range of possible values, say from 100 to 200 g, we can consider the prior distribution to be 'diffuse'; Figure 8.10 illustrates the situation. Referring to Equation (8.69), inference regarding the mean was based on the equation

$$f_A^{(u)}(a) = \frac{\exp\left[-\frac{n}{2\sigma^2}(\bar{x}-a)^2\right] \cdot f_A^{(i)}(a)}{N}.$$

The situation in Figure 8.10 suggests that the prior distribution be taken as uniform in the range of interest, with

$$f_A^{(i)}(a) = \text{constant}.$$

The updated distribution is then dominated by the likelihood function, and is of the form

$$f_A^{(u)}(a) = \frac{1}{\sqrt{2\pi}\,(\sigma/\sqrt{n})} \exp\left[-\frac{n}{2\sigma^2}(\bar{x}-a)^2\right], \tag{8.125}$$

which is normally distributed with a mean of $\bar{x}$ and a standard deviation of $(\sigma/\sqrt{n})$. This is identical to the classical result except that A is now a valid random quantity – see discussion of confidence intervals in Section 8.7.4 below. We may also achieve the same result as in Equation (8.125) by considering the conjugate prior distribution of Equation (8.72):

$$f_A^{(i)}(a) = \frac{1}{\sqrt{2\pi}\sigma_0} \exp\left[-\frac{1}{2}\left(\frac{a-\mu_0}{\sigma_0}\right)^2\right].$$

If we make σ_0 very large, then $w_0 = 0$, and $\mu_f = \bar{x}$ (for finite μ_0), and $w_f = n/\sigma^2$, as found in Equation (8.125).

8.6.2   Parameters of scale

When dealing, for example, with the standard deviation, we know that the parameter must be greater than zero, and that it scales the spread of the distribution. There are other scaling parameters that perform a similar function, for example the scaling parameter in the gamma distribution. As a result of this scaling the parameter covers the space from 0 to ∞. A uniform distribution over $(-\infty, \infty)$ as in the case of the mean does not make sense. Let us denote the scale parameter as A. A good way to look at this problem is to follow similar logic to that given in Chapter 4, in deriving utility functions using reasoning that originated with Daniel Bernoulli. Suppose that we have a small amount of probability dp to assign. We might reasonably decide that equal amounts dp should be assigned to equal *percentage* (or proportional) changes in A. In other words

$$dp \propto \frac{da}{a} \tag{8.126}$$

expresses this idea, which is fundamental to scaling.

Our assignment of probability is then proportional to the change in a, i.e. to da, and inversely proportional to the value of a. The form (8.126) is the characteristic form for **parameters of scale**. We can then write

$$f_A^{(i)}(a)\,da \propto \frac{da}{a} \quad \text{or}$$

$$f_A^{(i)}(a) \propto \frac{1}{a}. \tag{8.127}$$

Equation (8.127) is equivalent to that for the assignment of a uniform prior distribution to $(\ln A)$, since then

$$dp \propto d(\ln a) \quad \text{or} \tag{8.128}$$

$$dp \propto \frac{da}{a}.$$

Example 8.19 In the Poisson distribution of (8.116), the expected value ν is a scale parameter. This was denoted A in Equation (8.119). The prior distribution expressing vagueness is of the form (8.127):

$$f_A^{(i)}(a) \propto \frac{1}{a}.$$

This corresponds to $\alpha_i = 0$ and very large β_i in (8.119); the latter condition yielding $1/\beta_i = 0$ in (8.120). The posterior distribution of A is then such that the parameters in (8.120) are

$$\alpha_f = n\bar{r} \text{ and}$$
$$\frac{1}{\beta_f} = n. \tag{8.129}$$

This is the usual data-based interpretation. The mean of the distribution is $\alpha_f \beta_f = \bar{r}$, as would be expected.

8.6.3   Coherence

The expression $(1/a)$ cannot be integrated over the entire parameter space $[0, \infty)$; an infinite result appears. Much the same is found in the case of the prior to the mean (uniform over the space from $-\infty$ to $+\infty$). As a result, the reference prior distributions are often called 'improper'. Since we are attempting to specify 'complete ignorance' *a priori* – recall the mantra 'let the data set speak for itself' – it is not surprising that the distributions do not normalize. We become coherent only in the face of data! In reality, we will usually have some information, and then it is not difficult to say that our opinion is diffuse in a certain domain. This is why we prefer to term the prior distributions that give the data-based parameter values 'reference priors'. This does indeed represent a limiting case, and it is appropriate often to investigate it, and, coupled with sensitivity analysis, we can obtain results adequate for our decision-making. The difficulties with 'improper priors', which do not normalize, may be investigated; those who are mathematically minded may see, for example, the work of Jaynes.

But we can specify the prior distribution to be of the indifferent forms noted for the region of interest only, and then they can be normalized. For example, the uniform prior might only apply in a range $[a, b]$ that covers all values of interest. Or we might assume that the scale parameter $a \propto (\mathrm{d}a/a)$ for values $a_1 \leq a \leq a_2$, where this interval is reasonably wide. Then we can proceed with our analysis. The mathematical results for indifference are in reality limiting forms.

8.6.4   Unknown variance of normal distribution

If we specify the form (8.127) for the standard deviation, we can use it also for the variance or for the weight W. (Use the method for transformation of distributions, Section 7.2; see 'Transformations to different powers' just below. See also Exercise 8.16.) The prior distribution that leads to the classical result, and it must be emphasized, in an improved and corrected form, is a distribution that is, in the region of interest, of the following form:

$$f_W^{(i)}(w) \propto \frac{1}{w}.$$ (8.130)

If we combine expression (8.130) with the likelihood for the unknown variance – Equation (8.81) – we obtain

$$f_W(w) \propto w^{\frac{n}{2}-1} \exp\left[-\frac{w}{2}s_d^2\right].$$ (8.131)

This is of the same form as (8.89), with

$$W' = \frac{W}{w_f},$$

as before in Equation (8.90). We have again that W' follows a chi-square distribution, with $\nu_f = n$ degrees of freedom; in our previous notation – compare Equations (8.89) to (8.92) – we can see that

$$(1/w_f) = s_d^2 = \nu_f s^2 = ns^2.$$ (8.132)

These results imply that $\nu_0 \to 0$ in the prior distribution. This is a useful interpretation since the coefficient of variation of the chi-square distribution is $\sqrt{2/\nu}$, which tends to ∞ (becomes very diffuse) if $\nu \to 0$.

Application of these results follows the same method as in Example 8.16; it is now completely based on the new data summarized by s^2 and n. Classical use is taken up further in Sections 8.7.4 to 8.7.6 below.

Transformations to different powers

Had we used $\sigma^2 = \Theta$, say, as the parameter of interest, rather than W, we find that the prior distribution is again of the form

$$f_\Theta(\theta) \propto \frac{1}{\theta}.$$ (8.133)

This can be seen by making the transformation $\theta = 1/w$;

$$f_\Theta(\theta) = f_W(w) \cdot \left|\frac{dw}{d\theta}\right| \propto \frac{1}{w} \cdot \frac{1}{\theta^2} = \frac{1}{\theta}.$$ (8.134)

We may also consider the logarithmic form (8.128). By assigning a uniform density to $\ln w$ (again, over the region of interest), the same prior is achieved, since

$$d(\ln w) \propto \frac{1}{w}dw.$$ (8.135)

Transformation from w to w^k leads to a uniform distribution for the latter. One can see this by using the quantities $\ln w$ and $\ln(w^k)$, noting that $\ln(w^k) = k \ln w$.

8.6.5   Unknown mean and variance of normal distribution

We now consider the case where the mean and variance are unknown, as dealt with in Section 8.5.3. Based on the above reasoning, we assign a joint prior density for the parameters $\{A, \ln W\}$ such that they are independent, and with a uniform distribution (in the regions of interest); thus

$$f_{A,W}^{(i)}(a, w) \propto \frac{1}{w}. \tag{8.136}$$

This is combined with the likelihood function (8.101) to give

$$f_{A,W}^{(u)}(a, w) \propto w^{\frac{n}{2}-1} \exp\left\{-\frac{w}{2}\left[n(a-\bar{x})^2 + vs^2\right]\right\}. \tag{8.137}$$

We can compare this equation with (8.103) from the preceding section, repeated here for convenience:

$$f_{A,W}^{(u)}(a, w|n_f, \mu_f, v_f, \sigma_f) \propto w^{\frac{v_f}{2}-\frac{1}{2}} \exp\left\{-\frac{w}{2}\left[n_f(a-\mu_f)^2 + v_f\sigma_f^2\right]\right\}.$$

We can see that

$$v_f = v = (n-1); \tag{8.138}$$

$$\mu_f = \bar{x}, \tag{8.139}$$

and also that the only term in the exponent that is part of $(v_f\sigma_f^2)$ is (vs^2). The mechanical interpretation of Section 8.5.3, given after Equation (8.107), is that the only mass being considered is that due to the data. The moment of inertia referred to there is now (vs^2).

To obtain the marginal distributions, Equation (8.137) is integrated over the (unwanted) random quantity, as in Section 8.5.3. Thus

$$f_A^{(u)}(a) \propto \left(1 + \frac{t^2}{v_f}\right)^{-\frac{1}{2}(v_f+1)}, \tag{8.140}$$

where

$$t = \frac{a-\bar{x}}{s/\sqrt{n}}, \tag{8.141}$$

and the number of degrees of freedom is $v_f = n-1$. The marginal distribution of W is

$$f_W^{(u)}(w) \propto w^{\frac{v_f}{2}-1} \exp\left(-\frac{w}{2}v_f s^2\right), \tag{8.142}$$

where (to emphasize) $v_f = n-1$. This again is of the same form as (8.89), with

$$W' = \frac{W}{w_f}.$$

Then W' follows a chi-square distribution, with

$$v_f = n-1 \tag{8.143}$$

degrees of freedom; in our previous notation, we can see that

$$\frac{1}{w_f} = v_f s^2 = (n-1)s^2. \tag{8.144}$$

In the case of unknown mean and variance, the degrees of freedom are $(n-1)$, as compared to the case where the variance only is unknown (previous subsection), with n degrees of freedom. In both cases, the distribution of W *a priori* was $\propto (1/w)$, but in the case of mean and variance both unknown, the updating was carried out via the normal-gamma model, as described above, and in Section 8.5.3. When integrating out the unwanted random quantity (the mean), in order to obtain the distribution of the variance, the integral was proportional to $w^{-\frac{1}{2}}$; this combined with the term $w^{\frac{n}{2}-1}$ to produce $w^{\frac{n-1}{2}-1}$, and we can see the reason for the appearance of $(n-1)$ degrees of freedom.

One can again think of a mechanical analogy; if there are n normal modes in a vibration problem, these specify the configuration of the system. In the case of unknown variance, the calculation of sample variance is carried out with respect to the fixed mean μ. The n units can therefore vary independently, with n degrees of freedom, whereas in the case of the unknown mean and variance, the sample mean $\bar{x} = \sum x_i / n$ is used in the calculation of sample variance. This operates as a constraint, and the x_i then have $(n-1)$ degrees of freedom.

8.6.6   Binomial parameter

In the binomial process, if we have r successes in n trials, the natural inference in the absence of other information is that the frequency is (r/n). Uncertainty regarding this estimate will arise for small r and n. Referring to Equations (8.25) and (8.33), we see that this interpretation can be made if $r_0 = n_0 = 0$. Substituting this in Equation (8.21), we find that

$$f_A^{(i)}(a) \propto \frac{1}{a(1-a)}. \tag{8.145}$$

Applying Equation (8.145), we obtain

$$f_A^{(u)}(a) = \frac{a^{r-1}(1-a)^{n-r-1}}{B(r, n-r)}, \tag{8.146}$$

with mean

$$\langle A \rangle = \frac{r}{n}. \tag{8.147}$$

The proportionality (8.145) is the reference prior in the case of the binomial parameter. It is not possible to normalize this density, as is the case of other reference priors. As before, it represents a limiting case and we can normalize it by assuming that it is valid over a range of interest within the [0, 1] space of A.

The transition from the reference prior of proportionality (8.145) to Bayes' postulate of (8.34) and Section 8.3.3, i.e. $f_A^{(i)}(a) = 1$ for $0 \le a \le 1$, is quite interesting. By inspection of Equation (8.146), we can see that this is achieved if $r = 1$ and $n = 2$, that is, one success and

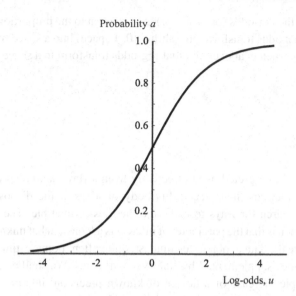

Figure 8.11   Probability and log-odds

one failure. One can then justify the use of the uniform prior rather than the improper one (8.145) by the knowledge that success and failure are possible, and that we judge these to be equally likely. This shows that we can move quite quickly away from the state of 'vagueness' implied by the reference prior.

Log-odds

Another interesting way to look at the vague prior of proportionality (8.145) is to consider the transformation into log-odds. Interpreting a as the probability of success on the next trial, then $[a/(1 - a)]$ represents the odds, and

$$u = \ln\left(\frac{a}{1 - a}\right) \tag{8.148}$$

represents the log-odds. This function has a long history; see Tribus (1969). It has, for example, been used to denote the 'evidence' for a proposition. Figure 8.11 shows the function graphically. We now take A and consequently U as random and assign a uniform distribution to the log-odds:

$$f_U(u) \propto \text{constant.} \tag{8.149}$$

Now

$$\frac{du}{da} = \frac{1}{a(1 - a)} \tag{8.150}$$

from (8.148). As a result, transforming the distribution (8.149),

$$f_A(a) \propto \frac{1}{a(1 - a)}. \tag{8.151}$$

Hence, a uniform distribution on the log-odds (or evidence) is equivalent to the proportionality (8.145). It is also worth noting that odds transform probability (0, 1 space) into a space from 0 to $+\infty$ (as in the case of scale parameters above) and that log-odds transform to a space from $-\infty$ to $+\infty$.

8.7   Classical estimation

8.7.1   Point estimation

Classical estimation is viewed and interpreted most effectively from a Bayesian perspective. The use of reference prior distributions then places this body of ideas at the disposal of the Bayesian, and we may then search for ways to make the theories compatible. The main assumption in the classical approach is that the parameter of interest is a 'constant but unknown' parameter, and that we have to use the data to obtain 'estimates', quite often 'point' estimates – single values – although sometimes supplemented by 'interval' estimates. We shall continue to use as an introductory example the case of a device of known precision, interpreted as meaning that the standard deviation σ is known and that readings follow a normal distribution. The classical analysis would consist of the estimation of the mean μ as being approximately equal to the sample mean $\bar{x}$, as noted in Section 8.6.1, Equation (8.124):

$$\mu \simeq \bar{x}.$$

We would be reasonably happy with this procedure if the data set (n) was very large. Otherwise, feelings of uncertainty arise from the 'approximate' part of this equation. To account for this, confidence intervals have been developed.

For the choice of a point estimate, a rule is needed. A Bayesian would look at the entire posterior distribution, and then make decisions following the methods of rational decision-making. In the classical approach, rules are set up often without any relation to the decision being made; see comments in Section 8.7.3 below. The criterion might constitute the following.

- The mean value.
- The most likely value.
- A value that shows no bias.

Examples of mean values that might be used as point estimates are the posterior means in Examples 8.14 and 8.15. These would also be unbiased since they are at the mean of the distributions representing the uncertainty regarding the parameter and therefore would constitute good 'average' estimators.

8.7.2   Maximum likelihood

One method of estimation is that of the maximum likelihood, which we shall briefly introduce. This was developed without the use of Bayes' theorem, mainly by R. A. Fisher, and therefore

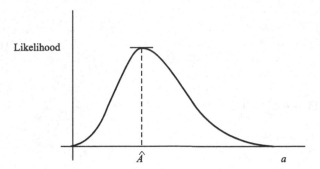

Figure 8.12   Most likely value

focusses on the information contained in the data. We repeat the definition (8.64) for the likelihood:

$$\ell = \prod_{i=1}^{n} f_{X|A=a}(x_i, a).$$

If we take the logarithm of the likelihood, we obtain a sum of terms rather than the product:

$$\ln \ell = \sum_{i=1}^{n} \ln f_{X|A=a}(x_i, a). \tag{8.152}$$

The idea is to obtain the estimator that leads to the maximum of the likelihood function, or of the equivalent for this purpose, the log-likelihood. Since it is based on data only, the method has strong application when there are large data sets and little prior information. In general, for a parameter A, the maximum likelihood function is shown schematically in Figure 8.12; this will approximate in shape the posterior density in cases where data dominates strongly. The mode of this distribution is the maximum likelihood estimator, as shown in Figure 8.12. This is denoted $\hat{A}$.

Example 8.20 We shall show an example right away so that the principle may be understood. Consider the normal distribution; leaving out inessential constants, from Equation (8.101) above,

$$\ell(a, w) = w^{\frac{n}{2}} \exp\left\{-\frac{w}{2}\left[vs^2 + n(\bar{x} - a)^2\right]\right\}.$$

We shall maximize $\ln \ell$. This is given by

$$\ln \ell(a, w) = \frac{n}{2} \ln w - \frac{w}{2}\left[vs^2 + n(\bar{x} - a)^2\right]. \tag{8.153}$$

In order to maximize $\ln \ell$,

$$\frac{\partial \ln \ell}{\partial a} = nw(\bar{x} - a) = 0,$$

or

$$a = \bar{x}.$$

Thus

$$\hat{A} = \bar{x}. \tag{8.154}$$

Also

$$\frac{\partial \ln \ell}{\partial w} = \frac{n}{2w} - \frac{1}{2} \left[vs^2 + n(\bar{x} - a)^2 \right] = 0,$$

or

$$w = \frac{n}{\left[vs^2 + n(\bar{x} - a)^2 \right]}.$$

If we substitute the value of $\hat{A} = a = \bar{x}$,

$$\widehat{W} = \frac{n}{vs^2} = \frac{n}{(n-1)s^2}. \tag{8.155}$$

Since W corresponds to $1/\sigma^2$, this estimate corresponds to

$$\frac{1}{\widehat{W}} = \widehat{\sigma^2} = \frac{(n-1)s^2}{n}. \tag{8.156}$$

The estimator $\hat{A}$ for the mean is the usual mean value but the estimator $\widehat{\sigma^2}$ differs from the usual value s^2. For large n, $(n-1)/n \simeq 1$. Lindley (1965, Part 2) has stressed the use of laws of large numbers in obtaining a good basis for the theory.

8.7.3 Comments on point estimates

The purpose of inference is to assist in decision-making. Returning to the basic decision problem of Figure 8.1(b), we should first idealize our problem in terms of the overall objectives. Let us take an everyday example: we wish to estimate the length of string that we need to tie up a parcel. There is a natural tendency to overestimate since there is a distinct disutility in having a piece of string that is too short, especially if we are coming to the end of the ball. The estimate, then, is not in general a mean, if there is some purpose intended. The length of string depends on our utility function and the disutility of being short. The same idea applies in ordering a length of pipe for a pipeline, to avoid penalty prices for re-ordering an additional length later. Or if we are borrowing money for an investment, we might overestimate to avoid short-term borrowing at a higher rate. Of course, we do not always want to overestimate – the point is simply that the utility also enters our thinking when we make an estimate. The general situation can result in bias in either direction away from the mean. See Exercise 8.25. We should not compose a rule in the void, without considering the specific context. The complete methodology has been explained in detail throughout this book.

Estimates at the centre of mass

The rationale for estimates that are as close to the mean value as possible is of interest. Suppose that we are considering the parameter A: this could be the mean value of a normal distribution, or the parameter of a binomial distribution. It is random, as we have insisted above. A good rule that leads to the mean for estimating a single value of A, say A^*, is to choose the value such that we lose an amount equal to a constant times $(A^* - a)^2$, for any value of $A = a$. This is a quadratic loss function and in implementing it we have to account for the randomness in A.

We have dealt with this situation before in Section 3.5.3. We wish to minimize the quantity

$$\int (A^* - a)^2 \, dF_A(a). \tag{8.157}$$

This expression represents the second moment of the probability distribution about A^*. It is well known in mechanics that the second moment is at a minimum when the axis about which moments are taken passes through the centre of mass: see Section 3.5.3. Therefore, expression (8.157) must be a minimum when A^* is at the centre of mass, or, in other words, at the mean. Then the best estimate of A is the mean value $\langle A \rangle$. Under these conditions, we do wish to use the mean as an estimator. The 'centre of mass' of our degree of belief expressed by a distribution of masses of probability is a logical point at which to place our 'best estimate'.

In fact, de Finetti uses the minimization of the quadratic loss as an alternative way of defining probability, and shows that it is equivalent to the 'fair bet' introduced in Section 2.1. For an event, say E, then $E^* = \langle E \rangle = \mathbf{Pr}(E)$. He goes further to define 'prevision' in the same way. As a result, means, expected values and probability have essentially the same meaning, with the latter being reserved for random events. The mean value of a parameter can therefore be said to represent a 'fair estimate' of the parameter, in the sense just described. In general, the choice of a particular value for some purpose depends on the situation at hand.

8.7.4   Confidence intervals; classical approach

As a result of the uncertainty in the parameter estimate that inevitably results when the data base is small, confidence intervals were developed. This arose in the context of classical data-based analysis but this uncertainty is quite naturally addressed in the Bayesian formulation. But the idea of confidence intervals can be applied to the Bayesian case too. The easiest way to bring out the points raised is to demonstrate them.

Example 8.21 In the analysis of the elevation of the bridge deck in Example 8.14, we found that the posterior mean elevation was 20.520 m with a standard deviation of 0.0202 m. We wish to express in a compact manner our uncertainty about the mean value. To do this, we construct the confidence intervals on the mean. We shall soon see that in the classical viewpoint, the data are treated as random but this runs into trouble once the (non-random) values are inserted. The Bayesian treats the data as non-random right from the start, and does not have to change the nature of the probabilistic statement.

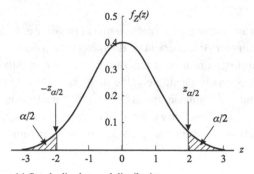

(a) Standardized normal distribution

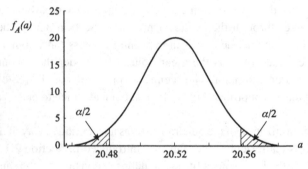

(b) Confidence intervals for bridge deck elevation

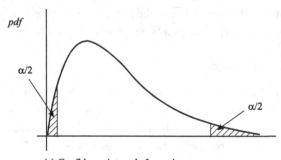

(c) Confidence intervals for variance

Figure 8.13 Confidence intervals

The following would be the initial statement. We denote by $\pm z_{\alpha/2}$ the points on the standard normal distribution containing the probability α; see Figure 8.13(a). Based on Equation (8.76) we can state that

$$\mathbf{Pr}\left[-z_{\alpha/2} \le \frac{(\mu_f - A)}{\sigma_f} \le z_{\alpha/2}\right] = 1 - \alpha,$$

where A is as before the quantity representing the (random and uncertain) mean. Thus

$$\mathbf{Pr}\left[\mu_f - z_{\alpha/2} \cdot \sigma_f \leq A \leq \mu_f + z_{\alpha/2} \cdot \sigma_f\right] = 1 - \alpha \qquad (8.158)$$

is the natural consequence of this reasoning. Figure 8.13(b) shows the confidence intervals with a probability density, following the Bayesian viewpoint (the classicist is, strictly speaking, forbidden to draw the probability distribution – see below). We choose $\alpha = 0.05$; then the values are

$$\mathbf{Pr}\left[20.520 - 1.96 \cdot 0.0202 \leq A \leq 20.520 + 1.96 \cdot 0.0202\right] = 1 - \alpha = 0.95 \text{ or}$$
$$\mathbf{Pr}\left[20.520 - 0.0396 \leq A \leq 20.520 + 0.0396\right] = 0.95 \text{ or}$$
$$\mathbf{Pr}\left[20.480 \leq A \leq 20.560\right] = 0.95, \qquad (8.159)$$

with the limits calculated to three decimal places. The values 20.480 and 20.560 in (8.159) enclose the **95% confidence interval**.

In general we use $(1 - \alpha)\,100\%$ confidence intervals; α is arbitrary to a considerable degree. It is stated that the level of 95% ($\alpha = 0.05$) had the attraction that it corresponded closely to ± 2 standard deviations of the normal distribution (a more precise value is 1.96, rather than 2). Under the assumption that prior information is vague, we might agree to use Equation (8.125) for a normal process with standard deviation σ. Then $\mu_f \simeq \bar{x}$ and $\sigma_f \simeq \sigma/\sqrt{n}$ so that (8.158) becomes

$$\mathbf{Pr}\left[\bar{x} - z_{\alpha/2} \cdot \frac{\sigma}{\sqrt{n}} \leq A \leq \bar{x} + z_{\alpha/2} \cdot \frac{\sigma}{\sqrt{n}}\right] = 1 - \alpha. \qquad (8.160)$$

The values

$$\bar{x} \pm z_{\alpha/2} \cdot \frac{\sigma}{\sqrt{n}} \qquad (8.161)$$

give the familiar classical confidence interval.

Classical approach

The classical method might present the following statement in connection with inferences regarding the mean of a normal distribution, where the standard deviation σ is known.

$$\mathbf{Pr}\left[-z_{\alpha/2} \leq \frac{\sqrt{n}\,(\bar{X} - \mu)}{\sigma} \leq z_{\alpha/2}\right] = 1 - \alpha,$$

in which $\bar{X}$ based on data has been presented as random. Then

$$\mathbf{Pr}\left[\bar{X} - z_{\alpha/2} \cdot \frac{\sigma}{\sqrt{n}} \leq \mu \leq \bar{X} + z_{\alpha/2} \cdot \frac{\sigma}{\sqrt{n}}\right] = 1 - \alpha.$$

We emphasize that the value of $\bar{X}$ continues to be considered as random in this equation.

Now we have pointed out that data can only be random when we don't know the values. The classicist is forced to retreat to the following statement, upon obtaining the data, from which $\bar{x}$ is calculated.

$$\bar{x} - z_{\alpha/2} \cdot \frac{\sigma}{\sqrt{n}} \leq \mu \leq \bar{x} + z_{\alpha/2} \cdot \frac{\sigma}{\sqrt{n}} \tag{8.162}$$

with $(1 - \alpha)\%$ confidence. The word 'confidence' replaces 'probability', since there are no probabilities left in expression (8.162). It is identical in numerical values to (8.160).

Variance

Another case was dealt with in some detail in Sections 8.5.2 and 8.6.4. This concerns the unknown variance. Here, we wish to develop the $(1 - \alpha)\,100\%$ confidence intervals for the variance σ^2. The Bayesian can include prior information or judge that the intervals be data-based (Sections 8.5 and 8.6 respectively). Further, the Bayesian can, as before, write a probabilistic statement based on Equation (8.89):

$$\mathbf{Pr}\left[\varkappa^2_{1-\alpha/2} \leq \frac{W}{w_f} \leq \varkappa^2_{\alpha/2} \right] = 1 - \alpha, \tag{8.163}$$

in which the random quantity W does not disappear. The symbol $\varkappa^2$ refers to values of the chi-square distribution as illustrated in Figure 8.13(c). The analysis applies to the case where the mean is either known or unknown, with appropriate degrees of freedom v_f. If we take the case of the reference prior expressing indifference for the case of unknown mean and variance, Equation (8.144), we find

$$\mathbf{Pr}\left[\frac{(n-1)\,s^2}{\varkappa^2_{\alpha/2}} \leq \sigma^2 \leq \frac{(n-1)\,s^2}{\varkappa^2_{1-\alpha/2}} \right], \tag{8.164}$$

in which we have (informally as regards notation!) treated σ^2 as random.

Figure 8.13(c) illustrates the posterior distribution in this case. In this we have shown the interval with equal areas $\alpha/2$ on either side; there are other ways to construct these limits; see Lindley (1965, Part 2).

Classical interpretation of chi-square distribution; biased or unbiased?

In the case of unknown variance σ^2, the usual estimator based on data is

$$s^2 = \frac{\sum_{i=1}^{n} (x_i - \bar{x})^2}{n - 1}. \tag{8.165}$$

We shall outline the background to the classical reasoning. It is common to refer to 'random data', a practice which (as previously indicated) I do not like since the data are not random once obtained. Let the 'population' consist of the random quantities $\{X_1, X_2, \ldots, X_i, \ldots, X_n\}$, normally distributed, and with common parameters μ and σ^2, respectively. We assume that the random quantities are also independent and therefore they are *iid*.

The chi-square distribution has been introduced previously. The name derives from the Greek symbol used in its analysis (χ^2). The distribution models sums of squares of deviations from the mean of n random quantities taken from normally distributed parents. These might have different means and standard deviations but are judged independent. We first normalize these deviations from the mean by dividing by the standard deviations, resulting in n independent random quantities following the standard normal N (0, 1) distribution. We then have

$$\chi^2 = Z_1^2 + Z_2^2 + \cdots + Z_i^2 + \cdots + Z_n^2, \tag{8.166}$$

where the Z_i are now *iid*.[4] The square of a normally distributed random quantity is chi-square (Exercise 8.21). It can be shown – see Exercises 8.21 and 8.22 as well as Example 7.21 – that χ^2 in Equation (8.166) follows a chi-square distribution with n degrees of freedom. Further, S^2 is now random:

$$S^2 = \frac{\sum_{i=1}^n \left(X_i - \bar{X}\right)^2}{n - 1}, \tag{8.167}$$

The quantity

$$\chi^2 = \frac{(n - 1)\, S^2}{\sigma^2} \tag{8.168}$$

has a chi-square distribution with $\nu = (n - 1)$ degrees of freedom. The number of degrees of freedom is in fact n if the known mean μ rather than $\bar{X}$ is used in Equation (8.167). See Exercise 8.23 and, for example, DeGroot and Schervish (2002).

In the classical method, based on Equation (8.168), the statement would be made that

$$\mathbf{Pr}\left[\varkappa_{1-\alpha/2}^2 \leq \frac{(n - 1)\, S^2}{\sigma^2} \leq \varkappa_{\alpha/2}^2\right] = 1 - \alpha;$$

see Figure 8.13(c) for the definition of terms. This leads to

$$\frac{(n - 1)\, s^2}{\varkappa_{\alpha/2}^2} \leq \sigma^2 \leq \frac{(n - 1)\, s^2}{\varkappa_{1-\alpha/2}^2}, \tag{8.169}$$

confidence intervals without a probability. The expression is the same as regards numerical values as (8.164). But we shall now investigate further; there are other differences.

Classical analysis of bias

It is noteworthy, first, that the mean of the distribution of (8.168) is $\nu = (n - 1)$, which corresponds to $\langle S^2 \rangle = \sigma^2$. We can also show this as follows. We know from Equations (3.107) and (3.106), regardless of the underlying distribution, that

$$\mu_{\bar{X}} = \mu \text{ and}$$

$$\sigma_{\bar{X}}^2 = \frac{\sigma^2}{n}.$$

---

[4] We have used a lower case letter χ in χ^2 to denote a random quantity. This deviates from our usual practice of using upper case letters for random quantities, and only to accord with established convention.

Using Equation (3.87) and these results, we can then conclude that

$$
\begin{aligned}
\langle S^2 \rangle &= \frac{1}{n-1} \left\langle \sum X_i^2 - n\bar{X}^2 \right\rangle \\
&= \frac{1}{n-1} \left(n\sigma^2 + n\mu^2 - \sigma^2 - n\mu^2 \right) \\
&= \sigma^2.
\end{aligned}
\tag{8.170}
$$

Thus a measurement of S^2 is, by this reasoning, an unbiased estimator of σ^2, and therefore a good estimator 'on average'. For this reason, s^2 from the sample is used with the divisor $(n-1)$ as in Equation (8.165) above. It does not follow from this that S is an unbiased estimator of σ, so that this property is not of compelling attraction. For large n, the value of $\langle S^2 \rangle$ tends to σ^2 and s will tend to σ also.

Equation (8.168) seems at first glance to provide us with a connection to our analysis of Sections 8.5 and 8.6 above, where we used chi-square distributions to model the prior and posterior distributions of $W = \sigma^{-2}$. Glancing over our shoulder as we leave the subject, we note that a chi-square distribution of W does not imply the same for σ^2, leading to Exercise 8.24, which we term '*Bayesian fly in the ointment*'. The classical result is the 'wrong way around'!

8.7.5   Classical method: hypothesis testing

Hypothesis testing is a formalized procedure that can be used as an aid to decision-making. There is a lot of common sense in the methodology but it has no real rigorous basis. Too often it is used in a spurious attempt to convert uncertainty to certainty, with conclusions of the kind 'the claim is true' or 'the mean is equal to (stated value)'. The role of utility in decision-making varies depending on the problem. The two extremes correspond to cases first where utility dominates the decision, whereas, in the second, we are trying to express a scientific opinion regarding probabilities. We have given an example of the first in the case of estimating the length of a piece of string to wrap a parcel; we overestimate to avoid being left short. In problems of safety, we use extreme values in estimates of load so as to reduce the probability and disutility of failure.

An example of a scientific judgement would be simply the best estimate of the strength of steel based on a set of tests. Or consider the question of a certain specified treatment on, say, the development of a seed to be used in agriculture. Tests are carried out, in which the crop from the new seed is compared to that from a traditionally used one. The scientist or engineer has to express an opinion on whether the new seed is an improvement on the existing one. (There are many problems of this kind; new treatments for disease, new manufacturing processes, compositions of metals, and so on.) The concept of a fair estimate was discussed in Section 8.7.3 'Comments on point estimates' above.

The difference between the two extremes just discussed is sometimes small; it comes down to the division between probability and utility, the question of judgement in terms of combining a 'fair bet' (probability) and what is desired (utility), a division which we have considered in great detail throughout this work. It is strongly advocated that any particular problem be analysed in a complete way using decision trees, so that the key elements in the problem can be clearly understood, as described in the earlier parts of this book. In my view, there is too much emphasis on hypothesis testing in standard texts and teaching; it is an 'ad hoc' procedure, an approximate form of decision-making, and should be taken with a pinch of salt.

Methodology

We now outline the methodology of hypothesis testing. The first step is to propose a hypothesis, which we then examine with regard to its consistency with regard to the data. A hypothesis is assertion regarding the parameter or parameters of the probability distribution under consideration. The basis hypothesis under test is termed the null hypothesis. This is often denoted as H_0, which is the notation that we shall use. As an example of a null hypothesis, we take the case of the engineer on the road construction site. The asphalt content of the paving material is measured by the normalized index A. It is desired that this index be 5. Let us assume further that experience has shown that the index measurements follow a normal distribution with a known standard deviation of 0.50.

Three samples are taken. The assertion made is that

Definition $H_0 : \mu = 5$.

In this, μ is the 'population' mean of the asphalt content in the batch being delivered. Suppose that the readings are

$\{4.81, 4.87, 5.15\}$.

It is usual also to make an **alternative hypothesis**. This could be as follows.

Definition $H_1 : \mu \neq 5$.

From a Bayesian point of view, the mean μ will be a random quantity, say A. The probability that A is equal to any specific value, for instance 5, is zero! What is meant in the present case is rather the following.

Definition (loose but more appropriate) $H_0 : A$ is reasonably close to 5.

Blind adherence to the formalism of hypothesis testing could obscure the real questions at hand: is the asphalt delivered up to the required standard for the proper performance of the road? Is the supplier trying to make a profit by reducing the asphalt content below specification?

The method of hypothesis testing usually follows the following six steps. They can be interpreted with regard to Figure 8.14. The null hypothesis is accepted if the sample statistic

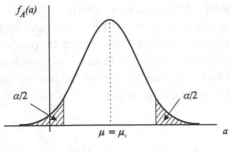

(a) Two-tailed test

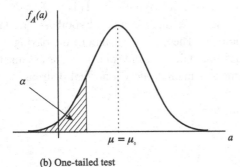

(b) One-tailed test

Figure 8.14   Hypothesis testing

falls within the central region of the probability distribution, which is drawn as if the null hypothesis were indeed true. The parameter α is termed the level of significance of the test.

Step 1 H_0 : A is reasonably close to 5 ($= \mu_0$).
Step 2 H_1 : A is not reasonably close to 5 (two-tailed),
        or H_1 : A is less than 5 (one-tailed).
Step 3 Specify the level of significance; $\alpha = 0.1$, $\alpha/2 = 0.05$.
Step 4 Critical region for rejection of H_0; $z < -1.645$, $z > 1.645$ (two-tailed), $z < -1.282$ (one-tailed). We would be more likely to consider a one-tailed test in this example since we are most interested in the adverse situation where the asphalt content is too low; a somewhat high value would not be of concern. Z is the standard normal random quantity, and

$$z = \frac{\bar{x} - \mu_0}{\sigma/\sqrt{n}},$$

where $\bar{x}$ is the sample mean, and n is the number of observations.
Step 5 From the data, $\bar{x} = 4.94$; $z = \frac{4.94 - 5.00}{0.50/\sqrt{3}} = -0.196$.
Step 6 Accept H_0, reject H_1.

The last step follows from the rule that H_0 is accepted if the sample value is in the central region of Figure 8.14. The practical decision is that the sample of road paving material is accepted. The actual decision should depend on the consequences, for example possible deterioration of the road and associated replacement costs if the material is substandard. If the asphalt content is only marginally less than the specification, this may be acceptable if the consequences are insignificant with regard to potential deterioration of the road. But the formalized procedure should be thoroughly investigated to ensure that the confidence intervals, and the procedure itself, reflect the objectives of the decision-maker. It is natural to ask the question: if there is a difference between the data and H_0, what is the probability that the difference arose by pure chance? This is a useful perspective to use in hypothesis testing.

While judgement is applied in the choice of α – the values commonly used range from 0.01 to 0.1 – some account of the consequences should be included in our thinking. The arbitrariness of α-values is one of the drawbacks of the method.

Type 1 and 2 errors

As noted, hypothesis tests could be one- or two-tailed. Figure 8.14(a) shows the two-tailed variety whereas (b) shows the one-tailed variety. Within the methodology of hypothesis testing, type 1 and 2 errors are defined as follows.

Definition *A type 1 error has been committed if the null hypothesis is rejected when it is true.*

We note that, using the formal definition of the null hypothesis, the probability of the null hypothesis being correct is zero! But we use the loose definition above, and we can then make progress at the expense of some precision in the definition.

Definition *A type 2 error has been committed if the null hypothesis is accepted when it is false.*

The same comments that are given after the definition of type 1 errors apply. To take this a little further, we illustrate the Bayesian approach in Figure 8.15(a). In this the parameter A is modelled as $f_A(a)$ with its mean at $\bar{x} = 4.94$, and a standard deviation of $\left(0.50/\sqrt{3}\right)$. The regions of acceptance of H_0 and H_1 are indicated for the one-tailed test. This is contrasted with the classical approach in (b), where the hypothesis H_0 (mean is equal to 5.0) is supposed to be true. But to use this for decisions, we have to understand that $\mu \simeq \mu_0$ really means 'the batch is acceptable', with all of the implications associated with this statement. The spirit of the definitions of type 1 and 2 errors can best be illustrated by means of the decision tree of Figure 8.16. A full decision analysis could be used to set the acceptance and rejection regions. With regard to decisions under uncertainty, we find the thoughts of Ambrose Bierce, recorded at the beginning of Chapter 4, to be apposite.

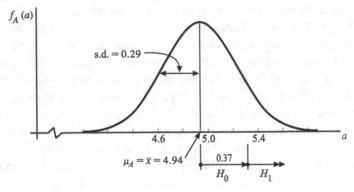

(a) Bayesian interpretation

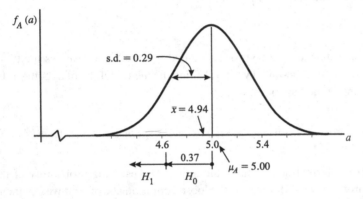

(b) Classical interpretation

Figure 8.15 One-tailed hypothesis test: comparison of Bayesian and classical interpretation. Critical interval $= (0.50/\sqrt{3}) \cdot (1.282) = 0.37$

8.7.6 Common classical inference problems

The techniques developed in the classical approach give good solutions under a variety of circumstances. But they have neither the rigorous basis of Bayesian theory nor the clear connection to decision theory stressed in this book. The development of a rational avenue to classical inference is largely associated with the work of de Finetti, who pointed out the assumptions that underly the classical approach. Fortunately the two approaches are consistent if one uses the reference prior distributions of the preceding section. This avenue permits the user to assess in a powerful way the merits of a particular procedure. Classical methods are widely used and disseminated; it is common in undergraduate education to introduce young engineers and scientists to classical inference problems. They are often referred to and used in practice and provide important common ground between the various viewpoints. The present section summarizes some applications of this kind. But the overriding need to place them in

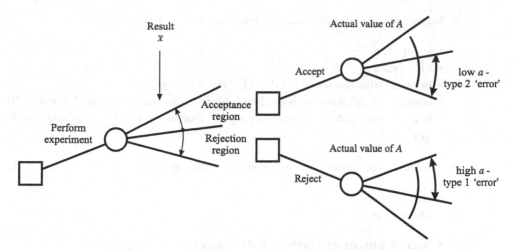

Figure 8.16 Illustration of hypothesis test as a decision problem. Consequences also enter the decision

a proper framework of decision-making is the task of the engineer or scientist using them. The following is a list of common inference procedures. Lindley (1965, Part 2), O'Hagan (1994), Bernardo and Smith (1994) and Lee (1997) give Bayesian interpretations of classical theory and in particular hypothesis testing.

Many of the common techniques rely on the assumption that the underlying random quantities (the X_i) are normally distributed. There are instances where it is known from past experience that a distribution is normal. In other cases, if the sample size is large enough, means and related random quantities will approximately follow the theory based on normal distributions. This follows from the Central Limit Theorem. A sample size of 30 or more for use of the normal distribution in these cases is often quoted. We shall denote the unknown parameter as A, in each case, as was done above. This will always be the parameter being estimated in the following.

The mean values below are generally interpreted as the 'point estimate' of classical theory. It is usually important to consider the distribution of A, not just $\langle A \rangle$, unless there is a large amount of data. One case where a single value might suffice is the 'most probable largest' value which appears in a very peaked distribution. See Chapter 9. The Bayesian interpretation of a confidence interval is that vague prior knowledge of the parameter leads to the probability $(1 - \alpha)$ of the parameter being in the interval.

1. Normal process or approximation. Estimate of $\mu \equiv A$; σ^2 known.

   - Using: $\bar{x}, \sigma^2$.
   - Mean value: $\langle A \rangle = \bar{x}$.
   - Interval estimate with probability $(1 - \alpha)$: $\bar{x} \pm z_{\alpha/2}\sigma/\sqrt{n}$.
   - Remarks: σ must be known. Normal distribution required for $n < 30$ (approximately). For $n \geq 30$, distribution of x need not be normal, also $s \simeq \sigma$.

2. Normal process or approximation. Estimate of $\mu \equiv A$; σ^2 unknown.

- Using: $\bar{x}, s^2$.
- Mean value: $\bar{x}$.
- Interval estimate with probability $(1 - \alpha)$: $\bar{x} \pm t_{\alpha/2} s / \sqrt{n}$.
- Remarks: σ unknown, estimated by s. Normal distribution assumed. For $n \geq 30$, tends to case above with $s \simeq \sigma$. Parameter for t-distribution $\nu = (n - 1)$ degrees of freedom (d.f.).

3. Two normal processes or approximations. Estimate of $\mu_1 - \mu_2 \equiv A$; σ_1^2 and σ_2^2 known.

- Using: $\bar{x}_1, \bar{x}_2$.
- Mean value: $\bar{x}_1 - \bar{x}_2$.
- Interval estimate with probability $(1 - \alpha)$: $(\bar{x}_1 - \bar{x}_2) \pm z_{\alpha/2} \sqrt{\dfrac{\sigma_1^2}{n_1} + \dfrac{\sigma_2^2}{n_2}}$.
- Remarks: σ_1, σ_2 known. Normal distribution assumed for $n_1, n_2 < 30$. For $n_1, n_2 \geq 30$, distribution need not be normal; also, then $s_1 \simeq \sigma_1, s_2 \simeq \sigma_2$.

4. Two normal processes or approximations. Estimate of $\mu_1 - \mu_2 \equiv A$; pooled estimate of variance.

- Using: $\bar{x}_1, \bar{x}_2, s_p^2$ based on s_1^2, s_2^2.
- Mean value: $\bar{x}_1 - \bar{x}_2$.
- Interval estimate with probability $(1 - \alpha)$: $(\bar{x}_1 - \bar{x}_2) \pm t_{\alpha/2} s_p \sqrt{\dfrac{1}{n_1} + \dfrac{1}{n_2}}$.
- $s_p^2 = \dfrac{(n_1 - 1)s_1^2 + (n_2 - 1)s_2^2}{n_1 + n_2 - 2}$.
- Remarks: σ_1, σ_2 unknown, but it is judged that $\sigma_1^2 \simeq \sigma_2^2$. For $n_1, n_2 < 30$, normal distributions for both variables are assumed. $t_{\alpha/2}$ with $\nu = n_1 + n_2 - 2$ d.f. $s_p = $ pooled estimate of variance.

5. Normal process or approximation. Estimate of $\mu_1 - \mu_2 \equiv A$; estimate of variance not pooled.

- Using: $\bar{x}_1, \bar{x}, s_1^2, s_2^2$.
- Mean value: $\bar{x}_1 - \bar{x}_2$.
- Interval estimate with probability $(1 - \alpha)$: $(\bar{x}_1 - \bar{x}_2) \pm t_{\alpha/2} \sqrt{\dfrac{s_1^2}{n_1} + \dfrac{s_2^2}{n_2}}$.
- $\nu = \dfrac{(s_1^2/n_1 + s_2^2/n_2)^2}{[(s_1^2/n_1)^2/(n_1 - 1)] + [(s_2^2/n_2)^2/(n_2 - 1)]}$ degrees of freedom.
- Remarks: σ_1, σ_2 unknown. Cannot assume $\sigma_1^2 = \sigma_2^2$ – use hypothesis test and F distribution to check. Normality has to be assumed for $n_1, n_2 < 30$.

6. Normal process or approximation for paired data (independence of pairs given mean and variance cannot be assumed; see comments below). Estimate of $\mu_D \equiv A$, where $D_i = X_{1i} - X_{2i}$.

- Using: $\bar{d}, s_d$.
- Mean value: $\bar{d}$.
- Interval estimate with probability $(1 - \alpha)$: $\bar{d} \pm t_{\alpha/2} s_d / \sqrt{n}$.
- Remarks: independence cannot be assumed: knowledge of one member of a pair causes us to change estimate of the other.

7. Estimate of $\sigma^2 \equiv A$; μ unknown.

- Using: $\bar{x}, s^2$.
- Mean value: classically s^2 but see Exercise 8.24.
- Interval estimate with probability $(1 - \alpha)$: $\dfrac{(n-1)s^2}{\chi^2_{\alpha/2}} < A < \dfrac{(n-1)s^2}{\chi^2_{1-\alpha/2}}$.
- Remarks: normal distribution of Xs assumed. χ^2 values from distribution with $(n-1)$ d.f.

8. Estimate of $\sigma_1^2 / \sigma_2^2 \equiv A$.

- Using: s_1^2, s_2^2.
- Mean value: classically s_1^2 / s_2^2.
- Interval estimate with probability $(1 - \alpha)$: $\dfrac{s_1^2}{s_2^2} \dfrac{1}{f_{\alpha/2}(\nu_1, \nu_2)} < A < \dfrac{s_1^2}{s_2^2} f_{\alpha/2}(\nu_2, \nu_1)$.
- Remarks: normal distribution for both samples assumed, $\nu_1 = n_1 - 1$, $\nu_2 = n_2 - 1$ d.f.

9. Estimate of proportion $p \equiv A$, for large samples.

- Using: $\bar{p} = x/n$.
- Mean value: $\bar{p}$.
- Interval estimate with probability $(1 - \alpha)$: $\bar{p} \pm z_{\alpha/2} \sqrt{\dfrac{\bar{p}\bar{q}}{n}}$.
- Remarks: $np, nq > 10$.

10. Estimate of difference in proportion $p_1 - p_2 \equiv A$, for large samples.

- Using: $\bar{p}_1 - \bar{p}_2 = x_1/n_1 - x_2/n_2$.
- Mean value: $\bar{p}_1 - \bar{p}_2$.
- Interval estimate with probability $(1 - \alpha)$: $\bar{p}_1 - \bar{p}_2 \pm z_{\alpha/2} \sqrt{\dfrac{\bar{p}_1\bar{q}_1}{n_1} + \dfrac{\bar{p}_2\bar{q}_2}{n_2}}$.
- Remarks: $np, nq > 10$.

The list just given gives the commonly used tests but there are others.

Example 8.22 Suppose that an engineer is considering two kinds of cable for use in an offshore operation, say from manufacturers A and B. There is a suspicion that manufacturer B is using an inferior grade of steel in the wires, and tests are conducted to determine whether this is a reasonable conclusion. Strands of wire taken from cables from each manufacturer are removed and tested. The values given now are in MN. There are 16 tests in each group; previous testing indicated that the assumption of normal distribution of strength is reasonable. It is desired first to test whether the variances are the same, and then to test the means.

	A	B
Mean (MN)	3456	3405
Variance (MN)^2	1190	3591

The tests will be conducted consecutively.

Step 1 $H_0 : \sigma_1^2$ is reasonably close to σ_2^2.

Step 2 $H_1 : \sigma_1^2$ is not reasonably close to σ_2^2.

Step 3 Specify the level of significance; $\alpha = 0.1$, $\frac{\alpha}{2} = 0.05$ (two-tailed).

Step 4 Critical region for rejection of H_0; $F > 2.40$, $F < 1/2.40 = 0.417$. F is the variance ratio; $f = s_1^2/s_2^2$.

Step 5 $f = 1190/3591 = 0.332$.

Step 6 Reject H_0, accept H_1.

Variances should not be pooled; case 5 above. The test on the means will now be conducted.

Step 1 $H_0 : \mu_1$ is reasonably close to μ_2.

Step 2 $H_1 : \mu_1$ is not reasonably close to μ_2.

Step 3 Specify the level of significance; $\alpha = 0.05$, $\frac{\alpha}{2} = 0.025$ (two-tailed).

Step 4 Critical region for rejection of H_0; $T < -2.064$, $T > 2.064$, where T is the student-t random quantity, and the degrees of freedom are

$$\nu = \frac{\left(\frac{1190}{16} + \frac{3591}{16}\right)^2}{\frac{1}{15}\left[\left(\frac{1190}{16}\right)^2 + \left(\frac{3591}{16}\right)^2\right]} = 23.96, \text{ say } 24.$$

Step 5 $t = \frac{51.0-0}{\sqrt{\frac{1190}{16} + \frac{3591}{16}}} = 2.95.$

Step 6 Reject H_0, accept H_1.

Of course, all the definite-sounding statements, 'accept', 'reject', and so on, cloak the fact that we are exercising judgment in the procedure above. It certainly appears unlikely that the test results A and B are different by 'pure chance'. In a scientific report, it is advisable to report all the statistics and probabilistic reasoning, with conclusions. Scientific background such as the chemical composition of the two steels is highly relevant. In the example above, we might decide to choose the 'A' cables but if the 'B' cables were a lot less expensive, and the strength nevertheless satisfactory, we might opt for this solution.

Paired observations

Sometimes there will be measurements taken on the same object or person, before and after a certain treatment. For example, the competence of group of persons at a particular skill (word processing, programming, playing squash, ...) might be tested before and after a training course. Or a set of engineered objects might be tested with and without a new treatment. The appropriate way to study these problems is to consider the *differences* between the pairs of readings on the same person, object, ..., as in case 6 in the list of common classical inference problems above. In this way, the effect of the treatment is not masked by the variation from one pair to another. For instance, an above-average performer will have higher-than-average readings compared to the rest of the group before and after the treatment. Contrast Exercises 8.30 and 8.32 below.

Consideration of *p*-values

One way to get a better appreciation of the results of a hypothesis test is to work out the '*p*-value'. For example, in the comparison of means in the second part of Example 8.22, we had the experimental value of t equal to 2.95. The critical value of 2.064 corresponded to $\alpha/2 = 0.025$ with $T > 2.064$ corresponding to rejection of H_0. We can also work out that $T > 2.95$ corresponds to $p = 0.0035$. This gives us a good idea of how unlikely is the result 2.95 or greater.

Concluding comment

The set of common methods in this section is not exhaustive and has only been outlined. Further detail can be found in many standard text books on the subject. See Exercise 8.27.

8.8   Partial exchangeability

We have discussed the use of exchangeability extensively in this chapter. This was implemented by means of Bayes' theorem. But Bayes' theorem by no means depends on this assumption. For example, consider Equations (8.76) to (8.78), in which we made inferences regarding the mean of the normal distribution (variance known). The updated distribution of the mean A was found to be

$$f_A^{(u)}(a) = \frac{1}{\sqrt{2\pi}\sigma_f} \exp\left[-\frac{1}{2}\left(\frac{a - \mu_f}{\sigma_f}\right)^2\right],$$

where

$$\mu_f = \frac{\bar{x}n\sigma^{-2} + \mu_0\sigma_0^{-2}}{n\sigma^{-2} + \sigma_0^{-2}},$$

and

$$\sigma_f^{-2} = n\sigma^{-2} + \sigma_0^{-2}.$$

Let us assume that there was only one reading x ($n = 1$), and that the prior information had corresponded to a single reading, x_0, on a device with standard deviation σ_0. Then

$$\mu_f = \frac{x\sigma^{-2} + x_0\sigma_0^{-2}}{\sigma^{-2} + \sigma_0^{-2}}$$

and

$$\sigma_f^{-2} = \sigma^{-2} + \sigma_0^{-2}.$$

The values $\{x_0, x\}$ are not exchangeable since they correspond to devices (say A and B) with different variances (σ_0^2 and σ^2 respectively). We have **partial exchangeability** if any set of readings **within** each group (A or B) are exchangeable. If we have n_0 and n_1 readings in each group respectively, then the weights in the above equations change from σ_0^{-2} and σ^{-2} to $n_0\sigma_0^{-2}$ and $n_1\sigma^{-2}$. We cannot exchange readings from one group to another since their variances are different.

In partial exchangeability, there are sets of circumstances which are perceived to be different, and these form groups. Readings are then taken within the groups. For example, the age of individuals in a set of measurements of the response to a certain drug might be considered a relevant piece of information, as might be the water content of concretes in a set of readings, or the location in measuring wave heights. An interesting account is given in de Finetti (1972), and we shall give a very brief outline of the reasoning. We consider k groups, for example age interval, water–cement ratio of the concretes or location, in the examples just given. There are n readings in each of the k groups; for group j the set of readings are

$$\{x_{1j}, x_{2j}, x_{3j}, \ldots, x_{nj}\}.$$

We then have exchangeability within each group. These could be treated separately, as in the example of readings with different standard deviations above, but there is the possibility of making inferences from one exchangeable group to another. For example, we might know the standard deviation of a certain steel, and be able to use the value to characterize the standard deviation of another steel. Often this kind of inference is done quite naturally by engineers, but at the same time, the use of the formal structure of partial exchangeability is potentially of great usefulness.

8.9   Concluding remarks

It is very common to have a set of measurements taken from an experiment, in which we judge that the order of the results is not important. It is commonly stated that the results are 'independent'. This is inappropriate if inferences are to be made using the data. Certainly when it comes to parameter estimation, the parameter and consequently the probabilities that ensue are highly dependent on the data. The judgement that the order does not matter is much simpler and appropriate. This constitutes exchangeability. The use of the concept of exchangeability then provides a sound basis for problems of inference. Bayesian methods provide a coherent framework within which inferences regarding parameters can be made.

Prior probabilities can often be represented by means of the conjugate families discussed in this chapter, but this is a particular choice that turns out to be convenient in terms of mathematical tractability. The case of reference priors, expressing vague prior opinion, in the common cases investigated, turns out to be a special (limiting) case of the conjugate families. These form a useful connection with inference based principally on the data, which we termed 'classical' inference. There is a desire amongst some analysts to achieve a Bayesian 'holy grail', in which all inferences follow automatically from a basic principle, including the choice of prior distribution. This rather contradicts the spirit of the Bayesian attitude, in which proper expression of an individual's opinion – based on available information – is a cornerstone of the approach. At the same time, there is no assurance that a conjugate prior distribution exists. It is most important that numerical methods for composing prior distributions and updating them are developed and become generally available.

The classical method, in which randomness regarding the parameter under investigation is not directly confronted, results in the parameter often being modelled by a single value, but the uncertainty surrounding this value is subsequently acknowledged by the use of 'confidence intervals' and related devices. Rational decisions should be derived from the **entire** probability distribution (see Chapter 4), which can be included in the Bayesian approach: then the development of simplified rules can be investigated from a better viewpoint. There may be cases where the posterior distribution is based on a large data set such that its distribution is a spike at a certain value. In this case, the classical inference and the Bayesian are very close to each other. Hypothesis testing must be approached with care.

We have dealt with only a few distributions in this chapter to illustrate the approach and to set a framework in which the various 'estimators' in the literature can be understood. Classical estimators are generally satisfactory in cases with little or no prior information; when combined with confidence intervals, the analysis is close to the one advocated here. It is but a small step in thinking to consider the whole distribution rather than just the intervals. Once non-informative reference priors are understood, then it is acceptable to use the classical approach where appropriate.

We take the approach that the probabilistic analyst has sufficient knowledge of the subject under consideration to assign a prior distribution, and that energy will not excessively be consumed in debates as to whether a uniform distribution should be assigned to the parameter A or to A^2, or to A^3, and so on. In fact, if the data dominates, there are quite simple rules to recapture the classical result, plus an infinity of others. At the same time there is a lot of interesting work on the subject of 'non-informative priors', invariance under parameter transformation, expression of complete ignorance and related subjects. The work of Jeffreys and Jaynes is particularly important. A further way to consider the reference priors is to consider 'data translation' (Box and Tiao, 1973). In this method, it is shown that the likelihood, if plotted in the appropriate metric (e.g. ln A for a scale parameter), translates without change of shape.

But the search for a Bayesian 'holy grail' is likely to be a vain search. We should learn instead to trust and use our judgement!

The parameters in a distribution give numerical expression to probabilities, and it could be argued that we are assessing 'probabilities of probabilities', or at least coming perilously close to this. The most appropriate way to consider the various values of the parameter is as corresponding to a set of 'hypotheses' or 'scenarios'. The parameter value is conditional on these scenarios, but we are uncertain regarding which scenario is true. In a simple piece of analysis, de Finetti (1977) showed that these hypotheses are arbitrary; in other words there are many possible sets of hypotheses that can be used in any particular case. This is as it should be.

8.10 Exercises

8.1 In a problem involving the binomial distribution, the uniform distribution is judged an appropriate measure of prior knowledge, i.e. $f_A^{(i)}(a) = 1, 0 \leq A \leq 1$. Ten observations of the process are made with x successes. Plot the *pdf* of the posterior distribution for all possible results, i.e. with $x = 0, 1, \ldots, 10$ successes.

8.2 Show that (8.32) reduces to (8.18) for large r', n'.

8.3 Show that $\omega_{x,n}$ in the case of Bayes' scholium is equal to $1/(n + 1)$ by integrating (8.43) with a uniform prior distribution. Also use (8.32) to verify this.

8.4 The voltage that can be carried by an integrated circuit is known to be normally distributed with a standard deviation of 0.10 volts. This is the 'system capacity' (see also Chapters 9 and 10). The mean capacity is to be studied in an application. Past measurements lead to the conclusion that the mean capacity is normally distributed ($\mu_0 = 5.50$ volts, $\sigma_0 = 0.05$ volts). Six tests are undertaken on prototype circuits, with the results of the measurements giving that $\bar{x} = 5.80$ volts.

(a) Update the distribution of the mean capacity using the prior information and the results of the measurements.
(b) Find the 99% interval estimate based on the Bayesian distribution found in (a). (This is the so-called 'Bayesian interval estimate'.)
(c) Contrast the result of (b) with the classical confidence intervals. Sketch all distributions.
(d) How many measurements need to be taken to reduce the standard deviation of the mean to 0.01? Use the result of (a) as the basis of calculation.

8.5 Assume now that the measurements on the prototype circuits in Exercise 8.4 have not been taken. It is being decided whether or not to take these measurements. Voltage surges occur in the circuit in practical applications. The maximum of these in a time period under consideration is termed the 'demand', which is to be compared to the capacity of Exercise 8.4. If demand exceeds capacity, failure occurs.

The basic question to be decided is whether to include the circuits in a design to be marketed. Draw a decision tree that models the problem. Include the option 'not to experiment' versus 'experiment with n prototype circuits'. Note on the decision tree the following consequences in units of utility:

design accepted and survival: 10 units;
design accepted and failure: -10 units;
design rejected: -3 units;
cost of testing: 0.5 units per circuit tested.

The failure probability can be worked out (in closed form for two normal or two lognormal distributions – see Chapter 10). This could be done with representative values of the parameters, and the problem solved fully. In the calculation of failure probability, a future value of capacity is required (see Exercise 8.6).

8.6 *Future values.* In many problems, we are interested in a future value. For example, we might have estimates of the parameters for a batch of steel specimens. Rather than the distribution of the parameter 'mean', we might wish to estimate the uncertainty in a future single value of strength, corresponding to a particular specimen. Describe how this might be accomplished.

8.7 Show how to obtain (8.89) from the prior distribution and the data.

8.8 Show that W in

$$f_{W'}\left(w'\right) = \frac{\left(w'\right)^{\frac{v_{\text{ref}}}{2}-1} \exp\left(-\frac{w'}{2}\right)}{2^{\frac{v_{\text{ref}}}{2}} \Gamma\left(\frac{v_{\text{ref}}}{2}\right)}, \tag{8.171}$$

where $W' = W/w_{\text{ref}}$ has a gamma distribution of the form (8.82) with parameters $\alpha = v/2$, and $\beta = 2w_{\text{ref}}$. The subscript 'ref' refers to the parameter of either the prior or posterior distribution (subscripts '0' or 'f' respectively). The chi-square random quantities in Equations (8.85) and (8.89) are of the form (8.171) and the result shows how to model W using either the chi-square or gamma forms.

8.9 Study the meaning of 'data-translated likelihood' with regard to non-informative prior distributions. See for example Box and Tiao (1973, p. 26ff).

8.10 *Geometric interpretations.*

(a) Show that Equation (8.105) represents the distance from the origin to the centre of mass of the combined masses n_0 and n.
(b) Show that Equation (8.107) represents the second moment of the distributed masses about the centre in (a). We have assumed, in writing $v_0 = n_0 - 1$, that the prior information is in the form of, or analogous to, a set of data with n_0 members. If this is not the case, one can think of the parameters n_0 and $(v_0\sigma_0^2)$ as the mass and moment of inertia of the prior information, without there being necessarily any link between v_0 and n_0.

8.11 Find the constant in the proportionality relationships (8.102) and (8.103).

8.12 It is very interesting to work through a problem with unknown mean and variance for a normally distributed process, Sections 8.5.3 and 8.6.5, along the lines of Example 8.17. It is suggested that a problem be composed in the reader's area of interest. It is instructive also to try various prior distributions including those that are non-informative. It is suggested to carry out the following.

(a) Use the mechanical analogy of Exercise 8.10.
(b) Plot the prior and posterior distributions; compare these to the cases where the mean and variance are known.
(c) Study in particular the change in shape of posterior distributions given first, vague prior information with 'sharp' data, and second, sharp prior information and vague data.

8.13 Plot the t-distribution for various degrees of freedom. Compare with the normal distribution.

8.14 Verify (8.120).

8.15 The occurrence of flaws in solid bodies can be modelled using the Poisson distribution, using length, area (see Exercise 3.8) or volume as the 'exposure' parameter. This can be used in failure analysis; for example the safety of steel pressure vessels in nuclear reactors, pipelines, ships, offshore rigs, and in aircraft.

(a) The occurrence of flaws in a welded joint is modelled as a Poisson process with a mean occurrence rate of Λ flaws per metre of weld. Prior information has show that this parameter has a mean of 0.5 flaws per metre with a coefficient of variation of 40%. Model this as a prior density using the appropriate conjugate family. Update the distribution using measurements with a reliable device indicating five flaws in a weld of length 9.2 metres. Calculate the posterior mean, variance and coefficient of variation.
(b) Structure a decision problem – such as that in (a) – using the Poisson process involving initial information regarding the expected value or rate of the process, which might be updated using Bayesian methods. Show decision trees in which the decision 'whether to experiment' is included (preposterior analysis). The problem could be concerned with events other than flaws in a material; arrivals of messages or packets of information could be considered, or any of the numerous uses of the Poisson process.

8.16 Show that the reference prior (8.127) for the scale parameter, applied to the standard deviation, also applies to the variance and to the weight W.

8.17 What is the reference prior for Example 8.18? State how would you interpret the prior

$$f_\Lambda (\lambda) = \frac{1}{\bar{\lambda}} \exp \left(-\frac{\lambda}{\bar{\lambda}} \right). \tag{8.172}$$

Is there a maximum-entropy interpretation of this prior distribution?

8.18 *Linear loss problems.* In problems of decision, there is often a linear loss associated with the parameter in the probability distribution. Raiffa and Schlaifer (1961) provide substantial background to this problem. For example, these authors describe a manufacturing process in which objects are mass-produced in lots of size n. The proportion of defectives is a binomial process with parameter

A. The lots can be sent directly for assembly, in which case the defective parts have to be replaced at a cost c_1 each. Or the lots can be screened at cost c_2 per lot and the defectives replaced at cost $c_3 < c_1$.

(a) Write the utility functions in terms of cost for the two strategies, assuming utility to be linear with money, in terms of the parameter $A = a$. Find the break-even point for a in terms of the constants.

(b) How would you take into account randomness in A? Research the linear loss integrals, integrated over two parts of the parameter space, as presented by Raiffa and Schlaifer. How could they be applied in case (a)?

(c) Sketch a Bayesian approach to the problem of determining the parameter A at minimum cost.

(d) Suggest conditions under which the cost of testing and experimentation is additive to the utility and independent of the current value of utility.

(e) Propose an application of the theory above in an area of interest to you. Note that Raiffa and Schlaifer have developed linear loss integrals for a variety of processes other than the binomial one.

8.19 Verify Equations (8.163) and (8.164).

8.20 Find 90% confidence intervals for the parameter W and for the corresponding standard deviation in Example 8.16.

8.21 Show that a random quantity that is the square of a standard normal distribution is chi-square with one degree of freedom.

8.22 We know (Example 7.21) that the gamma family is closed under convolution. Show that the distribution of χ^2 as given by Equation (8.166) is chi-square with n degrees of freedom.

8.23 Why in the classical approach are there $(n - 1)$ degrees of freedom in the chi-square distribution of (8.168) using (8.167)?

8.24 *Fly in the ointment: inverted gamma distribution.* Use the notation of Exercise 8.8. Make the transformation

$$Y = \frac{1}{W}, \tag{8.173}$$

where Y models the variance, to show that the distribution of Y follows the 'inverted gamma' (or 'inverted chi-square') distribution. Show that

$$\langle Y \rangle = \frac{\sigma_{\mathrm{ref}}^2 \nu_{\mathrm{ref}}}{\nu_{\mathrm{ref}} - 2}, \tag{8.174}$$

which differs from the 'unbiased' result of (8.170).

8.25 Write notes expressing your opinion on estimation from the point of view of decision-making, considering point estimates, centroids (mean values) and bias. We gave examples in Section 8.7.3 of cases where bias in the estimate is a desirable property of the estimate, and in particular an overestimate. Give further examples of this situation and give an example of the case where we might wish to underestimate.

8.26 Write notes expressing your opinion of the best estimator for a binomial process. Consider cases where there is strong and weak prior information. In particular, discuss the values r_0 and n_0 in (8.33):

$$\langle A \rangle = \frac{r + r_0}{n_1 + n_0}.$$

8.27 Research basis for members of the set of 'common classical inference problems', Section 8.7.6 above.

The following consist of a set of classical inference problems. Numerous others can be found in the many texts expounding classical methods, and these might be solved too. They contain a lot of useful thought on decisions: it is important to ponder the relevance of the level of significance α and the consequences of actions. It is suggested that decision trees be drawn in each case. Possible prior opinion should also be conjectured.

8.28 *Classical inference 1.* A manufacturer claims that the mean strength of a new type of strand for cable manufacture is more than 20 kN. The distribution of actual strengths is known to be normal to a good approximation, with a standard deviation of 1.4 kN. A random sample of five of the new cables has a mean of 22.5 kN.

(a) Is there sufficient evidence to support the claim?
(b) Now suppose that the standard deviation σ is unknown and that the measured standard deviation of the random sample is $s = 1.4$ kN. Is there sufficient evidence to support the claim?

8.29 *Classical inference 2.* Measurements of the output of power from a generator are known to be normally distributed with a standard deviation of $\sigma = 2.89$ kW. Find the smallest size of sample needed so that the mean of the 'random sample' lies within 0.10 kW of the unknown mean of the distribution 95% of the time (treat the sample as future unknown values).

8.30 *Classical inference 3.* A boat manufacturer claims that a new type of hull increases the average speed of a speedboat by more than 2 km/h over that of a competitor's hull design. Eleven speedboats, six with the new design and five with the old, are tested and their speeds (in km/h) are measured on a test course under identical environmental conditions.

New design (X_A): 36 42 38 36 37 39.
Old design (X_B): 31 33 30 37 34.

Conduct an appropriate hypothesis test at a level of significance of 5%. Is there sufficient evidence to accept the designer's claim? What assumptions have you made?

8.31 *Classical inference 4.* Samples are drawn at random from two separate populations, with no members in common, resulting in the following summary statistics:

$X : n = 50, \bar{x} = 643, s_X = 26$;
$Y : n = 90, \bar{y} = 651, s_Y = 32$.

Are these data consistent with the hypothesis that the two population means are equal? Calculate the p-value and comment.

8.32 *Classical inference 5.* Measurements of the torque output from an electric motor have been found to follow the normal distribution. The output is governed by a controller, and a new controller is being tested. Seven different motors are chosen at random and subjected to tests with the new and old controllers giving the following results. Carry out a hypothesis test on the null hypothesis that the torques using the new method are on average at least 2 units less for the new method, as compared to the old.

Motor	1	2	3	4	5	6	7
New method	6.01	3.16	4.43	5.25	6.12	5.75	2.21
Old method	8.88	6.22	7.46	7.83	8.83	8.19	5.64

See also Exercise 11.10 for the use of linear regression in this kind of analysis.

8.33 *Classical inference 6.* In the manufacture of a member carrying tension, two types of alloy are tested. It was found that 22 out of 40 specimens made of alloy A had strengths greater than a specified value, whereas 29 specimens out of 42 made with alloy B had strengths greater than the specified value. Do a test at the 5% level of significance to examine whether the difference between the two proportions is significant.

8.34 An engineer is studying the fatigue life, Y, in cycles of stress, of a steel connector device. It is convenient to consider the quantity

$X = \log Y$.

Previous measurements have shown that X is a normally distributed random quantity with a standard deviation of 0.20. The mean A of this distribution is the subject of study in the present investigation. Past experience has resulted in the assignment of a normal distribution to A with a mean of 5.80 and a standard deviation of 0.10. The following ten measurements have subsequently been taken of X:

5.61, 5.24, 5.12, 5.40, 5.14, 5.38, 5.40, 5.46, 5.41, 5.67.

Obtain the posterior distribution of A, and sketch the prior and posterior distributions. What is the name for the distribution of Y?

9   Extremes

The longer you fish, the bigger the fish you catch.

Old fisherperson's saying

9.1   Objectives

In the design of engineered systems, special account must be taken of the **extreme** conditions to which the system is exposed during its use. We wish our structure to resist the greatest load applied to it during its lifetime; telephone and computer systems should be able to serve extreme demands; biological systems, to survive over time, must be able to cope with the extreme environmental conditions – including those induced by humans – to which they are exposed; our manufacturing plant should, for maximum profit, be able to meet extreme demands at certain periods of its life. Other engineered systems include pipelines, transportation networks, the spillway on a dam, or a transmission line carrying electrical supply. Such systems will generally have a **demand** and a **capacity**. For example, the flow of water over a spillway, the loading applied to a structure, or the number of customers or users of a system, constitute demands on the system. Capacity represents the ability of the system to cope with the demand, for example the capacity of a spillway, resistance of the structural system, or the number of users that the system can serve.

Figure 9.1 shows schematically the demand and capacity, both being treated as random. The 'overlap' area is identified, in which there is a possibility of failure, in other words where demand can exceed capacity. The inability of the system to meet the demand is denoted as **system failure**. The calculation of failure probability is dealt with in Chapter 10 on Risk, safety and reliability: it is not equal to the overlap area but requires a convolution integral. Here, we wish to concentrate on aspects of the probability distributions for demand and capacity that require extremal analysis. In the case of capacity, we may wish to obtain the probability distribution of the smallest value, for example the weakest link in a chain, or the shortest path in a set of network links with random lengths of path, or the weakest cell in cancer research. In these examples, we are interested in smallest value of a set, rather than the largest, as is the case when considering demand.

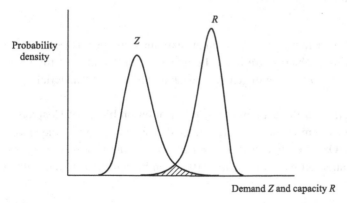

Figure 9.1   Possible exceedance of capacity by demand

The key point in the preceding is that often we wish to consider the **maximum** or **minimum** of a set of values. For example, an offshore installation has to resist wave loads generated by storms. For the period of time under consideration, we wish to obtain the distribution of the maximum of all the wave heights. Let the individual wave heights in a storm be represented by

$$\{X_1, X_2, X_3, \ldots, X_i, \ldots, X_n\},$$

which are random (usually future) values. The number of waves in a storm is n. Then we wish to find the distribution of the **extreme value** Z:

$$Z = \max(X_1, X_2, X_3, \ldots, X_i, \ldots, X_n),$$
$$= \max_n X_i. \tag{9.1}$$

Since the X_i are random, so is Z. It is important to note that **time** enters Equation (9.1) via the time period during which the storm lasts. The longer the time period, the greater is n, and as we shall see, this will affect the distribution of Z. We also need to consider the number of storms in our period of interest (for example, a year); see Example 9.13 in Section 9.8. Not all problems of extremes involve time, but those involving safety of human life generally do, and it is important to understand the manner in which time enters the problem.

A more general notion that includes time as a special case is that of '**exposure**'. An example where time is not involved is the 'weakest link' problem, mentioned above. In this, we have a chain (for example, the anchor chain in an offshore installation) with n links. The number of links in the chain represent the exposure in this case. The strength of each link is random, and, for the n links, the individual values are denoted again as

$$\{X_1, X_2, X_3, \ldots, X_i, \ldots, X_n\}, \tag{9.2}$$

for each of the $i = 1, \ldots, n$ links of the chain. The chain fails when the weakest link fails. We denote this value as W. Thus

$$W = \min(X_1, X_2, X_3, \ldots, X_i, \ldots, X_n)$$
$$= \min_n X_i. \tag{9.3}$$

This example shows that we may be interested in the extreme minimum as well as the extreme maximum in engineering problems. We might be interested in the shortest time to transmit information through a network. We will generally wish to purchase raw materials (or stocks!) at the minimum price.

It is noteworthy that, as in the deterministic optimization problems of Chapters 5 and 6, finding the extreme maximum is equivalent to a related problem of finding the extreme minimum, and *vice versa*. One merely considers the negatives of the values initially given. For example, the minimization problem of Equation (9.3) can be converted to the following one of maximization:

$$-W = \max(-X_1, -X_2, \ldots, -X_i, \ldots, -X_n). \tag{9.4}$$

The definitive work on extremes was written by Gumbel (1958), which represents a marvellous store of information. Galambos (1978), David (1981) and Leadbetter, Lindgren and Rootzén (1983) provide mathematical treatments while Castillo (1988) and Ochi (1990) focus on engineering applications. Maes (1985) is an excellent store of information for environmental extremes.

9.2   Independent and identically distributed (*iid*) random quantities; de Méré's problem

An assumption that we have encountered many times is that of independent and identically distributed (*iid*) random quantities. This applies if we have a set of a random quantities as in (9.2),

$$\{X_1, X_2, X_3, \ldots, X_i, \ldots, X_n\},$$

such that we judge the same distribution to apply to each X_i, which then have the same *cdf* $F_X(x)$. The random quantities are further judged to be stochastically independent – knowledge of one value does not lead us to change the probability distribution of another. Then the random quantities are *iid*, for example in die or coin tossing; the particles following the energy distributions of Chapter 6 were *iid*; this assumption formed the basis of the Central Limit Theorem in Chapter 7; it was a special case of exchangeability in Chapter 8. Using it, we can deduce particularly simple solutions.

Considering Equation (9.1), and the assumption that the random quantities X_i are *iid*,

$$F_Z(z) = \mathbf{Pr}\,(\text{all } X_i \leq z) = F_X^n(z). \tag{9.5}$$

Similarly, for *iid* random quantities, considering Equation (9.3),

$$F_W(w) = 1 - \mathbf{Pr}\,(\text{all } X_i > w) = 1 - [1 - F_X(w)]^n. \tag{9.6}$$

For continuous random quantities, we can obtain the probability density functions corresponding to these two equations by differentiation; respectively,

$$f_Z(z) = n F_X^{n-1}(z) f_X\,(z) \tag{9.7}$$

and

$$f_W(w) = n\,[1 - F_X(w)]^{n-1}\,f_X\,(w). \tag{9.8}$$

Example 9.1 We now consider an introductory example. Various two-dice problems have been used to discuss probabilistic issues, for example in Sections 4.3 and 4.6. For the present application, an interesting problem related to gambling was posed by the French nobleman, Antoine Gombauld, Chevalier de Méré. He consulted Pascal about it; Pascal wrote to Fermat, and the result was the famous Pascal–Fermat correspondence on the subject. The problem concerned gambling with dice, which subject interested de Méré considerably. In tossing a

die of six sides, the die must be tossed four times in order to achieve a six with a probability greater than one half. In gambling with two dice, it seemed natural to extrapolate from this result; in order to obtain a double-six, multiply the result of four by the ratio of the number of possible results (i.e. by 36/6). The answer, 24, is then supposed to indicate the number of throws required to achieve the double-six. De Méré found, probably by trial in gambling, that this logic did not work.

Let us solve the single-die problem first. The probability of obtaining no sixes in four throws is $(5/6)^4 = 0.482\ldots$ (remember *iid*!). Thus the probability of obtaining at least one six is the complement of this, $0.518\ldots$, which is greater than one half, and therefore a good long-run bet given even stakes. The corresponding probability of obtaining no double-sixes in 24 tosses is $(35/36)^{24} = 0.508\ldots \simeq 0.509$. The probability of obtaining at least one double-six is thus $0.491\ldots$, and we can see why this is not a money-making proposition for bets with even stakes. For the cases just discussed, we can give the flavour of extremal analysis by making the analogy that obtaining a six or a double-six could correspond to the event that the number of telephone calls at an exchange exceeds a certain number, or that the flood level in a river exceeds a specified value, with a probability of $1/6$ or $1/36$, respectively, on an annual basis. Then the probabilities of the number of telephone calls, or of floods exceeding the specified value at least once in four, and in 24, years is 0.518 and 0.491, respectively.

Formally, we may denote the outcome of the ith toss of a single die by the random quantity X_i, where each X_i may be one of the numbers $\{1, 2, 3, 4, 5, 6\}$ with equal probability. For two dice, we consider the sum of the two numbers, which can take the values $\{2, \ldots, 12\}$. The double-six corresponds to the value 12, and we denote the outcome again by the symbol X_i. Then, following the notation of the preceding section, the possible values in four and 24 tosses are $\{X_1, \ldots, X_4\}$ and $\{X_1, \ldots, X_{24}\}$ respectively. In a single toss, for any X_i,

$$F_X(5) = \mathbf{Pr}\,(X \le 5) = 5/6,$$

for the single die, and for the double-six,

$$F_X(11) = \mathbf{Pr}\,(X \leq 11) = 35/36.$$

Let the maximum value in a set of tosses be Z;

$$Z = \max\,(X_1, \ldots, X_4)\,, \text{ and}$$
$$Z = \max\,(X_1, \ldots, X_{24})\,,$$

for the two cases respectively. For there to be no sixes or double-sixes, respectively, each $Z \leq z = 5$, and $Z \leq z = 11$, and

$$F_Z(z) = \mathbf{Pr}\,(\text{all } X_i \leq z) = F_X^n(z) = \left(\frac{5}{6}\right)^4, \text{ or } \left(\frac{35}{36}\right)^{24},$$

for the two cases, again respectively, from Equation (9.5). The solution then follows, as discussed in Example 9.1 above.

9.3   Ordered random quantities; order 'statistics'

As in the case of inference, we distinguish between data, which are not random, and random quantities, which include future unmeasured values, the main subject of our uncertainty. The distinction can be represented as follows for the quantity X:

$$x_1, x_2, \ldots, x_{n_1}, X_{n_1+1}, X_{n_1+2}, \ldots, X_{n_1+n_2}.$$

The first n_1 lower case values represent data or measurements, and the next n_2 unknown (random) values, possibly in the future. We now concentrate attention on n of the second kind of quantities:

$$X_1, X_2, X_3, \ldots, X_i, \ldots, X_n. \tag{9.9}$$

These could be exchangeable or *iid* random quantities, or they could correspond to any other state of knowledge. We now imagine that the quantities are rearranged in order of increasing magnitude

$$X_{(1)}, X_{(2)}, X_{(3)}, \ldots, X_{(r)}, \ldots, X_{(n)}, \tag{9.10}$$

such that

$$X_{(1)} \leq X_{(2)} \leq X_{(3)} \leq \cdots \leq X_{(r)} \leq \cdots \leq X_{(n)}. \tag{9.11}$$

The subscripts in parentheses $(1), (2), (3), \ldots (r), \ldots, (n)$ therefore represent the 'order statistics', which is the term often used to describe data which are ranked by size. We shall use the term 'ordered random quantities' since we are not referring to data in the present context. We emphasize that we don't know which subscript i in the string (9.9) corresponds to which (r) in (9.10). The ordered random quantities are no longer exchangeable or *iid*; the position

in the set matters. We now wish to obtain details of the distribution of these ordered random quantities.

To do this, we assume first that the quantities under consideration in string (9.9) are judged to be *iid* with *cdf* $F_X(x)$. Further, we choose a particular value $X = x_{(r)}$, and we separate the ordered Xs of (9.10) into two groups, those less than or equal to $x_{(r)}$, r in number, and those greater than $x_{(r)}$, $(n - r)$ in number. We can at the same time construct an urn model, designating those less than or equal to $x_{(r)}$ as being red ('1' balls), and those greater than $x_{(r)}$ as being pink ('0' balls). Then the event $E_r = \{X_{(r)} \leq x_{(r)}\}$ corresponds to the drawing of a red ball. Since the distribution of the Xs is *iid*, drawing from the urn corresponds to a binomial process, with the probability of a red ball being $F_X\left(x_{(r)}\right)$.

The drawings result in $R = r$ red '1' balls and $n - r$ pink '0' ones. Then for a particular order of drawing, for example all red balls first, followed by all the pink ones,

$$\mathbf{Pr}(R = r) = p_{r,n} = F_X^r\left(x_{(r)}\right)\left[1 - F_X\left(x_{(r)}\right)\right]^{n-r}, \tag{9.12}$$

since we are dealing with a binomial process. Counting all orders,

$$\mathbf{Pr}(R = r) = \omega_{r,n} = \binom{n}{r} F_X^r\left(x_{(r)}\right)\left[1 - F_X\left(x_{(r)}\right)\right]^{n-r}; \tag{9.13}$$

see Equations (2.48) and (3.3).

We now wish to determine the probability that the rth ordered random quantity $X_{(r)}$ is less than or equal to $x_{(r)}$. This occurs if at least r of the n are less than or equal to $x_{(r)}$. For example, if $n = 10$ and $r = 8$, we are interested in the eighth in order, or the third largest value, $X_{(8)}$. If, for a particular $x_{(r)}$, there are eight, nine or ten values less or equal to than $x_{(r)}$, then the eighth value $X_{(8)}$ must be $\leq x_{(r)}$. As noted, there is a one-to-one correspondence between the events $E_r = \{X_{(r)} \leq x_{(r)}\}$ and the occurrence of red balls. It is possible then to construct the *cdf* of the rth ordered random quantity $X_{(r)}$:

$$F_{X_{(r)}}\left(x_{(r)}\right) = \sum_{t=r}^{n} \binom{n}{t} F_X^t\left(x_{(r)}\right)\left[1 - F_X\left(x_{(r)}\right)\right]^{n-t}. \tag{9.14}$$

We may differentiate Equation (9.14) to obtain

$$f_{X_{(r)}}\left(x_{(r)}\right) = \frac{n!}{(r-1)!\,(n-r)!} F_X^{r-1}\left(x_{(r)}\right)\left[1 - F_X\left(x_{(r)}\right)\right]^{n-r} f_X\left(x_{(r)}\right). \tag{9.15}$$

(See Exercise 9.1). Equation (9.15) may also be derived by observing that, if the rth quantity is in the interval from $x_{(r)}$ to $x_{(r)} + dx_{(r)}$, there are $(r - 1)$ values below $x_{(r)}$ and $(n - r)$ above $x_{(r)}$. Then, Equation (9.15), multiplied on each side by $dx_{(r)}$, represents the required probability that the rth quantity is in the interval noted.

We now have the basis for analysing the probability distributions of the central value, or the eighth out of ten, third largest, or other such quantities. This is useful in studying plotting positions, as we shall see in Chapter 11 (Section 11.4). For the remainder of this chapter we concentrate on the end points, with $r = 1$ or $r = n$, as was introduced in the preceding section, Equation (9.5) and in de Méré's problem. The first success in the vth trial is illustrated in Figure 9.2. We focus on *iid* random quantities; exchangeable ones are addressed in Section 9.9. Exercise 9.4 addresses the relationship between Equations (9.14) and (9.6).

Example 9.2 Equation (9.15) gives the distribution of the rth value from below out of n. Find the expression for the distribution of $_{(r)}X$, the rth value from above. The answer is

$$f_{_{(r)}X}\left(_{(r)}x\right) = \frac{n!}{(r-1)!\,(n-r)!}\left[1 - F_X\left(_{(r)}x\right)\right]^{r-1} F_X^{n-r}\left(_{(r)}x\right) f_X\left(_{(r)}x\right). \tag{9.16}$$

9.4   Introductory illustrations

Problems of extremes abound in engineering. Virtually all loadings – an exception being hydrostatic loading with fixed water level – must be dealt with probabilistically. This includes common situations, for instance surges in voltage which can cause failure of components, or loading on a floor in a building – although the background estimations for use in practice will likely have been made in the second example by a group of researchers and practitioners (for example a code committee) – as well as less common cases such as ice loading on a floating

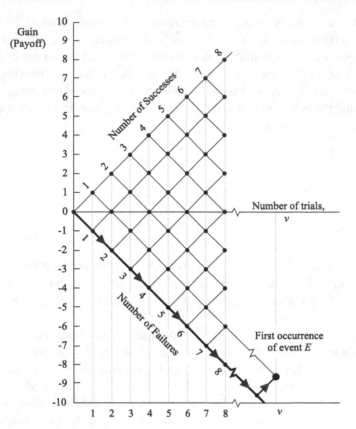

Figure 9.2   First success on the νth trial

structure that is operating in ice-covered waters. Floods are a common case involving extremes; spillway design and the design of systems to deal with electronic or vehicular traffic require analysis of extremes. We shall introduce some examples of maxima; minima can be treated similarly.

9.4.1   Random 'blows'

A stark situation is that of a structure that receives a certain number of randomly selected 'blows' during a time period. Structures buffetted (for example) by the wind, by waves, or by any naturally occurring phenomenon fall into this category. We choose the example of arctic shipping; this brings out the idea nicely but is quite general in principle. Here the vessel is buffetted by major interactions, termed 'rams', with heavy ice features. The methods of extremal analysis provided us with the key to classification of ships for arctic operations. Figure 9.3(a) shows a histogram of values of bow force experienced by the MV Arctic, an instrumented Canadian vessel that undertook a series of dedicated ice ramming trials (German

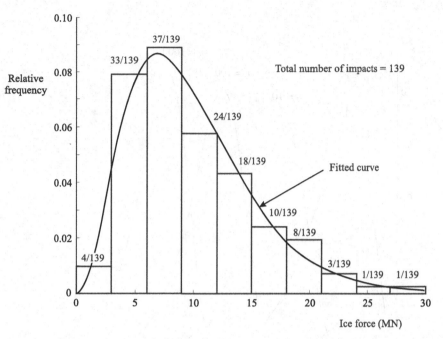

(a) Gamma distribution fitted to MV Arctic data

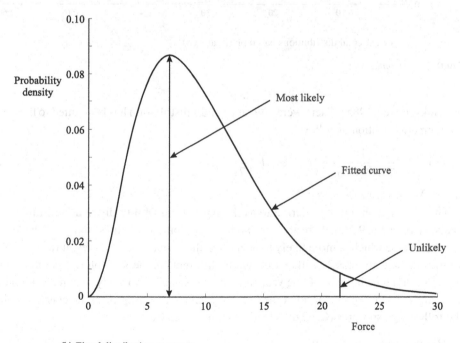

(b) Fitted distribution as parent

Figure 9.3   Obtaining extremal distributions (c) for the random 'blows' of (a)

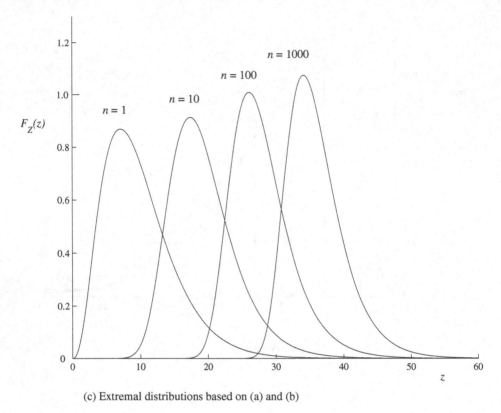

(c) Extremal distributions based on (a) and (b)

Figure 9.3   (*Continued*)

and Sukselainen, 1985). There were 139 rams, and a distribution has been fitted to these values, a gamma distribution as follows:

$$f_X(x) = \frac{1}{(2.77)^{3.58} \, \Gamma(3.58)} x^{2.58} \exp\left(\frac{-x}{2.77}\right), \tag{9.17}$$

where X represents force in MN.

This distribution will be referred to as the **parent** probability distribution in the extremal analysis. In Figure 9.3(a), there are some high values, some low values, and most cluster around a central value which is more likely to occur that the high (or low) values. It turns out that the various arctic class vessels will be exposed to different numbers of 'blows' per unit time. Let us consider a time interval of one year (it is often convenient to study the greatest load in one year, and then to take into account the lifetime of the structure). The arctic classes might have the following exposures (CAC denotes Canadian Arctic Class).

Highest class (CAC1) – thousands of rams per year.
High class (CAC2) – hundreds of rams per year.

Medium class (CAC3) – tens of rams per year.

Lowest class (CAC4) – several rams per year.

We wish to determine the maximum force for each class. The **extremal** distribution portrays the distribution of the maximum given the number of events. For ten events, we sample randomly ten values from the 'parent' of Figure 9.3(a), and throw away all values except the largest which is retained as a sample of the maximum value. We repeat this process many times, sampling ten at a time and keeping the maximum, throwing the rest away, and plot the new distribution of the set of maxima, one from each sample. It is not surprising that each value will tend to be higher than the average of the parent (although there is a small probability of being less than the average), and as a consequence the extremal distribution will be shifted to the right, as shown in Figure 9.3(c). We find the new distribution by simulation or by using Equations (9.5) and (9.17):

$$F_Z(z) = F_X^n(z).$$

If, instead of ten values, we sample 100 at a time, take the maximum, throw the others away, and repeat this process many times, it is not surprising that the resulting distribution shifts even further to the right, as shown in Figure 9.3(c). This is again repeated for 1000 and 10 000 repetitions. The distributions are seen to come closer together, and more 'peaked' as the number ($\log n$) increases. Exercise 9.2 will assist in understanding the process of going to extremes.

From a modelling standpoint, aside from the probabilistic process, the extremal distribution moves (as noted) to the right for a maximum, and eventually one might consider whether physical phenomena other than those characteristic of the parent distribution might come into play. For example, the higher forces that are given by the extremal distributions, particularly for CAC1, might lead to more large scale fracture processes in the ice, thereby alleviating the load.

9.4.2   Exponential parent distribution

As a second illustration we shall consider a phenomenon for which the parent follows the exponential distribution. This could be the waiting time for service in a computer application or communication network. We might be interested in the longest waiting time. Let X be the random quantity of interest. The parent distribution is as follows

$$f_X(x) = \frac{1}{\beta} \exp\left(-\frac{x-\alpha}{\beta}\right), \text{ for } x \geq \alpha, \tag{9.18}$$

and 0 elsewhere; α and β are constants with $\beta > 0$.

The *cdf* is as follows:

$$F_X(x) = 1 - \exp\left(-\frac{x-\alpha}{\beta}\right) \tag{9.19}$$

$$= 1 - \exp(-u_x), \tag{9.20}$$

where

$$U_x = \frac{X - \alpha}{\beta}. \tag{9.21}$$

Consequently the distribution of the maximum Z of n random quantities X is

$$F_Z(z) = \left[1 - \exp\left(-\frac{x - \alpha}{\beta} \right) \right]^n, \tag{9.22}$$

or

$$F_{U_z}(u_z) = \left[1 - \exp\left(-u_z \right) \right]^n, \tag{9.23}$$

where

$$U_z = \frac{Z - \alpha}{\beta}. \tag{9.24}$$

The *pdf* for the two forms (9.22) and (9.23) is obtained through differentiation:

$$f_Z(z) = \frac{n}{\beta} \exp\left(-\frac{z - \alpha}{\beta} \right) \left[1 - \exp\left(-\frac{z - \alpha}{\beta} \right) \right]^{n-1} \tag{9.25}$$

and

$$f_{U_z}(u_z) = n \exp\left(-u_z \right) \left[1 - \exp\left(-u_z \right) \right]^{n-1}. \tag{9.26}$$

Figure 9.4 illustrates the results using Equation (9.25). A feature of the extremes deriving from the exponential distribution is that their distributions quickly become stable in shape, and have even spacing for equal increments in $(\log n)$. This is dealt with further in Section 9.5, where useful descriptors of the extremes are introduced, and in the discussion of exponential-type distributions in Section 9.6.2.

9.4.3 Waves

The third illustration is concerned with a Rayleigh distribution of the amplitude of a Gaussian narrow band process. The distribution was first derived by Rayleigh in connection with sound waves. Another application is voltage from an AC generator. It has also been used in connection with wave heights in the sea; Cartwright and Longuet-Higgins (1956) presented the main results in a famous paper. Ochi (1990) gives a detailed introduction to the theory; see also Sarpkaya and Isaacson (1981) for a useful summary from a practical standpoint. The assumption made is of a sea state with contributions from different regions, of the same frequency but with random phase. This is essentially a Gaussian random process of surface elevation with a narrow band spectrum (Section 3.6, Example 3.39); see also Example 7.8 in Chapter 7. It was shown there that, under the assumptions noted, the probability distribution of the amplitude A of the process

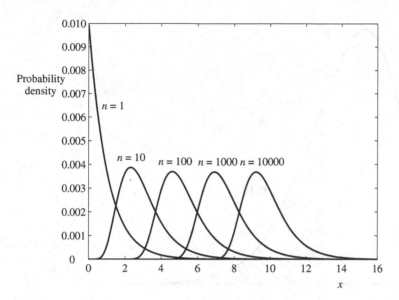

Figure 9.4 Extremes from exponential distribution

is given by

$$f_A(a) = \frac{a}{\sigma^2} \exp\left(-\frac{a^2}{2\sigma^2}\right), \text{ for } a \geq 0, \tag{9.27}$$

a Rayleigh distribution, where the parameter σ^2 is the variance of the narrow band Gaussian process describing the surface elevation; see Equation (7.48). This variance is also equal to both the area under the spectral distribution and the time average of the energy of the random process.

The expression for wave height in a storm is obtained by letting the crest-to-trough excursion H equal twice the amplitude A; since the determinant of the Jacobian $= 1/2$,

$$f_H(h) = \frac{h}{4\sigma^2} \exp\left(-\frac{h^2}{8\sigma^2}\right), \text{ for } h \geq 0. \tag{9.28}$$

For common spectral distributions such as the Bretschneider and Pierson–Moskowitz, the Rayleigh distribution can be fitted using the significant wave height h_s. This corresponds to visual observation of wave height, but is formally defined as the mean of the highest one-third of the wave heights ($h_s = h_{1/3}$ in Figure 9.5). It can be shown that, with good approximation (see Ochi, 1990, p. 282), that

$$h_s^2 \simeq 16\sigma^2, \tag{9.29}$$

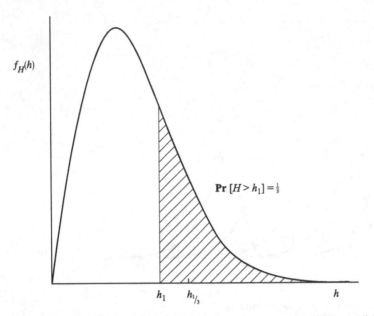

Figure 9.5   Significant wave height $h_s = h_{1/3}$, the mean of the highest 1/3 heights

and therefore we may write

$$f_H(h) = \frac{4h}{h_s^2} \exp\left[\left(-2\frac{h}{h_s}\right)^2\right].$$ (9.30)

The *cdf* of the random quantity H is

$$F_H(h) = 1 - \exp\left[\left(-2\frac{h}{h_s}\right)^2\right].$$ (9.31)

We are now in a position to compute the extremal distribution using Equation (9.5); thus the distribution of the extreme wave Z in a storm with n waves is given by

$$F_Z(z) = [F_H(z)]^n.$$ (9.32)

If we assume that $h_s = 9$ m, the Rayleigh individual wave height distribution, then

$$F_Z(z) = \left\{1 - \exp\left[-2\left(\frac{z}{9.0}\right)^2\right]\right\}^n.$$ (9.33)

The results are plotted in Figure 9.6 for $n = 1,10,100,1000,10\,000$. Again the distributions become more peaked and closer together.

A closed form solution is dealt with in Exercise 9.5.

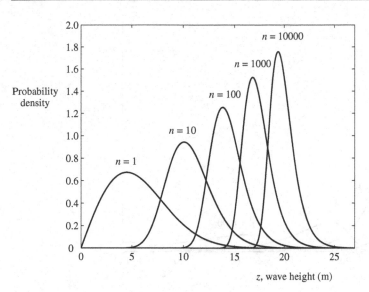

Figure 9.6   Extremes from Rayleigh distribution

9.4.4   Lognormal parent distribution

A strong contrast to the behaviour of the preceding introductory illustrations is found if a lognormal distribution is used as parent distribution. It has been used to model the luminosity of stars, the lifetime of electrical insulation (Castillo, 1988), the size distribution of particles produced upon the break-up of a solid, the daily discharges of a river, and various strength phenomena, including that of ice. The extreme values of these quantities might be of interest in an application. Figure 9.7 shows typical results obtained in finding the maximum of the indicated numbers of lognormally distributed random quantities. It is seen that, in contrast to the preceding examples, the distributions tend to become more spread out, and the means further apart, for equal increases in $(\log n)$. The lognormal distribution is not the only parent to exhibit this behaviour, and the generalization will be explored in the following subsection. The three types of behaviour are termed Class E1, Class E2 and Class E3 (Figure 9.8).

9.4.5   The three classes of extremes: E1, E2 and E3

We shall describe in Section 9.6 the fact that the distributions discussed above tend to an asymptotic distribution of the same mathematical form – the Gumbel distribution. This is sometimes expressed by saying that the **domain of attraction** is the Gumbel type. We have seen that the progression to the extreme distributions given in the illustrations shows different behaviour. In the first case – random blows – the extreme distributions become more and more peaked,[1] and the spacing between them decreases, as the number $(\log n)$ increases. This is

---

[1] The gamma distribution behaves in this way if the shape parameter $\alpha = 3.58$ in (9.17) is >1. See Table 9.1.

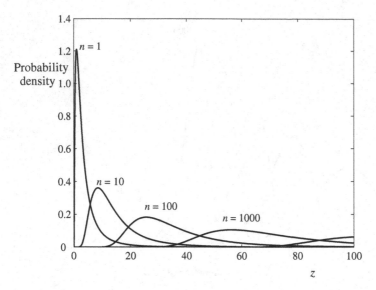

Figure 9.7   Extremes from lognormal distribution

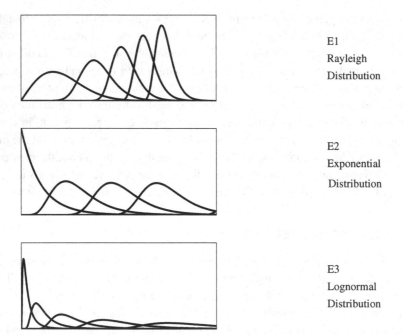

Figure 9.8   Classes of extremes; except for exponential (E2), distributions are examples

repeated in the case of waves, whereas the exponential parent results in stable behaviour. The lognormal example, by contrast, shows increased spacing and dispersion.

Figure 9.8 illustrates the three behaviours, which are Classes E1, E2 and E3. The behaviour of the extremal distribution is of great importance in design. The type of parent dictates the size of extreme; a lognormal parent may lead to much larger extremes than, say, a normal one. This emphasizes the importance of good modelling of the phenomenon under consideration. The convergence to the extremal distribution depends intimately upon the tail of the parent distribution. The larger n is, the further into the tail of the parent do the members of the extreme set originate. We have seen that the exponential tail in the parent results in stable behaviour,[2] and this is taken as the prototype of the parent distributions of the 'exponential type', and the defining behaviour for Class E2. For the present we note that Classes E1, E2, E3 originate in parents whose extreme tails converge, as the exposure $n \to \infty$, to the exponential type faster, the same as, or slower than, the exponential distribution. There is a general way to consider this which will be given in Section 9.6.

When considering the minimum of a set of *iid* random quantities, unlimited to the left, the extremal distribution shifts to the left, following Equation (9.6). This is a mirror image of results obtained for maxima. See Exercise 9.6.

9.5 Values of interest in extremal analysis

There are four main descriptors that are commonly used in describing extreme values and extreme behaviour. They are return periods and associated characteristic extremes, most probable largest (or smallest) values, and intensity functions.

9.5.1 Return period

We use the expression 'return period' frequently in considering design loads. For example the '100-year flood' or the '1000-year load' may be discussed in design situations. The idea is a good one and assists in formulating rules for design. The quantities just noted will correspond to exceedance probabilities of the annual maximum of 1% and one-in-a-thousand respectively. As an example, we might consider the maximum load in a year, and whether it does or does not exceed a certain value $Z = z$. The event E_i that it does, in the ith year, occurs with the exceedance probability p such that

$$p = 1 - F_Z(z). \tag{9.34}$$

The distribution $F_Z(z)$ is that of annual extremes; the fact that it is the same for each period derives generally from the *iid* assumptions, as in the derivation of Equation (9.5) or in the case of Poisson arrivals, as in Section 9.8. The assumption of independent trials is reasonable since

---

[2] There is a small decrease in the variance; see subsection 9.6.3 'Stability postulate'.

knowledge of events from one year would not usually cause one to change the estimate of probability for the next. The return period t_R is given by

$$t_R = \frac{1}{p}. \tag{9.35}$$

Gumbel (1958, p. 23) states that 'This result is self-evident: if an event has a probability p, we have to make, on the average $(1/p)$ trials in order that the event happen once.' Gumbel defines the return period as the average time to the first exceedance. We now explore the basis for Equation (9.35).

In general, one considers a series of events: exceedance or non-exceedance of a specified value for maxima; falling short (negative exceedance) or not for minima. Each event possibly occurs in each of a series of exposures or time periods (often a year). In defining the events, the distribution of the maximum (or minimum) for each exposure period is used. We therefore **define** the event of interest as the possible exceedance of a specified maximum or falling short of a minimum given the exposure; we shall use the word 'exposure', emphasizing the fact that we may be considering a variable other than time (for example, distance). The event in each period of exposure or time (E_i) will take values 1 or 0, a success or failure. Generally, independence between events and a constant probability of occurrence for each exposure is a reasonable assumption. This is the basic assumption that is made in the binomial distribution, the negative binomial and geometric distributions (Section 3.2). It is the last two of these distributions that are of interest here.

From Chapter 3, Section 3.2.4, the probability distribution of the number of trials V until the first success is given by the geometric distribution:

$$\mathbf{Pr}\,(V = v) = p_V\,(v) = pq^{v-1}, \tag{9.36}$$

for $v = 1, 2, 3, \ldots$ This is illustrated in Figure 9.2. The return period is usually defined (following Gumbel's idea above) as the expected value of V given by Equation (9.36):

$$t_R = \langle V \rangle = \sum_{v=1}^{\infty} vpq^{v-1} = \frac{1}{p}. \tag{9.37}$$

Although Gumbel (and many of us) might feel that this result is intuitively justified, it is in fact surprisingly difficult to prove given its simplicity. First we substitute $V = U + 1$; then

$$\langle V \rangle = \sum_{u=0}^{\infty} upq^u + \sum_{u=0}^{\infty} pq^u. \tag{9.38}$$

In the first summation the first term corresponding to $u = 0$ is equal to zero. The remaining terms of this summation are

$$\sum_{u=1}^{\infty} upq^u = q \sum_{u=1}^{\infty} upq^{u-1}$$
$$= q \langle V \rangle, \tag{9.39}$$

since the summation

$$\sum_{u=1}^{\infty} upq^{u-1}$$

is identical to the summation in Equation (9.37); it doesn't matter whether the positive integers $\{1, 2, 3, \ldots\}$ are indexed as u or v. We may write the second term of Equation (9.38) as

$$q \sum_{u=0}^{\infty} pq^{u-1};$$

the first term in the summation in this expression is (p/q) while the remaining terms are the probability masses in the geometric distribution, and sum to unity. Equation (9.38) may therefore be written as

$$\langle V \rangle = q \langle V \rangle + q \left(\frac{p}{q} + 1 \right), \text{ and consequently}$$

$$\langle V \rangle = \frac{1}{p}. \tag{9.40}$$

Another proof of this relationship is asked for in Exercise 9.3.

Example 9.3 In the design of a spillway for a dam, the distribution of the annual maximum flood Z (in $m^3 s^{-1}$) is given by the Gumbel distribution

$$F_Z(z) = \exp \left[- \exp \left(- \frac{z - 35}{8} \right) \right].$$

Find the return periods for discharges of 50 and 70 $m^3 s^{-1}$. The return periods are

$$t_{R50} = \frac{1}{1 - F_Z(50)} = 7.0 \text{ (years), and}$$

$$t_{R70} = \frac{1}{1 - F_Z(70)} = 79.9 \text{ (years).}$$

The flood levels of 50 and 70 $m^3 s^{-1}$ would be expected to occur, on average, every 7 and 80 years, respectively.

Example 9.4 The distribution function of an extreme voltage Z during an interval is given by

$$F_Z(z) = \exp \left[- \exp \left(- \frac{z - \alpha}{\beta} \right) \right]. \tag{9.41}$$

Determine the event (value of Z) with a return period of n intervals. Let z_n be the required value. Put

$$\frac{1}{n} = 1 - F_Z(z_n)$$

$$= 1 - \exp \left[- \exp \left(- \frac{z_n - \alpha}{\beta} \right) \right].$$

Solving for z_n:

$$z_n = \alpha - \beta \ln \left[- \ln \left(1 - \frac{1}{n} \right) \right]. \tag{9.42}$$

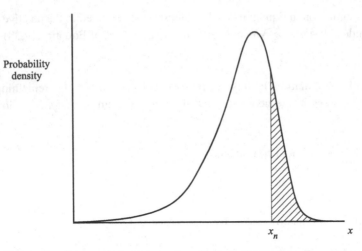

Figure 9.9   Characteristic largest value of parent distribution; E_i corresponds to $X_i > x_n$

This is an example of a characteristic value; see next subsection.

9.5.2   Characteristic values

The characteristic value is the value of the quantity corresponding to the return period. For a return period $t_R = n$ we denote it as x_n or z_n for the random quantities X or Z; thus the n-year event is z_n for the annual extreme Z. But the definition is often made so that it covers a broader range of applications than return periods; to this end we consider a set of n *iid* random quantities X with *cdf* $F_X(x)$. We use X as random quantity in this section since the n *iid* random quantities need not be annual maxima or minima (which we have denoted as Z and W). They could fit other applications as well, for example a set of n readings of electrical resistance, of which we are interested in the largest or smallest value. Using the largest as an example, we define E_i as the event that the ith of the set is greater than x_n, where x_n is a value that we choose, as described in the following (see Figure 9.9). The event E_i is defined as the event that $X_i > x_n$, and then $\tilde{E}_i$ is such that $X_i \leq x_n$. Hence $\mathbf{Pr}(\tilde{E}_i) = F_X(x_n)$ and

$$\mathbf{Pr}(E_i) = 1 - F_X(x_n). \tag{9.43}$$

Let

$$R = \sum_{i=1}^{n} E_i, \tag{9.44}$$

the number of 'successes' occurring when $E_i = 1$.

We define the **characteristic largest value** x_n such that, in n trials,

$$\langle R \rangle = 1. \tag{9.45}$$

We would expect on average one event in ten, one hundred, . . . , trials depending on the value of n. The process is again a binomial one, and the expected value of R is (np),

$$\langle R \rangle = np, \tag{9.46}$$

where $p \equiv \mathbf{Pr}(E_i)$. Therefore, from Equation (9.45),

$$\mathbf{Pr}(E_i) = \frac{1}{n}, \tag{9.47}$$

and (as desired)

$$\langle R \rangle = n \cdot \frac{1}{n} = 1.$$

In these equations, for practical purposes $n \geq 2$.

Consider now a series of intervals (such as a year), with one event (E_i or $\tilde{E}_i$) per interval. The return period is then n intervals. But the exposure period or time is not *necessarily* included in the notion of characteristic value. From the above

$$F_X(x_n) = 1 - \frac{1}{n}, \tag{9.48}$$

and considering the maximum Z of the n values,

$$F_Z(x_n) = F_X^n(x_n) = \left(1 - \frac{1}{n}\right)^n$$

$$\rightarrow \frac{1}{e} \tag{9.49}$$

as $n \rightarrow \infty$. From this we see that (for example) the probability of the n-year event being exceeded during n years tends toward $(1 - 1/e) = 0.6321$ as n increases. Thus there is a 63% chance of the 100-year flood or wave being exceeded in a large number of (say 100) years.

The characteristic smallest value $_n x$ is defined such that the expected number of occurrences of $B_i(\leftrightarrow X_i \leq {_n x})$ is unity. In this case

$$F_X({_n x}) = \frac{1}{n}. \tag{9.50}$$

Example 9.5 Find the characteristic extreme largest value for n iid quantities from an exponential distribution:

$$f_X(x) = \frac{1}{\beta} \exp\left(-\frac{x - \alpha}{\beta}\right), \quad \text{for } x \geq \alpha. \tag{9.51}$$

$$\frac{1}{n} = 1 - F_X(x_n) = \exp\left(-\frac{x_n - \alpha}{\beta}\right); \text{ thus}$$

$$x_n = \alpha + \beta \ln n. \tag{9.52}$$

The characteristic extreme increases in proportion to $\ln n$. The standardized random quantity

$$U = \frac{X - \alpha}{\beta} \tag{9.53}$$

has the characteristic extreme

$$u_n = \ln n \tag{9.54}$$

for the exponential distribution.

Example 9.6 Consider again the characteristic extreme of the Gumbel distribution (Example 9.4). Compare this value with the location parameter in the distribution of the maximum of n values taken from the distribution (9.41),

$$F_Z(z) = \exp\left[-\exp\left(-\frac{z - \alpha}{\beta}\right)\right].$$

The distribution of the maximum of the n values taken from this distribution is

$$[F_Z(z)]^n = \exp\left[-\exp\left(-\frac{z - \alpha - \beta \ln n}{\beta}\right)\right]. \tag{9.55}$$

The location parameter is

$$\alpha + \beta \ln n, \tag{9.56}$$

which is also the mode (see details in the next section). The characteristic extreme value z_n is, from Equation (9.42):

$$z_n = \alpha - \beta \ln\left[-\ln\left(1 - \frac{1}{n}\right)\right].$$

Since $\{\ln[-\ln(1 - 1/n)]\} \simeq -\ln n$ for large n, we see that the characteristic extreme approaches the location parameter $(\alpha + \beta \ln n)$ (mode) as n increases.

Example 9.7 Find the characteristic smallest value for the *cdf*

$$F_X(x) = 1 - \exp\left(-x^k\right), x \geq 0, k > 0 \tag{9.57}$$

(Weibull form), where X represents, for example, the time until the failure of a component in a series circuit, in which the failure of the first component is critical (these could also be the links in a chain). Equation (9.50) yields $F_X({}_n x) = 1/n$, or

$$_n x = \left[-\ln\left(1 - \frac{1}{n}\right)\right]^{\frac{1}{k}}. \tag{9.58}$$

For $k = 2$, and $n = 100$, this yields $_n x = (0.010\ldots)^{1/2} = 0.100\ldots$

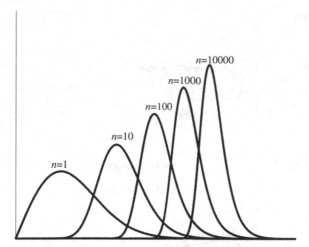

Figure 9.10   Class $E1$ extremal distribution; distributions become more peaked with $(\log n)$

9.5.3   Most probable largest values

For Class E1 parent distributions, the extreme distribution becomes more and more 'peaked' for many repetitions (Figure 9.10). When the peak of the distribution is very sharp, we may think of it increasingly as a spike of probability, and concentrate this near the mode of the distribution. This is the 'most probable largest' value, which is used in several codes[3] to specify values to be used in design. For the random quantity Z we denote it $\hat{z}$. It is a reasonable idea for the peaked distributions noted, but loses its appeal for those with a large coefficient of variation. This is particularly the case when rare events are considered, as discussed in Section 9.8 below.

The most probable largest value approximates closely the characteristic extreme if the number of periods in the exposure equals the number n in the set of quantities of Section 9.5.2 above. The resulting situation is illustrated in Figure 9.11. This can be shown for exponential-type distributions as follows. We start with a distribution of a random quantity Z that is, for example, an annual extreme. This is assumed to be exponential-type, not necessarily of Gumbel or other particular form. Its cdf is $F_Z(z)$. We then consider the maximum of n of these random quantities, which we denote Z_ℓ, with ℓ denoting 'largest'; the distribution of this is given by Equation (9.7):

$$f_{Z_\ell}(z_\ell) = n F_Z^{n-1}(z_\ell) f_Z(z_\ell).   \tag{9.59}$$

To obtain the mode of the extreme distribution, the derivative of (9.59) must be zero:

$$\frac{\mathrm{d}}{\mathrm{d}z_\ell} f_{Z_\ell}(z_\ell) = 0.   \tag{9.60}$$

Substituting for $f_{Z_\ell}(z_\ell)$ from (9.59), differentiating and dividing through by n,

---

[3] For example *DnV Rules for the Design, Construction and Inspection of Offshore Structures*, 1977, Appendix A, Environmental Conditions.

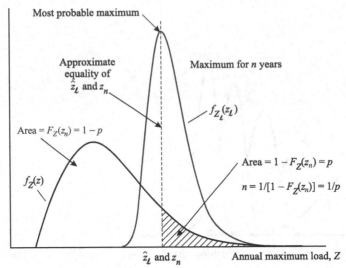

Figure 9.11   Most probable maximum; for example if $p = 0.01$, $n = 100$ (e.g. 100-year wave)

$$f_Z' (z_\ell) F_Z^{n-1} (z_\ell) + (n-1) f_Z^2 (z_\ell) F_Z^{n-2} (z_\ell) = 0, \tag{9.61}$$

where the prime denotes a derivative. The solution to this equation is the mode $z_\ell = \hat{z}_\ell$; using this notation and dividing through by $F_Z^{n-2} (\hat{z}_\ell)$, one gets

$$-f_Z' (\hat{z}_\ell) F_Z (\hat{z}_\ell) = (n-1) f_Z^2 (\hat{z}_\ell). \tag{9.62}$$

We write this in the abbreviated form

$$-f' F = f^2 (n-1), \tag{9.63}$$

or

$$\frac{f'}{f^2} = \frac{n-1}{F}. \tag{9.64}$$

We apply L'Hôpital's rule for large values of the quantity z, as follows:[4]

$$\frac{f}{1-F} \simeq -\frac{f'}{f}. \tag{9.65}$$

L'Hôpital's rule does not apply in all cases. Where it does, there are exponential-type parent distributions, which lead to the Gumbel form in the extreme; see Section 9.6.4 below.

For these cases (where L'Hôpital's rule does apply), substituting Equation (9.64) into (9.65) yields

---

[4] The left hand side of Equation (9.65) is the intensity function μ (see next subsection), and the ratio of the left and right hand sides is the 'critical quotient' introduced by Gumbel, further details of which are given in Section 9.6.4 below.

$$\frac{1}{1-F} \simeq \frac{n-1}{F},$$

(9.66)

or

$$F \simeq \frac{n-1}{n}.$$

(9.67)

Reverting to the full notation,

$$\frac{1}{1 - F_Z(\hat{z}_\ell)} \simeq n$$

(9.68)

for large n. We note that $[1 - F_Z(\hat{z}_\ell)]$ is the exceedance probability of the most probable largest value $\hat{z}_\ell$, and $1/[1 - F_Z(\hat{z}_\ell)]$ the return period. We have therefore shown that

$$\hat{z}_\ell \simeq z_n.$$

(9.69)

The result is illustrated in Figure 9.11. There is an interesting congruence of return period, characteristic extreme, and most probable largest extreme value. As we have discussed above, the peaked form of the extreme distribution occurs in many applications, particularly in the area of structural design for wind and waves, and therefore in naval applications. This makes it an attractive way to specify design values, for large n. An example is the application to gusty wind; see for example Davenport (1964). The approach works well where the parent distribution is Class E1. For other cases, it is not as useful, and for rare extreme events, the congruence disappears, as noted in Section 9.8.

Example 9.8 Find the mode of the standardized distribution for the maximum of n exponentially distributed random quantities, Equation (9.26), reproduced here without the subscript 'z':

$$f_U(u) = n \exp(-u) \left[1 - \exp(-u)\right]^{n-1}.$$

(9.70)

Differentiating and equating to zero, one finds that

$$\hat{u} = \ln n.$$

(9.71)

The result just obtained is most important; the mode of the distribution of the maximum of n exponentially distributed random quantities is equal to $(\ln n)$ for the standardized case, which is also the value of the characteristic extreme (see Example 9.5); thus

$$u_n = \hat{u} = \ln n.$$

(9.72)

It might be remembered that the exponential distribution is the prototype of the convergence of the Class E2 exponential-type distributions, which have equal spacing as they approach the extreme. Equation (9.69) is exact in this case, and for all n.

For a case where L'Hôpital's rule does not apply, see Exercise 9.11.

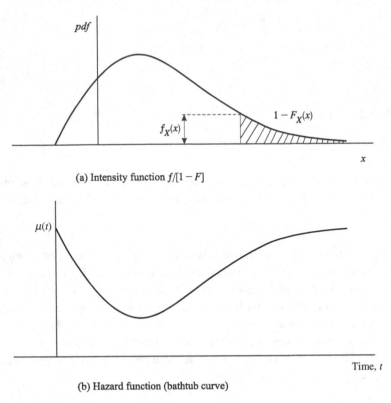

(a) Intensity function $f/[1-F]$

(b) Hazard function (bathtub curve)

Figure 9.12   Intensity and hazard

9.5.4   Intensity and hazard function

Given a continuous random quantity X, the probability that a value $X = x$ is exceeded is $[1 - F_X(x)]$. This probability of exceedance will decrease with increasing x, as will (in general) the density $f_X(x)$, at least for large X. This is shown in Figure 9.12(a). The ratio between the two has been termed the **intensity** $\mu(x)$ by Gumbel:

$$\mu(x) = \frac{f_X(x)}{1 - F_X(x)}. \tag{9.73}$$

This quantity will be discussed further when dealing with the definition of the exponential type of parent distributions (Section 9.6.4). It is intimately related to the dispersion of the extremal distribution that is produced from a given parent X; the greater the value of μ at an extreme point of the parent, the smaller the deviation of the extremal distribution at that point, and *vice versa*.

The intensity has also another interpretation when dealing with the lifetime of systems, their components, structures or people. Given that the age of the object or person is denoted T, with $F_T(t)$ being probability that the object (component) has failed, or that the person has died,

then

$$\mathbf{Pr}(t < T < t + dt | T > t) = \frac{f_T(t)\,dt}{1 - F_T(t)} = \mu(t)\,dt. \tag{9.74}$$

The function $\mu(t)$ presents a typical 'bathtub' shape, as shown in Figure 9.12(b). For a person, this gives one's mortality rate given that one has survived to the age $T = t$. One's prospects at first improve, with the rate decreasing, and then inevitably deteriorate.

We saw that in the case of the Poisson process, one could model occurrences in distance, area or volume, or in some other dimension than time. The same is true in the case of hazard functions; one could, for example, model distance to the first accident in an automobile. The intensity function $\mu(x)$ has, as noted a wider application, as will be explained in Section 9.6.4. It may increase, decrease or stay constant with increasing x. Gumbel (1958) has given an extended discussion of this quantity.

Example 9.9 Calculate the hazard function for the exponential distribution $f_T(t) = \lambda \exp(-\lambda t)$:

$$\mu(t) = \frac{\lambda \exp(-\lambda t)}{\exp(-\lambda t)} = \lambda. \tag{9.75}$$

Therefore the hazard function is a constant for the exponential distribution. Examples of application include physical phenomena such as large earthquakes or extreme winds. See also Exercise 10.11.

9.6   Asymptotic distributions

9.6.1   Overview

In the introductory illustrations given in Section 9.4, all dealing with maxima, the *cdf* of the parent distribution was raised to the power n, expressing the number of repetitions of the phenomenon within the exposure period. It is sometimes inconvenient to raise the parent to a power; in fact we might not even have knowledge of the parent. Commonly used parent distributions tend asymptotically as n increases to an extreme distribution of the same form, and this is a useful fact to exploit in analysis. We will focus again chiefly on maxima, but note that minima have analogous results to those given below; these will be introduced from time to time.

We have seen in Chapter 7 that, under quite wide circumstances, the distribution of the mean of a set of random quantities tends to a limiting distribution, the normal distribution, as $n \to \infty$ (Section 7.5). Further, the Central Limit Theorem deals with standardized random quantities in which a location parameter, the mean, is subtracted from the value of the random quantity, and the resulting difference is divided by a measure of dispersion, the standard deviation (Section 7.2.2). A striking analogy exists for the case of extremes. In this case, we consider *iid* random quantities X_i and their maximum

$$Z = \max_n (X_i).$$

As $n \to \infty$, the distribution of Y, given by

$$Y = \frac{Z - a_n}{b_n}, \tag{9.76}$$

tends to a limiting distribution. The normalizing values a_n and b_n are location and scale parameters.

For extremes, there are three limiting distributions, depending on the kind of parent distribution. In all of the cases used as introductory illustrations in Section 9.4, the parent tends to the same type of extremal distribution, the Gumbel (type I) distribution. While this is certainly the most important type, it is not the only one: there are, as noted, three, originating from three kinds of parent. These are the exponential type (type I), by far the most common; the Cauchy type (type II); and parent distributions limited so as to be less than a certain specified value (type III). We use the names Gumbel, Fréchet and Weibull to denote the three distributions, respectively. Fréchet developed the type II distribution; the other names recognize the extensive use by Gumbel and Weibull of the corresponding distributions in practice. The resulting extreme distributions are now given, following the usual convention. As noted, all of the illustrations in Section 9.4 correspond to the first type.

1. Type I. Exponential type, leading to the Gumbel or double exponential distribution, as follows:

$$F_Y (y) = \exp\left[-\exp(-y)\right] \text{ for} \tag{9.77}$$
$$-\infty < y < \infty.$$

2. Type II. Cauchy type, leading to the Fréchet distribution, as follows:

$$F_Y (y) = \exp\left(-y^{-\alpha}\right) \text{ with} \tag{9.78}$$
$$\alpha > 0, \text{ for}$$
$$-\infty < y < \infty.$$

3. Type III. Limited type, leading to the Weibull distribution, as follows:

$$F_Y (y) = \exp\left[-(-y)^{\alpha}\right] \text{ with} \tag{9.79}$$
$$\alpha > 0, \text{ for}$$
$$-\infty < y \leq 0.$$

The three limiting distributions are related. Consider Equation (9.77), but in terms of the quantity Z rather than Y:

$$F_Z (z) = \exp\left[-\exp\left(-\frac{z - a_n}{b_n}\right)\right]. \tag{9.80}$$

If we substitute

$$Z = \ln(V - \gamma) \tag{9.81}$$

and

$$a_n = \ln(d_n - \gamma), \tag{9.82}$$

then

$$F_V(v) = \exp\left[-\left(\frac{v - \gamma}{d_n - \gamma}\right)^{-\alpha}\right], \tag{9.83}$$

where $\alpha = 1/b_n$. This equation is of the same form as Equation (9.78). For other relationships, see Exercise 9.15. The logarithmic relationship (9.81) should be understood as a relationship between *extremes*. It does not imply that a logarithmic relationship between the *parent* distributions results in a similar relationship between the extremes. In fact both the normal and lognormal distributions as parents tend to the Gumbel (type I) distribution in the extreme. See Table 9.1 and also Hong (1994b).

The relationships may further be illustrated by the version, called the von Mises form (Castillo, 1988), which combines the three:

$$F_Z(z) = \exp\left\{-\left[1 + c\left(\frac{z - \gamma}{\delta}\right)\right]^{-\frac{1}{c}}\right\}, \tag{9.84}$$

with $0 \le 1 + c\left(\frac{z - \gamma}{\delta}\right)$. \tag{9.85}

We obtain the three extremal distributions with $c > 0$ (Fréchet), $c = 0$ (Gumbel) and $c < 0$ (Weibull). For $c = 0$, we must consider the limit $c \to 0$. Equation (9.84) is also known as the Generalized Extreme Value (GEV) distribution (Stedinger *et al.*, 1993). For $|c| < 0.3$, the shape is similar to that of the Gumbel distribution. The GEV has the possibility to model a richer variety of tail shapes.

Both types I and II deal with quantities that are unlimited to the right for either the parent or the extreme, so that these tails will decline to zero. In dealing with maxima, the type I distribution is by far the most useful, since it is the natural extreme for most parent distributions. In the case of the type II (Fréchet) distribution, one is concerned with 'pathological' parent distributions, for example the Cauchy or Pareto distributions. These either do not possess moments or only lower ones, as do the extremal distributions derived from them. For the Weibull distribution, one deals with the situation where there is a limitation on the maximum value that can be obtained. This is illustrated in Figure 9.13 (see left hand illustrations for maxima). This is not to say that the latter two types do not have uses; on the contrary, they do. In particular, the Weibull distribution has strong applications in the analysis of droughts and weakest-link theories (for example). See also Hong (1994a).

Yet in terms of maxima unlimited 'to the right', the Gumbel distribution has pride of place. Some writers have expressed concern that the distribution is unbounded to the left (when

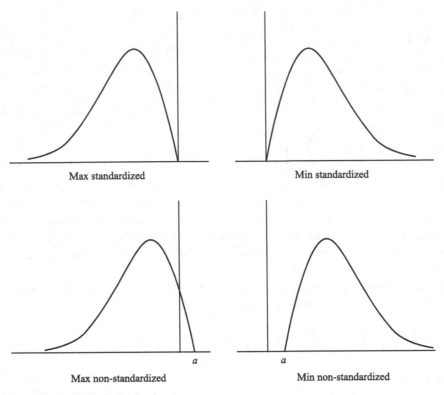

Figure 9.13 Weibull distributions

considering maxima). This is generally of little practical importance, given that the main mass of the distribution is taken from the right hand side of the tail of the parent, and the probability mass to the left of the origin in the extremal distribution will be negligible. In some cases, one might instead suppose a mass of probability at the origin corresponding to values less than zero; this arises naturally in Poisson processes, Section 9.8 below. There might also be grounds for truncating the distribution and renormalizing as described in Section 3.3.7 (Ochi, 1990).

We have introduced the theory in terms of **maxima**; a completely analogous theory can be derived for **minima**. Some aspects are summarized in Subsection 9.6.8 at the end of the present section, and in Section 9.7, Weakest link theories.

9.6.2   Gumbel distribution: type I

The Gumbel distribution has the well-known double-exponential form given in Equations (9.77) and (9.80). The distribution is illustrated in Figure 9.14 for the standardized form (9.77). We shall see it again in the discussion of Poisson processes, in Equation (9.160) below. For the standardized form of (9.77) the mode is at $y = 0$, the mean is equal to Euler's constant $\gamma = 0.5772$ (to four places of decimals), the standard deviation is $\pi/\sqrt{6} = 1.2825$ (to four

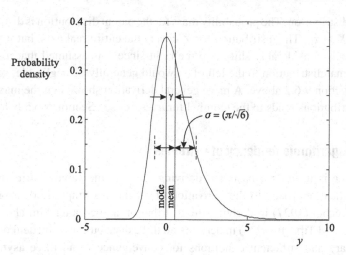

Figure 9.14 Standardized Gumbel distribution

places) and it has a skewness of 1.1396 (again to four places). For the form given in Equation (9.80), dropping the subscripts n applied in this equation to a and b, the mode $\hat{z} = a$, the mean is $a + \gamma b$, the variance $(\pi b)^2 /6$, and the skewness has the value just quoted.

We may derive the distribution directly by considering a parent distribution of the exponential form, with *cdf*

$$F_X(x) = 1 - \exp\left[-\lambda(x - x_0)\right]. \tag{9.86}$$

Since we are dealing with *iid* random quantities, the maximum of n of the Xs has the distribution

$$F_Z(z) = \{1 - \exp\left[-\lambda(z - x_0)\right]\}^n. \tag{9.87}$$

We can write the probability of exceedance p of x in the parent distribution as

$$p = \exp\left[-\lambda(x - x_0)\right], \tag{9.88}$$

and let

$$np = \eta. \tag{9.89}$$

Then Equation (9.87) becomes

$$F_Z(z) = \left(1 - \frac{\eta}{n}\right)^n. \tag{9.90}$$

If n becomes large, while keeping η constant, with p becoming small, then a close approximation is given by

$$\begin{aligned} F_Z(z) &= \exp(-\eta) = \exp(-np) \\ &= \exp\{-n\exp\left[-\lambda(z - x_0)\right]\} \\ &= \exp\{-\exp\left[-\lambda(z - x_0 - \ln n)\right]\}. \end{aligned} \tag{9.91}$$

This is the Gumbel distribution. The probability mass for the parent distribution is distributed over the real axis for $X \geq x_0$. The distribution for Z covers the entire real axis, but it should be noted that the mass is considerably shifted to the right since it is assumed that $n \to \infty$. Any mass in the extremal distribution to the left of x_0 would generally be negligible; see also comments at end of Section 9.6.2 above. A more general derivation shows that the maximum for most common distributions tends to the Gumbel for $n \to \infty$; see Section 9.6.5 below.

9.6.3   Stability postulate; logarithmic tendency of extremes

The **stability postulate** is used as a basis for deriving the extreme-value distributions; it represents a mathematical approach to the derivation of all three asymptotic distributions. It was first used by Fréchet (1927) in deriving the distribution named after him above, and subsequently by Fisher and Tippett (1928) in deriving all three distributions. Gnedenko (1941, 1943) derived necessary and sufficient conditions for convergence to all three asymptotic distributions. In these derivations it was postulated that the form of the extreme distribution stabilized as $n \to \infty$. This form was taken as

$$F_Z^n (z) = F_Z (g_n z + h_n).$$ (9.92)

In other words, the functional form of F_Z raised to the power n could be obtained from F_Z by a linear transformation of the argument of F_Z itself.

We take the Gumbel distribution (9.80)

$$F_Z (z) = \exp \left[-\exp \left(-\frac{z - a_n}{b_n} \right) \right]$$

as an example. In this case $g_n = 1$, and we can see that

$$
\begin{aligned}
F_Z^n (z) &= \left\{ \exp \left[-\exp \left(-\frac{z - a_n}{b_n} \right) \right] \right\}^n \\
&= \exp \left[-\exp \left(-\frac{z - a_n - b_n \ln n}{b_n} \right) \right] \\
&= F_Z (z - b_n \ln n).
\end{aligned}
$$ (9.93)

This represents a shift of the mode of the distribution by $(-h_n) = b_n \ln n$; the shape (and consequently g_n) remains constant. This is merely a demonstration; the original proofs resulting in the desired functional forms can be found in Gumbel (1958) and in other texts.

Equation (9.93) shows the logarithmic tendency of extremes; the mode increases in proportion to $\ln n$. We draw attention in this context to the exponential distribution; we saw in Example 9.5 that the standardized random quantity U has the characteristic extreme $u_n = \ln n$ for the exponential distribution. Further, the mode $\hat{u}$ is also equal to $\ln n$, as shown in Example 9.8.

The standardized exponential and Gumbel distributions are shown in Figure 9.15, together with extremes taken from these parents. It is seen that the exponential distribution rapidly

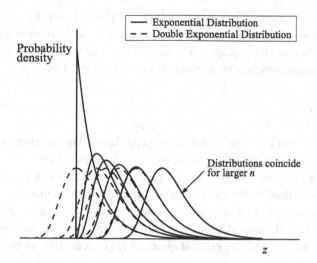

Figure 9.15   Extremal distributions from exponential and double exponential distributions

converges to the Gumbel form; the shape of the latter remains the same, as explained above. Gumbel (1958, p. 116) notes that the extremes of the exponential distribution exhibit a variance that, for $n = 100$, is only 0.6% less than the asymptotic value. We have seen in Section 9.4.5 above that parent distributions tend to the extreme distribution in different ways from those just discussed. Some become more sharply peaked and closer together, others more diffuse and widely spaced, as n increases.

9.6.4   Exponential type; critical quotient and classes E1, E2 and E3[5]

The prototype of the Class E2 convergence is the exponential distribution. For the standardized version, we have seen that the mode and the characteristic extreme are both *exactly* equal to $(\ln n)$, and the shape of the distribution of extremes converges quickly to the Gumbel type (Figure 9.15). The convergence and spacing of other distributions is therefore studied with reference to this distribution. The key aspect is how the tail of the parent behaves with regard to the prototype. Only those of the exponential type will converge to the Gumbel distribution. These are parent distributions whose tails approach the exponential form increasingly closely as $n \to \infty$. (We again consider the upper tails.)

The point of departure in our analysis is the intensity function $\mu(x)$ given in Equation (9.73) above. This was defined as

$$\mu(x) = \frac{f_X(x)}{1 - F_X(x)}.$$

It was found in Example 9.9 to be λ for the distribution $\lambda \exp(-\lambda x)$, and it therefore takes the

---

[5] The material in this subsection is not essential for the sequel.

value $\mu = 1$ for the standardized exponential distribution $[\exp(-x)]$ with $\lambda = 1$, for any value of the random quantity X. As we go further into the tail the numerator and denominator of the expression for μ become smaller, and indeed, approach zero. Hence μ becomes indeterminate. We can study this behaviour asymptotically by applying L'Hôpital's rule. Thus

$$\lim_{x \to \infty} \frac{f_X(x)}{1 - F_X(x)} = -\frac{f_X'(x)}{f_X(x)}, \tag{9.94}$$

and this will be approximately true for large x, the more so the larger. For our standardized exponential distribution, this ratio is again unity, and equal to λ for the *pdf* $[\lambda \exp(-\lambda x)]$.

We can calculate $\mu(x)$ for any parent distribution of interest, which includes for present purposes only distributions unlimited to the right. This will indicate the closeness to an exponential form for a given value of x. If μ is nearly constant, then one can deduce that it is approximately exponential. The right hand side of Equation (9.94) will indicate the tendency in the extreme for large x. Gumbel defines the **critical quotient** $Q(x)$ as the ratio of these two quantities:

$$Q(x) = \frac{\mu(x)}{-f'(x)/f(x)} = \frac{-f^2(x)}{f'(x)[1 - F(x)]}. \tag{9.95}$$

This function will tend to unity for distributions of the exponential type, as shown in Figure 9.16. Whether it approaches from the top, giving 'peaked' convergence to exponential type, Class E1; or is equal to unity, Class E2; or approaches from the bottom, giving 'diffuse' convergence with wide spacing, Class E3; is dictated by the form of the parent. The three kinds of convergence are summarized in the following behaviours of Q:

$$Q(x) = \begin{cases} 1 + \epsilon(x) \\ 1 \\ 1 - \epsilon(x), \end{cases} \tag{9.96}$$

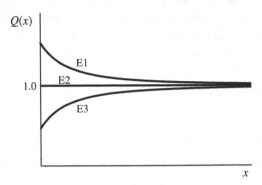

Figure 9.16   Critical quotient $Q(x)$

where $\epsilon(x)$ is a small positive quantity that tends to zero as $n \to \infty$. Equation (9.96) defines Class E1 (upper expression), Class E2 (middle expression) and Class E3 (lower expression).

In Exercise 9.12 it is asked to calculate and plot various intensity functions and critical quotients.

Example 9.10 Consider the *cdf*

$$F_X(x) = 1 - \exp\left(-x^k\right), x \geq 0, k > 0. \tag{9.97}$$

Find the intensity, critical quotient, and class as a function of k. Note that $k > 0$. First we deduce that

$$f_X(x) = kx^{k-1} \exp\left(-x^k\right)$$

and

$$f'_X(x) = kx^{k-1} \exp\left(-x^k\right)\left[-kx^{k-1} + (k-1)x^{-1}\right].$$

Substitution into Equation (9.73) gives

$$\mu(x) = kx^{k-1}, \tag{9.98}$$

and into Equation (9.95):

$$Q(x) = \frac{kx^{k-1}}{kx^{k-1} - (k-1)x^{-1}}$$
$$= \frac{1}{1 - \left(1 - \frac{1}{k}\right)\frac{1}{x^k}}.$$

We note that, for large x, $1/x^k$ is small, so that we can write

$$Q(x) \simeq 1 + \left(1 - \frac{1}{k}\right)\frac{1}{x^k}. \tag{9.99}$$

We see that $Q(x) \to 1$ for large x; the quantity ϵ in Equation (9.96) takes a positive sign for $k > 1$, is zero for $k = 1$, and takes a negative sign for $k < 1$, leading to Class E1, E2 and E3 behaviour in the three cases, respectively. Equation (9.97) is the Weibull distribution for minima; see Table 9.1 and Section 9.6.8 below.

9.6.5   General derivation of the Gumbel distribution

We obtained the Gumbel distribution from the exponential parent in Section 9.6.2. In Section 9.8 below, we obtain it again using the Poisson process. We shall now derive the distribution in a general way that focusses attention on the characteristic extreme as well. Gumbel's book (1958) contains a detailed treatment of the different derivations; he attributes the following to Cramér and von Mises. One considers a *cdf* that can be expressed as follows:

$$F_X(x) = 1 - \exp\left[-g(x)\right], \tag{9.100}$$

where $g(x)$ is a monotonically increasing function of x, without limit. It is most important to capture this behaviour in the upper tail, as we are considering again maxima. The characteristic extreme x_n is given by

$$1 - F_X(x_n) = \exp[-g(x_n)] = \frac{1}{n}. \tag{9.101}$$

Therefore

$$\exp[g(x_n)] = n \tag{9.102}$$

and

$$F_X(x) = 1 - \frac{\exp\{-[g(x) - g(x_n)]\}}{n}. \tag{9.103}$$

We take as before

$$Z = \max_n X_i,$$

and then

$$
\begin{aligned}
F_Z(z) &= \lim_{n \to \infty} \left(1 - \frac{\exp\{-[g(z) - g(x_n)]\}}{n}\right)^n \\
&= \exp\left(-\exp\{-[g(z) - g(x_n)]\}\right).
\end{aligned} \tag{9.104}
$$

We now use the Taylor expansion of $g(z)$ in the neighbourhood of x_n. This is

$$g(z) = g(x_n) + g'(x_n)(z - x_n) + g''(x_n)\frac{(z - x_n)^2}{2!} + \cdots, \text{ or}$$

$$g(z) - g(x_n) = g'(x_n)(z - x_n) + \cdots \tag{9.105}$$

For small differences $(z - x_n)$, we substitute only the first term of the right hand side of this expression into Equation (9.104) and obtain approximately

$$F_Z(z) = \exp\left(-\exp\{-[g'(x_n)(z - x_n)]\}\right). \tag{9.106}$$

Compare this to Equation (9.80), reproduced here:

$$F_Z(z) = \exp\left[-\exp\left(-\frac{z - a_n}{b_n}\right)\right].$$

We see that the characteristic extreme of the parent, x_n, becomes – in the limit – the mode a_n of the extreme distribution at that point, as previously found in Section 9.5.3. We can also see from Equation (9.100) that

$$g'(x_n) = \frac{f_X(x_n)}{1 - F_X(x_n)} = n f_X(x_n) = \mu(x_n). \tag{9.107}$$

This provides a more complete interpretation of the intensity; it provides a measure of the dispersion of the extremal distribution; comparing again to Equation (9.80),

$$g'(x_n) = \mu(x_n) = \frac{1}{b_n}. \tag{9.108}$$

9.6.6   Common distributions of exponential type

The various parent distributions with tails of the exponential type, leading to the Gumbel distribution, approach the limit at various speeds. Our criterion for this convergence has been the 'prototype', the exponential distribution; we saw in Example 9.5 that the standardized random quantity has the characteristic extreme and the mode equal to $(\ln n)$. Furthermore, the Class E2, defined by the middle expression of Equation (9.96), is the exponential distribution. At the same time, the shape of this distribution changes and converges to the Gumbel form quite rapidly, as we saw in Figure 9.15 above. The Gumbel distribution does not change shape but, as seen in Example 9.4, the characteristic extreme is given by

$$z_n = \alpha - \beta \ln[-\ln(1 - 1/n)],$$

which tends to

$$\alpha + \beta \ln n$$

as $n \to \infty$. This is shown in Figure 9.17, in which

$$d_n = -\ln[-\ln(1 - 1/n)], \tag{9.109}$$

so that $z_n = \alpha + \beta d_n$. The tail of the Gumbel distribution converges to the exponential form quite quickly. It is actually of Class E1 in the classification above. Some distributions, like the normal distribution, are very slow to converge; others, like the Gumbel (for characteristic extreme) and exponential (for shape) are very fast.

To set the scene for Table 9.1 below, we study this question using the characteristic extreme a_n given by

$$1 - F_X(a_n) = \frac{1}{n}$$

or

$$n = \frac{1}{1 - F_X(a_n)}, \tag{9.110}$$

and the dispersion b_n given as

$$b_n = \frac{1}{\mu(a_n)} = \frac{1 - F_X(a_n)}{f_X(a_n)} = \frac{1}{n f_X(a_n)},$$

using Equations (9.108) and (9.110).

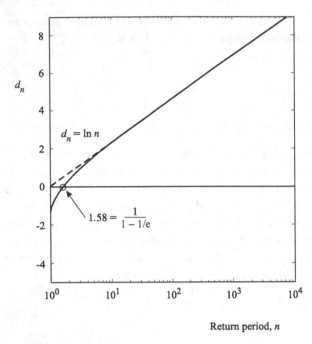

Figure 9.17 Factor d_n as a function of return period

We know that the characteristic extreme for the exponential distribution is $z_n = \alpha + \beta \ln n$ and the dispersion can be verified to be β. These values and other common ones are shown in Table 9.1, based on Maes (1985), Gumbel (1958), Galambos (1978), Leadbetter *et al.* (1983) and Castillo (1988). Han Ping Hong made several suggestions. The various parent distributions converge in different ways towards the asymptotic (Gumbel) distribution. Asymptotic expressions are valid for large n. Gumbel (1958) notes that the approximation $\mu + \sigma \sqrt{2 \ln(0.4n)}$ was used by Frank to analyse the velocity of the quickest gas molecule, given normally distributed individual velocities.

We may also study in a general way the behaviour of a_n and b_n with increasing n. First, we note that

$$\frac{\mathrm{d}a_n}{\mathrm{d}n} = \frac{\mathrm{d}a_n}{\mathrm{d}F_X(a_n)} \cdot \frac{\mathrm{d}F_X(a_n)}{\mathrm{d}n}$$

$$= \frac{1}{f_X(a_n)} \cdot \frac{1}{n^2} \tag{9.111}$$

$$= \frac{b_n}{n}; \tag{9.112}$$

and thus

$$\frac{\mathrm{d}a_n}{\mathrm{d}\ln n} = b_n. \tag{9.113}$$

Table 9.1 *Common probability distributions with Gumbel domain of attraction*

Parent; cdf $F_X(x)$	Characteristic extreme a_n; Dispersion factor b_n	Class E1, E2 or E3
Exponential; $1 - \exp\left(-\frac{x-\alpha}{\beta}\right)$	$\alpha + \beta \ln n$; β	E2
Logistic; $1/\left[1 + \exp(-x)\right]$	$\ln(n-1)$; $\frac{n}{n-1}$	E1
Gumbel; $\exp\left\{-\exp\left[-\left(\frac{x-\alpha}{\beta}\right)\right]\right\}$	$\alpha + \beta \ln n$; β	E1
Gamma $\displaystyle\int_0^x \frac{u^{\alpha-1}\exp(-u/\beta)\,du}{\Gamma(\alpha)\beta^\alpha}$	$\beta \ln\left[\dfrac{n}{\Gamma(\alpha)}\right]$ or $^{(1)}\beta\left\{\ln\frac{n}{\Gamma(\alpha)} + (\alpha-1)\ln\left[\ln\frac{n}{\Gamma(\alpha)}\right]\right\}$; $\beta\left[\ln\frac{n}{\Gamma(\alpha)}\right]^{1-\alpha}$ for $n > 10$, or $^{(1)}\dfrac{a_n}{a_n/\beta + 1 - \alpha}$	$\begin{cases}\alpha > 1 : \text{E1} \\ \alpha = 1 : \text{E2} \\ \alpha < 1 : \text{E3}\end{cases}$
Normal $N(\mu, \sigma^2)$; $\frac{1}{\sqrt{2\pi}\sigma}\displaystyle\int_{-\infty}^x \exp\left[-\left(\frac{u-\mu}{\sigma}\right)^2\right]du$	$\mu + \sigma\left[(2\ln n)^{\frac{1}{2}} - \dfrac{(\ln\ln n + \ln 4\pi)}{2(2\ln n)^{\frac{1}{2}}}\right]$; $\sigma(2\ln n)^{-\frac{1}{2}}$	E1
Lognormal; see Appendix 1 $\frac{1}{\sqrt{2\pi}\sigma}\displaystyle\int_0^y \frac{1}{w}\exp\left[-\frac{(\ln w - \mu)^2}{2\sigma^2}\right]dw$	$\exp\left\{\mu + \sigma\left[(2\ln n)^{\frac{1}{2}} - \dfrac{(\ln\ln n + \ln 4\pi)}{2(2\ln n)^{\frac{1}{2}}}\right]\right\}$; $a_n\sigma(2\ln n)^{-\frac{1}{2}}$	E3
Rayleigh $1 - \exp\left(-\frac{x^2}{2\sigma^2}\right)$	$\sigma\sqrt{2\ln n}$; $\sigma/\sqrt{2\ln n}$	E1
2 parameter Weibull $1 - \exp\left[-\left(\frac{x}{\beta}\right)^\alpha\right]$	$\beta(\ln n)^{1/\alpha}$; $\frac{\beta}{\alpha}(\ln n)^{\frac{1-\alpha}{\alpha}}$	$\begin{cases}\alpha > 1 : \text{E1} \\ \alpha = 1 : \text{E2} \\ \alpha < 1 : \text{E3}\end{cases}$

Asymptotic expressions are approximate, valid for large n.
(1): relationship kindly suggested by Han Ping Hong.

If we know the variation of b_n with n, we can determine the behaviour of a_n. We explore this.

$$\frac{db_n}{dn} = \frac{d}{dn}\left(\frac{1}{nf_X(a_n)}\right) = -\frac{1}{n^2 f_X(a_n)} - \frac{f_X'(a_n)}{nf_X^2(a_n)}\frac{da_n}{dn}$$

$$= -\frac{1}{n^2 f_X(a_n)} - \frac{f_X'(a_n)}{nf_X^2(a_n)}\frac{1}{f_X(a_n)}\cdot\frac{1}{n^2},$$

using Equation (9.111). Then

$$\frac{db_n}{dn} = \frac{1}{n^2 f_X(a_n)}\left[\frac{1}{Q(a_n)} - 1\right], \tag{9.114}$$

using Equation (9.95). The term before the square brackets is a positive quantity, while that within is negative for Class E1, zero for Class E2, and positive for E3. The result is plotted schematically in Figure 9.18(a). From Equation (9.113), the behaviour of a_n with increasing values of $(\ln n)$ is sketched in Figure 9.18(b), based on 9.18(a).

9.6.7   Fréchet distribution: type II

The main result for the type II distribution was given in Equation (9.78) in terms of the standardized quantity Y. For the random quantity $(Z - a_n)/b_n$, this can be written as

$$F_Z(z) = \exp\left[-\left(\frac{z - a_n}{b_n}\right)^{-\alpha}\right], \tag{9.115}$$

for $\alpha > 0$ and $-\infty < z < \infty$.

Unlike the Gumbel distribution, the shape will depend on the parameter α. The distribution has limited moments and, as stated earlier, distributions such as the Cauchy and Pareto fall into the Fréchet domain of attraction. The tail of the distribution is rather fat, compared to the Gumbel. This has implications with regard to calculation of failure probability. The mean and standard deviation are given in Appendix 1. The distribution has been used, for example, in analysis of extreme wind speeds and floods.

9.6.8   Weibull distribution: type III; minima

The Weibull distribution of Equation (9.79) is readily converted to the linearly transformed random quantity $(Z - a_n)/b_n$:

$$F_Z(z) = \exp\left[-\left(-\frac{z - a_n}{b_n}\right)^{\alpha}\right] \tag{9.116}$$

for $\alpha > 0$ and $-\infty < z \leq a_n$.

As in the preceding subsection, the parameter α dictates the shape of the distribution (see Appendix 1, also for mean and standard deviation). The distribution is based on parent distributions that are limited in value, to the right for maxima; see Figure 9.13. Often we consider that there is a physical maximum, yet its definition is vague and difficult to define, for example the

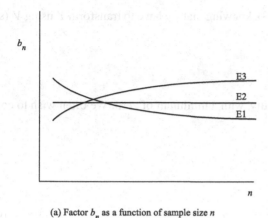

(a) Factor b_n as a function of sample size n

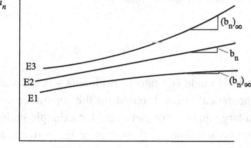

(b) Characteristic extreme as a function of logarithm of sample size n

Figure 9.18   Dispersion factor and characteristic extreme as functions of n

maximum wave height that is physically possible. It is generally better in such instances to treat the quantity as being unlimited, and to check that probabilities generated, for example with the Gumbel distribution, are acceptably small in the regions where a physical limit is considered possible. See also Feller (1968, Chapter 1). A physical limit may be set in an engineered system by devices such as fuses for maximum current, or weak links that fail under specified conditions.

So far in this subsection we have concentrated on maxima. We now transform Equation (9.79) to the case of a minimum. We might be considering droughts, the shortest time to transmit electronic information, or failure of the weakest link in a set of connected links under uniaxial tension. From Equation (9.3)

$$W = \min_n X_i.$$

Then, from Equation (9.4),

$$-W = \max_n (-X_i). \tag{9.117}$$

We use the normalized Equation (9.79), knowing that we have to transform Y using V (say) $= -Y$. Then, using Equation (7.8),

$$F_V(v) = 1 - \exp(-v^\alpha) \tag{9.118}$$

with $0 \leq v < \infty$.

This is a normalized random quantity, with a minimum of zero. We often wish to consider the transformed quantity W, in

$$V = \frac{W - a}{b_n}.$$

Equation (9.118) then becomes

$$F_W(w) = 1 - \exp\left[-\left(\frac{w - a}{b_n}\right)^\alpha\right] \tag{9.119}$$

with $a \leq w < \infty$.

In these equations, we use a rather than a_n since this indicates a physical limit; see Figure 9.13. This is the factor associated with the Weibull domain of attraction.

9.7   Weakest link theories

The strength of materials – even carefully made laboratory specimens – is generally several orders of magnitude less than the theoretical strength based on the separation of atoms in the crystal lattice. Solids consist of a large quantity of elements, for example grains in poly-crystalline materials. The size, orientation and shape of these may differ from one to another, together with the morphology of the grain boundaries. The presence of dislocations will result in movements under stress which will lower the stresses at failure. Heterogeneities and stress concentrations cause failures locally in the material. Defects and flaws in the material will intervene to cause fracture at lower stresses than the theoretical values. These comments are especially relevant to geophysical materials forming under natural conditions, since the flaw structure will be quite significant (Jordaan, 2000).

We now consider an initial model of the 'weakest-link' variety (Weibull, 1951; Bolotin, 1969; Maes, 1985). This is the chain – or, more generally, a structure that may be considered as a series of n elements such that the structure fails when one of the elements fail. If T_i is the strength of the ith element or link, and the strengths of the elements are considered to be *iid* with distribution function $F_T(t)$, the strength of the assembly, $R = \min_n T_i$, and from Equation (9.6),

$$F_R(r) = 1 - [1 - F_T(r)]^n.$$

This can be written

$$F_R(r) = 1 - \exp\{n \ln[1 - F_T(r)]\}. \tag{9.120}$$

In the case where the structure may be considered to be composed of elements that are each of volume v_0, with the total volume $v = nv_0$, then

$$F_R(r) = 1 - \exp\left\{\frac{v}{v_0} \ln\left[1 - F_T(r)\right]\right\}. \tag{9.121}$$

The assumption that continues to be made is that the structure fails when the weakest element fails.

Weibull suggested the use of a material function $m(r)$ to represent the expression $\{-\ln[1 - F_T(r)]\}$ within Equation (9.121), and in particular the power-law equation

$$m(r) = \left(\frac{r - r_0}{r_1}\right)^\alpha, \tag{9.122}$$

where α, r_0 and r_1 are constants with r_0 representing a lower limit on strength. Substituting this into Equation (9.121), it is found that

$$F_R(r) = 1 - \exp\left[-\frac{v}{v_0}\left(\frac{r - r_0}{r_1}\right)^\alpha\right]. \tag{9.123}$$

We compare this with the type III extreme value distribution, given in Equation (9.119), reproduced here:

$$F_W(w) = 1 - \exp\left[-\left(\frac{w - a_n}{b_n}\right)^\alpha\right].$$

The equations are in fact identical in form, with

$$R \equiv W, \tag{9.124}$$

$$r_0 \equiv a_n \text{ and} \tag{9.125}$$

$$\left(\frac{v_0}{v}\right)^{\frac{1}{\alpha}} r_1 \equiv b_n. \tag{9.126}$$

There is thus considerable basis for Weibull's theory: we can interpret it as the asymptotic distribution of the minimum of a set of random strengths with a lower minimum value r_0.

Often it is reasonable to set $r_0 = 0$ in Equation (9.123):

$$F_R(r) = 1 - \exp\left[-\frac{v}{v_0}\left(\frac{r}{r_1}\right)^\alpha\right]. \tag{9.127}$$

9.7.1   Inhomogeneous stress states

Where the state of stress is inhomogeneous, it is still supposed that failure can take place in any link or element. The failure probability of each element is a function of the stress state, and in order to make the analysis tractable, we attempt to write a failure criterion that is a simple function of this stress state.

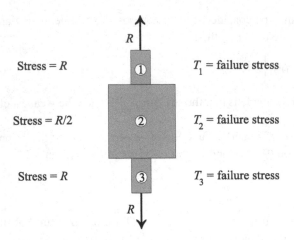

Figure 9.19   Three elements in series

Example 9.11 As an introductory example, consider the situation illustrated in Figure 9.19. This shows three elements in series subjected to uniform tension, the cross sectional area of the elements being a, $2a$ and a, for the elements 1, 2 and 3 respectively. Uniform tensile stress in each element is assumed, stress concentrations being ignored. These stresses are then r, $r/2$ and r respectively for the three elements. We assume that the material fails according to the same distribution function as above, written here as $F_T(t)$, where T_i is the stress in the ith element, $i = 1, 2, 3$, implying that the same assumption as before, that the T_i are *iid*, is made. The failure stress is random since the T_i are, and the common value is denoted R (as before). This value has to be associated with one of the stresses, and we choose the stress in element 1, i.e. we use r (recall that the stresses in elements 1, 2 and 3 are r, $r/2$ and r respectively); thus r is the value of R at which failure of the assembly takes place. The failure of the assembly is then given by

$$R = \min\left(T_1, T_2, T_3\right), \tag{9.128}$$

and

$$F_R(r) = 1 - \left[1 - F_T(r)\right]^2 \left[1 - F_T(r/2)\right]. \tag{9.129}$$

The consequence of the inhomogeneity is reflected in the inclusion of the different stress values in the three elements.

For more general states of stress, for instance a varying three-dimensional stress in a continuum given by the stress tensor σ_{ij}, Weibull proposed that one consider a homogeneous stress state in a small volume Δv around a point with cartesian coordinates $\mathbf{x}$. In order that the stress σ_{ij} at this point be represented by a single scalar value, say s, it is supposed that one can write

$$s = r_R \cdot \phi(\mathbf{x}), \tag{9.130}$$

where $\phi(\mathbf{x})$ is a function of position only, and r_R represents a reference value, for instance the maximum value in the body. See Exercise 9.18 for application of this to Example 9.11.

A more general example is given by the failure criterion represented by the maximum elastic strain (Bolotin, 1969):

$$\epsilon = \frac{1}{E}\left[\sigma_1 - \mu\left(\sigma_2 + \sigma_3\right)\right], \tag{9.131}$$

where $E =$ Young's modulus and $\mu =$ Poisson's ratio. This strain can be written as a function of position, expressed by a non-dimensional function $\phi(\mathbf{x})$ as

$$\epsilon \equiv s = r_R \phi(\mathbf{x}) = r\phi(\mathbf{x}). \tag{9.132}$$

The value $r_R = r$ is, as noted, a reference value, for example the maximum strain at a certain point in the body. The strength T_i of the ith element is then described by the distribution function

$$F_T(s) = F_T\left[r\phi(\mathbf{x})\right]. \tag{9.133}$$

The function $F_T(t)$ is the same as that used previously, for example in Equation (9.121); in effect s replaces t, with ϕ representing a deterministic variation in stress. Randomness is embodied in $F_T(\cdot)$, taking particular values $r\phi$. The reduction of a stress state to a single value, as just described, can also be used with the maximum-principal-stress or the Coulomb–Mohr failure criteria, or others. See also Exercise 9.19.

The volume v_0 of the body is divided into k small elements Δv_m, $m = 1, \ldots, k$. The supposition is made that the elements are small enough to ensure a homogeneous stress state, and yet the assumption of independence of strength between elements is maintained (exchangeability would often make a more natural assumption; see Section 9.9). Again we study $(\min T_i)$. The randomness is (as noted above) represented in $F_T(\cdot)$, which takes an independent value for each element. Equation (9.121),

$$F_R(r) = 1 - \exp\left\{\frac{v}{v_0}\ln\left[1 - F_T(r)\right]\right\},$$

becomes

$$F_R(r) = 1 - \prod_{m=1}^{k}\exp\left(\frac{\Delta v_m}{v_0}\ln\left\{1 - F_T\left[r\phi_m(\mathbf{x})\right]\right\}\right). \tag{9.134}$$

This is a generalization of Equation (9.129). The value v_0 should be interpreted as a reference volume; it has been described, for example, as the volume of a standard test specimen (e.g. Bolotin, 1969). The function $\phi_m(\mathbf{x})$ applies to the small volume Δv_m. Equation (9.134) is written as

$$F_R(r) = 1 - \exp\left[\frac{1}{v_0}\sum_{m=1}^{k}\left(\Delta v_m \ln\left\{1 - F_T\left[r\phi_m(\mathbf{x})\right]\right\}\right)\right]. \tag{9.135}$$

Replacing the sum in Equation (9.135) with an integral,

$$F_R(r) = 1 - \exp\left\{\frac{1}{v_0}\int_v \ln\{1 - F_T\left[r\phi\left(\mathbf{x}\right)\right]\}\,dv\right\}. \tag{9.136}$$

Now a similar simplification as before is made, writing the material function $m\,(r)$ as

$$m\,(r) = -\ln\{1 - F_T\left[r\phi\left(\mathbf{x}\right)\right]\} = \left(\frac{r\phi\left(\mathbf{x}\right) - r_0}{r_1}\right)^{\alpha}, \tag{9.137}$$

so that

$$F_R(r) = 1 - \exp\left\{-\frac{1}{v_0}\int_v\left(\frac{r\phi\left(\mathbf{x}\right) - r_0}{r_1}\right)^{\alpha}dv\right\}. \tag{9.138}$$

A simplification can be made if we let $r_0 = 0$:

$$F_R(r) = 1 - \exp\left\{-\frac{1}{v_0}\left(\frac{r}{r_1}\right)^{\alpha}\int_v\phi^{\alpha}\left(\mathbf{x}\right)dv\right\}. \tag{9.139}$$

The integral in this equation defines a 'reduced volume'

$$v_* = \int_v\phi^{\alpha}\left(\mathbf{x}\right)dv, \tag{9.140}$$

so that

$$F_R(r) = 1 - \exp\left[-\frac{v_*}{v_0}\left(\frac{r}{r_1}\right)^{\alpha}\right]. \tag{9.141}$$

Comparing this with Equation (9.127), we see that the equations are identical if the reduced volume v_* replaces the volume v of the body under homogeneous stress.

Example 9.12 Consider a beam of span l, with a rectangular cross section (depth h and width b), subjected to pure bending moment m. The material is brittle and can be modelled using Weibull's theory with a maximum-tensile-stress criterion. Find the reduced volume v_*. Let y be the distance from the neutral axis. We need to consider stresses only in the tensile half of the cross section. The maximum stress is taken as the reference value of stress, i.e.

$$r_R = r = \frac{6m}{bh^2}. \tag{9.142}$$

Then the stress varies only with y and is equal to

$$r\cdot\frac{2y}{h} \tag{9.143}$$

and

$$\phi\,(y) = \frac{2y}{h}. \tag{9.144}$$

The reduced volume is

$$
v_* = bl \int_0^{\frac{h}{2}} \left(\frac{2y}{h}\right)^\alpha dy
$$

$$
= \frac{blh}{2(\alpha + 1)}.
$$

(9.145)

This is the entire volume divided by $2(\alpha + 1)$.

In Exercise 9.20, other loading cases are considered. We should note the simplifying aspects involved in this analysis. The shearing stress is zero in Example 9.12, and shear is neglected in the exercises involving bending. It might be reasonable to neglect shear in slender beams, but its presence is of considerable importance in many instances, particularly in reinforced and prestressed concrete beams. Shear is important in many other cases too; the spalling activity in compressive failure in ice-structure interaction, leading to mitigation of load severity, may often be associated with shear acting across weak grain boundaries.

9.7.2   Scale effects

The mean and variance of the distribution (9.123) are

$$
\langle R \rangle = r_0 + r_1 \left(\frac{v}{v_0}\right)^{-\frac{1}{\alpha}} \Gamma \left(1 + \frac{1}{\alpha}\right)
$$

(9.146)

and

$$
\sigma_R^2 = r_1^2 \left(\frac{v}{v_0}\right)^{-\frac{2}{\alpha}} \left[\Gamma \left(1 + \frac{2}{\alpha}\right) - \Gamma^2 \left(1 + \frac{1}{\alpha}\right)\right].
$$

(9.147)

From the mean value, when $r_0 = 0$, we may compare the strength of two volumes v_1 and v_2:

$$
\frac{\langle R \rangle_1}{\langle R \rangle_2} = \left(\frac{v_2}{v_1}\right)^{\frac{1}{\alpha}};
$$

(9.148)

this may also be used with reduced volumes.

The standard deviation of R is

$$
\sigma_R = r_1 \left(\frac{v}{v_0}\right)^{-\frac{1}{\alpha}} \left[\Gamma \left(1 + \frac{2}{\alpha}\right) - \Gamma^2 \left(1 + \frac{1}{\alpha}\right)\right]^{\frac{1}{2}},
$$

(9.149)

and the coefficient of variation γ is given by

$$
\gamma = \frac{r_1 \left(\frac{v}{v_0}\right)^{-\frac{1}{\alpha}} \left[\Gamma \left(1 + \frac{2}{\alpha}\right) - \Gamma^2 \left(1 + \frac{1}{\alpha}\right)\right]^{\frac{1}{2}}}{r_0 + r_1 \left(\frac{v}{v_0}\right)^{-\frac{1}{\alpha}} \Gamma \left(1 + \frac{1}{\alpha}\right)}.
$$

(9.150)

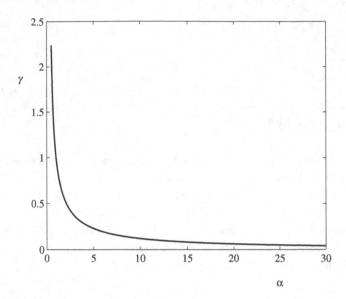

Figure 9.20 Coefficient of variation γ as a function of α

For $r_0 = 0$, this becomes

$$\gamma = \frac{\left[\Gamma\left(1 + \frac{2}{\alpha}\right) - \Gamma^2\left(1 + \frac{1}{\alpha}\right)\right]^{\frac{1}{2}}}{\Gamma\left(1 + \frac{1}{\alpha}\right)}$$

$$= \left[\frac{\Gamma\left(1 + \frac{2}{\alpha}\right)}{\Gamma^2\left(1 + \frac{1}{\alpha}\right)} - 1\right]^{\frac{1}{2}}, \tag{9.151}$$

a function of α only. This equation can also be derived from Equation (9.141) using reduced volume v_*. Figure 9.20 shows a plot of γ versus α. It is seen that the coefficient of variation decreases considerably for large α.

9.7.3   Applications and discussion

First, we consider the geometry of the solid. If the cross-sectional area is constant, for instance in a bar under tension, volume in Equation (9.123) can be replaced by length. In other cases, the stressed volume of a solid might be proportional to a linear dimension cubed. For instance, the 'bulb' of stressed material in a semi-infinite solid with contact area proportional to the square of a linear dimension l will have a stressed volume proportional to l^3. Figure 9.21 shows the results from tensile tests by Griffith (1920) on glass threads, while Figure 9.22(a) shows values for tensile (flexural) tests (Jordaan and Pond, 2001). Certainly scale effects are in evidence, and lower values of α, in evidence for ice, support also the high variability, as suggested by Figure 9.20. Figure 9.22(b) shows a collection of results from ice compressive

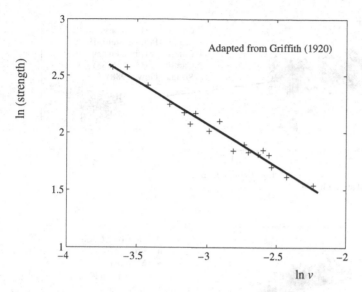

Figure 9.21   Griffith's results for scale effect

failure. In the latter case, the stressed volume is proportional to l^3, and area to l^2. The observed decrease, on average, of strength with $(area)^{-1/2}$ is supported by measured values of $\alpha \simeq 3$ in Equations (9.127) and (9.141); see Sanderson (1988).

The methods above have been applied to 'bundles'. Freudenthal (1968) introduces this concept, which originated in the work of Daniels in 1945. In this analysis, instability of the dominant flaw no longer leads to failure of the entire specimen or structure under consideration. The bundle consists of a large number of parallel threads or fibres with the same length and cross-sectional area. Failure first takes place in the weakest fibre, and the load is then distributed amongst the remaining ones, but does not necessarily lead to failure of the assembly. Another important area of research lies in the attempt to link the failure process to the flaw structure in the material, taking into account the state of stress at the point under consideration. The approaches of Hunt and McCartney (1979) and Maes (1992) represent examples of this important direction of research.

9.8   Poisson processes: rare extreme events

A model of arrivals of events that is often a good abstraction is the Poisson distribution (Sections 3.2.6 and 3.6.1). It is especially useful for cases of rare events, but can be used more widely. Arrivals of electronic communications, packets of data, telephone calls, errors, accidents, floods, earthquakes, and exceedance of wind speeds during typhoons, all of these subjects can be analysed as Poisson processes. There are cases where the 'arrival rate' for the extreme events is quite small, for example where the events are bolts of lightning, accidents,

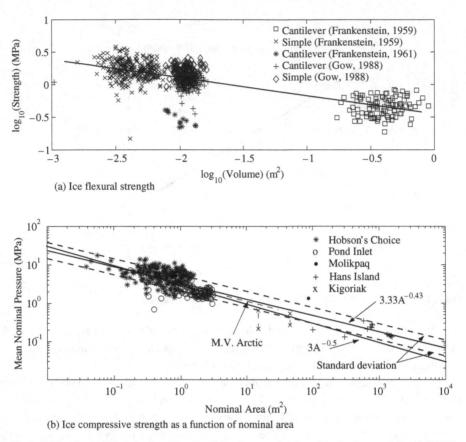

(a) Ice flexural strength

(b) Ice compressive strength as a function of nominal area

Figure 9.22   Scale effect for ice. Details of the data can be found in Jordaan and Pond (2001)

hurricanes, earthquakes or collisions of icebergs with structures at certain offshore locations. Figure 9.23 illustrates the situation, which we shall explain in detail in the following.

We denote the events under consideration (messages, accidents, floods, typhoons,...) by the symbol E_i, $i = 1, 2, 3, \ldots$ It should be recalled (Section 3.2.6) that, for many applications, the arrival rate does not have to be constant. For our analysis we need the expected number of arrivals ν during the period of interest t (often a year):

$$\nu = \int_0^t \lambda(\tau)\,d\tau,$$

where the rate $\lambda(\cdot)$ can vary with (for example) season; if λ is constant, then $\nu = \lambda t$. We consider again the case where the parent distribution of the random quantity of interest, X, is $F_X(x)$. The magnitude X_i is associated with each event E_i, and again we make the assumption that the events are independent and identically distributed (*iid*).

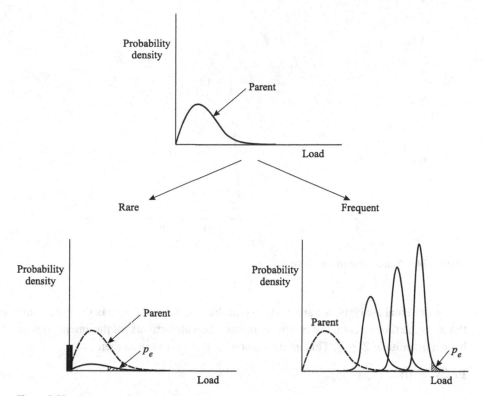

Figure 9.23 Rare and frequent processes

Within the Poisson process, we consider those occurrences with a magnitude X greater than a specified value $Z = z$. These events are denoted as F_i, a subset of the E_i, such that $\{X_i > z\}$. The quantity Z denotes, as before, the extreme value of interest during the period of interest. The number of events E_i during the period is a random quantity, denoted N, and this can be modelled by the Poisson distribution:

$$p_N(n) = \frac{\exp(-\nu)\,\nu^n}{n!}, \text{ for } n = 0, 1, 2, \ldots \tag{9.152}$$

Now the events F_i correspond to those events within the group E_i that exceed in magnitude $Z = z$, so that the expected number of events F_i is

$$\nu' = \nu\,[1 - F_X(z)]. \tag{9.153}$$

The F_i therefore constitute another Poisson process, with the amended parameter ν'. The arrivals in this case number N' such that

$$\mathbf{Pr}\left(N' = n'\right) = \frac{\exp\left(-\nu'\right)\left(\nu'\right)^{n'}}{n'!}. \tag{9.154}$$

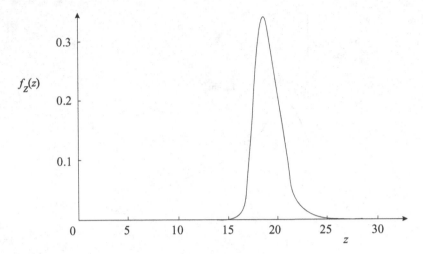

Figure 9.24 Annual maximum wave height

For extremal analysis, we are interested in the case where $Z = z$ is the maximum value of the X_i in the time interval under consideration. Then all 'arrivals' of the amended process must have a magnitude $Z \leq z$. This implies that $n' = 0$ in (9.154) and then

$$\mathbf{Pr}\left(N' = 0\right) = \exp\left(-\nu'\right).$$
(9.155)

This corresponds to the case that $Z \leq z$ so that (9.155) is equal to

$$\mathbf{Pr}\left(Z \leq z\right) = F_Z\left(z\right) = \exp\left\{-\nu\left[1 - F_X\left(z\right)\right]\right\}.$$
(9.156)

This gives us the probability distribution of the extreme value Z.

Example 9.13 We may apply this idea to the problem introduced in Section 9.4.3, where we considered a number of waves per storm to obtain the distribution of the extreme wave during the storm (Equations (9.32) and (9.33) and Figure 9.6). If there are $\nu = 5$ storms per year then the distribution of the annual maximum, on the assumption of a Poisson distribution of the number of storms per year, is

$$F_Z\left(z\right) = \exp\left[-5\left(1 - \left\{1 - \exp\left[-2\left(\frac{z}{9.0}\right)^2\right]\right\}^n\right)\right].$$
(9.157)

This is shown in Figure 9.24 for five storms per year and $n = 1000$ waves per storm.

If the distribution of X is exponential, let us say the shifted kind

$$f_X\left(x\right) = \frac{1}{\beta}\exp\left(-\frac{x - \alpha}{\beta}\right), \ x \geq \alpha$$
(9.158)

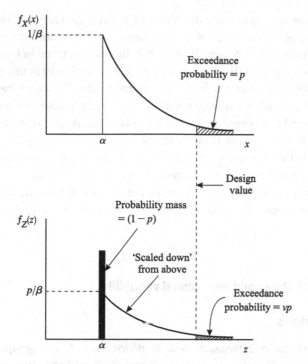

Figure 9.25 Simple scaling

(and zero elsewhere), then

$$F_X(x) = 1 - \exp\left(-\frac{x - \alpha}{\beta}\right). \tag{9.159}$$

Equation (9.156) now takes the double exponential form

$$F_Z(z) = \exp\left\{-\nu\left[\exp\left(-\frac{z - \alpha}{\beta}\right)\right]\right\}$$

$$= \exp -\left\{\exp\left[-\left(\frac{z - \alpha - \beta \ln \nu}{\beta}\right)\right]\right\}. \tag{9.160}$$

The contrast between rare and frequent events is illustrated in Figure 9.23. In the case where ν is sufficiently small, Equation (9.156) becomes

$$F_Z(z) \simeq 1 - \nu[1 - F_X(z)] = 1 - \nu p, \tag{9.161}$$

where p is the probability that X exceeds the value z. This distribution for rare events is shown in Figure 9.25, taking the exponential form of Equations (9.158) and (9.160) as an example. This case of 'simple scaling' has been used for the specification of design loads for rare environmental processes, such as earthquake and some ice loading events (Jordaan and Maes, 1991), and in studies of spills of hazardous materials by the author.

The Poisson process will automatically give a spike of probability at the lower bound (where it exists) of the quantity X. Suppose that $F_X(z) = 0$ at $x = z = \alpha$ in Equation (9.156) – an example of such a lower bound below which there is no probability mass is given in Equation (9.158). Then Equations (9.156) and (9.160) tell us that $F_Z(z = \alpha)$ is $\exp(-\nu)$ at this point. This has to consist of a probability mass at $z = \alpha$, corresponding to the Poisson probability of $N = 0$ in Equation (9.152). It is approximately equal to $(1 - \nu)$ in the case of Equation (9.161). There is therefore no need to be concerned about the occurrence of quantities that are less than α in Equation (9.158) for the case of Poisson arrivals.

The methodology described in this section has been applied in studies of floods, wave heights and extreme wind speeds during typhoons. It is described in the literature in these areas as the 'peak over threshold' (POT) model. These applications are concerned with the analysis of environmental events which exceed a certain threshold, modelled as $Z = z$ in Equation (9.155).

9.9   Inference, exchangeability, mixtures and interannual variability

9.9.1   Exchangeable random quantities

In Chapter 8, we pointed out that making inferences about the distribution of a random quantity using data invariably leads to changing one's assessment of probabilities. The same situation exists with regard to extremes: the use of data will lead to inferences regarding parameters of the distributions under consideration, in this case an extremal distribution. The consequence is that probabilities will change, and independence will no longer apply. Then our *iid* assumption used so far in this chapter is no longer valid. The simplest and most natural way out of this quandary is to consider again exchangeable random quantities. Then independence will again be embedded in the structure of the analysis as part of exchangeability (conditional on the parameter value). The analysis is then aimed at modelling the uncertainty regarding the parameter(s) of the extremal distribution.

Under the hypothesis of exchangeability, if we consider a set of values, possibly observations, we might make the judgement that the order of the values in the set is of no consequence, as discussed in Chapter 8. The subject of study might be the time between arrival of packets of data in a network during a peak transmission period, or the heights of individual waves during a storm. We may again avail ourselves of an urn model, as in Chapter 8 and in Section 9.3 above. Suppose again that it contains $R = r$ red '1' balls and $n - r$ pink '0' ones. But the urn model will be of random composition, reflecting the fact that we are uncertain about the value of the parameter(s) of our distribution, exactly as we were in Section 8.2. We denote the parameter(s) by the Λ, taking particular values λ. To focus thinking, Λ might represent the parameters a_n and b_n in the extremal distribution of wave heights in the design of an offshore rig.

We now develop an equation analogous to (9.12). Instead of a known composition of the urn we suppose that the proportion of red balls represents the *cdf* $F_\Lambda(\lambda)$ of the parameter.

Then we write $F_{X,\Lambda}(x|\lambda)$ as the probability that $X \leq x$ given that $\Lambda = \lambda$. The probability of a particular composition being in the interval $\lambda < \Lambda < \lambda + d\lambda$ is then

$$dF_\Lambda(\lambda) = f_\Lambda(\lambda)\,d\lambda$$

for a continuous distribution $f_\Lambda(\lambda)$. For any particular order of drawing, for example all red balls first, followed by all the pink ones,

$$\mathbf{Pr}(R = r) = p_{r,n} = \int_{-\infty}^{\infty} F_{X,\Lambda}^r(z|\lambda)\left[1 - F_{X,\Lambda}(z|\lambda)\right]^{n-r}\,dF_\Lambda(\lambda). \tag{9.162}$$

Counting all orders,

$$\mathbf{Pr}(R = r) = \omega_{r,n} = \binom{n}{r}\int_{-\infty}^{\infty} F_{X,\Lambda}^r(z|\lambda)\left[1 - F_{X,\Lambda}(z|\lambda)\right]^{n-r}\,dF_\Lambda(\lambda). \tag{9.163}$$

Using the same arguments as we developed leading up to Equation (9.14), the *cdf* of the rth ordered random quantity $X_{(r)}$ is:

$$F_{X_{(r)}}(z) = \sum_{t=r}^{n}\binom{n}{t}\int_{-\infty}^{\infty} F_{X,\Lambda}^t(z|\lambda)\left[1 - F_{X,\Lambda}(z|\lambda)\right]^{n-t}\,dF_\Lambda(\lambda). \tag{9.164}$$

The theory can be further developed into consideration of order statistics, but we shall concentrate on the case where $r = n$ or $r = 1$, that is

$$F_Z(z) = \int_{-\infty}^{\infty} F_{X,\Lambda}^n(z|\lambda)\,dF_\Lambda(\lambda) \tag{9.165}$$

for a maximum, and

$$F_W(w) = 1 - \int_{-\infty}^{\infty}\left[1 - F_{X,\Lambda}(w|\lambda)\right]^n\,dF_\Lambda(\lambda) \tag{9.166}$$

for a minimum. Compare the corresponding *iid* equations, (9.5) and (9.6). The *pdf* of the maximum (for example) is

$$f_Z(z) = n\int_{-\infty}^{\infty} F_{X,\Lambda}^{n-1}(z|\lambda)\,f_{X,\Lambda}(z|\lambda)\,dF_\Lambda(\lambda), \tag{9.167}$$

corresponding to Equation (9.7).

We see that, in order to implement exchangeability, we may, as in Chapter 8, treat the random quantity as being *iid*, conditional on the parameter value, that $\Lambda = \lambda$. One then integrates over the possible parameter values to obtain the required distribution. It is always important to take into account uncertainty in parameter values in probabilistic analysis, and this can be very important in estimating extreme values in design.

Example 9.14 We consider a random quantity X ($0 \leq X \leq 1$) representing the proportion of devices that successfully fulfil their function during the lifetime of an assembly. The uncertainty is modelled using the power-law relationship

$$f_{X,\Lambda}(x|\lambda) = (\lambda + 1)x^\lambda, \tag{9.168}$$

with uncertainty in the parameter Λ expressed by

$$f_\Lambda(\lambda) = \frac{1}{\beta}\exp - \left(\frac{\lambda}{\beta}\right), \tag{9.169}$$

where β is a constant.

From Equation (9.168),

$$F_{X,\Lambda}(x|\lambda) = x^{\lambda+1},$$

and the distribution of the maximum of n of the Xs is, by Equation (9.165),

$$\begin{aligned}
F_Z(z) &= \int_0^\infty \left(z^{\lambda+1}\right)^n \frac{1}{\beta}\exp - \left(\frac{\lambda}{\beta}\right)\, d\lambda \\
&= \frac{z^n}{\beta}\int_0^\infty \exp\left[\left(n\ln z - \frac{1}{\beta}\right)\lambda\right]d\lambda \\
&= \frac{z^n}{1 - \beta n \ln z}.
\end{aligned} \tag{9.170}$$

Conjugate priors

Generally, the extremal distributions do not possess conjugate prior distributions, as were introduced in Chapter 8; see Exercise 9.23. Both classical and Bayesian inference present difficulties. There are approximate techniques to deal with this, for example Evans and Nigm (1980) dealt with the Weibull distribution. Maes (1985) developed a special distribution (the exchangeable-extremal distribution) that does have sufficient statistics and a conjugate prior. Numerical techniques are also of importance; Monte Carlo techniques are also of great usefulness; see Chapter 11. It is noteworthy that the parameter of the Poisson distribution may be modelled using conjugate prior distributions of the gamma family (Section 8.5.4). This can be used with the Poisson distribution in the extremal processes of Section 9.8. This interpretation offers one way forward.

9.9.2 Parameter uncertainty; fixed- and variable-in-time; return periods; inter-period variation

Uncertainty in the parameter of an extreme distribution will generally result when estimating from a data set; there will be more or less uncertainty depending on the size of the data base. Inference proceeds via prior information combined with data (likelihood). We now discuss some aspects that relate to the source of uncertainty. In Nessim *et al.* (1995) a classification was proposed for environmental loads, that was designed to be as simple as possible. This included the distinction between uncertainties that are fixed-in-time and variable-in-time. Some related aspects are briefly discussed.

One might find that, as data comes in, the parameter for extreme loading is at the upper end of the supposed range. There may be physical reasons why a parameter takes a high value at a certain physical location, for instance as a result of exposure to wind, or because of other construction nearby a structure might become exposed to above-average wind loading. In these cases, knowledge of the parameter leads to higher estimates of loading. We have advocated the use of Equation (9.165) for the extremal distribution $F_Z(z)$. In calculating return periods, Equation (9.35),

$$t_R = \frac{1}{1 - F_Z(z)},$$

the uncertainty associated with the parameter must be taken into account. It is fixed-in-time under the circumstances just discussed, and will correctly lead to longer return periods.

If we assume that the final uncertainty represents a mixture, with **different parameter values for different structures**, but that the different structures are exchangeable, there will be different but unknown safety levels for the different structures. The inclusion of parameter uncertainty in the estimate of return period in this case will result in an overall estimate, averaged over the distribution of the parameter, and the resulting reliability will not be uniform. This might be of minor significance if the parameter uncertainty between structures is relatively small, but is a valid subject of study.

A second distinct situation occurs when there is variability **between exposure periods**. The modelling of a wave or wind regime does not usually acknowledge explicitly the different conditions from year to year. Higher-than-average seismic activity might result during periods when rock masses are in a highly stressed state – but this might be unknown to the analyst. One is forced to acknowledge this in considering the severity of threat from icebergs; there are years on the Grand Banks when no icebergs appear so that the probability of collision is zero. In other years the areal density is considerable, up to 50 per degree-rectangle, and then there is a probability of collision greater than zero. Similarly, ice thickness in temperate zones or the arctic can vary from year to year depending on the accumulated freezing degree-days for the year in question. Biological systems also have to contend with changes from year to year that are detectable and measurable. Drought and the presence or absence of ice are examples. Demand in data transmission (for example) will vary from one period to another for a variety of reasons. These examples are variable-in-time.

In this second situation, the use of the expression for t_R above needs special attention. Let us continue with the question of iceberg populations as an example. To simplify the discussion, let us assume that in our area of concern, two kinds of iceberg 'years' occur. In the kind 'A', there are no icebergs in the area, so that the probability of collision is zero. In the kind 'B', the iceberg population is such that the probability of collision is 0.2. The return period corresponding to the condition A is $t_R = 1/p = \infty$, and for B it is $t_R = 1/p = 5$ years. If we now say that the return period is the expected value of these possibilities, both of which occur with a probability of $1/2$, we obtain a result equal to infinity. This is clearly unreasonable. A better result is obtained if one judged the yearly conditions to be exchangeable between years.

The probability of collision for a given year for which we have no information as to conditions would be the weighted probability of collision given the two possible scenarios. This is

$$p = 0.5\,(0) + 0.5\,(0.2) = 0.1.$$

The return period is then

$$\frac{1}{p} = 10 \text{ years.}$$

This makes sense in terms of the usual understanding of the expression 'return period', and the result corresponds to the expression $t_R = 1/[1 - F_Z(z)]$ with the weighted probability.

We see that the definition of return period as the expected number of years to the first event does not work if we make explicit analysis of the changing probabilities from year to year. If we use the 'average' probability, things work out when there is inter-period variability. The overall analysis might be viewed as corresponding to the Raiffa urn problem, as discussed in Chapter 2 (Figure 2.10) in the context of Bayes' theorem. If the parameter value corresponds to the composition of one urn only (of the two), but we don't know which, we have the situation at the beginning of the present subsection (fixed-in-time). If we have a mixture of urns, with a (randomly) different urn each period, we have the situation just discussed, with different iceberg concentrations corresponding to different urns in different years. Exchangeability implies that we have no means of distinguishing *a priori* between the periods, and details of the 'composition' would be hidden. The reader might wish to construct the Raiffa urn corresponding to this situation.

9.9.3   Mixtures

One might wish to model explicitly the inter-period variability. Thus the value of $F_X(x)$ could be given by a mixture

$$F_X(x) = \int F_{X|A}(x|a) f_A(a)\,da. \tag{9.171}$$

As a mixture, the distribution $f_A(a)$ does not converge to a 'spike' with accumulated data as in the case of exchangeability but expresses an unchanging reality – for example a reflection of nature, with some interannual variation such as mean temperature, or accumulated rainfall or freezing degree-days. To use this as a model, some additional information has to be included in the idealization, a parameter that characterizes the physics such as those just mentioned. It could then be used for updating probabilities in a particular year (or other period).

9.10   Concluding remarks

Analysis of extremes has an important role in the design of engineering systems. The application usually follows one of two approaches. The first is evaluation of the parent distribution, and subsequent generation of the extremal distribution. One of the difficulties that might present itself is that of extrapolation for large n. Do the same physical phenomena carry over into higher values of the quantity under consideration? The other approach involves direct estimate of the extremal distribution based on data. In either case, understanding of the generation of extreme-value distributions is essential, as this will guide us in our modelling of physical systems.

9.11 Exercises

9.1 Verify that Equation (9.15) follows from Equation (9.14).

9.2 *Basic exercise in extremes to become familiar with the process.*

(a) Carry out Monte Carlo simulations (Chapter 11) to reproduce the extremal distributions obtained in Section 9.4. The Monte Carlo procedure should follow the procedure outlined under 'Random blows'. Histograms and fitted distributions should be produced.

(b) Repeat (a) by using the relationship $F_Z(z) = F_X^n(z)$ based on the parent *cdf*, thereby obtaining graphical representations of extremal distributions as in Section 9.4.

9.3 Find a second proof of the return period relationship $1/p$.

9.4 Show that Equation (9.14) reduces to the same form as (9.6) for the lowest endpoint ($r = 1$).

9.5 Find a closed form solution for *pdf* of the extreme wave (Section 9.4.3) given a Rayleigh parent. Plot behaviour with respect to $\log n$.

9.6 Repeat Exercise 9.2 but using the minimum of a set of random quantities, first using Monte Carlo methods and second by implementing Equation (9.6). Consider parent distributions both unlimited and limited to the left.

9.7 Repeat Exercise 9.2 but considering now a situation based on Equation (9.15), for instance study the distribution of the fifth out of eight for a given parent distribution such as the exponential.

9.8 An event (rams of an arctic vessel, voltage surges of a certain magnitude) occurs 20 times in a year. The distribution of its magnitude X is exponential, $\lambda \exp(-\lambda x)$. Find the magnitude in terms of the mean $1/\lambda$ such that the return period is 100 years. Assume *iid*.

9.9 An approximation that is often useful for very small p ($\ll 1/n$) is

$$(1 - p)^n \simeq 1 - np. \tag{9.172}$$

(a) Apply this in Exercise 9.8.

(b) Equation (9.49) states that

$$F_Z(x_n) = F_X^n(x_n) = \left(1 - \frac{1}{n}\right)^n \to \frac{1}{e},$$

in other words the n-year load on a structure has a 63% chance of being exceeded in n years for large n. A designer wishes therefore to choose a more conservative value. He then considers a value $\check{z}$ with a small probability of exceedance of the n-year load p_{ne} such that

$$p_{ne} = 1 - F_Z(\check{z}) \tag{9.173}$$

rather than the mode. Then if X is the parent,

$$[F_X(\check{z})]^n = 1 - p_{ne}, \text{ or} \tag{9.174}$$

$$F_X(\check{z}) = (1 - p_{ne})^{1/n} \simeq 1 - \frac{p_{ne}}{n}, \tag{9.175}$$

using a variation of the rule (9.172). Obtain p_{ne} in terms of a_n, b_n and α for a Weibull parent distribution based on Equation (9.116), with X replacing Z:

$$F_X(x) = \exp\left[-\left(-\frac{x - a_n}{b_n}\right)^\alpha\right]. \tag{9.176}$$

An engineer in offshore practice would realize that a load factor is usually applied to the 100-year load, giving a result that approaches the situation described.

9.10 Find the mode of the Rayleigh distribution

$$f_X(x) = \frac{2x}{\sigma^2} \exp\left(-\frac{x^2}{\sigma^2}\right).$$

9.11 Consider the following form of the Pareto distribution:

$$F_X(x) = 1 - x^k,$$
$$x \geq 1, k > 0,$$

which has been used, for example, to model relative income distributions. Show that L'Hôpital's rule does not apply, and that the critical quotient is equal to

$$\frac{k}{k + 1}.$$

9.12 Compute and plot the intensity function and critical quotient for the following distributions: normal, logistic and lognormal. Compare to that of the exponential distribution.

9.13 Show that the mode of the Gumbel distribution

$$f_Z(z) = \frac{1}{\beta} \exp\left[-\exp\left(-\frac{z - \alpha}{\beta}\right) - \frac{z - \alpha}{\beta}\right]$$

is $\hat{z} = \alpha$. Show also that this is approximately equal to the characteristic extreme z_n.

9.14 Compare the results of using Table 9.1 with distributions obtained from the parent using Equation (9.5).

9.15 Derive the type III (Weibull) distribution from type I and type II. Explain which parameters depend on n.

9.16 Calculate the hazard function for the distribution $F_X(x) = 1 - \exp(-x^k)$, where k is a constant.

9.17 Plot Equation (9.114),

$$\frac{db_n}{dn} = \frac{1}{n^2 f_X(a_n)} \left[\frac{1}{Q(a_n)} - 1 \right],$$

for the distribution in Example 9.10, and discuss the behaviour of b_n with a_n.

9.18 Consider Example 9.11. Let the vector

$$\mathbf{x} = \begin{Bmatrix} 1 \\ 2 \\ 3 \end{Bmatrix}$$

represent the position of the three elements. We took r_R as r (the stress in element 1). Find a matrix ϕ that transforms this into

$$r \begin{Bmatrix} 1 \\ \frac{1}{2} \\ 1 \end{Bmatrix},$$

giving the relative values of s in each element.

9.19 Consider the failure criterion

$$s = \left(\frac{\sigma_{11}}{\sigma_0} \right)^d + \left(\frac{\sigma_{22}}{\sigma_0} \right)^d + \left(\frac{\sigma_{33}}{\sigma_0} \right)^d,$$

where σ_{11}, σ_{22} and σ_{33} are principal stresses, and d and σ_0 are constants. Show how this can be used in Weibull's strength distribution using an equation of the kind (9.122).

9.20 Find the reduced volume v_*, as in Example 9.12 for a cylindrical bar under torsion; for a prismatic beam simply supported under point load; and for a prismatic beam, simply supported, loaded symmetrically by two point loads. Treat the reduced stress as the greatest principal stress.

9.21 Consider the following non-homogeneous failure condition. A bar is stressed under uniform tension. Two aspects are to be considered. The first relates to a population of surface flaws, and the second to flaws within the volume of the material. How would you approach this problem?

9.22 Show that the extremal distributions do not possess conjugate prior distributions or sufficient statistics.

9.23 Analyse how you would take into account parameter uncertainty in Equation (9.148) for the scale effect, and in (9.151) for the variance.

9.24 Consider an extremal process that is modelled as a Poisson process. Treat the arrival rate as being random. Develop a 'peak over threshold' (POT) model that includes parameter uncertainty in the arrival rate and an application of the model.

10 Risk, safety and reliability

The R-101 is as safe as a house, except for the millionth chance.

Lord Thomson, *Secretary of State for Air, shortly before boarding the airship headed to India on its first flight,*
October 4, 1930

If you risk nothing, then you risk everything.

Geena Davis

10.1   What is risk? Its analysis as decision theory

It is commonplace to make statements such as 'life involves risk', or 'one cannot exist without facing risk', or 'even if one stays in bed, there is still some risk – for example, a meteorite could fall on one, or one could fall out and sustain a fatal injury', [1] and so on. The Oxford dictionary defines risk as a 'hazard, chance of . . . bad consequences, loss, . . . ' ; Webster's definition is similar: 'the chance of injury, damage or loss' or 'a dangerous chance'. Two aspects thrust themselves forward: first, chance, and second, the unwanted consequences involved in risk. We have taken pains to emphasize that decision-making involves two aspects: probabilities (chance), and utilities (consequences). Protection of human life, property, and the environment are fundamental objectives in engineering projects. This involves the reduction of risk to an acceptable level. The word 'safety' conveys this overall objective.

In Chapter 4, aversion to risk was analysed in some detail. The fundamental idea was incorporated in the utility function. We prompt recollection of the concept by using an anecdote based on a suggestion of a colleague. A person is abducted by a doctor who is also a fanatical decision-theorist. He offers the person the following two options before release from captivity. First, removal of one kidney, then release. Second, toss an honest coin: if heads, release of the captive (with both kidneys intact); if tails, removal of both kidneys, followed by release. Our attribute, to use the notation of Chapter 4, would be 'number of kidneys lost'. The expected loss for each option is the same; one kidney. We would be inclined to choose the first option to avoid the possible dire consequences of the second. The utility would then be a concave

---

[1] or, less commonplace, in the last instance: 'one could suffer the fate of Howard Hughes!'

514

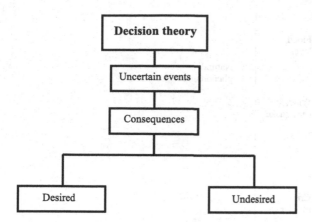

Figure 10.1   Risk analysis as a branch of decision analysis dealing with undesired consequences

decreasing function of the attribute (kidneys lost) with the loss of two corresponding to the lowest utility (death). It is most important that the non-linearity of utility with attributes in the case of risk aversion be borne in mind when reading this chapter.

The direct relationship between the structure of rational decision-making and the concepts embodied in the definition of risk implies that we can conduct risk analysis with the tools already developed in all of the preceding chapters. There is one condition that must apply to risk analysis: the consequences are unwanted, that is, of low utility. This is emphasized in Figure 10.1, which summarizes the relationship with decision theory. Some decision trees with implications for risks of various kinds are given in Figure 10.2. The risks are classified broadly into those associated with natural phenomena, societal issues, and technological aspects. But some risks do not fall clearly into one of these; for example taking risks on the stock market might involve all three aspects.

The subjects covered by the term 'risk analysis' are extensive, and it would be futile to attempt to review them all. Our purpose is to expose those general principles that might be applied in idealizing and solving problems in any area. The material reviewed in this chapter will serve the purpose of illustration and no claim is made to comprehensiveness.

10.1.1   Safety Index (*SI*)

We may summarize the concepts in Figure 10.3, which illustrates both the probabilities and consequences of actions. Taking a scale of utility from 0 (least desirable) to 1 (most desirable), the figure shows the area of interest in risk analysis. It is convenient for many purposes to use probabilities calculated on an annual basis, so that, for example, the annual risk to an individual is assessed. Often it is useful to express this probability on a logarithmic scale (lower part of Figure 10.3). Usually the range under consideration is from about 10^{-2} to 10^{-7}, giving a range on the (negative) log scale of 2 to 7. This logarithm to the base 10 will be

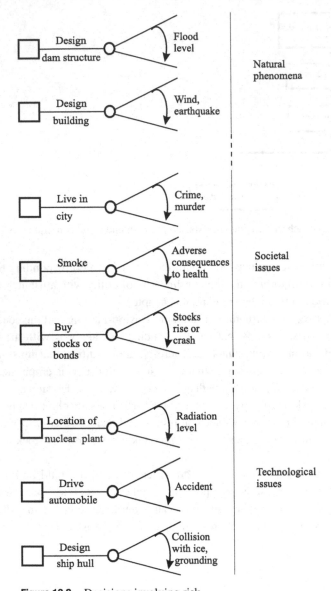

Figure 10.2   Decisions involving risk

termed the '**Safety Index**' (*SI*) which we abbreviate to *SI* (see Jordaan, 1988, and also Paulos, 1988)[2].

The consequence scale in Figure 10.3 in general deals with events that might include single, several or many deaths and injuries, minor or major damage to the environment (for example,

---

[2] This is not the Safety Index of structural reliability theory, which we have termed the 'Reliability Index' (*RI*): see Section 10.5.

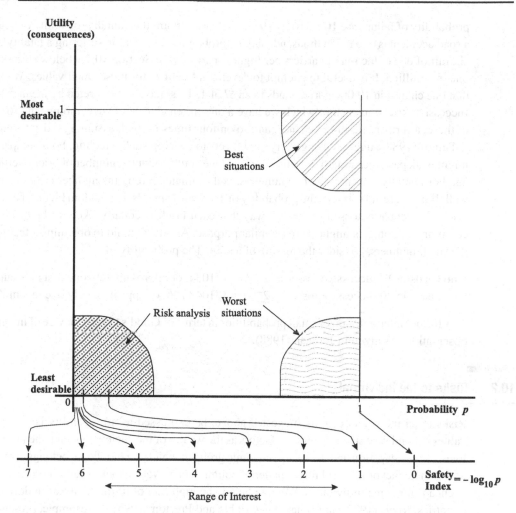

Figure 10.3 Domain of risk analysis

small or large oil spills), or combinations of these. If the set of two consequences {loss of life, survival} for a single person is considered, this may be modelled by the two points {0, 1}. One is considering the risks to an individual which can be expressed as a probability per unit time, generally a year. This is a very useful measure of safety and can be a good guide to acceptable risks. Section 10.4 deals with this question. Further discussion of the basis for the logarithmic measure is given in Jordaan (1988).

10.1.2 Small probabilities

The values quoted in the preceding paragraph show that it is necessary to deal with small probabilities. This is true also of extreme values used in design (Chapter 9) as well as probabilities of failure of engineering systems – the risk of failure. Common target values for the probability of failure are 10^{-3}, 10^{-4}, 10^{-5}, 10^{-6} per annum; the annual probability of death in a road accident is one in ten thousand, and the probability of a person suffering a fatality on an aircraft of one of the major carriers per flight departure (see Section 10.2.3 below) is less than one in a million. It is useful to attempt to develop a 'feeling' for these small values. We know that one chance in 10 000 corresponds to an SI of 4. This may be interpreted by means of four successive 'trials' in each of which we have a one-in-ten chance of succumbing, and for each of these, a further one-in-ten chance, and so on four times in total, assuming independence.

Paulos (1988) suggests using everyday objects to develop such a feeling. For example, one might walk past a certain brick wall every day. One could count the number of bricks vertically and horizontally, multiply the two numbers, and estimate thereby the number of bricks in the wall. If there are 1000 bricks the probability of 10^{-3} corresponds to a random hit on a particular brick if a pebble is tossed (in such a way that each brick is equally likely to be hit). Other commonplace objects might serve a similar purpose. As a further aid to obtaining a feeling for the small numbers, consider the tossing of a coin. The probability of

- no heads in 10 successive tosses $= 2^{-10} = 1/1024$, or approximately one in a thousand;
- no heads in 20 successive tosses $= 2^{-20} = 1/1048\,756$, or approximately one in a million.

A further interpretation of small probabilities in terms of Good's (1950) 'device of imaginary observations' is given in Jordaan (1988).

10.2   Risks to the individual

Risks to an individual can, as suggested above, be expressed as a probability per year. The values depend very strongly on such factors as the wealth of the society in which the individual lives, their occupation and the socio-economic group to which the individual belongs. This is an unfortunate fact of life, and it is a matter of common observation that in certain societies, 'life is cheap' for some individuals. We note below the high rate of traffic fatalities in developing countries. Rowe (1977) and Bulatao in Gribble and Preston (1993), for example, present data on risks to persons in the developing world as against those in industrial countries. In the latter reference, it is found that the age-adjusted rates of death in developing countries are double those in industrial ones. Some extreme happenings can be noted. For example, Melchers (1993) mentions the risks in the Indian nuclear industry, whereby low-caste workers are employed at attractive daily rates to perform maintenance work under conditions of high exposure to radiation.

We shall present data mainly from the perspective of a Canadian, but similar risks would be applicable to other countries with similar GDP per capita and social structure. To set the scene, we first present data from Statistics Canada, in Table 10.1. The values of annual risk (AR) per 10 000 persons is a useful number and it is a good idea to remember several key values. Table 10.1 shows that the annual risk of death for a Canadian ('other things being equal') is about 0.7%. The SI is a little more than 2, an average value for all ages, both sexes, and for all reasons.

Table 10.1 *Risks for Canadians based on data from Statistics Canada*

Cause or activity	Annual risk (AR) per 10 000 persons		Safety Index (SI) $[\log_{10}(AR/10\ 000)]$	
	Male	Female	Male	Female
All causes	66		2.2	
All causes (male and female)	84	52	2.1	2.3
All causes				
Age 20–24	10	3	3.0	3.5
40–44	22	13	2.7	2.9
60–64	137	77	1.9	2.1
Heart disease	23	13.0	2.6	2.9
Cancer (all types)	23	14.9	2.6	2.8
Lung cancer	8.1	4.5	3.1	3.4
HIV	0.4	0.1	4.4	5.3
All accidents	4.2		3.4	
Motor vehicle	1.0		4.0	
Suicide	2.0	0.5	3.7	4.3
Homicide	0.2		4.7	

The table shows that risk is a lot less for younger people and that females are far less prone to accidents than males. As a person ages, the cause of death is likely to be associated more with disease, but again women are less liable to succumb. This is manifested in the greater life expectancy of women. Disease is a significant cause of death, particularly as noted with ageing. Accidents take their toll, especially on the road. It pays to be careful to avoid falls. The average citizen, especially if male, is more likely to succumb to suicide than to murder. Figure 10.4 summarizes various risks covering a range of values of the Safety Index.

Another risk measure that is useful when the activity under consideration is the fatality accident rate (*FAR*), which is the risk of death per 10^8 person-hours that the person is occupied in the activity under consideration. Table 10.2 gives a set of values of *FAR*, several of which have been taken from Hambly and Hambly (1994). Recall that there are 8760 hours per year; in cases where the risk is given as, say 1/1000 per year at work, this annual risk is divided by the number of hours at work, taken as 50 weeks × 40 hours = 2000 hours to give a *FAR* of 50 × 10^{-8}. Vinnem (1998) shows how a risk profile for a worker in the offshore industry can be constructed, including variations of the risk level during a working day.

10.2.1 Smoking

Smoking is a voluntary activity or, more properly, an addiction that is very destructive of human life. In Canada, and in many countries, it is the leading cause of preventable deaths.

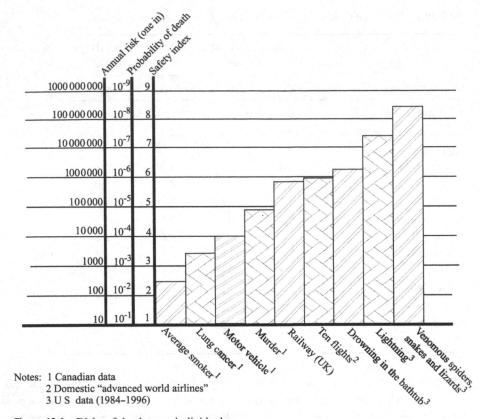

Figure 10.4   Risks of death to an individual

Notes: 1 Canadian data
2 Domestic "advanced world airlines"
3 U S data (1984–1996)

The percentage of the population who smoke has declined from about 50% in the 1960s to the present rate of less than 30%; yet in 1996, over 45 000 Canadians died from smoking-related diseases. This translates into an annual risk (*AR*) per smoker of more than 5×10^{-3}, giving a Safety Index of 2.3. These values were used to deduce the average *FAR* in Table 10.2. This rate would be higher for heavier smokers.

10.2.2   Traffic accidents

According to an article in the Guardian Weekly (July 5, 1998), accidents associated with motor vehicles have claimed 30 million lives since the first death in 1898. During the 1990s, the annual total rose to 500 000 deaths, most occurring in developing countries such as India. This human hazard is projected to overtake war, tuberculosis and HIV as one of the world's biggest killers by 2020. At the same time the death rate has been improving in most developed countries. Increased emphasis on the hazards of drinking and driving, road and car safety, use of seat belts – all of these have assisted in reducing the death rate. In France, the number of deaths has

Table 10.2 *Fatality accident rate (FAR)*

Cause, activity of condition: risks are given for time while activities are undertaken	Risk of death per 10^8 person hours (*FAR*)
1. Plague in London in 1665	15 000
2. Rock climbing while on rock face	4000
3. Travel by helicopter	500
4. Travel by motorcycle and moped	300
5. Average man in 60s (all causes)	160
6. Average woman in 60s (all causes)	90
7. Construction, high-rise erectors	70
8. Smoking (average; see Section 10.2.1)	60
9. Gold mining (Witwatersrand, South Africa)	55
10. 1 in 1000 per year at work	50
11. Motor vehicle accidents	23
12. Average man in 20s (all causes)	11
13. Travel by air	10
14. 1 in 10 000 per year at work	5
15. All accidents	5
16. Offshore oil and gas extraction (mobile units)	5
17. Travel by train	5
18. Construction worker, average	5
19. Average woman in 20s (all causes)	4
20. Metal manufacturing	4
21. Travel by bus	1
22. 1 in 10 000 per year living near major hazard	1
23. 1 in 100 000 per year at work	0.5
24. Radon gas natural gas radiation, UK average	0.1
25. 1 in 100 000 per year living near nuclear plant	0.1
26. Buildings falling down	0.002

been reduced from about 17 000 in 1972 to 8000 in 1997. This remains a high figure. Similar high figures are found in other countries. There were about 3400 road deaths in Great Britain in 1998, with a population of about 58 million.

In the USA, over 41 000 people were killed on the road in 1999 at which time the population was about 273 000 000. The totals in the late 1960s and early 1970s exceeded 50 000 per year. During this period, the Vietnam war raged with the main US involvement from 1965 to 1973. The total number of Americans killed in the war was 58 209, according to a website devoted to this subject. It might be a good thing if the attention devoted to road deaths were commensurate with the risks, yet society is content to accept a callous attitude regarding the number of deaths that occur at present. In Canada the number of deaths per year in the late 1990s was about 3000, about 1 per 10 000 persons (risk of 10^{-4} per person per year). This rate has been halved during the last twenty years. The rate for Great Britain (based on the figures above) is about 0.6×10^{-4}; differences in distance travelled and vehicle ownership (number per capita) make detailed comparisons difficult.

A better figure for broad comparison of road safety is the annual death rate per vehicle; the Guardian article referred to above reports that in many 'developed' countries the rate is less than five per 10 000 vehicles and as low as two in Japan and Australia. In India, the annual rate is 40 per 10 000 vehicles; 77 in Bangladesh; 111 in Ghana, and 192 in Ethiopia. A general conclusion is that the driving of automobiles involves high risks, in countries that are considered developed or not.

The Twin Towers

10.2.3 Aircraft safety

The following was written in the period just before September 11, 2001. 'I know of two cases where it is reasonable to conclude that a pilot of an important airline took the entire aircraft, including passengers and other crew to their death in a suicide mission. This is a minor if alarming possibility, minor given the number of aircraft taking off each day.' The element of alarm has increased since September 11, but the main conclusion is still valid. The number of such incidents is still proportionately small.

The risk of death in aircraft accidents has diminished considerably over the last decades (Lancaster, 1996). Most of the risk during normal flying occurs during takeoff, climb, descent and landing (Boeing, 2000). While statistics per hour in the activity are useful (given for air travel in Table 10.2), it is most interesting to consider the risk per flight, since this would be approximately constant if most of the risk occurs on landing and takeoff. It is also of interest for travellers to know the risk of death if they choose a flight at random. Some years ago, a colleague and I estimated the probability of death per flight on major Canadian air carriers to be less than one in a million (Wirasinghe and Jordaan, 1985). This is a considerably better rate than the average for all aircraft. The answer appears to lie in the type of air carrier considered. Barnett and Wang (2000) analysed the risk of death per passenger per flight on various airlines for the years 1987–96. For US domestic airlines with scheduled routes (trunklines) the risk averaged at 1 in 6.5 million for jet carriers. For domestic flights in 'advanced-world' scheduled jet operations, the risk was estimated at 1 in 11 million, whereas in the developing world the risk rose to 1 in 500 000. Slightly higher risks were found for international flights for either category of development. It would seem appropriate to choose one of the 'advanced-world' carriers if at all possible, yet the risks are still low for either category, 'advanced-world' or developing. A final point is of interest: the majority of accidents are associated with human error (Lancaster, 1996; Stewart and Melchers, 1997). Smaller aircraft, microlights and helicopters pose significantly increased risks.

10.2.4 Risks at work

Risks at work vary considerably from country to country and from industry to industry. Some occupations are particularly hazardous, for example flying helicopters, deep-sea diving and fishing, and mining. Fatality rates at work vary from less than 10^{-6} per year to 10^{-5} in vehicle manufacturing to 10^{-4} in construction to 10^{-3} in some occupations (see also Table 10.2). Miners face high risks. Wilson (1979) notes annual death rates in the 1970s of $1.6 \cdot 10^{-3}$ for coal miners. There is the additional hazard of lung disease. Fatality rates in the range 1 to $1.25 \cdot 10^{-3}$ are found in the South African gold mines (Webber, 1996), many associated with rockbursts. Often it is useful to collect data for a particular system. Risks in the offshore industry are often quoted as being 10^{-3} per annum. The following example illustrates the usefulness of searching for data.

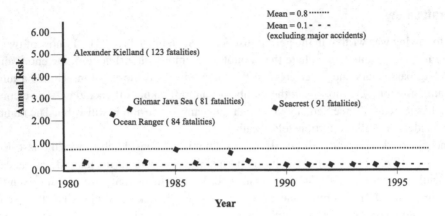

Figure 10.5   Annual risk for 1000 workers on mobile offshore units (WOAD)

Example 10.1 *Offshore oil and gas industry.* An engineer is investigating the safety of mobile units for production in the offshore industry so as to formulate safety provisions for future applications. The engineer considers that current estimates of the order of 10^{-3} per year for the risk to life of worker are too high. The main source of data consulted is WOAD, the Worldwide Offshore Accident Databank of Det norske Veritas. For mobile offshore units, the record for the years 1970 to 1979 indicates an annual risk of death to the worker of close to 10^{-3}. This includes exploration structures, for example jackups, drill ships and semisubmessibles. Many accidents are associated with drilling operations, and more than half of fatalities were associated with this activity in the WOAD database for the years 1980–1995.

An analysis of recent trends, from 1980 to 1996, yields the results in Figure 10.5. The results were obtained by dividing the number of deaths per year by the corresponding number of personnel-years. It can be seen that the average level of risk is again close to $1 \cdot 10^{-3}$, of the order of $0.8 \cdot 10^{-3}$. But it is noteworthy that this value includes several large occurrences, such as the Alexander Kielland and the Ocean Ranger incidents. If these are omitted from the data base, the remaining risk level is in fact close to $1 \cdot 10^{-4}$. It is certain that serious lessons were learned from these incidents: it is generally the case, for example in the airline industry, that lessons from failures are learned, and that the subsequent safety record is improved. It is therefore reasonable to assume that, with careful design, a target of 10^{-4} or better is obtainable. It is emphasized that the data upon which Figure 10.5 is based are derived for all kinds of offshore mobile units. An inspection of the data shows that most of the accidents occur in drilling, as noted, and none were recorded in WOAD in the category 'production' for the years 1980–1995. The Petrobras P-36 accident of March 2001 involved 11 deaths in production using a semisubmersible.

As noted above, the values in this example have been calculated by dividing the number of deaths per year by the personnel-years for the year in question. The latter is based on the number of rigs and the cabin capacities. The resulting risk level then covers personnel at risk

24 hours per day, as is usual for offshore work. The risk for a particular individual is less than this since the person will work only on certain shifts.

10.2.5 Perception of risk

We often read in the newspapers about aircraft accidents, but seldom do we read about traffic accidents unless there are a large number of people involved. Yet far more lives are lost on the roads than in accidents associated with air travel. Large accidents attract the most attention, mainly as a result of media coverage, which quite naturally focusses on such happenings. The crash of an aircraft, resulting in over a hundred lives lost, is a tragic occurrence. There will inevitably be a large amount of time and space in the media devoted to such events. There is nothing wrong with this, yet we know that in Canada, the death toll on the roads each year is of the order of 3000, and of 40 000 in the United States. The tally of 3000 deaths corresponds to 20 major airline crashes, each involving 150 people; Canadians would be horrified by such an annual occurrence. If there were $40\,000/150 = 267$ major crashes in the USA in a year, almost one per day, there would certainly be a major reaction!

In the UK and elsewhere, a large amount of media space and attention has been devoted to BSE – bovine spongiform encephalopathy, or mad cow disease. This is linked to CJD, Creutzfeldt–Jakob disease, or a variant, vCJD. The rate of occurrence of CJD worldwide is about one in a million people; for the period 1996 to 2000, there were 87 cases of vCJD in the United Kingdom, resulting in death. These deaths are tragic, particularly in young children, and the cause should be addressed. The point being made here is there are far more road deaths, equally tragic in their circumstances, which do not receive proportionate attention in the media. We noted above that far more Americans were killed on the road during the period of the Vietnam war than in the war itself. The attention given by the media results to a large extent from the size of the incidents, but there is certainly an element related to the novelty, as in the case of new diseases such as BSE and HIV.

There are other instances where humans will accept risks which can be seen to be quite high. There is the question of proneness to risk – the exhilaration which some people might feel when exposed to danger. Yet a person who is not especially risk-prone might fly microlight aircraft. A vessel navigating on the Labrador coast might not reduce speed in foggy conditions, even though the chance of encounter with an iceberg is relatively high. People smoke while knowing of the risks; here an addiction is involved. Why do we accept high risks on the road? The answer seems to lie partly in the convenience of automobile use, but also the experiences of the person involved and the consequent perception of risk. Even though we might consider an annual risk of 10^{-4} on the road to be high given the number of people killed, the return period in years to the death of an individual, given present conditions is 10 000. One hears remarks such as 'it is the number of years to the next ice age', which gives the misleading impression of a very low risk.

We do not often encounter cases of persons close to us who are killed on the road. The captain of the vessel on the Labrador coast might have had no experience or report of interaction

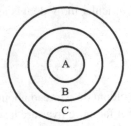

Figure 10.6　Circles of risk perception for an individual
　　　　　　A = close family and friends
　　　　　　B = distant family and acquaintances
　　　　　　C = persons known: friends of friends, movie stars, sports heros

between ships and icebergs, since the density of shipping is very low. In terms of personal perception, Figure 10.6 illustrates a possible interpretation of an individual in society. The figure shows 'circles of perception', in which there are first (A) a number of people close to the person: close family and friends, say 20 in total. Then there are (B) more distant relatives, and casual acquaintances, say 500 in total. Finally, there are (C) people known to the person: friends of friends, celebrities, movie stars, leaders in the community, say 5000 in total.

Given the annual probability of 10^{-4}, the chance of a person in the 'inner circle' being killed in a year is (20/10 000) or one in five hundred. The majority of people will not experience this kind of tragedy personally. It becomes increasingly probable that an individual knows somebody in circles B and C who has experienced a tragedy involving death on the road. The analysis using risk circles corresponds to a person's experience. More empathy with regard to the plight of others needs to be developed before we can solve the problem of road deaths. Of course this might never happen.

10.3　Estimation of probabilities of system or component failure

The failure of structural components and systems is of central interest to engineers. This was introduced in Chapter 9, Section 9.1 and Figure 9.1. We shall first deal with demand possibly exceeding capacity. The term system usually refers to an entire collection of objects and components, such as an assemblage of mechanical or electronic components or the structural elements – beams, columns, slabs perhaps – which make up an entire structure. Demand and capacity might refer to the entire system; for instance a gravity-based structure offshore might be analysed for possible sliding or rotation under the action of wave loading. Failure would imply system failure. The problem could be approached using a single demand and capacity. In other cases, this is not possible, and the analysis might then be applied to a component of the system, for example a particular structural component such as a beam or a single electrical component. Failure of the component might not lead to failure of the system, only to local failure. The failure modes for the entire system might be rather complex and high in number, involving various component failures.

Failure rates for many standard industrial components and pieces of equipment are given in data banks; Wells (1996) gives typical values for various items of equipment. Stewart and Melchers (1997) review various data bases. There are also proprietary data sets compiled by various organizations. Applications are many and varied, and the engineer must search for the best data in any particular application.

In the next subsection, we give methodology for estimating the failure probability given probability distributions for demand and capacity. This approach is common in civil and structural engineering, where the failure rates are rather low because of the implications of failure for human life. Often in these disciplines it is possible – as discussed above – to model only the failure of a component of a system, supplemented by approximate techniques and judgement when dealing with the entire system. The failure of the entire system might be difficult or impossible to model accurately; structural redundancy is an important example. Bounds or approximations might be the best that one can achieve. The term 'system' should be interpreted rather loosely in the following subsection; the subsequent Section 10.3.2 gives more detail of the relation between components and systems.

10.3.1 Demand and capacity

We now add flesh to the bones of the introduction of Section 9.1. Figure 10.7(a) shows detail of the 'overlap' area between demand and capacity, introduced in Figure 9.1. The demand (load on a structure, number of users of a transportation network, . . .) is denoted as S, and the capacity (structural resistance, capacity of the transportation network, . . .) as T. As always in the use of probability theory, the possible states must be clearly defined. Failure could be modelled as collapse of a system or component, but it could also if desired be defined as some state of damage or malfunction other than total collapse or failure.

We assume that S and T are modelled as continuous random quantities, and that they are stochastically independent. This follows the usual definition: one judges that knowledge of the load does not affect one's probabilistic estimate of resistance or *vice versa*. We are considering the case where the calculation is made separately from any actual application of the demand or load, so we do not have evidence of the system response. This may be contrasted with the case of proof loading of a system. If we actually load the system to a certain level and observe its response, for instance in the proof loading or testing of a component, the probability distribution of capacity or resistance must be updated to reflect this new information. If the system does not fail under a certain load, we know that its resistance is greater than this value; see Exercise 10.2.

There are other cases where load and resistance might be correlated for some physical reason. For example, a certain temperature might be applied to a structural component causing stresses in the component. The loading corresponds to the temperature yet the response or resistance of the material might be temperature-dependent. Such correlations must be taken into account. The present analysis is concerned with independent events.

To focus ideas we shall use the expressions 'load' for S and 'resistance' for T. Generally the quantity S will be an extreme value, for instance the annual maximum load or demand

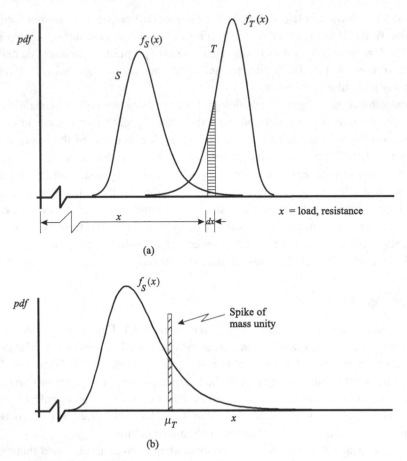

Figure 10.7 Calculation of probability of failure

during a given period of time. This is how time enters into the problem. The distribution of extreme values has been dealt with extensively in Chapter 9. In Figure 10.7(a) we consider the overlap area between load and resistance. Failure, a random event F, is defined as 'load exceeds resistance' ; then $F = 1$ in the Boolean notation of Section 2.2. Its probability $\mathbf{Pr}(F)$ is denoted p_f. No failure, or survival corresponds to $F = 0$. The survival of the structure is written as R; in Boolean notation

$$R = \tilde{F} = 1 - F. \tag{10.1}$$

The probability $\mathbf{Pr}(F)$ can be written in several ways:

$$
\begin{aligned}
p_f &= \mathbf{Pr}(T \leq S) \\
&= \mathbf{Pr}(T - S \leq 0) \\
&= \mathbf{Pr}(\frac{T}{S} \leq 1) \\
&= \mathbf{Pr}[(\ln T - \ln S) \leq 0]. \tag{10.2}
\end{aligned}
$$

In structural engineering these are often written generally as

$$p_f = \mathbf{Pr}\left[g\left(S, T\right) \leq 0\right], \tag{10.3}$$

where $g\left(\cdot\right)$ is termed a **limit state function**. The probability of survival is written as

$$\mathbf{Pr}\left(R\right) = r = 1 - p_f, \tag{10.4}$$

and is called the **reliability**.

We now consider ways of evaluating p_f from the distributions of S and T. Referring to Figure 10.7(a), load and resistance are considered as being in the same units (e.g. MN, V, ...), with probability densities $f_S\left(s\right)$ and $f_T\left(t\right)$. The common value of s and t is denoted x. Then a small increment in p_f corresponds to the event (logical product)

$$(T \leq x)(x \leq S \leq x + dx),$$

or

$$dp_f = F_T\left(x\right) \cdot f_S\left(x\right) dx.$$

Therefore

$$p_f = \int_{-\infty}^{\infty} F_T\left(x\right) f_S\left(x\right) dx. \tag{10.5}$$

This can also be formulated as (see Exercise 10.3):

$$p_f = \int_{-\infty}^{\infty} f_T\left(x\right)\left[1 - F_S\left(x\right)\right] dx. \tag{10.6}$$

A special case of this result can be obtained if the resistance shows little randomness, as in Figure 10.7(b) with mean μ_T. The probability of failure is then

$$p_f = 1 - F_S\left(\mu_T\right). \tag{10.7}$$

Similarly, if we can neglect the randomness in S,

$$p_f = F_T\left(\mu_S\right), \tag{10.8}$$

where μ_S is the mean (deterministic) value of S.

Another way to calculate p_f is to consider the **margin of safety** M (see Figure 10.8) from Equation (10.2), a special case of the limit state function $g\left(\cdot\right)$,

$$g\left(T, S\right) = M = T - S. \tag{10.9}$$

If M is negative, failure occurs and $F = 1$. If the margin of safety is positive, $F = 0$. Then

$$p_f = \mathbf{Pr}\left(F\right) = \mathbf{Pr}\left(M \leq 0\right). \tag{10.10}$$

We attempt to derive the distribution of M from the distributions of T and S. Then we can solve for the probability of failure using (10.10).

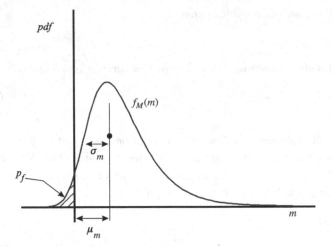

Figure 10.8 Margin of safety

Closed form solutions for p_f are available only for a limited number of distributions, and the calculation of failure probability is commonly carried out using numerical integration; see Melchers (1999). We now give two examples: where R and S are exponentially and normally distributed. Lognormal distributions are also commonly used: see Exercises 10.4 to 10.7. A practice that should be avoided is that of using distributions to model load and resistance that are convenient only for computational purposes. Computer packages that perform the numerical integration for a variety of distributions are useful in this regard.

Example 10.2 The case of exponentially distributed random quantities was used in Example 7.7 to illustrate the transformation of distributions. This is a convenient illustration, but real-world problems involve distributions such as the Gumbel. The following forms are then used for illustration:

$$f_T(t) = \lambda_1 \exp(-\lambda_1 t) \text{ and}$$
$$f_S(s) = \lambda_2 \exp(-\lambda_2 s), \tag{10.11}$$

for $s, t \geq 0$, and 0 otherwise. The resulting distribution of M is

$$f_M(m) = \frac{\lambda_1 \lambda_2}{\lambda_1 + \lambda_2} \exp(\lambda_2 m) \text{ for } M \leq 0, \text{ and}$$

$$f_M(m) = \frac{\lambda_1 \lambda_2}{\lambda_1 + \lambda_2} \exp(-\lambda_1 m) \text{ for } M \geq 0, \tag{10.12}$$

with failure probability

$$p_f = \int_{-\infty}^{0} \frac{\lambda_1 \lambda_2}{\lambda_1 + \lambda_2} \exp(\lambda_2 m) \, dm = \frac{\lambda_1}{\lambda_1 + \lambda_2}. \tag{10.13}$$

Example 10.3 If S and T are normally distributed and independent with parameters $\left(\mu_S, \sigma_S^2\right)$ and $\left(\mu_T, \sigma_T^2\right)$ respectively, then M is also normally distributed with parameters $\left(\mu_T - \mu_S, \sigma_T^2 + \sigma_S^2\right)$. Then

$$p_f = \Pr\left(M \le 0\right) = \Pr\left(Z \le \frac{0 - \mu_M}{\sigma_M}\right)$$

$$= F_Z\left(\frac{-\mu_M}{\sigma_M}\right) = F_Z\left(-\beta\right), \tag{10.14}$$

where Z is the standard normal random quantity, and

$$\beta = \frac{\mu_M}{\sigma_M}$$

is the Reliability Index discussed below (Section 10.5).

Extreme values usually follow non-normal distributions, as discussed in the preceding chapter. These are generally appropriate for the random quantity S. The failure probability is very sensitive to the distributions used in modelling. This illustrates the need for careful modelling and, in particular, attention to the appropriate tails of the distributions of T and S.

Generalization

The theory developed in the preceding subsection works well under the conditions of independence of demand and capacity for failure that arises from a single action (demand) exceeding a corresponding capacity. Rather than single actions T and S, we might need to consider sets of these:

$$T_1, T_2, T_3, \ldots \text{ and } S_1, S_2, S_3, \ldots, \tag{10.15}$$

related through the joint density function

$$f_{S_1, S_2, S_3, \ldots, T_1, T_2, T_3, \ldots}\left(s_1, s_2, s_3, \ldots, t_1, t_2, t_3, \ldots\right) \equiv f_{S_i, T_i}\left(s_i, t_i\right).$$

Then the failure probability for failure mode j can be considered to be the integration over some domain d_j of the space occupied by the random quantities under consideration, for which the limit state function, $g_j\left(s_i, t_i\right) < 0$; the latter is a generalization of (10.9). This is illustrated in two dimensions in Figure 10.9. The failure probability can then be written as

$$p_f = \int_{S_i, T_i \in d_j} f_{S_i, T_i}\left(s_i, t_i\right) \mathrm{d}s_1 \mathrm{d}s_2 \ldots \mathrm{d}t_1 \mathrm{d}t_2 \ldots; \tag{10.16}$$

see Melchers (1999). Evaluation of Equation (10.16) would generally be carried out by some form of numerical integration; there are several commercially available software packages available for this purpose. See Melchers (1999) and Madsen *et al.* (1986), for examples.

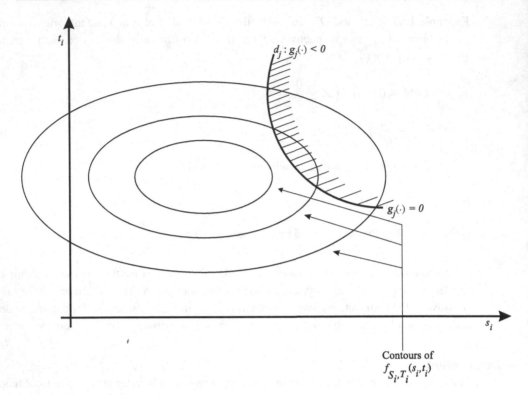

$d_j : g_j(\cdot) < 0$

$g_j(\cdot) = 0$

t_i

s_i

Contours of

$f_{S_i, T_i}(s_i, t_i)$

Figure 10.9　Failure domain

10.3.2　Systems of components; redundancy

There is often a need to combine results for various components into system failure or reliability. For many components, data bases exist; see the introductory comments to Section 10.3 and Exercise 10.1. Alternatively, the probability of failure of various components can be estimated by the methods of the preceding subsection. We shall introduce the analysis of systems of components in this subsection; see McCormick (1981) and Billinton and Allan (1992), for example, for further detail. The most common systems are series and parallel ones (Figure 10.10). We assume in the following that the units (components) act independently of each other. Recall from Equation (10.1) that $R = 1 - F$. We shall use F, R, p_s and r_s for events and probabilities pertaining to the system, and F_i, R_i, p_i and r_i for events and probabilities pertaining to the units ($i = 1, 2, \ldots, n$). The ps denote failure probability and the rs the probability of survival. From Equation (10.4),

$$p_s + r_s = 1 \text{ and} \tag{10.17}$$

$$p_i + r_i = 1 \text{ for } i = 1, 2, \ldots, n. \tag{10.18}$$

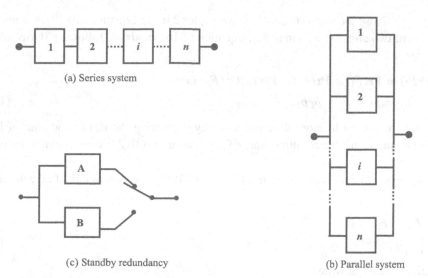

(a) Series system

(c) Standby redundancy

(b) Parallel system

Figure 10.10   Series and parallel systems

For the series system of Figure 10.10(a) failure occurs if any one – or more – of the units fails. Then

$$F = F_1 \vee F_2 \vee \cdots \vee F_i \vee \cdots \vee F_n$$
$$= 1 - \prod_{i=1}^{n} (1 - F_i) \tag{10.19}$$
$$= 1 - \prod_{i=1}^{n} R_i, \tag{10.20}$$

using the Boolean notation; see Equation (2.17). We can also write

$$R = \prod_{i=1}^{n} R_i; \tag{10.21}$$

the system operates if all components operate. From Equation (10.21) and using also Equation (2.40) based on independence,

$$r_s = \prod_{i=1}^{n} r_i. \tag{10.22}$$

If the component failures are *iid*,

$$r_s = r^n, \tag{10.23}$$

where $r_i = r$ is now common to all components. This is the 'weakest link' model; see Sections 9.1 and 9.2 and Equation (9.6). Note that $p_s = 1 - r_s$. If one wishes to calculate p_s using the p_is, (10.22) becomes

$$p_s = 1 - \prod_{i=1}^{n} (1 - p_i). \tag{10.24}$$

One may also use the method described in Chapter 2 using Equation (10.19). For instance, in the case of two units ($n = 2$) in series, Equation (2.16) yields the following (based also on independence):

$$\mathbf{Pr}(F_1 \vee F_2) = \mathbf{Pr}(F_1) + \mathbf{Pr}(F_2) - \mathbf{Pr}(F_1)\mathbf{Pr}(F_2) \text{ or}$$

$$p_s = p_1 + p_2 - p_1 p_2. \tag{10.25}$$

This may be extended to more than two units by expanding the right hand side of Equation (10.19) and using the resulting linear form. Equation (10.22) does provide the easiest solution.

Dealing now with the parallel system of Figure 10.10(b), the system fails if all components fail:

$$F = F_1 F_2 \cdots F_i \cdots F_n$$
$$= \prod_{i=1}^{n} F_i. \tag{10.26}$$

Alternatively, one can write

$$R = 1 - \prod_{i=1}^{n} F_i$$
$$= 1 - \prod_{i=1}^{n} (1 - R_i). \tag{10.27}$$

The failure probability of the system is

$$p_s = \prod_{i=1}^{n} p_i \tag{10.28}$$

and

$$r_s = 1 - \prod_{i=1}^{n} (1 - r_i). \tag{10.29}$$

Again in the *iid* case,

$$p_s = p^n, \tag{10.30}$$

where $p_i = p$ is common to all units. This might also be thought of as the failure of the weakest link in a parallel system. One can use an equation analogous to (10.25) with the ensuing sentence to develop equations for r_s.

The term 'active-parallel' is often used to describe this situation; it means that no unit is held on standby until other units fail; see Figure 10.10(c). Billinton and Allan (1992) discuss this issue in some detail. Figure 10.11 shows the effect of increasing the number of components in series and parallel systems. As would be expected, the parallel systems show increasing reliability since more and more units must fail to effect system failure. The converse is true for series systems. The question of redundancy is important in many engineering systems and can be most beneficial.

Example 10.4 In the design of a system, it is required that the system reliability should be at least 0.99. It comprises ten identical units in series. What should be the minimum reliability

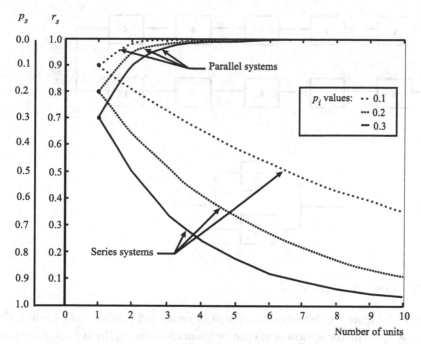

Figure 10.11   Reliability of series and parallel systems

of each component? From Equation (10.23),

$$0.99 = r^{10},$$

or

$$r_i = r = (0.99)^{\frac{1}{10}} = 0.9990,$$

assuming independence.

Example 10.5 A system comprises three units in parallel, with reliabilities of 0.98, 0.99 and 0.97. Find the system reliability. Assuming independence and using Equation (10.28), we find that $p_s = (0.02)(0.01)(0.03) = 6 \cdot 10^{-6}$ and therefore $r_s = 1 - 6 \cdot 10^{-6} = 0.999\,994$.

Example 10.6 Figure 10.12(a) shows a system of identical series and parallel units. Deduce the reliability of the system. The system in (a) can be reduced to that in (b), where

$$r_{11} = r_{12} = r_i^5,$$

and consequently

$$\begin{aligned}
r_s &= 1 - (1 - r_{11})(1 - r_{12}) \\
&= 1 - \left(1 - r_i^5\right)\left(1 - r_i^5\right) \\
&= 1 - \left(1 - r_i^5\right)^2.
\end{aligned}$$

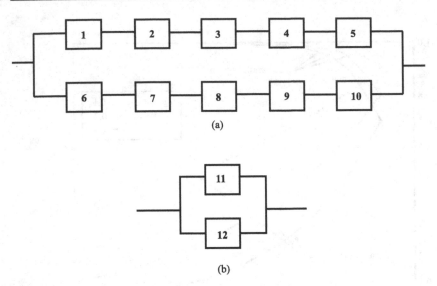

Figure 10.12 Units in Example 10.6

Example 10.7 Figure 10.13(a) shows a system of series and parallel units. Deduce the reliability of the system. The system in (a) can be reduced to that in (b) as an intermediate step towards a solution. Then

$r_8 = r_1 r_2,$

$r_9 = 1 - (1 - r_3)(1 - r_4)(1 - r_5)$ and

$r_{10} = 1 - (1 - r_6)(1 - r_7).$

The system reliability is then

$r_s = r_{10}[1 - (1 - r_8)(1 - r_9)].$

10.3.3 Human factors

We have noted that most accidents to aircraft result from human error. In structural systems, the rôle of human error is also dominant; Palmer (1996) analysed pipeline failures and concluded that most result from human errors. He, as well as Nessim and Jordaan (1985), Madsen *et al.* (1986) and Melchers (1999), reviewed literature on failures in structural engineering; all reached the same conclusion that human error is the principal cause of failure. Williamson (1997) lists incidents resulting from human error related to software. These include problems with air traffic control systems, ambulance dispatching and in nuclear power systems. Finnie and Johnston (1990) list uncontrolled inflight closure of throttles on Boeing 747-400s, malfunction of linear accelerators leading to death during hospital treatment, flood gates opening in Norway during maintenance, and people killed by robots in Japan. Stewart and Melchers

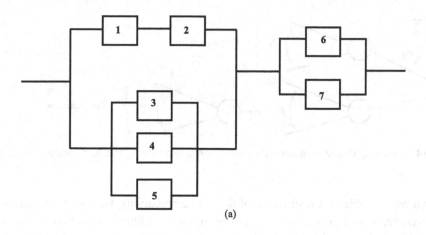

(a)

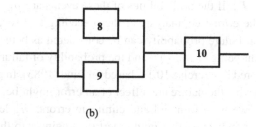

(b)

Figure 10.13   Units in Example 10.7

(1997) survey the significant effect of human error in the offshore industry, in shipping and in power plants. In particular areas of human activity there are studies of rates of error occurrence and their cause, for example on human performance in inspection and quality control; these may be searched and reviewed (if available) in any particular application.

Bea (1998) studied actual failures in the offshore industry and concluded that most accidents (80%) are associated with human and organization factors. 'Human error' and 'operator error' are simplifications insofar as the causes of the incidents might be related to organization, communication and regulation in a complex manner. The Piper Alpha accident is a case in point. The subsequent Cullen report (1990) spurred the move towards the 'safety case', in effect objective-based regulation rather than prescriptive regulation. A healthy balance between the two is needed. Perrow (1999) notes that 60–80% of accidents are related to 'operator error' but that the situation is much complicated by unexpected and complicated interactions amongst failures.

Since human error is a significant contributor to failure of systems, the calculation of probability of failure based only on the previous two subsections, in which errors are ignored, may

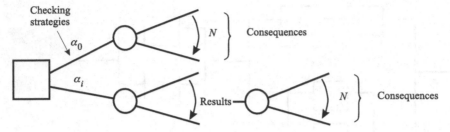

Figure 10.14   Checking strategies in design: α_0 corresponds to no check, and α_i represents the set of possible strategies

represent a considerable underestimation of the failure probability. Gauss divided errors into random, systematic and gross. Gross errors are mistakes or blunders such as misreading a number or misplacing a decimal point. Misunderstanding the behaviour of a system also constitutes gross error. It is the gross kind of error that is associated with structural and probably most other failures. One can consider that a structure or system either is based on an error (E) or is error-free $(\tilde{E})$. If the probabilities of these events are p_E and $(1 - p_E)$ respectively, one can consider the error-free analysis of the preceding two sections as being conditional on $\tilde{E}$. Then the final failure probability can be considered as being the weighted sum of this analysis – the weight being $(1 - p_E)$ – and the probability of failure given an error, weighted by p_E. This is explored in Exercise 10.14, based on Lind (1983). In practice the main difficulty lies in determining what the nature and effect of an error might be.

It is natural that we wish to avoid and eliminate errors. While one cannot predict error occurrence, there are indicators that might alert the engineer to the increased probability of error. Much of the difficulty in structural engineering arises from the use of new or unusual materials, methods of construction and types of structure (Pugsley, 1973). Misunderstanding of principles also is important. These indicators of increased probability of error might usefully be generalized to other areas of endeavour; certainly it is observed that changes in, for example, computer or control systems are liable to result in increased chance of malfunction. Checking represents the most important method of error control, particularly for errors that can be anticipated. For errors in the design process, several studies have been made. Melchers (1999) reviews this area. Nessim (1983) and Nessim and Jordaan (1985) conducted a Bayesian analysis in which error occurrence is modelled as a Poisson process. Figure 10.14 illustrates the overall approach, which is useful in analysing the relative merit of various strategies together with associated cost. The analysis was aimed at those errors that the person or method of checking can detect.

10.3.4   Probability and frequency

Failure probabilities estimated on the basis of the theoretical exercises of this section turn out often to differ from the measured rates based on data. The role of errors and structural redundancy in making estimates difficult has been noted. Some authors advocate the use of 'nominal' probabilities of failure. A better approach is to acknowledge those factors that are not included in the calculations, such as errors and redundancy, and to improve modelling so as to close the gap. Probabilities are always conditional on certain events or assumptions, and this might be a good way to interpret 'nominal' values. We have advocated a subjective conception of probability different from that of frequency. This should not be interpreted to mean that the probabilities are somehow less 'accurate' than those based only on frequencies or that one can then accept discrepancies with regard to data.

We have often used frequencies and data to estimate probabilities. The idea is to 'massage' frequencies where necessary, to estimate probabilities. For example, in estimating the probability of death in a scheduled flight on a major airline, we would look to accident records for similar aircraft. In cases where the aircraft were different or were subjected to different maintenance procedures, we might adjust the data. Certain classes or ages of aircraft might be eliminated from our data base. This is a question of our probabilistic judgement. We cannot use the excuse that our probabilities are subjective and therefore somehow they need not be in accordance with measured data or that we can ignore the question of measurement.

The relationship between probability and frequency has been studied by de Finetti (1937). Let the events $\{R_1, R_2, \ldots, R_i, \ldots, R_n\}$ represent survival in a given time period, say a year, of the objects of interest, for example a set of n electronic components or a set of n offshore rigs. We consider a future year so that these values are not known. The R_i can, as usual, take the values 0 or 1. The probabilities associated with these events are $\{r_1, r_2, \ldots, r_i, \ldots, r_n\}$ so that they represent reliabilities. Then

$$r_i = \langle R_i \rangle$$

and $\sum_{i=1}^{n} R_i$ represents the number of 'successes' in n trials. Now we let $\{\omega_0, \omega_1, \omega_2, \ldots, \omega_j, \ldots, \omega_m\}$ represent the probabilities that $\{0, 1, 2, \ldots, j, \ldots, n\}$ of the events occur.

We can write the expected number of survivals in two ways:

$$\left\langle \sum_i R_i \right\rangle = \sum_i \langle R_i \rangle = r_1 + r_2 + \cdots + r_i + \cdots + r_n \tag{10.31}$$

$$= (0)(\omega_0) + (1)(\omega_1) + (2)(\omega_2) + \cdots + (j)(\omega_j) + \cdots + (n)(\omega_n); \tag{10.32}$$

these two ways must agree in order that the assignments $\{r_i\}$ and $\{\omega_j\}$ be consistent. Equations (10.31) and (10.32) represent, in different ways, the expected number of events, for example, the expected number of components or rigs that survive. Equation (10.32) may be divided by n, giving

$$\frac{\sum_{i=1}^{n} r_i}{n} = \bar{r} = \left(\frac{0}{n}\right)(\omega_0) + \left(\frac{1}{n}\right)(\omega_1) + \cdots + \left(\frac{j}{n}\right)(\omega_j) + \cdots \left(\frac{n}{n}\right)(\omega_n). \tag{10.33}$$

Now (j/n), with $j = 0, 1, 2, \ldots, n$, represents the frequency of occurrence of the events.

Since this is unknown in the present analysis, because we are considering the events in a future year, it is a random quantity. Let us denote the frequency as G, taking possible values j/n. Then by Equation (10.33)

$$\bar{r} = \langle G \rangle ; \tag{10.34}$$

the expected frequency equals the average probability of occurrence of the events R_i. This reasoning must apply both to reliability and probability of failure.

Some particular cases are of use in dealing with risk. They deal with the use of frequency in assisting us to estimate probabilities. Two situations will be discussed.

(1) If the events $\{R_1, \ldots, R_i, \ldots, R_n\}$ are considered to be equiprobable, and if the frequency is well defined by past records, then the probabilities $r_i = r = \bar{r}$ must be consistent with the frequencies.

(2) There are cases where the frequency is known, but not the individual probabilities. De Finetti gives the example of the case where a ballot is taken with n voters. There are r favourable votes in support of a certain candidate, or proposition. We do not know how any of the individuals voted; we may express our uncertainty probabilistically as $\{r_1, r_2, \ldots, r_i, \ldots, r_n\}$, where these express our probabilities that individuals $\{1, 2, \ldots, i, \ldots, n\}$ voted in favour of the candidate or proposition. Yet we know the frequency in this case since we know the total number (j) who voted for the proposition, and the total number of voters (n). By virtue of Equation (10.34) our assignment must be constrained by the condition $\bar{r} = j/n$. In assigning the r_i we would use any knowledge we might have regarding the propensities of the individuals concerned.

Let us now apply these ideas to engineering components and systems. Often we find that theoretical estimates of failure probability do not accord with observed failure rates, as was noted above. The most constructive approach to this situation is to take the view that it points out deficiencies in our analysis. For example, calculations of failure probabilities of structural systems may not account completely for structural indeterminacy and redundancy or for errors. Instead, a judgemental allowance may be made for such factors. We should, at the same time, strive to improve our estimates of failure probability so that they match reality and that they capture the essential points. The key point is that the probabilities and frequencies must be consistent. The use of conditional probabilities, for example for a limit state at first yield, may represent a temporary way out, but the long term objective should be consistency.

If the failure rates for the system or component are judged equiprobable, then these should be similar to observed frequencies, possibly suitably 'massaged'. In this, we might make the judgement that the present systems are similar to those used in the past, or that there have been changes affecting the failure rates. For example, there are observable improvements in aircraft safety records over time. It is of course never true that the future is the same as the past. Yet we may make this judgement as being sufficiently accurate. For example, in estimating wave heights or iceberg populations, we use statistical parameters (means, variances, ...) to estimate future conditions. We know that climate changes with time, but we might ignore this for a period of short duration so as to make a reasonable approximation for the commercial life of (say) an oil production facility. For other purposes, involving the longer term, we might want to include models of climate change, greenhouse effects, and so on. Common sense should guide us: a new dam constructed affects conditions downstream, and we would adjust our flood estimates accordingly.

We might judge that failure rates are not equiprobable; components from two different manufacturers might have different failure rates. The safety record for fixed and mobile offshore units is different. For a given proportion of the two kinds of objects, our assignment of probability must be consistent with the overall average. Often, in practice, a set of data has been collected, yet, as a result of proprietary ownership, the results are not freely available. How tantalizing such a set of data might appear to us! We might, at the same time, have some knowledge of the average values collected, and our assignment of individual values must be consistent with averages known. For example, we might know that, in a certain city or region, 42% of the population use a particular transportation mode or use a certain product. We might not have data for each socio-economic group. Assignment of probabilities for these groups must also be constrained by $\bar{p} = 0.42$. If the city contained three groups 'low income', 'medium income', and 'high income', each with the same number of members, we might validly assign $p_1 = 0.60$, $p_2 = 0.40$, $p_3 = 0.26$, with a mean of 0.42, for the probabilities that an individual from each of the three groups uses a certain transit mode.

10.4   Risk analysis and management

Risk analysis is a branch of decision theory; all of the background developed in this book is therefore of use in risk analysis. This provides an intellectual structure on which to base the analysis. Often approximate methods are used, for example the use of target levels of probability for safety as a 'cutoff', as described in Section 10.4.2. In cases where decisions are clearcut, these approximate methods might be sufficient. In other situations, more detailed decision analysis is needed, and factors such as risk aversion need to be addressed. Even in extended works on risk analysis, fundamental aspects such as risk aversion, as presented in Chapter 4, are frequently neglected if the connection to decision theory is not appreciated.

Many engineering problems are concerned with the analysis and mitigation of risk. The siting of industrial facilities such as those involved in natural gas processing, or nuclear power stations, transportation of hazardous goods, avoidance of computer system error or failure of control systems – these represent typical areas of application. **Management** of risk is an important activity. Periodic inspection and maintenance are necessary to the safety of many engineering systems. Common engineered objects where safety is an important aspect are aircraft, offshore rigs, ships, vehicles used in road transport, and pipelines. Inspection in these cases will include, for example, the checking of the appearance and growth of cracks. (We have noted in Chapter 2 the importance of cracks in the fracture and possible failure of structural systems.) In industrial systems generally, inspection and quality control are important parts of the production process. The best way to formulate a strategy for maintenance or inspection is by means of decision theory. Financial risk can also be analysed: uncertainty in interest rates, inflation, the price of oil, stock and bond prices – these can all be modelled using probability theory.

Engineers will naturally attempt to answer questions posed by clients in industry or government. Most problems will be concerned with specific areas of interest, examples of which have been given in the preceding paragraph. A commendable case study in which multiattribute utility has been combined with probability is the siting of the Mexico City airport (de Neufville and Keeney, 1972; Keeney and Raiffa, 1976). The utility functions included attributes concerned with cost, safety, access and noise pollution.

Full use of probabilistic decision theory is not always necessary; sometimes cost is used as a 'surrogate', as, for example in a study of flood control and reservoir dam gates (Putcha and Patev, 2000), or individual outcomes such as possible fatalities and oil spill size (Tveit *et al.*, 1980) can be considered in designing offshore production platforms for oil and gas. In some cases, analysis reveals that the probabilities of undesired consequences are small, and further analysis is unnecessary. In others, the consequences turn out to be unimportant, and again further analysis is not necessary. Often we develop approximate rules, especially for small repetitive problems. Cost–benefit approaches are frequently used.

In all of these instances, decision theory as developed in this book provides the necessary guidance in understanding the nature of the problem and associated risks, and in making appropriate approximations. Occasionally we deal with those having large potential consequences; the design of nuclear power stations or air traffic control systems are examples. Certainly design rules and codes of practice for structures are of this kind. The result is generally that the probability of undesirable events such as major adverse consequences to human life or the environment is reduced to a very small value.

Risk mitigation can often be assisted by analysis of risks prior to initiation of the project. This can be achieved by probability-reducing or consequence-reducing means. To take the question of transport of hazardous goods as an example, the probability of an accident near a populated area can be reduced by re-routing the transportation system; consequences can be reduced – say in the case of rail transport of hazardous goods – by the design of special rail cars that prevent the release of their contents.

10.4.1 Tools: event and fault trees

We have used decision trees extensively in this book. If we consider a series of chance forks, this is often referred to as an **event tree**. The probabilities of the various events should be estimated and added in an analysis. If these are conditional upon each other, or are independent, they can be multiplied to obtain the final event probability. Event trees should be viewed as a natural part of the decision process and of the broader use of decision trees that include decision forks as well.

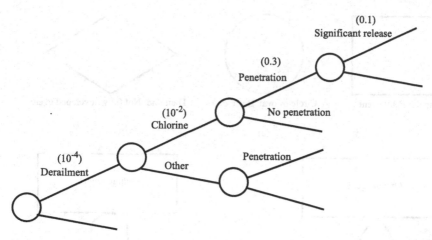

Figure 10.15   Event tree for release of hazardous goods from a railroad car for Example 10.8

Example 10.8 Figure 10.15 illustrates the case of transportation of hazardous goods by rail. Notional values of probability in the figure are added as an example. If we assume that the probability that a car contains chlorine is independent of the event of a derailment, and that the probability of penetration of the containment is conditional on the car containing chlorine – for example if the car is specially designed for the purpose – and that the probability of a significant release is conditional upon penetration, then the final probability of a significant release is

$$\left(10^{-4}\right) \cdot \left(10^{-2}\right) \cdot (0.3) \cdot (0.1) = 3 \cdot 10^{-8}.$$

Fault trees were developed in the aerospace industry in the early 1960s (Barlow and Lambert in Barlow, Fussell and Singpurwalla, 1975). Their use is now widespread, for instance in analysis of chemical plants (Wells, 1996) and nuclear plants (McCormick, 1981). Fault trees can be used as a design tool or in diagnostic analysis of existing systems and failures. In effect, they identify the Boolean and–or situations – the series and parallel system attributes analysed in Section 10.3.2 above. There are two types of symbols, those describing events as in Figure 10.16(a) and logic gates as in 10.16(b). The events include rectangles denoting failure that results from the logic gates. The focus of fault tree analysis is the 'top event', invariably an undesirable event, usually system failure. This will be explained by means of an example.

Example 10.9 Figure 10.17 shows an electrical circuit. The top event is 'motor does not operate'. This is shown in Figure 10.18(a), together with primary faults contributing to failure and the use of an 'or' gate. Figure 10.18(b) shows a fault tree which incorporates an 'and' gate.

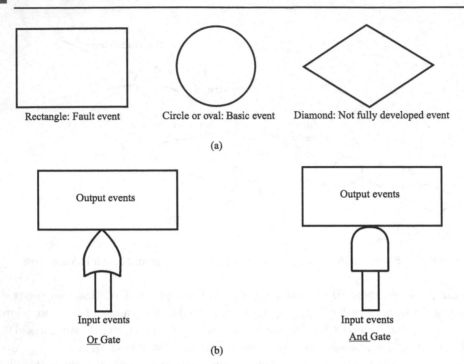

Rectangle: Fault event   Circle or oval: Basic event   Diamond: Not fully developed event

(a)

Output events

Input events

Or Gate

Output events

Input events

And Gate

(b)

Figure 10.16   (a) Event symbols (b) Symbols for logic gates

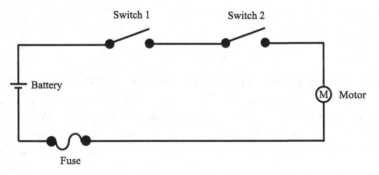

Switch 1   Switch 2

Battery

M Motor

Fuse

Figure 10.17   Electric circuit

There exist other event and gate symbols; further, the failure of a system can always be analysed in more depth if desired. The literature is rich in examples. See, for example, Barlow, Fussell and Singpurwalla (1975), McCormick (1981), Wells (1996), Billinton and Allan (1992).

Example 10.10 *Boolean analysis.* Figure 10.19 shows the relation of fault tree analysis to the reliability of components as expressed in a reliability block diagram. Boolean analysis can be applied; for example in Figure 10.19(a)

$$F = F_1 F_2 F_3 F_4,$$ (10.35)

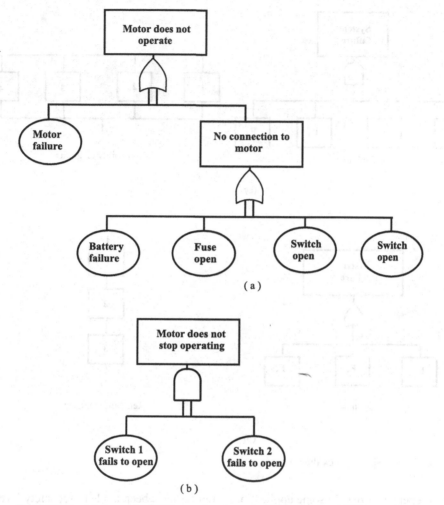

Figure 10.18   (a) Example of OR gate development (b) Example of AND gate development

whereas in Figure 10.19(b)

$$F = 1 - (1 - F_1)(1 - F_2)(1 - F_3).$$ (10.36)

10.4.2 Safety and target probabilities: as low as reasonably practicable (*ALARP*)

In many cases, for instance in the design of nuclear power plants, and the siting of facilities dealing with explosive materials such as liquefied natural gas, a detailed decision analysis might be considered necessary. In many other cases, the safety is addressed by specifying 'target probabilities', for instance, of failure of a system. These will be set at a sufficiently low level to ensure a desired level of safety. We shall discuss this issue mainly from the point of

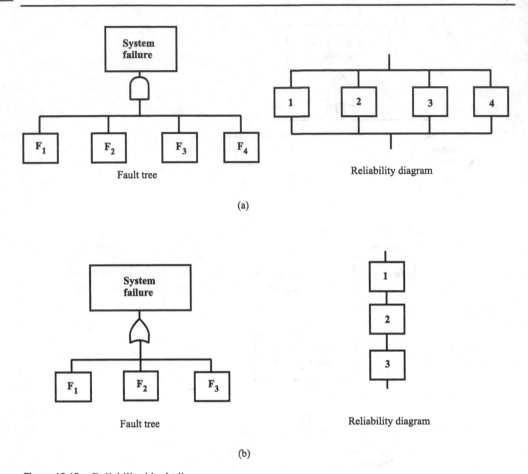

Figure 10.19 Reliability block diagrams

view of a person at risk. In some applications attempts have been made to set safety levels by using a cost–benefit approach, setting a monetary equivalent for loss of life. Baker (1977) and Flint and Baker (1976) review some applications in the offshore industry, and a calculation is given in Jordaan and Maes (1991). The conclusion is that this calculation is quite arbitrary and that often exceedingly high and unrealistic risks to humans are indicated. There is confusion over the perspective: the owner might have an entirely different point of view from the worker on the installation.

It is the conviction of the writer that the person at risk should be the predominant factor in one's thinking. An analysis by Nessim and Stephens (1995) illustrates the point. In considering a tradeoff between safety and economy, Figure 10.20 illustrates various solutions to a design problem, based on economic analysis, exemplified by curves A, B and C. A minimum safety level has been specified corresponding to a Safety Index of 4 (10^{-4} per year). The situation where the point corresponding to an optimal solution from the economic standpoint violates the

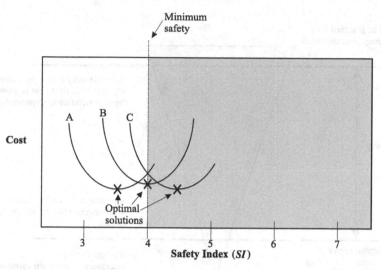

Figure 10.20   Tradeoff between optimal cost and safety: $SI = -\log_{10} p_f$

minimum safety requirement is shown in the case of curve A. It is considered very reasonable then to set a 'cutoff' corresponding to minimum safety.

In setting targets, there will often be a range of values that need to be considered; some occupations, even within a given industry, are more hazardous than others. One might want to express a target which one wishes over time to achieve. The *ALARP* (as low as reasonably practicable) principle has been formulated and used in the UK in considering risks from nuclear power stations (Wells, 1996; Melchers, 1993). This is illustrated in Figure 10.21. One can think of the diagram as representing an arrow pointing in the direction in which one wishes things to improve. This idea can be related to target values, as in Figure 10.22. In this figure, a target of 10^{-5} is proposed, yet values of 10^{-4} are countenanced. See Moan (2004) for a discussion of offshore floating platforms.

Basic risk

The idea of basic risk is useful in deriving target risks in industrial projects. An individual in society will face the following risks. (See Chapter 9 in Lind, 1972; also Starr, 1969.)

- Basic risk r_0, defined as the risk that is impossible to avoid in living in society.
- Voluntary risk r_v, under the person's control, that the person is prepared to accept.
- Liability risk r_ℓ, where blame or liability attaches to a third party.

An estimate of basic risk is often taken as a proportion (about one-third) of the risks due to all accidents. Taking the value from Table 10.1, it is found that $r_0 \simeq 1.4 \cdot 10^{-4}$ per annum. This is somewhat more than the value for motor vehicle accidents. The level of 10^{-4} is often given as a desired maximum risk per year in industrial activities. This is taken here to be a reasonable value for this purpose.

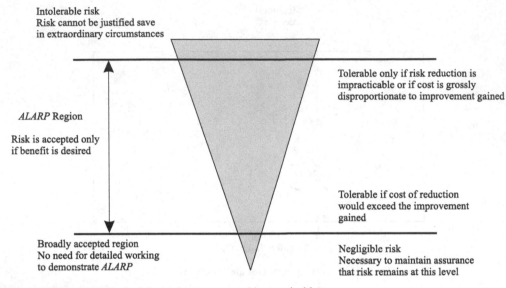

Intolerable risk
Risk cannot be justified save
in extraordinary circumstances

Tolerable only if risk reduction is
impracticable or if cost is grossly
disproportionate to improvement gained

ALARP Region

Risk is accepted only
if benefit is desired

Tolerable if cost of reduction
would exceed the improvement
gained

Broadly accepted region
No need for detailed working
to demonstrate *ALARP*

Negligible risk
Necessary to maintain assurance
that risk remains at this level

Figure 10.21 *ALARP* principle (as low as reasonably practicable)

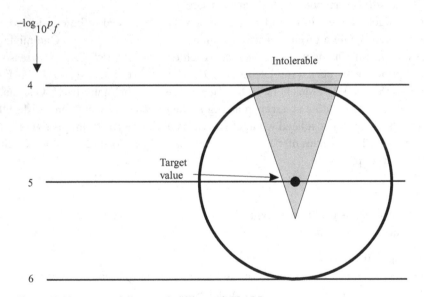

$-\log_{10} p_f$

Intolerable

Target
value

Figure 10.22 Target failure probability and *ALARP*

Target values

Generally risk will be considered with regard to all causes and also with regard to a single cause. Risks to an offshore worker, apart from structural failure, include numerous causes such as helicopter crashes, explosions, blowouts, fires, ship collisions and dropped objects. The same

Table 10.3 *Some target values of tolerable risk to the individual*

	Target risk per year	Safety Index
At work/all causes	10^{-4}	4
Specific causes	10^{-5}	5
At home/all causes	10^{-5}	5
Specific causes	10^{-6}	6

is true of other industries. One then has to think of 'all causes' as against 'specific causes'. In the calibration of offshore standards (Jordaan and Maes, 1991) we considered this aspect and arrived at values of 10^{-4} and 10^{-5} per annum for the two cases, respectively, the latter value pertaining to structural safety. Wells (1996) proposed similar values from the perspective of the designer of process plants, particularly in the chemical industry. Table 10.3 suggests some values of tolerable risk. The values for those at work correspond to occupations where there is above-average risk. Values for society (at home) are based on Wells (1996).

A final point can be made. If the target value of failure probability is p_t, we may suppose that this should be comfortably lower than the basic risk r_0 (Jordaan, 1988):

$$p_t = r_0 \cdot \Delta, \tag{10.37}$$

where Δ is a multiplier. Taking logarithms,

$$(-\log_{10} p_t) = (-\log_{10} r_0) + (-\log_{10} \Delta). \tag{10.38}$$

Arguments are presented Jordaan (1998), that suggest that the logarithm of failure probability is a reasonable utility measure. If this is accepted, one can illustrate the situation as in Figure 10.23 where the change in utility associated with Δ is shown. A reasonable value for an offshore structure is of the order of $p_t = 10^{-5}$ per annum where human life or significant damage to the environment is involved. One can also use another factor, Δ_1 say, in (10.37) to account for the effect on human safety of (for example) evacuation of a rig in advance of a hurricane.

10.4.3 Acronyms: PPHA, HAZOP, FMEA and FMECA

The activities described under these names summarize methods that have been of use in practice. Our approach has been to stress the approach and methods that one would use from first principles, but for completeness we summarize the meaning of the acronyms; see Wells (1996) and Stewart and Melchers (1997) for further details. Preliminary process hazard analysis (PPHA) is used at the preliminary design stage, and is aimed at identifying the major hazards, their cause and consequences. The method has been used in the analysis of process plants. Hazard and operability (HAZOP) studies were developed in the chemical and process industries. It was originally oriented towards flow lines in process plants, together with associated equipment, and aims at a systematic analysis of hazards and other potential problems. Wells (1996) gives excellent case studies. Failure modes and effects analysis (FMEA),

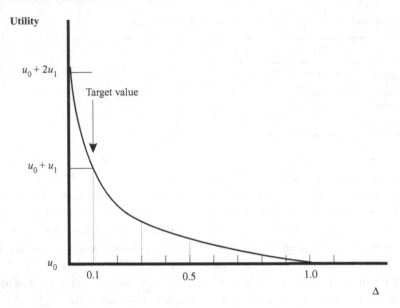

Figure 10.23   Variation of utility with Δ, a multipilier of basic risk R_0. Utility is in relative units

developed originally for aerospace applications, and failure mode, effect and criticality analysis (FMECA) aims at systematic and extensive identification of failure modes in systems. Stewart and Melchers (1997) review literature and summarize some applications.

10.5   Structural systems

The use of **limit states** is natural for structural engineers. A structure may collapse, an oil storage facility may be breached, a high-rise building may suffer excessive deflections to the discomfort of its occupants, vibrations may make the use of machinery difficult or impossible. The different limit states that make the structure unfit for purpose are a natural focus in design. In the context of the present analysis, if a limit state is reached, this is termed 'failure'. As can be seen in the list of examples just given, the consequences of this failure vary considerably. Palmer (1996) discusses the development of structural engineering, and points to the thinking that must have gone behind the construction of arches in Roman times. These stone structures would be prone to failure due to instability resulting from external loads. The arch geometry might result in the formation of a mechanism. Similar arguments apply in the case of early cathedrals and arch bridges. Pugslcy (1951) refers to the failure as one of loss of 'block stability'. This limit state would not involve failure of the stone. Dead weight of the structure was the main concern; wind and the movement of people on the structures were considered relatively unimportant. As Pugsley points out the Achilles heel of this construction was foundation settlement. Limit states corresponding to this failure mode were not considered in masonry construction.

Wood had been extensively used in construction, and particularly in railway bridges. Strength was an issue, and the variability led to the introduction of load factors that were greater than, say, those for wrought iron (Pugsley, 1966). In bridge construction, wood was superseded by cast iron during the expansion of railways in the nineteenth century. Again, material response played an important role because of cast iron's brittleness and vulnerability to fatigue. Many failures of iron bridges occurred, including famous examples such as the Tay bridge. Its failure during a storm alerted engineers to the importance of wind loads, and subsequent designs included allowance for lateral loads due to wind.

Because of the failure of the material, testing of beams and columns was initiated. This led to the introduction of a **factor of safety** as the ratio of the load required to break the girder or component to the load that it would be expected to carry. The possible presence of large flaws in castings led to the use of large safety factors (4 to 6; Pugsley, 1951). The development of steels then permitted the use of its ductility in design. The understanding of beam theory allowed the introduction of stress and the definition of factor of safety as the ratio of the stress at failure to maximum working stress. In aeronautical engineering the use of histograms of peak acceleration, and in building design the plotting of histograms for floor loading represent examples of the introduction of probabilistic ideas and risk from the point of view of determination of loadings. This, when combined with uncertainty in strength, led to the early determination of probability of failure. Pugsley in the UK and Freudenthal in the USA (Freudenthal *et al.*, 1966 for example) were leaders in these developments.

One can see the step-by-step process of learning by experience. A facet of this experience was the need to deal with uncertainty, for example in allowing larger factors of safety when there was uncertainty in strength. The format used in current limit-states design is

$$\gamma l^* \leq \phi r^*, \tag{10.39}$$

where l^* is the specified load with load factor γ, and r^* is the specified resistance with resistance factor ϕ. This has been generalized to state that 'the **effect** of the factored loads does not exceed the factored resistance' (for example, Canadian Standards Association, 1992).

Specifications of load for design purposes reflect the seriousness of consequences; the more serious, the smaller the exceedance probability – and the longer the return period – on the design loading event. In particular, there may be implications of failure for human life. In such cases where the consequences of failure are serious and can be anticipated, we effectively 'engineer out' the unwanted events by using loads and resistance that ensure that the probability of exceedance of load over capacity becomes very small. An example of the introduction of consequences into the design setting is the use of safety classes which govern the selection of target levels of safety. An example is given in the Canadian Standards Association (1992)[3] for fixed offshore structures in which two safety classes are defined for the verification of the safety of the structure or any of its structural elements:

---

[3] The 2004 edition is now available.

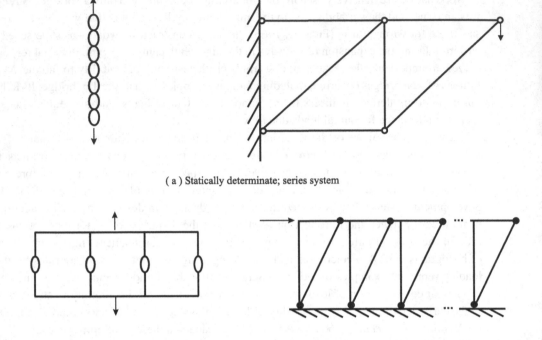

(a) Statically determinate; series system

(b) Statically indeterminate; parallel system

Figure 10.24   Structural systems

(a) Safety Class 1 – failure would result in great risk to life or a high potential for environmental damage, for the loading condition under consideration;

(b) Safety Class 2 – failure would result in small risk to life and a low potential for environmental damage, for the loading condition under consideration.

The target annual probabilities of failure are given in the standard as 10^{-5} and 10^{-3} respectively.

As noted in the above (Section 10.3.3) in reality most structural failures result from errors and human factors. An opposite tendency to that of errors is that of redundancy and reserve capacity in structural systems. Redundancy is often not included in the methodology of structural engineering. Figure 10.24 illustrates simple structural systems that correspond to the assumptions of series and parallel behaviour, as introduced in Section 10.3.2. In the case of parallel systems, the usual mode of plastic collapse is illustrated. It is required that sufficient ductility be provided so that the postulated modes of failure can occur. Analysis of system ductility is often complex and not fully calculated in design, it being possible only to model the failure of a component of a system, supplemented by approximate techniques involving reserve strength ratios or judgement when dealing with the entire system. An additional level of safety against collapse results. The calculation of failure probability, or estimates based on code calibrations or comparisons, are then only approximate.

A last issue that should be mentioned is that of **model uncertainty**. The models used for estimation of loads often contain bias and uncertainty. For example, measurements of water drag forces might show scatter when compared to theory (Morison's equation). Nessim *et al.* (1995) review this area and propose a framework for analysis. Models of structural resistance are also subject to uncertainties of this kind. Experienced engineers are generally familiar with model uncertainties and make allowances for them in the design process.

10.5.1 Tail sensitivity: Reliability Index

There can be real difficulties in developing confidence in estimates of probability of failure. This results partly from the fact that structural design will generally entail extremely low probabilities of failure given that safety of human life is involved. Failure will then correspond to the upper and lower tail regions of the load and demand curves (respectively) of Figure 10.7(a). There are difficulties in inferring details of distributions in the tails where failure will take place – this might be several standard deviations from the mean. Observations might not be available, and extrapolation might add uncertainty. It is often, at the same time, possible to extrapolate on the basis that the physical processes are similar in the extremes as for the central part of the distribution, for example for waves, and to use asymptotic distributions that are well justified on the basis of the process itself. In some cases of rare events, the rareness itself results in the key part of the distribution not being far in the tail. These factors might assist, but modelling the tails of distributions does require special attention. One occasionally sees particular distributions chosen merely for ease of computation; the lognormal has been a choice on occasion for this reason.

Failure probabilities are sensitive to the distributions of the tails, just discussed; different distributions with similar moments can give results that differ considerably. Failure rates can be difficult to observe since they are – *ipso facto* – rare. As a result of these small failure rates it is not easy to obtain statistical data to match the theoretical estimates. Palmer (1996) used a published estimate of reliability for the yielding limit state in pipelines to show that for 50 North Sea pipeline systems of length 10^4 km each, the time until one failure on average would correspond to the lifetime of the universe, estimated at $2 \cdot 10^{10}$ years. Similar small values were obtained in estimates of the probability of rupture in the plates and panels in arctic shipping design (Carter *et al.*, 1992). The latter results from the fact that design concentrates more on a serviceability limit state – the desire to avoid excessive denting – than on rupture. If one concentrated on rupture only, there would inevitably be unacceptable denting since the ultimate membrane strength of the plates is many times the value corresponding to the three-hinge bending failure. Furthermore, there are not many vessels in the arctic, so that a failure data base is not expected, although 'hungry horse' ship plates are commonly observed. At the same time, the failure probabilities for the second example just given (ship plates) are sufficient; we need only to know that they are very small, since, as stated, serviceability dominates.

One of the ways of avoiding the problems of dealing with the tails of distributions is the use of first and second moments only. This is based on the margin of safety $M = T - S$; see

Equation (10.9) and Cornell (1969). From this, a measure of structural safety has been derived: the 'β-factor', or **Reliability Index** defined as

$$\beta = \frac{\mu_M}{\sigma_M},\tag{10.40}$$

where μ_M and σ_M are the mean and standard deviation of the margin of safety M (Section 10.3.1). The motivation is that one would wish the mean value of M, μ_M, (the average margin of safety) to be greater if there is more uncertainty, which occurs in the case of greater σ_M. Keeping β constant then keeps the ratio between the safety margin and the level of uncertainty, as measured by the standard deviation, constant. If one judges that the uncertainty can be modelled using normal distributions, the probability of failure is as given in Equation (10.14):

$$p_f = F_Z(-\beta).\tag{10.41}$$

The simplicity of dealing only with first and second moments in reliability theory is appealing and is important in many applications.

It should be borne in mind that this is indeed only an approximation, and was so intended by the originator. Recall that in Section 4.4, in Equation (4.2), we discussed the use of such approximations in making decisions. The idea was to use a single value to represent the lottery. This was taken as being equal to the mean less a constant times the standard deviation; using the notation of the present section,

$$\mu_M - \beta\sigma_M,\tag{10.42}$$

where β is a constant depending on our degree of risk-averseness. We showed in Section 4.4 using Equation (4.2) – essentially the same as (10.42) – that there are cases where we cannot discriminate between two actions since μ_M and σ_M are the same for each action, yet we have a clear preference for one of the two actions. Figure 10.25, which mimics Figure 4.5, shows

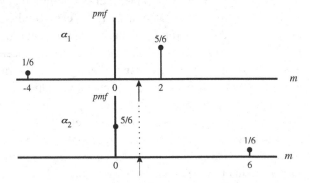

Figure 10.25 Probability mass functions for strategies α_1 and α_2. The first choice leads to possible collapse (negative M) whereas the second one does not

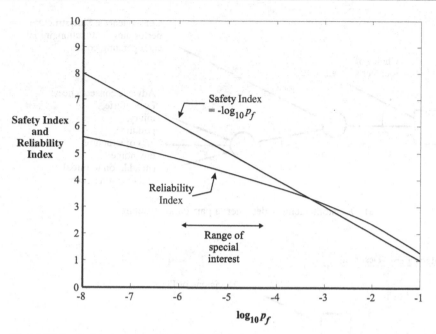

Figure 10.26 Safety Index and Reliability Index

two simple probability distributions resulting from two actions α_1 and α_2. The use of (10.42) does not permit one to discriminate between α_1 and α_2. A utility function was needed to do this.

But the β-factor was always intended as an approximation to take into account the very insufficiency of information in estimating probabilities and should therefore not be used for calculating p_f if this is the case. For unidimensional S and T, there is a simple relationship between β and p_f; see Equation (10.14) or Exercise 10.8. Figure 10.26 shows the relationship between our Safety Index $(-\log_{10} p_f)$ and the Reliability Index β. It is seen that this is essentially linear over the range of interest. We have proposed the Safety Index as a utility function for safety (Jordaan, 1988). The index β, being a linear transformation, could serve the same purpose. But the calculation of probability of failure from normal distributions of load and resistance, implied in β, might not be accurate.

If one has detailed information on the probability distributions involved, and if one wishes to make correspondingly more detailed evaluations of risk, it would be better to proceed to detailed probabilistic design. Over the last decades the development, indeed redefinition of β so as to make a better estimate of p_f, has been a central part of structural reliability (see Madsen *et al.*, 1986). In this approach, the definition of β has been amended so that random quantities representing load and resistance are transformed into normal distributions. Then β for the transformed system gives accurate values of p_f. This development has been good in the sense of providing consistency with the past.

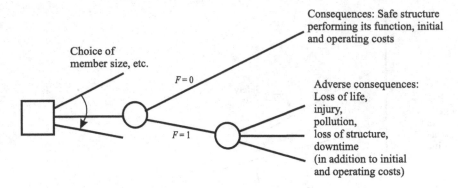

(a) Decisions facing a designer: a particular structure

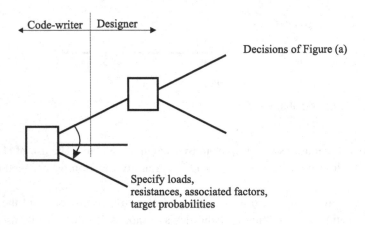

(b) Decisions facing the code-writer: all structures in the class covered by the code.
Vertical line shows division between two groups of decision makers

Figure 10.27   Code-writing decisions

10.5.2   Codes of practice

Codes of practice are formulated with the intention of setting minimum levels of safety. The decisions facing a designer are illustrated in Figure 10.27(a). There is no way that a code-writer can, or should, account for all the economic decisions in designing a structure; these are the domain of the owner, designer and operator. See, for example, Rosenblueth (1987). The interested parties are

(i) the owner and operator (who might be the same);
(ii) government and regulatory bodies responsible for safety;

(iii) the designers and contractors;

(iv) not least, the future personnel who will work on the facility.

The purpose of the code is to ensure adequate safety and performance of the structure. The code-writer therefore attempts to make a judgement on behalf of society, to set a safety level as a constraint on the design process. It therefore should focus on the interests of the fourth of the interested parties above. The role of the code as a constraint on design is shown in Figure 10.27(b). The responsibility of protecting persons at risk was emphasized in Section 10.4.2.

  The code may include rules of thumb that assist the designer in obtaining quick and reliable solutions. Fully probabilistic design is not a necessary or reasonable objective; the code should relieve the designer of having to perform intricate probabilistic calculations. What is needed is a judicious use of probability theory (Palmer, 1996), with intense effort for large projects involving possible loss of life or damage to the environment.

10.6   Broad issues

Attempts have been made to develop methods to deal with broad issues facing humankind as a whole. Our response to risks and dangers in society is often irrational; several examples have been given in the section on 'Risks to the individual', above. A more rational allocation of resources could certainly result in less suffering and in greater human happiness. A good example of this kind of analysis has been presented by Lind, Nathwani and Siddall (1991). Their analysis weighs the benefits against the costs of various possible actions. They draw attention to what they term 'misallocation of resources'. Blind reduction of risk is not a good thing; resources should be allocated where they can do the most good. They give the example from the nuclear industry, where the expenditure on the saving of a life by installing hydrogen recombiners to prevent explosions amounts to about $7 billion. At the same time, it is estimated that detection and reduction of radon in homes is 1000 to 100 000 times more cost-efficient. They discuss the lack of consistency in safety standards whereby pollutants in the outdoors from industrial sources are controlled yet the risk from a variety of organic chemicals indoors is 100 times greater. On the other hand, some of us would wish to aim at air in the outdoors that is clean as possible, for all sectors of society!

  In formulating attributes to use in studying human happiness, careful thought must be paid to the parameters proposed to measure the attribute in question. These should reflect the attribute that is desired. As an example where an attribute might be misleading, consider the use of Gross National Product (GNP) as a measure of the happiness of individuals in a nation. Goldsmith (1994) argues that, despite the quadrupling and trebling of the GNPs, in real terms, of the USA and UK respectively, in the last fifty years, both nations are 'profoundly troubled'. GNP measures the extent of economic activity, and calculations based on it do not include, for example, the contribution of mothers or fathers who spend their time looking after their families without pay, thus not contributing to the official economy. Policies such as globalization, which lead to an increase in GNP, do not necessarily lead to increased happiness. Billions of people from countries with high unemployment – for instance India or Vietnam – have entered the world economy, and they are able to work at a rate of pay approximately one-fiftieth that of a person in the developed economies. Enterprises naturally move those factories requiring extensive labour to the developing countries where 50 people can be employed for the price of one in the developed world. The result is the high level of unemployment and associated misery found in the developed country despite strides in the level of GNP.

Lind *et al.* (1991) discuss other indices, for example ones that include life expectancy, literacy and purchasing power based on GDP per capita. This is the human development index (HDI) developed by the United Nations Development Programme (UNDP). The main message in the study is the need for frameworks at the national level to make rational decisions regarding the allocation of national revenue and the development of resources so as to maximize benefits and the avoidance of inefficient risk-reduction schemes. These efforts at analysis are worthwhile, yet whether they will succeed in the long run remains to be seen. There is a considerable element of irrationality in attempts by humankind to come to terms with national, and more so, international planning. For the foreseeable future, we can expect to see the desecration of the environment and the disregard for social issues as well as the disrespect and contempt for other animals and species to continue.

10.7   Exercises

There are many reference case studies in the area of risk analysis. It is a useful educational exercise to search for and consult those in your area of interest.

10.1 It is a useful exercise to conduct a search on data bases available in your area of interest.

10.2 Consider the case where proof loading is applied to a system. Describe how this may affect the distribution of resistance $f_T(t)$. Consider cases where the proof loading does and does not cause signs of distress in the system (for example cracking in a structure).

10.3 Prove Equation (10.6).

10.4 Derive p_f for the case where T and S are uniformly distributed.

10.5 A component has random resistance that is modelled by a symmetrical triangular distribution in the interval (10, 15) in arbitrary units. The loading is uniform in the interval (4, 12) using the same units. Calculate the probability of failure.

10.6 Derive p_f for the case where T and S are modelled by chi-square distributions.

10.7 For lognormally distributed load and resistance, show that

$$p_f = F_Z \left(-\frac{\ln\left[\frac{\mu_T}{\mu_S}\left(\frac{1+v_S^2}{1+v_T}\right)^{\frac{1}{2}}\right]}{\left\{\ln\left[(1+v_T^2)(1+v_S^2)\right]\right\}^{\frac{1}{2}}} \right), \tag{10.43}$$

where μ_T and μ_S are the means of T and S whereas v_T and v_S are the coefficients of variation. Also show that the following approximations are reasonable.

$$p_f \simeq F_Z \left(-\frac{\ln\left[\frac{\mu_T}{\mu_S}\right]}{(v_T^2 + v_S^2)^{\frac{1}{2}}} \right) \tag{10.44}$$

for $v_T, v_S < 0.3$. Also

$$\frac{\mu_T}{\mu_S} \simeq \exp\left[\beta\left(v_T^2 + v_S^2\right)^{\frac{1}{2}}\right]. \tag{10.45}$$

10.8 Show that p_f for the case of normal distributions can be written as

$$\mathbf{Pr}\left(-\beta\right) = \frac{1 - \mathrm{erf}\left(\frac{\beta}{\sqrt{2}}\right)}{2}. \tag{10.46}$$

Use this or (10.14) to verify Figure 10.26.

10.9 Verify Equations (10.19), (10.21), (10.26) and (10.27) by substituting 1s and 0s using $1 = \mathrm{TRUE}$, $0 = \mathrm{FALSE}$ in the F_i and R_i.

10.10 Extend Equation (10.25) to find p_s for three units in series, using Equation (2.17).

10.11 The failure rate generally follows the analysis of the intensity function of Section 9.5.4 and Figure 9.12(b). For the special case of a constant intensity as in Example 9.9, show that the reliability can be written as

$$R(t) = \exp(-\lambda t),$$

where t is time and λ is a parameter representing the failure rate. Use this to develop the mean-time-to-failure ($MTTF$) for series and parallel systems.

10.12 Find the reliability of the system in the figure, given a component reliability of 0.9.

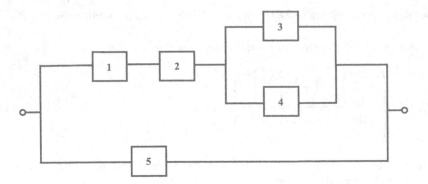

10.13 The error rate in a design/programming process has been found to be one per 100 calculations or programming statements.

    (a) Given a total of 1000 statements, find the probability that the number of errors exceeds 20.

    (b) Given a checking process in which one in ten errors is detected, calculate the probability that the number of errors exceeds 20 after checking.

    (c) Formulate a model in which the error and detection rate are treated as random. See Figure 10.14 and Nessim and Jordaan (1985).

10.14 Research the effect of human error on the capacity T of a system. In particular, consider the model of human error due to Lind (1983).

10.15 Give examples of practical probability- and consequence-reducing measures in risk analysis in your area of interest.

'I checked it very thoroughly,' said the computer, 'and that quite definitely is the answer. I think the problem, to be quite honest with you, is that you've never actually known what the question is.'

Douglas Adams, *The Hitchhiker's Guide to the Galaxy*

Entities should not be multiplied unnecessarily.

Occam's Razor

11.1 Introduction

Engineers continually face the necessity to analyse data and records from the past. These might consist of measurements of material strength, flood records, electronic signals, wave heights in a certain location, or numbers per unit area of icebergs. The intention is to use these records of the past to assess our uncertainty regarding future events. The overall aim is, as always, to make decisions for design at an acceptable level of risk. Apart from good analysis, data provide the most significant means of decreasing uncertainty and improving our probabilistic estimates. Data are obtained from experiments. These always represent an available strategy, although money and effort are required to carry them out. It is the author's experience that in cases where there is considerable uncertainty regarding a physical process, one real experiment is worth a thousand theories.

In the use of probabilistic models two aspects present themselves: the choice of the model itself, and the estimation of model parameters. We showed in Chapter 8 how to use data to estimate parameters of *known distributions*. The framework described in Section 8.1.1 involved the knowledge of a distribution or process. In the present chapter, we describe the use of data to fit and compare distributions. The choice of a particular model can be made by an analysis and understanding of the underlying process involved, or on an empirical analysis of data, or some combination of the two. The quantity under investigation might be the sum of many other quantities leading to the use of the normal distribution. Or it might be multiplicative in character – bacteria multiply, and possibly fracture processes might show this characteristic, leading to the lognormal distribution. Or the strength of an element might be the smallest of elements in its composition, leading to the Weibull distribution. The occurrence of accidents in time might be random, leading to the Poisson distribution.

Alternatively, the data might be compared empirically to various distributions and a choice made on that basis. A combination of knowledge of the process and empiricism might be used. In some cases, one might use a set of empirical data itself as a model, in which case there might not be any need to estimate model parameters. We summarize the approach noted with a Bayesian relationship, to be considered as a pragmatic guide in one's thinking:

$$\mathbf{Pr}\,(C|\text{data}) \propto \mathbf{Pr}\,(\text{data}|C) \cdot \mathbf{Pr}\,(C),\qquad\qquad(11.1)$$

where C indicates the choice of a particular distribution. A common-sense result is that with diffuse prior knowledge, the closer the data are to the fitted distribution, the higher the chance that the fitted distribution is the best hypothesis. The importance of prior knowledge is also apparent. The equation is introduced to show that Bayesian thinking can guide the work of the present chapter. But the work does relate to real world data, which often do not fit the ideal theoretical model.

The present chapter also deals with simulation. Often convenient closed-form mathematical models are not available, necessitating the use of numerical methods. Often data are not available or are found for certain parameters that are of indirect relevance to the problem at hand. In these cases, Monte Carlo simulation is employed to generate results in a manner that is in some ways analogous to an experiment, and certainly to a 'thought-experiment'. This forms the second related subject of the chapter.

11.2 Data reduction and curve fitting

We shall introduce the ideas using an extended example. This concerns a problem of measurement. An engineer is using a survey instrument to measure the difference in elevation from a reference point of known elevation to the underside of a bridge deck. This problem was introduced in the context of inference in an example at the end of Section 8.5.1. It could be an important one for determining clearance under the bridge, for instance for a rail system that passes underneath. In order to assess the accuracy of the measuring system, the engineer takes 100 repeated measurements. These are given in Table 11.1. The problem is to determine the distribution that best fits the data. The engineer suspects that the data follows the normal distribution. This distribution of 'errors' in measurement was used extensively by Gauss in studies of astronomy and geodesy.

Calculations have traditionally been done manually. Now, computer-based methods provide an effective and powerful means of dealing with data. For a beginner, it is useful for complete understanding to do some of the calculations by hand at least once. We shall lead the way through the analysis of the data set. The first thing to do is to obtain representative statistics. It should be noted that the treatment of the data set is based on the concept that 'order doesn't matter' thus being consistent with the assumption of exchangeability.

Table 11.1 *Measurements of bridge elevation in metres from reference point*

27.893	27.832	27.959	27.927	27.929
27.914	27.925	27.929	27.988	27.830
27.960	27.859	27.887	28.001	27.763
27.799	27.826	27.898	27.883	27.924
27.958	27.883	27.859	28.000	27.854
27.882	27.882	27.853	27.892	27.940
27.900	27.847	27.916	27.868	27.971
27.939	27.850	27.940	27.925	27.855
27.926	27.913	27.957	27.897	27.945
27.827	27.814	27.894	27.952	27.907
27.873	27.888	27.866	27.975	27.884
27.970	27.838	27.819	27.905	27.872
27.876	27.842	27.920	27.929	27.881
27.955	27.859	27.926	27.993	27.836
27.971	27.926	27.938	27.884	27.910
27.805	27.982	27.888	27.932	27.909
28.032	27.960	27.881	27.913	27.867
27.961	27.877	27.890	27.861	27.934
27.947	27.802	27.892	27.785	27.826
27.865	27.910	27.909	27.792	27.852

11.2.1 Sample descriptive measures

We are given a set of measured values

$$x_1, x_2, x_3, \ldots, x_i, \ldots, x_n. \tag{11.2}$$

Note that these values cannot be random – they correspond to actual deterministic values such as those in Table 11.1 above. A first task is to obtain minimum, maximum and the range, defined as the maximum minus the minimum. It is also useful for several purposes to rank the data; this is usually done in ascending order and represents a task that is much better done by a computer than by hand.

Example 11.1 For the data set of Table 11.1, the ranked data is shown in Table 11.2. The minimum and maximum are 27.763 and 28.032 m respectively, with a range of 0.269 m.

Central tendency

Moments are generated for data in a manner analogous to those for distributions. One can imagine, in a general way, that each data point has a probabilistic 'weight' of $1/n$. But we will see that for variance and standard deviation, usage suggests $1/(n - 1)$. The most common and

Table 11.2   *Data of Table 11.1 ranked in ascending order*

27.763	27.854	27.884	27.913	27.945
27.785	27.855	27.884	27.914	27.947
27.792	27.859	27.887	27.916	27.952
27.799	27.859	27.888	27.920	27.955
27.802	27.859	27.888	27.924	27.957
27.805	27.861	27.890	27.925	27.958
27.814	27.865	27.892	27.925	27.959
27.819	27.866	27.892	27.926	27.960
27.826	27.867	27.893	27.926	27.960
27.826	27.868	27.894	27.926	27.961
27.827	27.872	27.897	27.927	27.970
27.830	27.873	27.898	27.929	27.971
27.832	27.876	27.900	27.929	27.971
27.836	27.877	27.905	27.929	27.975
27.838	27.881	27.907	27.932	27.982
27.842	27.881	27.909	27.934	27.988
27.847	27.882	27.909	27.938	27.993
27.850	27.882	27.910	27.939	28.000
27.852	27.883	27.910	27.940	28.001
27.853	27.883	27.913	27.940	28.032

useful measure of central tendency is that of the sample **mean** (commonly referred to as an average):

$$\bar{x} = \frac{\sum_{i=1}^{n} x_i}{n}.$$ (11.3)

Thus

$$\sum_{i=1}^{n} (x_i - \bar{x}) = 0.$$ (11.4)

Example 11.2 The mean of the data set of Table 11.1 is found to be 27.898 m.

Strictly, the mean presented in Equation (11.3) is the arithmetic mean. Weighted means apply different weights to particular data points (in which case order might matter). Geometric and harmonic means can also be calculated in particular instances, as in the following example.

Example 11.3 The geometric mean of n observations is the nth root of their product. For example, if a stock in a particular exchange increases by 5%, 10% and 12% in three successive years, we would be interested in the geometric mean

$$\sqrt[3]{1.05 \cdot 1.10 \cdot 1.12} = 1.0896 \text{ (to four decimal places).}$$

The geometric mean is always less than the arithmetic one and gives better results when multiplicative compounding is the issue, as in the present example.

Measures of central tendency include the **mode** and the **median**. The mode corresponds to the most frequently occurring value, while the median represents the value in the middle of the ranked data set. For a data set with an odd number of observations, the median is the $[(n + 1)/2]$th value, for example the fifth if $n = 9$. If the number of data points is even, then one averages the $(n/2)$th and the $[(n + 2)/2]$th values.

Example 11.4 In our data set of Table 11.2, several values occur twice (for instance 27.826, 27.881, 27.909 and 27.913; see Table 11.2). Three values occur three times (27.859, 27.926 and 27.929). There are too few repetitions of values to obtain a good estimate of central tendency. The use of the mode works better for grouped data, using the peak value in the frequency distribution; see Subsection 11.2.2 below.

Example 11.5 The median of the data set of Table 11.2 is the mean of the 50th and the 51st ranked data points, i.e. 27.8955 m.

Dispersion

The first measure of dispersion of the data is the **range**, given above as 0.269 m for the data of Tables 11.1 and 11.2. By far the most common measure of spread is the **sample variance** (s^2) and sample **standard deviation** (s). These values are given by

$$s^2 = \frac{\sum_{i=1}^{n} (x_i - \bar{x})^2}{n - 1} \tag{11.5}$$

and

$$s = \sqrt{\frac{\sum_{i=1}^{n} (x_i - \bar{x})^2}{n - 1}} \tag{11.6}$$

respectively. The use of $(n - 1)$ rather than n in Equations (11.5) and (11.6) is common and is sometimes termed 'Bessel's correction'. We have adopted this definition because of its common usage.

But the reason for this choice is more historic than compelling. It appeared naturally when considering some inferences using the normal distribution where the mean and variance were being studied, for example in Equations (8.99), (8.100) and (8.143), in Section 8.6.5, and there the factor $(n - 1)$ was termed the 'degrees of freedom', ν. The use of $\nu = n$ appeared in Sections 8.5.2 and 8.6.4 where inferences regarding the variance only are being made. Hence there does not seem to be a strong reason to choose $(n - 1)$ rather than n, except that the problems of inference involving $(n - 1)$ are more common. It is also stated in the literature that the estimator s^2 is 'unbiased' but there are good reasons to question this, as described in

Section 8.7.4 and in Exercise 8.24 ('fly in the ointment'). Discussion of degrees of freedom is given in Section 11.3.2 below.

The **sample coefficient of variation** is defined in a manner analogous to that for probability distributions, as given in Equation (3.66):

$$\frac{s}{\bar{x}} \cdot 100, \text{ in per cent.} \tag{11.7}$$

Example 11.6 The standard deviation of the sample of Table 11.1 is found to be 0.0535 m and the coefficient of variation is

$$\frac{0.0535}{27.898} \cdot 100 = 0.19\% \text{ (two significant figures).}$$

The mean elevation is measured on an interval rather than a ratio scale (Section 4.7) so that there is a certain arbitrariness in the coefficient of variation.

Note that

$$s^2 = \frac{\sum_{i=1}^{n}(x_i - \bar{x})^2}{n-1}$$

$$= \frac{\sum_{i=1}^{n} x_i^2 - 2\bar{x}\sum_{i=1}^{n} x_i + n\bar{x}^2}{n-1}$$

$$= \frac{\sum_{i=1}^{n} x_i^2 - n\bar{x}^2}{n-1}, \text{ or}$$

$$s^2 = \frac{n\sum_{i=1}^{n} x_i^2 - \left(\sum_{i=1}^{n} x_i\right)^2}{n(n-1)}. \tag{11.8}$$

The last form is useful for computation.

A measure of dispersion can be based on **quantiles** (quartiles, percentiles, deciles, ...). These can be defined for an ordered data set in a similar way to that used for the median above. The basic idea is to find values such that a specified probability lies below the value in question. Data do not of themselves represent probabilities; one would likely want to smooth the data by fitting a distribution and to consider uncertainties in the parameters of the distribution (Chapter 8). The transition from data to probability is one that requires some introspection, and one should not automatically make the association that data represent probability (Section 11.2.3 below).

We shall now give the rules for obtaining quantiles based only on the data (in other words before the 'transition to probability'). The method is much the same as for medians above. Suppose that one wishes to obtain the point corresponding to the first quartile. We then calculate $(n/4)$. If this value is not an integer, round up to the next integer and take the corresponding ordered value of the data as the quartile. If the value is an integer, calculate the mean of the corresponding ordered value of the data and the next highest. For third fractiles we use $(3n/4)$ with a similar calculation procedure. For the (i/r)th fractile, we use $(i/r) \cdot n$ and either choose a value or average as just described. When a non-integer is found, one chooses the next integer as the division point giving the quantile, thus dividing the data set into two groups of size $(i/r) \cdot (n-1)$ and $[(r-i)/r] \cdot (n-1)$. When an integer is found, the division point is halfway between this integer and the next, thus dividing the data set into two groups of size $(i/r) \cdot n$ and $[(r-i)/r] \cdot n$. Rather than taking the midpoint, linear interpolation could be used.

Example 11.7 Suppose the data set has $n = 75$ values. We wish to find the first quartile so that $r = 4$ and $i = 1$. Then $n/4 = 18.75$ and we take the 19th value as the quartile. For the third quartile $i = 3$ and $3n/4 = 56.25$ so that we use the 57th value.

Example 11.8 For the data set of Table 11.1, $(n/4) = 25$, so that the first quartile is halfway between the 25th and the 26th values in the ranked data. This gives $(27.859 + 27.861)/2$, or 27.860. Similarly, the third quartile is 27.933.

The method just described is based only on the data; a better method might be required based on the fitted probability distribution. See Subsection 11.2.3 (transition to probability) below.

Skewness and kurtosis

We give the usual classical definitions, which are useful for assessing skewness and flatness of distributions. The **coefficient of skewness** is defined as

$$\bar{g}_1 = \frac{\frac{1}{n}\sum_{i=1}^{n}(x_i - \bar{x})^3}{s^3}. \tag{11.9}$$

The value will be positive for distributions skewed to the right and negative for distributions skewed to the left (see also Figure 3.20).

Example 11.9 For the data of Table 11.1, the skewness is $\bar{g}_1 = -0.0742$. This low value indicates a symmetric distribution.

The **coefficient of kurtosis** is given by

$$\bar{g}_2 = \frac{\frac{1}{n}\sum_{i=1}^{n}(x_i - \bar{x})^4}{s^4}. \tag{11.10}$$

This value is often compared to $\bar{g}_2 = 3$, which is the value for a normal distribution. The kurtosis is a measure of 'flatness' and tail heaviness of distributions.

Example 11.10 For the data of Table 11.1, we find $\bar{g}_2 = 2.69$, which is close to the value 3 for a normal distribution.

The definitions given in Equations (11.9) and (11.10) should be used with large samples. For small samples, the estimates will become unreliable, and a small bias will become evident;

see Exercise 11.2 There are also other ways to define skewness and kurtosis, for example using the values $\sqrt{\bar{g}_1}$ and $(\bar{g}_2 - 3)$. In using computer packages, it should be checked which version of the definitions is used.

11.2.2 Grouped data and histograms

Histograms represent an excellent means of judging the form of the probability distribution. We shall focus on the data of Table 11.1. Figure 11.1 shows the result; the shape of the distribution is bell-shaped and a good appreciation of the distribution of the measurements is obtained. The first step in producing this display is to develop a tally chart, as shown in Table 11.3. Computer programs compute these automatically, but it is useful to complete at least one by hand. Most programs will have a default number of class intervals and these are often unsatisfactory; it is better oneself to specify the class boundaries. Too few intervals may lead to loss of information regarding the shape of the distribution and too many lead to erratic shaped results, hard to interpret. It is worth experimenting with the number of intervals – using a computer package! – so as to get a feeling for these effects.

A useful approximate rule that has been found by experience for the number of intervals is

$$1 + 3.3 \log_{10} n. \tag{11.11}$$

For $n = 100$, this gives 7.6 intervals. The following was decided upon: start at 27.761 m and end at 28.033 m, with eight intervals of size 0.034 m. The tally chart is shown in Table 11.3. The 'cut points' are the class boundaries whereas 'class marks' indicate the midpoints of each

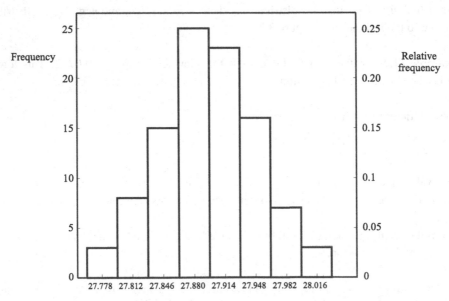

Figure 11.1 Histogram of elevation data

Table 11.3 *Tally chart for data of Tables 11.1 and 11.2*

Cut points (class boundaries)	Midpoints (class marks, $\acute{x}_j$)	Tally (frequency)	Relative frequency, g_j
27.761–27.795	27.778	3	0.03
27.795–27.829	27.812	8	0.08
27.829–27.863	27.846	15	0.15
27.863–27.897	27.880	25	0.25
27.897–27.931	27.914	23	0.23
27.931–27.965	27.948	16	0.16
27.965–27.999	27.982	7	0.07
27.999–28.033	28.016	3	0.03

interval. The usual assumption is that if a value coincides with the upper class boundary it is taken as being in that class, not the next. The intervals are often simply referred to as 'bins'.

The tally from Table 11.3 is plotted as the frequency in Figure 11.1. The relative frequency, which is the frequency divided by the number n of data points, is given in the last column of Table 11.3, and is shown on the right hand ordinate of Figure 11.1. The most likely value, the mode, can now successfully be calculated; it corresponds to the class mark 27.880 in Table 11.3 and in Figure 11.1.

One can calculate the mean, standard deviation and other statistics from a set of grouped data. By using grouped data, one does lose some accuracy as compared to the use of the original data set. The method is analogous to that above. For a set of grouped data with class marks

$$\acute{x}_1, \acute{x}_2, \acute{x}_3, \ldots, \acute{x}_j, \ldots, \acute{x}_k \tag{11.12}$$

and associated frequencies

$$g_1, g_2, g_3, \ldots, g_j, \ldots, g_k, \tag{11.13}$$

the sample mean and variance are

$$\bar{x} = \frac{\sum_{j=1}^{k} g_j \acute{x}_j}{n} \tag{11.14}$$

and

$$s^2 = \frac{\sum_{j=1}^{k} g_j \left(\acute{x}_j - \bar{x}\right)^2}{n-1} \tag{11.15}$$

$$= \frac{n \sum_{j=1}^{k} g_j \acute{x}_j^2 - \left(\sum_{j=1}^{k} g_j \acute{x}_j\right)^2}{n(n-1)}, \tag{11.16}$$

respectively.

11.2.3 Cumulative frequency plots, fitting distributions and the transition to probability

We consider initially the grouped data of Table 11.3. In Table 11.4, these values are used to generate cumulative frequency and cumulative relative frequency (crf) values; these are the sum of frequencies and relative frequencies respectively for those intervals (or bins) below the value (boundary) in question. The data can be used to plot the **ogive** given in Figure 11.2. Usually these curves are S-shaped as in the figure.

Table 11.4 *Data for ogive and probability plot*

Less than or equal to	Cumulative frequency	Cumulative relative frequency $= crf$	Normal $z = F_Z^{-1}(crf)$
27.761	0	0	$-\infty$
27.795	3	0.03	-1.881
27.829	11	0.11	-1.227
27.863	26	0.26	-0.643
27.897	51	0.51	0.025
27.931	74	0.74	0.643
27.965	90	0.90	1.282
27.999	97	0.97	1.881
28.033	100	1.0	∞

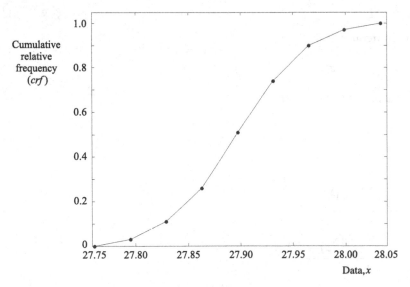

Figure 11.2   Frequency ogive for data of Table 11.4

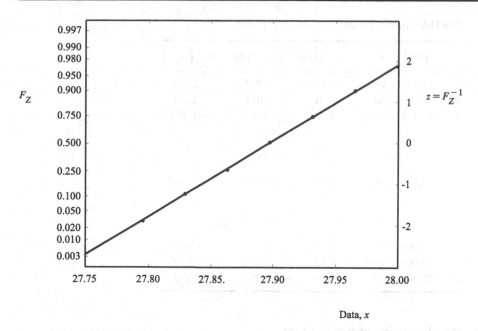

Figure 11.3 Probability plot of binned data of Table 11.4

To check for normality of a distribution, 'arithmetic probability paper' has traditionally been used. This has in the past been done by hand and again it is useful for a beginner to complete one or two plots by hand. The plots are now done very effectively by computer for a variety of distributions. The idea is to produce a graph, which if linear, would suggest that the distribution under consideration is the appropriate one for use. The 'arithmetic probability paper' could in the past be purchased at stationers but can be produced as follows. We shall complete the calculation for the normal distribution, but the method can be applied to any distribution which can be normalized using location and scale parameters. The method for those that cannot is given below.

The cumulative relative frequencies range from 0 to 1, so that it is natural to use these as 'first estimates' of cumulative probabilities. Since we are considering the normal distribution, for convenience the comparison is made with the standard normal distribution; see Equations (3.26) and (3.27), the cumulative distribution of which is denoted F_Z. We calculate the inverse values, F_Z^{-1}, as in the last column of Table 11.4. A straight line in a plot of data against these values supports the assumption of a normal distribution. Figure 11.3 shows the result in the present case. One can see that the data follows approximately a straight line, suggesting normality. It should be noted that the first and last points with relative frequencies of 0 and 1 correspond to $-\infty$ and $+\infty$ on the inverse cumulative axis for all distributions that range from $-\infty$ to $+\infty$.

In the example, the standard normal distribution has been used, but the analysis can be done for other chosen distributions; see Figures 11.6 and 11.8 and related text below.

Table 11.5 *Ranked data for probability plot*

r	Value	r/n	$(r-1)/n$	$(r-\frac{1}{2})/n$	$F_Z^{-1}\left(\frac{r-\frac{1}{2}}{n}\right)$
1	27.763	0.01	0	0.005	−2.576
2	27.785	0.02	0.01	0.015	−2.170
3	27.792	0.03	0.03	0.025	−1.960
⋮	⋮	⋮	⋮	⋮	⋮
⋮	⋮	⋮	⋮	⋮	⋮
50	27.894	0.5	0.49	0.495	−0.0125
⋮	⋮	⋮	⋮	⋮	⋮
⋮	⋮	⋮	⋮	⋮	⋮
98	28.000	0.98	0.97	0.975	1.960
99	28.001	0.99	0.98	0.985	2.170
100	28.032	1	0.99	0.995	2.576

Probability plots using entire data set

We now use the ranked data of Table 11.2 rather than the grouped data. This is again shown in Table 11.5, with r denoting the position in the ordered data set, giving a realization

$$x_{(1)}, x_{(2)}, x_{(3)}, \ldots, x_{(r)}, \ldots, x_{(n)}$$

of the set of ordered random quantities

$$X_{(1)}, X_{(2)}, X_{(3)}, \ldots, X_{(r)}, \ldots, X_{(n)}; \tag{9.10}$$

see also Chapter 9 and Equation (9.10). We now wish to associate a value with each entry that gives a good idea of the associated probability of the random quantity being less than the value in question – equivalent to the cumulative relative frequency of Table 11.4. For this purpose, three columns are shown in Table 11.5. The first gives (r/n) which gives values ranging from $(1/100)$ for the first $(r = 1)$ to $(50/100) = 0.5$ for $r = 50$ to $(100/100) = 1$ for the last $(r = 100)$. At first glance this seems reasonable, but we note that the last point should be plotted at $F_Z^{-1}(1) = \infty$ in the case of distributions unlimited to the right. The next column showing $[(r-1)/n]$ corresponds to the case where the data are ranked in *decreasing* order, say with indices $\{1, 2, 3, \ldots, r', \ldots, n\}$. Then the frequency of exceedance of a value is (r'/n) and, that of non-exceedance is

$$1 - \frac{r'}{n} = \frac{r-1}{n}$$

since $r + r' = n + 1$. This has the disadvantage that the first point has to be plotted at $-\infty$.

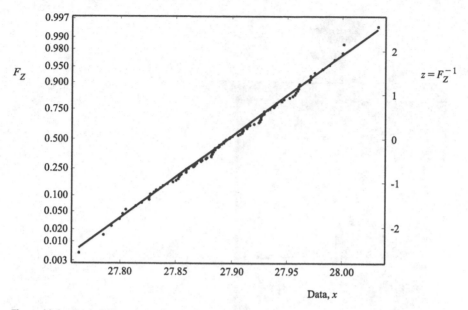

Figure 11.4   Probability plot using entire data set of Table 11.5

Hazen suggested a compromise plotting position between (r/n) and $[(r-1)/n]$:

$$\frac{r-\frac{1}{2}}{n}.$$

(11.17)

This allows the first and last points to be plotted. It is the default plotting position in MATLAB®, and has been used in Figure 11.4; see Section 11.4 below for further discussion. The vertical axes in the figure show

$$F_Z^{-1}\left(\frac{r-\frac{1}{2}}{n}\right)$$

(11.18)

and the corresponding value of F_Z.

It is a matter of judgement to move from data to probabilities. The straight line in Figure 11.4 is convincing and one might accept that a normal distribution is an acceptable probabilistic model in the present case. It is useful also to present the final result of the analysis in the form of the histogram together with the proposed probability distribution. Figure 11.5 shows the result. The fitted normal curve was plotted using the sample mean and variance noted in Examples 11.2 and 11.6 above. The illustration shows Francis Galton's apparatus for generating bell-shaped distributions. Balls, beads or particles are inserted into the receptacle at the top and fall through the maze of spikes, bouncing off them, with the resulting distribution at the bottom.

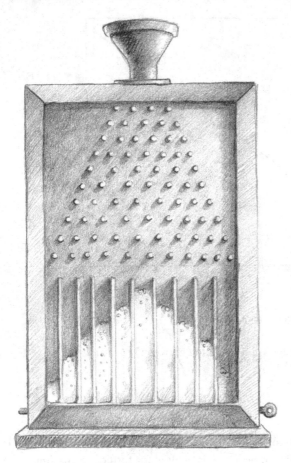

Francis Galton's apparatus

Distributions not normalized by location and scale parameters

Figure 11.6 shows data from dedicated rams of the vessel MV Arctic into multiyear ice, discussed in Section 9.4.1 above, with Table 11.6 giving the cumulative frequency data. Figure 11.7 shows the ogive. The data is unsymmetrical, and after some experimentation with different distributions, it is decided to fit a gamma distribution. There is no generally convenient way to normalize distributions that do not have location and scale parameters. The mean and variance from the histogram are found to be 9.85 and 28.33 respectively, giving $\alpha = 3.4266$ and $\beta = 2.8753$ in the gamma distribution:

$$f_X(x) = \frac{x^{\alpha-1} \exp(-x/\beta)}{\beta^\alpha \Gamma(\alpha)};$$

see Appendix 1 for details of distributions.

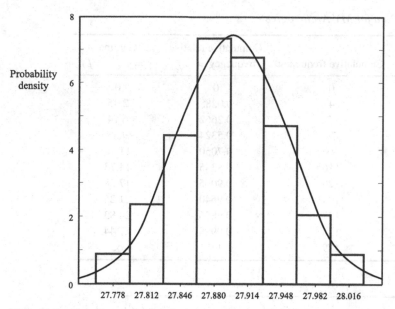

Figure 11.5   Fitted normal curve for bridge elevation

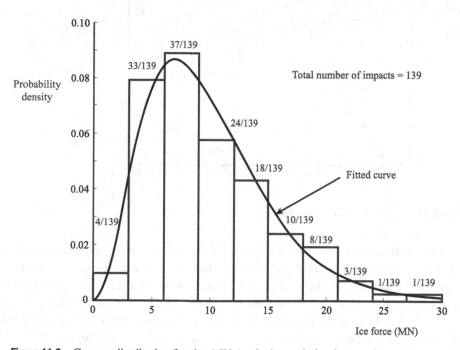

Figure 11.6   Gamma distribution fitted to MV Arctic data; relative frequencies as shown on histogram

Table 11.6 *Data for MV Arctic ogive*

Less than or equal to	Cumulative frequency	Cumulative relative frequency $= crf$	Gamma $x = F_X^{-1}(crf)$
0	0	0	0
3	4	0.0288	2.45
6	37	0.2662	6.14
9	74	0.5324	9.33
12	98	0.7050	11.90
15	116	0.8345	14.73
18	126	0.9065	17.28
21	134	0.9640	21.25
24	137	0.9856	24.83
27	138	0.9928	27.44
30	139	1.0	∞

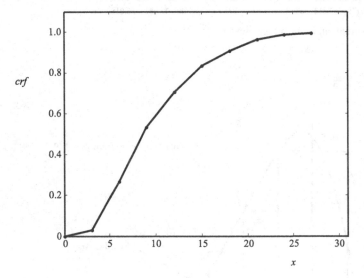

Figure 11.7   Ogive for MV Arctic data

The fourth column in Table 11.6 shows values of the inverse gamma distribution, using the parameter values just noted and using the cumulative probability values given in the third column. For example, the value in the third row (6.14) is the inverse gamma value for the *cdf* equal to 0.2662 and with the parameters (α, β) noted. The resulting probability plot is given in Figure 11.8 and a reasonable fit is obtained. Figure 11.6 shows the histogram and fitted curve. (Note that the parameters used in Chapter 8 – Equation (9.17) – were slightly different as they were based on slightly different data.)

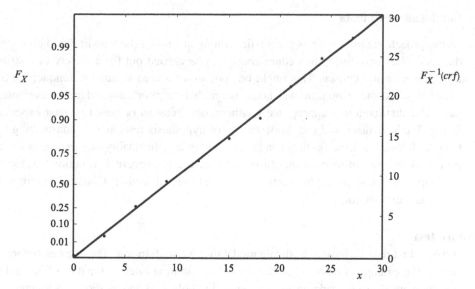

Figure 11.8 Probability plot for gamma distribution (from Table 11.6)

In order to find the values of F_X^{-1}, parameters (α, β in the present case) had to be estimated to deduce the value of the random quantity; the curve fitted was not a 'standard' curve in the sense of the standard normal which is universally the same. But the procedure is very similar. In the present case, some iteration on the parameter values might be undertaken to find the best fit. Since the product of analysis is a straight line (Figures 11.3, 11.4 and 11.8), the tools of linear regression are often used to find the least-square fit (Section 11.3).

Transition to probability

Data when presented in histograms, ogives or probability plots will show 'humps and bumps', variations with regard to the postulated distributions. This is to be expected, as data sets will show scatter. The fitted distribution might represent much better one's judgement regarding probability since the irregularities will have been ironed out. The assignment of probability using data is a matter of the analyst's best judgement. Generally, if a probabilistic model is appropriate with regard to the process and if the data show a good fit, then the fitted curve will be used to estimate the probabilities.

Quantiles can also be deduced from the fitted distribution rather than from the data themselves. Then the quantiles give probability- rather than the data-based values given in Section 11.2.1 above. Uncertainty in parameters is also important in giving final estimates of probability, as outlined in Chapter 8. Once the distribution is decided upon that represents one's probabilities, one can read off from the *cdf* the various quantiles that give one's best estimate of probabilities less than or greater than a specified value or of values with a specified value of probability above or below. This is in distinction to the quantiles obtained earlier that represented percentages of data points less than or greater than a specified value.

11.2.4　Goodness-of-fit tests

Our approach in this section is pragmatic, aiming at finding the best-fit distributions given a data set. The curve-fitting procedure above can be carried out for a variety of distributions and the best result chosen. This might be done using a least-squares technique; see Section 11.3 below. At the same time we should be guided by prior knowledge, for example that a particular distribution is appropriate for theoretical reasons or based on past experience of fitting. We have discussed the shortcomings of hypothesis tests as decision-making tools in Chapter 8 (Section 8.9). Yet they can be useful to gauge the quality of a particular fit, and the procedure will be outlined for the chi-square test. The chi-square distribution was introduced in Chapter 8, see in particular Section 8.7.4, and the Subsection 'Classical interpretation of chi-square distribution'.

Chi-square test

Assume first that we have a candidate model that we wish to test. There are as before, n data points. For example, this could be the normal distribution fitted to the data of Table 11.1. To use the method, we consider grouped data, as in Table 11.3 and as shown in Figure 11.3. We compare with the model class marks (11.12) the observed frequencies (11.13),

$$g_1, g_2, g_3, \ldots, g_j, \ldots, g_k, \tag{11.13}$$

used in constructing the histogram, an example being the last column of Table 11.3. We assume that the model yields probabilities

$$p_1, p_2, p_3, \ldots, p_j, \ldots, p_k, \tag{11.19}$$

corresponding to these values. The p_j can be calculated by taking the value of the probability distribution at the class mark and multiplying this value by the class interval or by integrating over the class interval. In effect we consider a partition and compare the measurements with the fitted curve. The same procedure is followed for any type of distribution that we are fitting.

The chi-square test is based on the quantity

$$\sum_{j=1}^{k} \frac{(g_j - np_j)^2}{np_j}, \tag{11.20}$$

which for large n is approximately chi-square with $(k - 1)$ degrees of freedom. We may interpret Equation (11.20) as representing

$$\sum \frac{(\text{observed frequency minus expected frequency})^2}{(\text{expected frequency})}. \tag{11.21}$$

The probability of the result (11.13) follows the multinomial distribution

$$\frac{n!}{g_1! g_2! \cdots g_j! \cdots g_k!} p_1^{g_1} p_2^{g_2} \cdots p_j^{g_j} \cdots p_k^{g_k}. \tag{11.22}$$

The trend to a limiting chi-square distribution can be proved in two ways. These correspond to the inductive and classical approaches. First, a Bayesian might start by considering Equation (11.1) with the prior term $\mathbf{Pr}(D)$ as being uniform. In other words we have vague opinions *a priori* regarding the choice of distribution. Then the term $\mathbf{Pr}(D|\text{data})$ is proportional to the likelihood

$$p_1^{g_1} p_2^{g_2} \cdots p_k^{g_k} \tag{11.23}$$

with the p_j random. In the classical approach, the g_j are considered to be random. We can write (11.20) as

$$\sum_{j=1}^{k} \frac{(g_j - np_j)^2}{np_j} = \sum_{j=1}^{k} \frac{g_j^2}{np_j} - n. \tag{11.24}$$

If the number of data points n from which frequencies are calculated is large then the distribution of (11.20) becomes chi-square with $(k-1)$ degrees of freedom. In rough outline, the proofs result in the distribution of (11.20) or (11.21) as being the integration of squares of normally distributed random quantities over the volume of the annulus of a shell in $(k-1)$ dimensions. For example, the g_j might be considered as binomial or multinomial random quantities which tend to be normally distributed as n increases. The proofs are considered in more detail in Exercise 11.7.

If the distribution is discrete, the class marks coincide with the possible values of the random quantity, and it becomes a question of taking the value of the *pmf* at the point under consideration – see Exercise 11.6. We noted that, in the case of continuous distributions, one calculates the integral of the probability density over the bin between the class boundaries (approximately the density at the class mark multiplied by the class width).

If the data are used to estimate parameters of the distribution, note that there is a reduction in the degrees of freedom. If r parameters are estimated from the data, there are

$$k - r - 1 \tag{11.25}$$

degrees of freedom; see Example 11.11 for an application.

Example 11.11 We use the data of Table 11.3 above, and the fitted normal distribution with $\bar{x} = 27.898$ m and $s = 0.0535$ m. The data in the first and last two bins have been combined to be consistent with the practical rule: use five or more bins with at least five samples in each bin. The number of degrees of freedom is then six (number of bins) minus one minus two (for the two parameters estimated with the data), resulting in three degrees of freedom.

The 95% range of the chi-square distribution is from 0.216 to 9.35. The observed value of 0.32 is within this range but near the lower end, indicating a better-than-average fit.

One can obtain curious results if one does not have reasonably well-populated bins; hence the rule quoted in this example. It is important to put goodness-of-fit tests into perspective. For example, the test just conducted says very little about the tails of the distribution. All of the

Table 11.7 *Chi-square goodness-of-fit test*

Cut points	Frequency (tally)	Relative frequency	Estimated normal probability	Normal frequency	Equation (11.21)
< 27.829	11	0.11	0.0986	9.86	0.1324
27.829–27.863	15	0.15	0.1579	15.79	0.0397
27.863–27.897	25	0.25	0.2361	23.61	0.0824
27.897–27.931	23	0.23	0.2388	23.88	0.0323
27.931–27.965	16	0.16	0.1635	16.35	0.0073
> 27.965	10	0.10	0.1052	10.52	0.0259
Totals	100	1.00	1.0001	100.01	0.3200

displays and results discussed in the preceding should be used in making judgements about probabilistic models.

Other goodness-of-fit tests

There are numerous goodness-of-fit tests which are readily available in the standard literature and which appear in computer packages. These can be researched for individual applications; the overall idea has been conveyed in the previous section of the chi-square test. See also D'Agostino and Stephens (1986).

11.2.5 Estimating model parameters

In Chapter 8, we dealt with the problem of parameter estimation. The Bayesian prescription was advocated as the most rational approach. The literature and computer packages abound in different methods, without guidance as to which is appropriate. For example, point estimates can be based on the method of moments (calculating moments from the data), best-fit lines in probability plots, or maximum likelihood methods. The methods can be assessed using the Bayesian approach. If there are many data points, there will not be a large difference between classical methods and the Bayesian approach.

In the final analysis, it not so much the estimates of position and spread of a distribution that matter as incorporating all of the uncertainty – the entire probability distribution including its parameters – in decision-making. But there are also many cases where the parameter uncertainty is minor, or where the decision only depends in a general way on the probabilities, and then a large amount of detail is not necessary.

11.3 Linear regression and least squares

Engineers often have to deal with relationships between two quantities, for example temperature and reaction rate, pressure and volume, load and deflection, voltage and current, fatigue life and load range. The term 'regression' arose from early measurements by Francis Galton of the size of sweet peas – as suggested to him by Charles Darwin – and then of human beings. The studies focussed on the relationship between the size of parents and that of their offspring. In both cases – peas and humans – small parents lead to offspring that are below average in size while large parents lead to offspring that are above average in size. Yet he found that the smaller-than-average offspring of small parents were closer to the mean than were their parents; likewise the larger-than-average offspring of large parents were also closer to the mean than were their parents. There is a tendency to inherit small or large characteristics but also a tendency to 'regress to the mean'. This led to the use of the term 'regression' in studying relationships between two quantities. It has also found application in other areas. In investments (Bernstein, 1996) overvalued stocks certainly have a tendency to regress, as any investor in technology in the first part of this century will have come to realize. Galton also found that the data on sizes of offspring fitted the normal distribution, an assumption that is usually made in regression studies.

In present-day usage, the term 'regression' in statistical analysis does not refer necessarily to actual regression to the mean. It is used in a general way to describe the study of the relationship between two or more quantities. The term **linear regression** refers to the study of linear relationships; Figure 11.9 shows a typical subject of study in which there are indications of a linear relationship with scatter about the mean line. We commence our analysis by considering two quantities, y and x. Let x be interpreted as the 'independent variable' and y the dependent one. It is natural to consider $y|x$ or 'y given x'. Figure 11.10 shows the line with the variation around it indicated with a distribution at x_i. For example, x could represent the height of parents in the example given in the preceding paragraph. The offspring would not all have the same height; there would be scatter as in Figure 11.9. Let us assume that there are sufficient offspring to model the distribution about the line. At point x_i in Figure 11.10 we indicate a normal distribution of values with the line $(\alpha + \beta x)$ as the mean. It is then quite natural to model

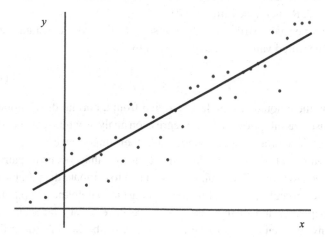

Figure 11.9   Typical linear regression

Table 11.8 *Regression analysis data pairs*

Independent quantity	x_1	x_2	x_3	$\cdots$	x_i	$\cdots$	x_n
Dependent quantity	y_1	y_2	y_3	$\cdots$	y_i	$\cdots$	y_n

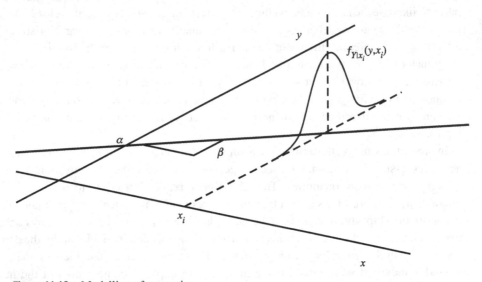

Figure 11.10   Modelling of regression

the random quantity $Y|x$ as being normally distributed. Often we can take the variance σ^2 about the line as being constant, not a function of x. (This is sometimes termed homoscedastic; if σ^2 varies with x, heteroscedastic. Rather awkward terms!)

In summary, we assume a normal distribution about a straight line with constant variance for all the x of interest. A measured value of y, say y_i given by

$$y_i = \alpha + \beta x_i + \epsilon_i, \tag{11.26}$$

will depart from the line by the amount ϵ_i which is sampled from a normal distribution with variance σ^2. These are the usual assumptions in linear regression analysis which can and should be checked against available data, or assessed with regard to their importance.

We always commence with a set of data, in which the x and y values occur in pairs as in Table 11.8. Before we proceed to probabilistic analysis, we need to estimate the parameters of the line $(\alpha + \beta x)$. The values, α and β, correspond to the 'constant but unknown' parameters that were discussed in Chapter 8, and to the case where we have a very large amount of data, enough to remove any uncertainty regarding the position of the line. As described in

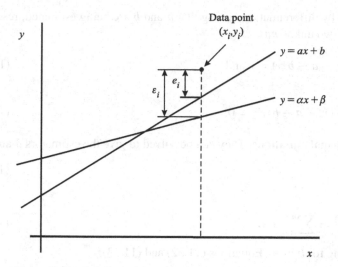

Figure 11.11 Relation of data to regression lines

Chapter 8, we can never determine them; we can only estimate them and account for associated uncertainties. We write the estimated line as

$$y = a + bx. \tag{11.27}$$

The situation is illustrated in Figure 11.11. A particular data point will deviate a distance e_i from the estimated line $(a + bx)$:

$$y_i = a + bx_i + e_i. \tag{11.28}$$

11.3.1 Obtaining the regression line: method of least squares

How do we estimate the regression line? (Recall that 'regression' no longer refers to 'regression to the mean' as discovered by Galton.) The method generally used is to minimize the squares of the deviations from the mean. If we regard the data points as entities with equal mass, this will give the line with minimum variance about it – see also Section 3.5.3. We can see that there is uncertainty in a and b, since the database is a sample limited in size. We will denote these as A and B, which are then random. We shall obtain two estimators for a and b, which we denote $\hat{a}$ and $\hat{b}$ respectively. Following this strategy, we wish to minimize the 'sums of the squares of the errors', or SSE, equal to $\sum_{i=1}^{n} e_i^2$, that is to find

$$\text{Min} \left(SSE = \sum_{i=1}^{n} e_i^2 \right) = \text{Min} \left(\sum_{i=1}^{n} [y_i - (a + bx_i)]^2 \right). \tag{11.29}$$

The solution is found by differentiation, noting that a and b are being estimated, resulting in two equations for the two unknowns:

$$\frac{\partial (SSE)}{\partial a} = \sum_{i=1}^{n} -2 (y_i - a - bx_i) = 0 \text{ and} \tag{11.30}$$

$$\frac{\partial (SSE)}{\partial b} = \sum_{i=1}^{n} [-2x_i (y_i - a - bx_i)] = 0. \tag{11.31}$$

These are termed the **normal equations**. They can be solved to give the estimators $\hat{a}$ and $\hat{b}$:

$$n\hat{a} + \hat{b} \sum_{i=1}^{n} x_i = \sum_{i=1}^{n} y_i \tag{11.32}$$

and

$$\hat{a} \sum_{i=1}^{n} x_i + \hat{b} \sum_{i=1}^{n} x_i^2 = \sum_{i=1}^{n} x_i y_i. \tag{11.33}$$

Eliminating $\hat{a}$ and solving for $\hat{b}$ using Equations (11.32) and (11.33):

$$\hat{b} = \frac{n \sum_{i=1}^{n} x_i y_i - \left(\sum_{i=1}^{n} x_i\right)\left(\sum_{i=1}^{n} y_i\right)}{n \sum_{i=1}^{n} x_i^2 - \left(\sum_{i=1}^{n} x_i\right)^2}. \tag{11.34}$$

Equation (11.32) can be written as

$$n\hat{a} + \hat{b}n\bar{x} = n\bar{y}, \text{ or}$$
$$\bar{y} = \hat{a} + \hat{b}\bar{x}. \tag{11.35}$$

Therefore

$$\hat{a} = \bar{y} - \hat{b}\bar{x}. \tag{11.36}$$

We can see from this equation that the least-squares line passes through the centroid of the data points $(\bar{x}, \bar{y})$.

In conducting regression analysis, the following notation is useful.

$$S_{xx} = \sum_{i=1}^{n} (x_i - \bar{x})^2 = \sum_{i=1}^{n} x_i^2 - \frac{\left(\sum_{i=1}^{n} x_i\right)^2}{n}, \tag{11.37}$$

$$S_{yy} = \sum_{i=1}^{n} (y_i - \bar{y})^2 = \sum_{i=1}^{n} y_i^2 - \frac{\left(\sum_{i=1}^{n} y_i\right)^2}{n}, \tag{11.38}$$

$$S_{xy} = \sum_{i=1}^{n} (x_i - \bar{x})(y_i - \bar{y}) = \sum_{i=1}^{n} x_i y_i - \frac{\left(\sum_{i=1}^{n} x_i\right)\left(\sum_{i=1}^{n} y_i\right)}{n}. \tag{11.39}$$

These equations can result in small differences in two large numbers corresponding to the two terms on the right hand side. Care must therefore be taken in completing the calculations. One can treat each data pair as a 'point mass' and then consider any of the Equations (11.37) to (11.39) as representing the moment of the masses about the centre-of-mass (left hand side).

This is equal to the moment about the origin of the masses (first term on the right) minus the moment of the sum of the masses applied at the centroid about the origin (second term; note that these are equal to $n\bar{x}^2, n\bar{y}^2$ and $n\bar{x}\bar{y}$). This results from the parallel axis theorem in mechanics; the conclusion from a computational point of view is that the closer the origin of computation is to the centroid, the smaller the problem resulting from calculating small differences between large quantities. In the next subsection, the origin is taken at the mean of the x_i; calculating with the centroid as origin is a good computational strategy.

Comparison of Equation (11.34) with Equations (11.37) and (11.39) results in

$$\hat{b} = \frac{s_{xy}}{s_{xx}}.$$
(11.40)

We may also show that

$$SSE = s_{yy} - \hat{b}s_{xy}$$
(11.41)

as follows. Since

$$SSE = \sum_{i=1}^{n} (y_i - \hat{y}_i)^2,$$

where

$$\hat{y}_i = \hat{a} + \hat{b}x_i,$$
(11.42)

$$SSE = \sum_{i=1}^{n} \left(y_i - \hat{a} - \hat{b}x_i\right)^2$$

$$= \sum_{i=1}^{n} \left[(y_i - \bar{y}) - \hat{b}(x_i - \bar{x})\right]^2, \text{ using } \bar{y} = \hat{a} + \hat{b}\bar{x}, \text{ and consequently}$$

$$SSE = \sum_{i=1}^{n} (y_i - \bar{y})^2 - 2\hat{b}\sum_{i=1}^{n} (y_i - \bar{y})(x_i - \bar{x}) + \hat{b}^2 \sum_{i=1}^{n} (x_i - \bar{x})^2$$

$$= s_{yy} - 2\hat{b}s_{xy} + \hat{b}^2 s_{xx}$$

$$= s_{yy} - \hat{b}s_{xy}, \text{ using Equation (11.39)}.$$

Example 11.12 Fatigue data are often studied using the methods of regression analysis. The data generally result from cyclic testing to failure of specimens loaded to a specified stress level s, with the number n_c of cycles to failure being counted. Various stress levels are used; the higher the stress level, the smaller the number of cycles to failure. Table 11.9 shows some typical results. Experience has shown that $s-n_c$ curves are non-linear, so that log-scales are used. We shall use

$$x = \log s, \text{ and}$$

$$y = \log n_c.$$

This is one way of dealing with non-linearities – to transform the original quantities by means of non-linear functions.

Table 11.9   *Data for regression analysis*

Stress, s MPa	Cycles n_c	$x = \log s$ (3 dp)	$y = \log n_c$ (3 dp)
160	301 740	2.204	5.480
160	200 400	2.204	5.302
160	425 210	2.204	5.629
200	58 890	2.301	4.770
200	82 020	2.301	4.914
200	81 810	2.301	4.913
300	38 070	2.477	4.581
300	28 840	2.477	4.460
300	39 931	2.477	4.601
300	37 270	2.477	4.571
300	24 520	2.477	4.390
300	31 790	2.477	4.502
300	68 360	2.477	4.835
400	13 150	2.602	4.119
400	8 830	2.602	3.946
400	22 370	2.602	4.350
400	15 920	2.602	4.202
400	16 930	2.602	4.229
400	21 970	2.602	4.342
400	15 440	2.602	4.189

The results of the regression analysis are as follows.

$S_{xx} = 0.4147$, $s_{yy} = 3.9284$, $s_{xy} = -1.1981$, $\hat{b} = -2.8891$, $\hat{a} = 11.7044$, $\bar{x} = 2.453$, $\bar{y} = 4.6161$, SSE $= 0.4669$, $\sigma = 0.1611$, $n = 20$.

The data and the fitted line are shown in Figure 11.12.

11.3.2   Uncertainty in model parameters, mean and future values

All of the parameters and quantities in this section are modelled as being random, and the knowledge that we use is the data set of Table 11.8 (Table 11.9 is an example). In other words, we do not include prior knowledge regarding the parameter values so that the analysis corresponds to the prior distributions expressing 'indifference' of Chapter 8. Exercises 11.16 and 11.17 deal with conjugate-prior analysis. The only knowledge then is the set of data of Table 11.8. Further, the assumption of exchangeability will be made, such that the order of the n data pairs of Table 11.8 is not deemed important. There is a fair bit of mathematical detail in the following; the theory is very similar in structure to that regarding the normal distribution in Chapter 8. We have stressed particularly in Chapter 8 that all important uncertainties should

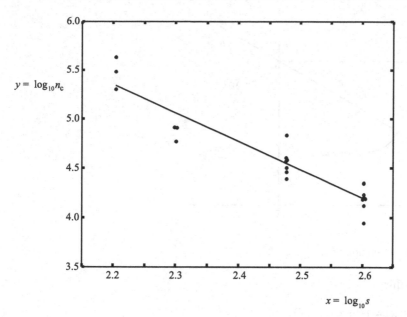

Figure 11.12   Example regression plot

be taken into account. While the uncertainty of a future value with regard to the regression line will be taken into account by analysing the scatter about the line, the parameters of the regression line will themselves include uncertainties. As in Chapter 8 it is inherent in the present approach to treat these parameters as being random. We have just demonstrated how a and b in the regression equation can be estimated by the method of least squares yielding the values $\hat{a}$ and $\hat{b}$. For convenience, we consider the quantity a' rather than a as the constant of the regression line. This is depicted in Figure 11.13; a' is the intercept at the value $x = \bar{x}$ rather than the value a at $x = 0$ and then

$$\hat{a}' = \bar{y} \tag{11.43}$$

from Equation (11.36).

The parameters a' and b of the regression are now modelled as random quantities A' and B respectively. We assume further that the scatter about the regression line is normally distributed with variance σ^2 (to be estimated). We follow the strategy adopted in Chapter 8 and associate w with $(1/\sigma^2)$ as the 'weight' representing the (inverse) spread of the distribution. In our Bayesian analysis, W is a random quantity. The distribution of Y_i is then normal:

$$f_Y\left(y|a', b, w = \frac{1}{\sigma^2}, x\right) = \frac{1}{\sqrt{2\pi}\sigma} \exp\left\{-\frac{1}{2\sigma^2}\left[y - a' - b\left(x - \bar{x}\right)\right]^2\right\}$$

$$\propto w^{\frac{1}{2}} \exp\left\{-\frac{w}{2}\left[y - a' - b\left(x - \bar{x}\right)\right]^2\right\}. \tag{11.44}$$

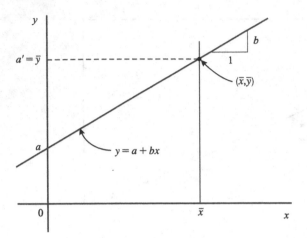

Figure 11.13 Definition of a'

Also from this

$$\langle Y|a', b, w, x\rangle = a' + b(x - \bar{x}).$$ (11.45)

Our basic assumption is that the prior distributions of A', B and $\ln W$ are independent and uniform expressing little prior knowledge. Then

$$f^{(i)}_{A', B, W}(a', b, w) \propto \frac{1}{w}.$$ (11.46)

The independence often makes sense from a judgement of the problem, and the uniform distributions express initial 'indifference' regarding the values concerned (see also Section 8.6). Given this initial vagueness, we should not be surprised that the assumptions result in equations similar to the classical data-based analysis. But the structure and interpretation are more rational. Conjugate prior analysis can also be undertaken; see Exercise 11.16.

We consider now the likelihood of the data set, which is considered as an exchangeable set of n data pairs. The likelihood of the y_i given the x_i is

$$\ell(a', b, w) \propto \prod_{i=1}^{n} w^{\frac{1}{2}} \exp\left\{-\frac{w}{2}\left[y_i - a' - b(x_i - \bar{x})\right]^2\right\}$$
$$= w^{\frac{n}{2}} \exp\left\{-\frac{w}{2}\sum_{i=1}^{n}\left[y_i - a' - b(x_i - \bar{x})\right]^2\right\}.$$ (11.47)

Now

$$\sum_{i=1}^{n}\left[y_i - a' - b(x_i - \bar{x})\right]^2 = n(\bar{y} - a')^2 + \sum_{i=1}^{n}\left[(y_i - \bar{y}) - b(x_i - \bar{x})\right]^2$$
$$= n(a' - \bar{y})^2 + s_{yy} - 2bs_{xy} + b^2 s_{xx}$$
$$= n(a' - \hat{a}')^2 + s_{xx}(b - \hat{b})^2 + SSE,$$ (11.48)

using (11.37) to (11.39), (11.43), (11.40) and (11.41). It is noteworthy that $\hat{a}'$ and $\hat{b}$ enter naturally into Equation (11.48) and we can see that $a' = \hat{a}'$ and $b = \hat{b}$ minimize Equation (11.48), the left hand side of which is the same as (11.29). Since Equation (11.48) is minimized, the parameter estimates just noted constitute the maximum of the likelihood ℓ of (11.47).

The posterior distribution is obtained by multiplying the prior, Equation (11.46) by the likelihood, Equation (11.47) into which Equation (11.48) has been substituted:

$$f_{A',B,W}^{(u)}\left(a', b, w\right) \propto w^{\frac{n}{2}-1} \exp\left\{-\frac{w}{2}\left[n\left(a' - \hat{a}'\right)^2 + s_{xx}\left(b - \hat{b}\right)^2 + SSE\right]\right\}. \tag{11.49}$$

We now determine the marginal distributions of A', B and W by integration.

Uncertainty in W

Integration of Equation (11.49) with respect to a' yields

$$f_{B,W}^{(u)}\left(b, w\right) \propto w^{\frac{n}{2}-1} \exp\left\{-\frac{w}{2}\left[s_{xx}\left(b - \hat{b}\right)^2 + SSE\right]\right\}$$
$$\times \int_{-\infty}^{\infty} \exp\left\{-\frac{w}{2}\left[n\left(a' - \hat{a}'\right)^2\right]\right\} \mathrm{d}a',$$

in which the integral has the form of the normal distribution; see also Section 8.5.3 and Equation (8.113). It is proportional to $w^{-1/2}$, and consequently

$$f_{B,W}^{(u)}\left(b, w\right) \propto w^{\frac{n}{2}-\frac{3}{2}} \exp\left\{-\frac{w}{2}\left[s_{xx}\left(b - \hat{b}\right)^2 + SSE\right]\right\}. \tag{11.50}$$

We obtain the integral of Equation (11.50) with respect to b; again it is of normal form, proportional to $w^{-1/2}$, and we find that

$$f_W^{(u)}\left(w\right) \propto w^{\frac{n}{2}-2} \exp\left[-\frac{w}{2}\left(SSE\right)\right]. \tag{11.51}$$

Thus $[W \cdot (SSE)]$ follows a chi-square distribution with $(n-2)$ degrees of freedom; see Equation (8.83). The mean is

$$\langle W \cdot (SSE) \rangle = n - 2, \text{ or}$$

$$\frac{1}{\langle W \rangle} = \frac{SSE}{n-2} = s^2 = \frac{\sum_{i=1}^n e_i^2}{n-2}. \tag{11.52}$$

This does not imply that the mean of the probability distribution of variance, i.e. the mean of $(1/W)$ is s^2; see Section 8.7.4, and in particular Exercise 8.24 and Equation (8.174). In fact,

$$\left\langle \frac{1}{W} \right\rangle = \frac{SSE \cdot \nu}{\nu\left(\nu - 2\right)} = \frac{SSE}{n-4} \tag{11.53}$$

since $\nu = n - 2$.

Degrees of freedom

The term 'degrees of freedom' used in statistics is similar to the usage in mechanics. In the present instance, we have $(n - 2)$ degrees of freedom. In estimating the variance from a data set, with mean and variance unknown – see the discussion after Equation (8.142) – the number of degrees of freedom in the distribution of the variance was $(n-1)$. We can think of the number of independent elements in the sample used in the calculation of the parameter of interest: one equation constraining the sample values has been used in calculating the mean, resulting in $(n-1)$ degrees of freedom remaining. In the case above there are n data pairs but two equations constraining them have been used in calculating the parameters $\hat{a}$ and $\hat{b}$, resulting in $(n-2)$ degrees of freedom remaining. In mechanics, the number of normal modes of vibration dictates the degrees of freedom, the number of variables needed to specify completely the configuration of the system.

Uncertainty in B

We now wish to find the distribution of B. We therefore integrate (11.50) with respect to w. We follow the strategy given in Section 8.5.3 by making the substitution, as in (8.109),

$$z = \frac{w}{2}\left[S_{xx}\left(b-\hat{b}\right)^2 + SSE\right] \tag{11.54}$$

in Equation (11.50). Then

$$f_B^{(u)}(b) \propto \left[S_{xx}\left(b-\hat{b}\right)^2 + SSE\right]^{-\frac{1}{2}(n-1)} \int_0^\infty z^{\frac{n}{2}-\frac{3}{2}}\exp\left(-z\right)dz$$

$$= \Gamma\left(\frac{n}{2}-\frac{1}{2}\right)\left[S_{xx}\left(b-\hat{b}\right)^2 + SSE\right]^{-\frac{1}{2}(n-1)}. \tag{11.55}$$

We now make the substitution

$$t = \frac{\left(b-\hat{b}\right)}{s/\sqrt{S_{xx}}} \tag{11.56}$$

and find that

$$f_T^{(u)}(t) \propto \left[1 + \frac{t^2}{n-2}\right]^{-\frac{1}{2}(n-1)} \tag{11.57}$$

since the Jacobian is a constant. We conclude that the posterior distribution of

$$\frac{B-\hat{b}}{s_B} \tag{11.58}$$

with

$$s_B^2 = \frac{s^2}{S_{xx}} = \frac{SSE/(n-2)}{S_{xx}} \tag{11.59}$$

is a t-distribution with $(n-2)$ degrees of freedom. If we have a very large amount of data, B will model the idea of a 'population' parameter β which has little uncertainty associated with it.

Uncertainty in A'

In order to estimate this uncertainty we must integrate Equation (11.49) first with respect to b and then with respect to w. Then

$$f_{A,W}^{(u)}(a, w) \propto w^{\frac{n}{2}-1} \exp\left\{-\frac{w}{2}\left[n\left(a'-\hat{a}'\right)^2 + SSE\right]\right\} \int_{-\infty}^{\infty} \exp\left\{-\frac{w}{2}\left[S_{xx}\left(b-\hat{b}\right)^2\right]\right\} db.$$

Again the integral is of the form of the normal distribution and is proportional to $w^{-1/2}$; then

$$f_{A,W}^{(u)}(a, w) \propto w^{\frac{n}{2}-\frac{3}{2}} \exp\left\{-\frac{w}{2}\left[n\left(a'-\hat{a}'\right)^2 + SSE\right]\right\}. \tag{11.60}$$

Following the method used for B, we make the substitution

$$z = \frac{w}{2}\left[n\left(a'-\hat{a}'\right)^2 + SSE\right]$$

in Equation (11.60). Then

$$f_A^{(u)}(a) \propto \left[n\left(a'-\hat{a}'\right)^2 + SSE\right]^{-\frac{1}{2}(n-1)} \int_0^{\infty} z^{\frac{n}{2}-\frac{3}{2}} \exp(-z)\,dz$$

$$= \Gamma\left(\frac{n}{2}-\frac{1}{2}\right)\left[n\left(a'-\hat{a}'\right)^2 + SSE\right]^{-\frac{1}{2}(n-1)}. \tag{11.61}$$

Following the method as before, we substitute

$$t = \frac{\left(a'-\hat{a}'\right)}{s/\sqrt{n}} \tag{11.62}$$

and find the same result as in (11.57),

$$f_T^{(u)}(t) \propto \left[1 + \frac{t^2}{n-2}\right]^{-\frac{1}{2}(n-1)},$$

since the Jacobian is a constant. The posterior distribution of

$$\frac{A'-\hat{a}'}{s_{A'}} \tag{11.63}$$

with

$$s_{A'}^2 = \frac{s^2}{n} = \frac{SSE/(n-2)}{n} \tag{11.64}$$

is a t-distribution with $(n-2)$ degrees of freedom.

Again, for a very large amount of data, the uncertainty in A becomes negligible, thus modelling the idea of a 'population parameter' α, as does B model β.

Uncertainty in mean response

We have now modelled A' and B as random quantities. This means that there is uncertainty in both the position of the line (A') and its slope (B). Figure 11.14 illustrates the fact that the uncertainty regarding these parameters results in a spread of the range about the mean response

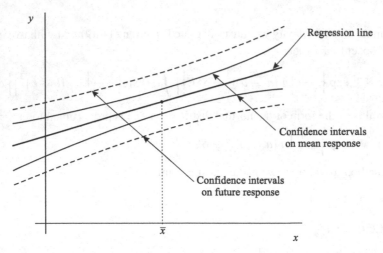

Figure 11.14   Illustration of confidence intervals in regression; narrowest intervals are at $\bar{x}$

that widens as the x-value moves away from the mean $\bar{x}$. We shall now demonstrate this fact. First, we consider $W = w = (1/\sigma^2)$ as a constant in Equation (11.49); then

$$f_{A',B}^{(u)}\left(a', b | w = \frac{1}{\sigma^2}\right) \propto \exp\left\{-\frac{1}{2\sigma^2}\left[n\left(a' - \hat{a}'\right)^2 + s_{xx}\left(b - \hat{b}\right)^2\right]\right\}$$

$$= \exp-\left[\frac{n\left(a' - \hat{a}'\right)^2}{2\sigma^2}\right]\exp-\left[\frac{s_{xx}\left(b - \hat{b}\right)^2}{2\sigma^2}\right].$$

(11.65)

We see from this that given σ^2, A' and B are independent and $N\left(\hat{a}', \sigma^2/n\right)$ and $N(\hat{b}, \sigma^2/s_{xx})$ respectively. This independence occurs at the centroid of the data – see Figure 11.14. It is also one of the reasons for using a' (at the centroid) rather than a in our uncertainty analysis.

We now consider the mean value, Equation (11.45), at a particular value $x = x_0$:

$$\langle Y | a', b, w, x_0 \rangle = a' + b\left(x_0 - \bar{x}\right).$$

(11.66)

We denote the random quantity $\langle Y \rangle$ as M. Since A' and B are random, independent and normally distributed (given σ^2) and since $(M | \sigma^2)$ is a linear function of a' and b as in Equation (11.66), consequently it is

$$N\left\{\hat{a}' + \hat{b}\left(x_0 - \bar{x}\right), \sigma^2\left[\frac{1}{n} + \frac{\left(x_0 - \bar{x}\right)^2}{s_{xx}}\right]\right\}.$$

(11.67)

We see why the bands of uncertainty of Figure 11.14 enlarge as x_0 moves away from $\bar{x}$; the variance in (11.67) increases with a term proportional to $(x_0 - \bar{x})^2$.

We now have to take into account uncertainty in σ^2. We know that W is distributed as in Equation (11.51):

$$f_W^{(u)}(w) \propto w^{\frac{n}{2}-2} \exp\left[-\frac{w}{2}(SSE)\right],$$

and therefore we can write

$$f_M(m) = f_{M|W=w}(m, w) f_W^{(u)}(w)$$

and integrate over w. Following the same procedure as before (see Exercise 11.12), we find that

$$T = \frac{M - [\hat{a}' + \hat{b}(x_0 - \bar{x})]}{s\sqrt{\frac{1}{n} + \frac{(x_0-\bar{x})^2}{S_{xx}}}}$$
(11.68)

follows a t-distribution with $(n-2)$ degrees of freedom.

Uncertainty regarding future response

Although it is often necessary to have information regarding uncertainty of the mean line, it is also common in engineering to consider a future value. In our example of the fatigue of a specimen, we would very likely wish to use the data to estimate the fatigue behaviour of a specimen included in the actual structure. In effect, we consider now the random quantity Y rather than the mean $\langle Y \rangle$, and expect more uncertainty in this particular value than in the mean. The proof follows similar lines as in the case of $\langle Y \rangle$.

From Equation (11.44), we know that by assumption $f_Y(y|a', b, w = \frac{1}{\sigma^2}, x)$ is a normal distribution. We consider the value of Y at a particular value $x = x_0$. Then $(Y|a', b, w, x_0)$, is $N[a' + b(x_0 - \bar{x}), \sigma^2]$. We know that $(A'|b, w, x_0)$ is $N(\hat{a}', \sigma^2/n)$ from Equation (11.65). Since

$$f_{Y,A'}(y, a'|b, w, x_0) = f_Y(y|a', b, w, x_0) \cdot f_{A'}(a'|b, w, x_0),$$
(11.69)

we can deduce that $(Y|b, w, x_0)$ is

$$N[\hat{a}' + b(x_0 - \bar{x}), \sigma^2 + \sigma^2/n];$$
(11.70)

see Exercise 11.17(c). Again using the same method, and the fact that $(B|w, x_0)$ is $N(\hat{b}, \sigma^2/s_{xx})$, one can show that $(Y|w, x_0)$ is

$$N\left[\hat{a}' + \hat{b}(x_0 - \bar{x}), \sigma^2 + \frac{\sigma^2}{n} + \sigma^2\frac{(x_0 - \bar{x})^2}{S_{xx}}\right].$$

Finally, taking into account the uncertainty in W, standing for $(1/\sigma^2)$, using the same method as before, we find that

$$T = \frac{Y - [\hat{a}' + \hat{b}(x_0 - \bar{x})]}{s\sqrt{1 + \frac{1}{n} + \frac{(x_0-\bar{x})^2}{S_{xx}}}}$$
(11.71)

follows a t-distribution with $(n-2)$ degrees of freedom.

Confidence intervals

The calculation of confidence intervals assists in making judgements in regression analysis. It is useful to plot them on the graphs showing the regression analysis so as to obtain an immediate feeling for the uncertainty. The uncertainty is consistent with all of the models for decision-making presented in previous chapters, and this presents the most rigorous approach. We shall write down below the confidence intervals which can be derived in the manner described in Chapter 8.

Hypothesis testing is also commonly applied in regression analysis. For example, we might want to make a judgement regarding the slope of the regression line; is it significantly greater than zero? We have discussed before the limitations of hypothesis testing and the attempt to convert uncertainty into certainty. We advise the calculation of confidence intervals and the use of judgement from that point onwards.

Confidence intervals for A'

Using Equations (11.63) and (11.64) we find that

$$\Pr\left[\hat{a}' - t_{\alpha/2} \cdot s_{A'} \le A' \le \hat{a}' + t_{\alpha/2} \cdot s_{A'}\right] = 1 - \alpha, \tag{11.72}$$

where the t-distribution has $(n - 2)$ degrees of freedom and

$$s_{A'}^2 = \frac{s^2}{n} = \frac{SSE/(n-2)}{n}. \tag{11.73}$$

The intervals can be expressed simply as

$$\hat{a}' \pm t_{\alpha/2} \cdot s_{A'}. \tag{11.74}$$

If one wishes to obtain confidence intervals for A rather than A', then we use

$$s_A^2 = s^2 \frac{\sum x_i^2}{n s_{xx}} \tag{11.75}$$

rather than (11.73); see Exercise 11.14.

Confidence intervals for B

Equations (11.58) and (11.59) are used to show that

$$\Pr\left[\hat{b} - t_{\alpha/2} \cdot s_B \le B \le \hat{b} + t_{\alpha/2} \cdot s_B\right] = 1 - \alpha, \tag{11.76}$$

where the t-distribution has $(n - 2)$ degrees of freedom and

$$s_B^2 = \frac{s^2}{s_{xx}} = \frac{SSE/(n-2)}{s_{xx}}. \tag{11.77}$$

Again we can write the intervals as

$$\hat{b} \pm t_{\alpha/2} \cdot s_B. \tag{11.78}$$

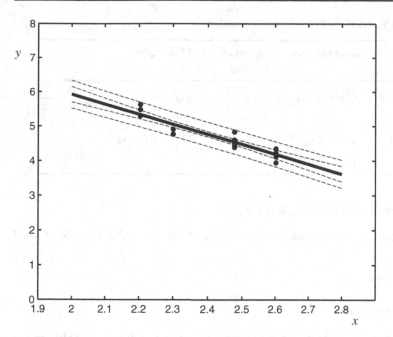

Figure 11.15 Regression line and 95% confidence intervals for mean and future response in Example 11.12

Mean response M at $x = x_0$

The confidence intervals can be derived as before using Equation (11.68):

$$\left[\hat{a}' + \hat{b}\,(x_0 - \bar{x})\right] \pm t_{\alpha/2} \cdot s \sqrt{\frac{1}{n} + \frac{(x_0 - \bar{x})^2}{S_{xx}}}. \tag{11.79}$$

Future response Y at $x = x_0$

Again proceeding as before based on Equation (11.71), the intervals are

$$\left[\hat{a}' + \hat{b}\,(x_0 - \bar{x})\right] \pm t_{\alpha/2} \cdot s \sqrt{1 + \frac{1}{n} + \frac{(x_0 - \bar{x})^2}{S_{xx}}}. \tag{11.80}$$

In regression analysis, confidence intervals for the variance are not usually done; see Exercise 11.14.

Example 11.13 The results discussed have been applied to Example 11.12. Figure 11.15 shows the widening of the intervals as one moves away from the mean $\bar{x}$.

Table 11.10   *ANOVA Results for example problem*

Source	Sum of squares	Degrees of freedom	Mean square	F
Due to regression	$SSR = 3.461$	1	$SSR = 3.461$	$\dfrac{SSR}{SSE/(n-2)}$ $= 133.4$
Residual (SSE)	$SSE = 0.467$	$n-2 = 18$	$\dfrac{SSE}{n-2} = 0.02594$	–
Total	$SST = 3.928$	$n-1 = 19$	–	–

11.3.3 Analysis-of-variance (ANOVA) approach

The calculations can also be written in an analysis-of-variance (ANOVA) table. From Equation (11.41)

$$s_{yy} = \hat{b}s_{xy} + SSE.$$

This may be written as

Total sum of squares = Sum of squares due to regression + Residual (SSE), or   (11.81)

$$SST = SSR + SSE.$$

To see this we write the left hand side of (11.81) as

$$\sum_{i=1}^{n} (y_i - \bar{y})^2 = \sum_{i=1}^{n} (y_i - \hat{y})^2 + \sum_{i=1}^{n} (\hat{y}_i - \bar{y})^2 + 2\sum_{i=1}^{n} (y_i - \hat{y}_i)(\hat{y}_i - \bar{y})$$

$$= \sum_{i=1}^{n} (\hat{y}_i - \bar{y})^2 + \sum_{i=1}^{n} (y_i - \hat{y}_i)^2,$$ (11.82)

which is the same as (11.81). To prove (11.82) we write

$$(y_i - \bar{y}) = (y_i - \hat{y}_i) + (\hat{y}_i - \bar{y})$$

and note that

$$\sum_{i=1}^{n} (y_i - \hat{y}_i)(\hat{y}_i - \bar{y}) = \sum_{i=1}^{n} (y_i - \hat{a} - \hat{b}x_i)(\hat{a} + \hat{b}x_i - \bar{y})$$

$$= \hat{a}\sum_{i=1}^{n} (y_i - \hat{a} - \hat{b}x_i) + \hat{b}\sum_{i=1}^{n} x_i (y_i - \hat{a} - \hat{b}x_i)$$

$$- \bar{y}\sum_{i=1}^{n} (y_i - \hat{a} - \hat{b}x_i).$$

These last three terms are all equal to zero by the normal equations, (11.30) and (11.31).
Equation (11.81) gives expression to (11.82); the results are summarized in Table 11.10 for Example 11.12. We note that

$$SST = s_{yy}, \text{ and}$$
$$SSR = \hat{b}s_{xy}.$$

Other values in the table include mean squares (sums of squares divided by degrees of freedom) and the F-value, the ratio of the mean squares, as noted in the table. This value assists in assessing whether the sum of squares due to regression is a significant component of the total. The 5% exceedance value for F with 1 and 18 degrees of freedom is 4.41. This is considerably below the value of 133.4 which must then correspond to a very small exceedance probability (actually of the order of 10^{-9}). One can conclude that the regression is very strong. For background on the F-tests, see exercise 11.13.

11.3.4 Correlation

The correlation coefficient ρ_{XY} was introduced for two-dimensional probability distributions in Chapter 3, defined in Equation (3.93) by

$$\rho_{X,Y} = \frac{\sigma_{X,Y}}{\sigma_X \sigma_Y},$$

where $-1 \leq \rho_{X,Y} \leq 1$, always. It measures the extent of linear relationship between X and Y. Corresponding to this, the **sample correlation coefficient** is defined as

$$r = \frac{S_{xy}}{\sqrt{S_{xx}S_{yy}}}. \tag{11.83}$$

Another expression for this is

$$r = \hat{b}\sqrt{\frac{S_{xx}}{S_{yy}}}. \tag{11.84}$$

A related quantity is r^2, **the sample coefficient of determination**:

$$r^2 = \frac{S_{xy}^2}{S_{xx}S_{yy}}; \tag{11.85}$$

this can also be calculated using the square of (11.84); other forms can be deduced. An interesting version is

$$r^2 = \frac{SSR}{SST} = 1 - \frac{SSE}{SST}, \tag{11.86}$$

which expresses the proportion of the variation in Y that is explained by the regression line. We then know that r^2 indicates in a better way the effectiveness of the regression in terms of the 'sums of squares', or variance in Y. If $r = 0.90$, then $r^2 = 0.81$, a less impressive number; this disparity is more evident when $r = 0.70$, seemingly a good correlation yet only $r^2 = 0.49$ – less than half – is the proportion of variance explained by the regression.

For a two-dimensional normal distribution given by Equation (A1.140): in Appendix 1, an analysis using Bayesian methods treats ρ, a parameter in the distribution, as a random quantity (Lindley, 1965). Prior information is taken as uninformative. Let us denote this random quantity as R. Then the result is that $\Omega = \tanh^{-1} R$ is approximately ($n \to \infty$) normal with mean equal to tanh^{-1} r and variance $(1/n)$. This approach is particularly useful when we are considering the relationship between two quantities in cases where there is no preference for $(y$ on $x)$ as against $(x$ on $y)$. We might be interested, for example, in the relationship between tensile and compressive strengths of a material without wanting to consider either as the independent or dependent quantity. (But see the next section for a case where the reverse is wanted.) Or we might be considering the relationship between wind speeds at two locations or between the foot size and arm length of individuals, with no asymmetry suggesting $(y$ on $x)$ rather than $(x$ on $y)$. Jaynes (1976) discusses a strategy for determining regression coefficients that would not reflect this asymmetry. This avenue is promising and would be useful to complete. Where one quantity is used as an indicator, the asymmetry returns, as is discussed in the following section.

11.3.5 Uncertainty in X: Bayesian approach

We now consider the case where knowledge of one quantity, say Y, is used to improve the estimate of the other. We mentioned in the preceding paragraph the case where pairs of measurements are made without any preference for one or the other as the independent quantity. An example was given concerning the pairs of tensile and compressive strengths of a material. This set of pairs of measurements can provide background information for another analysis. We might subsequently be given one value, say $Y = y$, and wish to infer the distribution of X given this information together with a prior distribution $f_X(x)$. It is quite common in building projects to measure only compressive strengths of concrete – denoted Y – and we might be interested during a design to obtain estimates of tensile strength, say X, for example if crack opening could present a problem, such as in storing fluids.

Then we can write

$$f_{X|Y=y}(x, y) \propto f_{Y|X=x}(x, y) \cdot f_X(x), \qquad (11.87)$$

using Bayes' theorem. The term $f_{Y|X=x}(x, y)$ is obtained from regression in the present analysis. We also see in this equation why we took $Y = y$ as given rather than $X = x$; the conditional distribution on the right hand side is reversed, as is always the case in Bayes' theorem. In the example of determining tensile from compressive strength, we would perform a regression of compressive strength (Y) on tensile strength (X).

We take as an example the case where the assumption of a normal distribution applies in the regression analysis as well as in the prior distribution $f_X(x)$. We assume further that the parameters of the distributions are known, on the basis of sufficient prior evidence and data, so that there are no parameter uncertainties to be taken into account. Then $Y|x$ is

$$N\left(\alpha + \beta x, \sigma^2\right) \qquad (11.88)$$

based on a linear regression analysis. As noted, we assume further that $f_X(x)$ is

$$N\left(\mu_X, \sigma_X^2\right). \qquad (11.89)$$

The method of Exercise 11.17 can be used; the result is that $X|y$ is

$$N\left[\mu_X + (y - \mu_Y)\frac{\rho\sigma_X}{\sigma_Y}, \sigma_X^2\left(1 - \rho^2\right)\right],$$ (11.90)

where $\mu_Y = \alpha + \beta\mu_X$, $\sigma_Y^2 = \sigma^2 + \beta^2\sigma_X^2$ and $\rho = \beta\sigma_X/\sigma_Y$ is the correlation coefficient. The analysis results in a two-dimensional normal distribution $f_{X,Y}(x, y)$ with the conditional distribution (11.90), consistent with the assumptions of linear regression since the mean value of $X|y$ is a linear function of y. In applications, the parameters α, β, σ^2 and ρ are estimated from the regression data of Y on X. In general, we would wish to use the variance of a future response from (11.80) in modelling the uncertainty in Y given X. In the present case, we have assumed that parameters are well known so that $\sigma^2 \cong s^2$ is used in the analysis.

Example 11.14 Maddock and Jordaan (1982) used the approach described in an analysis of decisions in structural engineering.

11.3.6 Regression in several dimensions; non-linear regression

For background to analysis in several dimensions, see Raiffa and Schlaifer (1961), Box and Tiao (1973), or O'Hagan (1994) for example. Problems are often non-linear in character; we dealt with this in Example 11.12 by transforming the quantities under consideration. Non-linear problems are dealt with in many texts, including the one by Box and Tiao just quoted.

11.4   Plotting positions

In obtaining empirical fits to data, we showed in Section 11.2 above how the appropriate use of a transformed ordinate, for example in Figures 11.4 and 11.8, leads to a straight line representation, indicating that the chosen distribution provides a reasonable fit to data. Having now dealt with least squares and linear regression, we can also use these techniques to assist in fitting a straight line. In this section we return to the question of plotting positions, on which there appear to be almost as many opinions as there are statisticians. We shall adopt a pragmatic approach, and rely principally on Cunnane (1978) who has brought clarity to the issue, with a certain amount of character. In many cases, particularly where the data set is large, the choice of plotting position has little effect on the results.

We discussed in Section 11.2 briefly the choice of value for the plotting position, and used

$$\frac{r - \frac{1}{2}}{n}$$ (11.17)

as a compromise position between (r/n) and $[(r - 1)/n]$. These two had the disadvantage that either the last or the first points in the data cannot be plotted in the case of unlimited

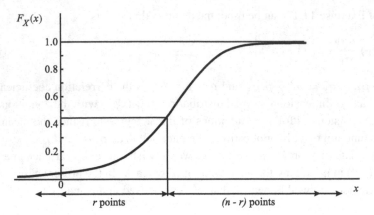

Figure 11.16   Ogive and separation into r and $(n - r)$ points. Note linear ordinate scale.

distributions since they are unity or zero and correspond in plots such as Figure 11.4 to $+\infty$ or $-\infty$. Weibull (1939) suggested

$$\frac{r}{n+1};\tag{11.91}$$

see Cunnane (1978) for a brief history. The expression (11.91) has endpoints that are not 0 or 1.

For most purposes where there is a fair amount of data, Equation (11.17) is a reasonable choice. For small data sets, further care is appropriate. Cunnane (1978) has developed several rules to deal with this situation. To study the question further, we divide the problem into two stages.

Stage 1

In the first stage, we consider the construction of the ogive, without yet considering the question of fitting a particular distribution. We therefore analyse the random quantity (X) and its *cdf* $F_X(x)$ as in Figure 11.16. This constitutes then an analysis of the quantity under consideration before obtaining data. In stage 2 we will study the question of linear presentations which are a transformation of the ordinate, as in the ordinate of Figure 11.4. We consider the data ranked in ascending order, as above and in Section 9.3. There, we derived the distribution of the rth ordered random quantity $X_{(r)}$ in terms of the parent $f_X(x)$ in Equation (9.15):

$$f_{X_{(r)}}\left(x_{(r)}\right) = \frac{n!}{(r-1)!\,(n-r)!} F_X^{r-1}\left(x_{(r)}\right)\left[1 - F_X\left(x_{(r)}\right)\right]^{n-r} f_X\left(x_{(r)}\right)$$

$$= r\binom{n}{r} F_X^{r-1}\left(x_{(r)}\right)\left[1 - F_X\left(x_{(r)}\right)\right]^{n-r} f_X\left(x_{(r)}\right).\tag{11.92}$$

One can see that

$$r\binom{n}{r} F_X^{r-1}\left(x_{(r)}\right)\left[1 - F_X\left(x_{(r)}\right)\right]^{n-r}\tag{11.93}$$

constitutes a beta-distribution of the random quantity F_X with parameters r and $(n - r + 1)$. Blom (1958) generalized this as the transformed β-quantity.

One approach that appears in the literature is as follows. Noting that

$$\frac{dF_X\left(x_{(r)}\right)}{dx_{(r)}} = f_X\left(x_{(r)}\right),$$
(11.94)

we may substitute into Equation (11.92) or (9.15) to obtain

$$f_{X_{(r)}}\left(x_{(r)}\right) dx_{(r)} = r\binom{n}{r} F_X^{r-1}\left(x_{(r)}\right)\left[1 - F_X\left(x_{(r)}\right)\right]^{n-r} dF_X\left(x_{(r)}\right).$$
(11.95)

But we can weight the *cdf* F_X by $f_{X_{(r)}}$ to get a mean position on the F_X axis. We do this by multiplying each side of (11.95) by $F_X\left(x_{(r)}\right)$ and integrating,

$$\int_{-\infty}^{\infty} F_X\left(x_{(r)}\right) f_{X_{(r)}}\left(x_{(r)}\right) dx_{(r)} = \int_0^1 r\binom{n}{r} F_X^r\left(x_{(r)}\right)\left[1 - F_X\left(x_{(r)}\right)\right]^{n-r} dF_X\left(x_{(r)}\right)$$

$$= \frac{r}{n+1}.$$
(11.96)

In other words

$$\left\langle F_X\left(x_{(r)}\right)\right\rangle = \frac{r}{n+1}.$$
(11.97)

This is a distribution-free result but is in fact hard to interpret. It has been used to lend support to the use of (11.91) as a plotting position, the argument being that the 'average' position of $x_{(r)}$ on the F_X axis will be correct, as given by (11.97). But the vertical axis that we use in our fitting is not a linear F_X scale (or F_Z scale in the case of the standard normal distribution) – it is a transformed non-linear scale as in Figure 11.4 (see F_Z values on vertical scale). Again, in Figure 11.8, the vertical scale is linear in F_X^{-1}; the corresponding F_X-values are distinctly non-linear. A uniform distribution of f_X in (11.94) only will result in a uniform F scale in curve fitting. Equation (11.91) might then be expected to work well for ogives.

Stage 2

The intention in data reduction is often, as outlined above, to determine empirically the best-fit distribution or to confirm or deny a distributional assumption. These presentations will generally be carried out using transformations of the ordinate in Figure 11.16, examples of which were just discussed, using the ordinates of Figures 11.4 and 11.8 above. A linear fit in plots such as Figures 11.4 or 11.8 will be of the form

$$Y = \frac{X - c}{d},$$
(11.98)

where X models the data and Y stands for the transformed quantity, termed F_X^{-1} and F_Z^{-1} in the examples above. This kind of transformation was introduced in Chapter 7, Equation (7.14).

We can then deduce that

$$\langle Y \rangle = \frac{\langle X \rangle - c}{d} \tag{11.99}$$

and

$$\langle Y_{(r)} \rangle = \frac{\langle X_{(r)} \rangle - c}{d}. \tag{11.100}$$

The second of these two equations means that, if we can obtain $\langle X_{(r)} \rangle$, we can estimate $\langle Y_{(r)} \rangle$.

Rather than concentrate on the mean value of $F_X \left(x_{(r)} \right)$, as in Equation (11.97) we consider instead $X_{(r)}$, the random quantity under question, knowing that the transformation (11.100) will result in an appropriate mean value since the linearity ensures an equivalence. Using Equation (9.15), repeated as (11.92) above, the analysis suggests that

$$\langle X_{(r)} \rangle = \int_{-\infty}^{\infty} x_{(r)} f_{X_{(r)}} \left(x_{(r)} \right) \, dx_{(r)}$$

$$= r \binom{n}{r} \int_{-\infty}^{\infty} x_{(r)} F_X^{r-1} \left(x_{(r)} \right) \left[1 - F_X \left(x_{(r)} \right) \right]^{n-r} f_X \left(x_{(r)} \right) \, dx_{(r)} \tag{11.101}$$

be solved.

There is no convenient and universal solution to Equation (11.101). We know that Equations (11.96) and (11.97) are solutions in the case where the possible points on the Y axis are uniformly distributed, as would be the case for a uniform *pdf*. Approximate numerical solutions to (11.101) have been obtained (Cunnane, 1978). These involve the generalization of (11.17) and (11.91) to

$$\frac{r - \alpha}{n + 1 - \alpha - \beta} \tag{11.102}$$

and

$$\frac{r - \alpha}{n + 1 - 2\alpha}, \tag{11.103}$$

where the equations have to be solved for α and β. Table 11.11 summarizes the results of these endeavours. As noted, for cases with a reasonable amount of data, the Hazen result works well and in Cunnane's paper has a good overall performance.

Recommendations vary: the Associate Committee on Hydrology of the National Research Council of Canada (1989) recommends the Cunnane compromise position below. In my view, the Hazen or Cunnane (compromise) positions represent simple and effective all-purpose rules, with further investigation needed in cases of small data sets.

Table 11.11   *Main results for plotting positions*

Source	Formula	α, β in (11.102)	Comments
Increasing rank	$\dfrac{r}{n}$	$\alpha = 0, \beta = 1$	
Decreasing rank	$\dfrac{r-1}{n}$	$\alpha = 1, \beta = 0$	
Hazen	$\dfrac{r-\frac{1}{2}}{n}$	$\alpha = \frac{1}{2}, \beta = \frac{1}{2}$	Works well in many cases
Weibull	$\dfrac{r}{n+1}$	$\alpha = \beta = 0$	Uniform or ogive
Blom	$\dfrac{r-\frac{3}{8}}{n+\frac{1}{4}}$	$\alpha = \beta = \frac{3}{8}$	Developed for normal
Gringorten	$\dfrac{r-0.44}{n+0.12}$	$\alpha = \beta = 0.44$	Developed for Gumbel
Cunnane	$\dfrac{r-\frac{2}{5}}{n+\frac{1}{5}}$	$\alpha = \beta = 0.40$	Compromise

11.5   Monte Carlo simulation

In analysis of engineering problems more often than not it is difficult to find the required solution in closed form. A powerful way to solve problems numerically is the Monte Carlo method, which is very flexible and has been widely used in industry. It offers a way out of intractable problems. At the same time it is good advice to keep any closed form solutions as part of the algorithm developed. This can reduce considerably the complexity of the program and the computer time required, and at the same time increase the elegance of the solution. Hammersley and Handscomb (1964) state that 'Each random number is a potential source of added uncertainty in the final result, and it will usually pay to scrutinize each part of a Monte Carlo experiment to see whether that part cannot be replaced by exact theoretical analysis contributing no uncertainty.'

11.5.1   Basis of method and historical comments

Monte Carlo simulation is essentially a method of integration (Hammersley and Handscomb, 1964; Kalos and Whitlock, 1986). Our applications will be principally in the use of random numbers in problems of decision and probability, but let us commence with the problem of determining areas such as those in Figure 11.17. In the *Old Testament* it was observed that the girth of the columns of King Solomon's temple were about three times as great as the diameter. If we do not know the value of π we might wish to determine the area of the circle in Figure 11.17(a) by means of an experiment. In our case let us set up such an experiment. We can construct a model such as that in Figure 11.17(a) on a larger scale, with a circle inscribed in a square. Assume that we know how to calculate the area of the square; to find the area of the circle, we conduct a random experiment. Let us drop an object with a sharp point such as a dart in such a manner that it is equally likely to penetrate at any point in the square with length

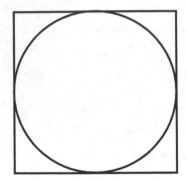 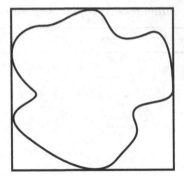

Figure 11.17   Areas for hit-and-miss

of side d. If we repeat this many times we can count the number of times that the dart arrives in the circle and divide this by the number of times it penetrates the square. This ratio will approximate the ratio of the area of the circle and the area of the square, or

$$\frac{\pi d^2/4}{d^2} = \frac{\pi}{4}.$$

This experiment enables us to determine an approximation to the value of π. Example 11.15 below shows how this experiment can be carried out in MATLAB.

One interpretation of the problem is in terms of 'geometric probabilities', outlined in Sections 2.3.1 and 2.3.3, used also in the Titanic illustration of Chapter 2. In Example 11.15 below, the probability of an object arriving randomly within the circle is $(\pi/4)$. The solution also approximates the integral

$$\int_{-1}^{1} \int_{-\sqrt{1-y^2}}^{\sqrt{1-y^2}} dx dy.$$

It should be noted that other techniques of integration, for instance based on quadrature, might prove more effective than Monte Carlo with better convergence. But we are interested in Monte Carlo because of its flexibility in solving complex problems possibly with many dimensions.

Another way of obtaining an approximation to π is the problem in Exercise 11.20 introduced by the Comte de Buffon in the eighteenth century. In this, a needle of length d is dropped at random onto a horizontal surface ruled with straight lines a distance s apart. It can be shown – see Exercise 11.20 – that the probability that the needle touches one of the lines is $(2d/\pi s)$. Laplace later suggested that this might constitute a way to determine an approximation to π by repeated throws of the needle. It might be worth noting that the scour of an ice feature on the seabed forms fairly straight lines which might intersect pipelines causing damage and possibly oil spills. This problem has been treated in a number of applications.

Drawing coloured or numbered balls from urns might be viewed as Monte Carlo experiments. It is recorded that Kelvin in 1901 used advanced Monte Carlo techniques. Student (W. S. Gosset) early in the twentieth century used random sampling in his studies of distributions. The use of Monte Carlo techniques achieved impetus by von Neumann and Ulam in the development of the atomic bomb during the second World War. These two together with Fermi and Metropolis continued this usage after the war. During the intervening years, the method has become an established part of engineering analysis.

11.5.2 Random numbers

Random numbers may be generated by physical processes such as coin- or die-tossing, drawing objects from jars or urns, or drawing lotto chips. The name 'Monte Carlo' conjures roulette wheels and other gambling devices. A few decades ago it was common to publish tables of random numbers derived on the basis of physical experiments. Copies of these tables can still be found. The most common practice today is to use computer-generated sequences. In the sense that the sequences can be reproduced exactly by repeating the same algorithm on a computer, these sequences are not strictly random and are therefore termed *pseudorandom* numbers (*prns*). Provided the method works well, the result is a string of *iid* numbers. We can then obtain support for the methods from other theory involving *iid* quantities, and in particular that of Chapter 7 including the Central Limit Theorem and the laws of large numbers.

The most common method of generating *prns* is the congruential method with the numbers generated by a relation such as

$$x_{i+1} \equiv a x_i + c \pmod{m}, \tag{11.104}$$

in which x_i is the current number, x_{i+1} the next in the recurrence relation and a, c and m are constants. The expression 'mod' or 'modulo' is used in the context of 'congruence modulo m' relationships. The relationship could involve differences between numbers; then x_{i+1} and $(ax_i + c)$ would differ by a multiple of the chosen natural number m. In random number generation it relates to the remainder term. Specifically, the relationship (11.104) indicates that x_{i+1} is the remainder when $(ax_i + c)$ is divided by m. The value a is the *multiplier*, c the *increment* and m the *modulus*. For example, let $a = 3, c = 3, m = 16$, and $x_1 = 2$. Then the relationship (11.104) results in the sequence

$$2, 9, 14, 13, 10, 1, 6, 5, 2, 9, \ldots$$

After reaching again the initial value or 'seed', 2 in the present case, the sequence repeats itself. This is always the case; eventually the initial seed recurs. In the example, the period is 9. The maximum period would be m. The objective is to obtain a period as close as possible to the maximum by judicious choice of a, c and m. In reality we use much larger values of a, c and m than those used in the example above; typically $m = 2^{31} - 1, a = 7^5, c = 0$ (due to Park and Miller in Press *et al.*, 1992). Kalos and Whitlock (1986) state that the choice of c, if not zero, has little influence on the quality of the generator, a being most important. But the example emphasizes that the period is finite and the choice of constants important in a pseudorandom number generator. The objective should be to achieve a period equal to m. If this is achieved, the choice of seed does not matter since the sequence includes all possible integers, from 0 to $(m - 1)$ if $c \neq 0$. In the case where $c = 0$ in (11.104) the seed must never equal zero or it simply perpetuates itself. If we divide the random numbers by m, the result is a set of *prns* in the range [0,1); to obtain numbers in the range [0,1], we divide by $m - 1$.

We mention that *quasirandom* numbers are sometimes used. These fill out the space of integration according to a pattern filling the volume of integration in a predetermined fashion. This is an example of a variance-reduction technique (see below and Press *et al.*, 1992).

Lagged Fibonacci generators

These generators have become increasingly used in recent years. The name derives from the Fibonacci sequence which is as follows:

$$1, 1, 2, 3, 5, 8, 13, 21, \ldots$$

Each number apart from the first two is the sum of the two preceding numbers. This may be written as

$$x_n = x_{n-1} + x_{n-2}.$$

The first two numbers are given. In the case of lagged Fibonacci generators (*lfgs*)

$$x_n = x_{n-\ell} + x_{n-k}; \tag{11.105}$$

k and ℓ are the 'lags'. The current value of x is determined by the value of x at k and ℓ places ago. Equation (11.105) is used with a modulo relationship in the algorithm. It is claimed that periods greater than 2^{1000} can be obtained.

Statistical tests

Press *et al.* (1992) give a detailed analysis of *prn* generators, including algorithms other than the linear congruential method above. They detail the shortcomings of many system-supplied generators. My suggestion is to check your algorithm against this good advice. There are many statistical tests for *prns* (see for example Kalos and Whitlock, 1986). The simplest one is the chi-square test of Equation (11.21); one can use this

$$\sum \frac{(\text{observed frequency minus expected frequency})^2}{(\text{expected frequency})}$$

to test for uniformity or against any desired distribution; see for example Exercise 11.19. One can also test for correlations between successive values, runs-up and runs-down, as well as other aspects. It is a good idea to do appropriate tests on your *prn* generator.

11.5.3 Hit-or-miss versus expected value methods

Rubinstein (1981) made the division suggested in the title to this subsection. We start with a hit-and-miss example.

Example 11.15 We can write a computer program in MATLAB to simulate the dropping of darts in a circle inscribed in a square. One solution is as follows. Only Pythagoras' theorem is used.

```
x=unifrnd(-1,1,1,100000);y=unifrnd(-1,1,1,100000);
z=(x.^2)+(y.^2);
(Pythagoras)
w=(z<=1);
v=sum(w);
4*v/100000
```

Estimates of π can be made with accuracy that we can calculate, increasing with n, the number of simulations. The process is close to a binomial set of trials, n in number with a probability of success of $(\pi/4)$ in each trial. The expected value is $np = (n\pi/4)$ with a standard deviation of $\sigma = \sqrt{npq} \simeq 0.411\sqrt{n}$. The coefficient of variation is then

$$\frac{\sqrt{npq}}{np} = \frac{\sqrt{q}}{\sqrt{np}}.$$

This gives a value of 0.00 165 for $n = 100\,000$, easily confirmed by the simulations.

Given the large n, the normal distribution approximates very well the spread of values. The standard deviation and 95% confidence intervals for $n = 100\,000$ are then 130 and $\pm 1.96 \cdot 130 \simeq \pm 250$ (in round numbers) respectively. The confidence intervals represent $(250/78\,500) = 0.32\%$ of the mean. The mean will in general have a standard deviation that is proportional to

$$\frac{\sqrt{npq}}{n} = \frac{\sqrt{pq}}{\sqrt{n}} \propto \frac{1}{\sqrt{n}}. \tag{11.106}$$

Example 11.16 The use of expected or mean value is simpler. One is attempting to solve the integral

$$\int_0^1 \sqrt{1 - x^2}\, dx, \tag{11.107}$$

which we know has the value $\pi/4$. (See also Example 11.18 below.) This can be simulated as

```
x=rand(1,100000);
y=sqrt(1-x.^2);
4*mean(y)
```

The variance σ^2 of this method can be assessed as in Section 11.5.5 below. For n simulations, it is σ^2/n, resulting in a coefficient of variation of

$$\sqrt{\frac{0.05}{100\,000}} / (0.785) = 9 \times 10^{-4}.$$

This decreases according to $1/\sqrt{n}$ as above, but is less than the value 0.00 165 obtained for the 'hit-and-miss' method. This shows that the variance depends on the Monte Carlo method chosen for computation. This has important implications in design methods; a wise choice of methodology can lead to results that do contain the minimum variance. See Section 11.5.5 below.

11.5.4 Sampling from distributions

To focus the discussion, we observe that (x_{i+1}/m) in the use of (11.104) results in a uniform distribution in the range from 0 to 1. Most *prn* generators will produce such a distribution (note that some algorithms may exclude one or both endpoints). Now we introduce a most useful result.

All cdfs are uniformly distributed.

This statement can be understood by considering the *cdf* $F_X(x)$ of any random quantity X. Since the *cdf* represents probability on a uniform scale from zero to unity, equal increments of any size along the vertical axis represent equal increments of probability. *Therefore all points are equally likely.* To amplify, taking a continuous distribution as an example,

$$dF_X(x) = f_X(x)\, dx,$$

and we see that the shaded areas in Figure 11.18 are equal for equal increments in $dF_X(x)$. Similar arguments can be used for discrete distributions. The result just described was contained

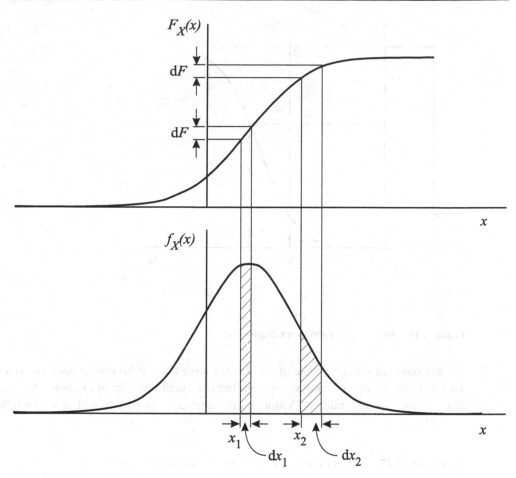

Figure 11.18   Equal increments dF of probability

in the *probability integral transform* given in Chapter 7 in Example 7.2 and in Figure 7.1(b). These statements are also consistent with our conclusion above in Section 11.4 regarding plotting positions, that there is an even spacing of points in the ordinate for ogives, but not for the linearized plots in which the scale has been transformed.

Most computer packages will give random numbers tailored to particular distributions. For example, one can automatically generate random numbers to simulate Poisson or normal distributions with given parameters. But it is useful to outline methods for carrying out these simulations. The use of *cdf*s is the most common. Suppose that we wish to sample random values of the quantity X from a distribution of which we know the *cdf* $F_X(x)$. We first sample from a uniform distribution of the random quantity U on the interval from 0 to 1. Denote this value u. But u represents a possible realization of any point of $U = F_X(x)$. We now calculate the inverse value $x = F_X^{-1}(x)$, as shown in Figure 11.19. This can be repeated

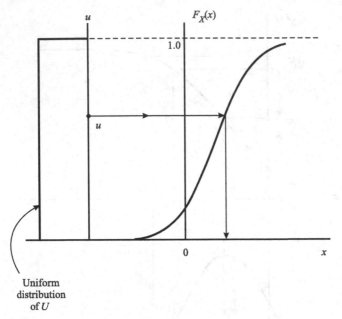

Uniform
distribution
of U

Figure 11.19   Sampling using uniform distribution

for any required size of sample of the Xs. We term this the **inverse transform technique**.
One can use empirical data in simulations but one should be careful to ensure that one does
not exclude any possibilities. Continuous distributions can be sampled as in the following
example.

Example 11.17 Consider the Gumbel (double exponential) distribution

$$F_Z(z) = \exp\left[-\exp\left(-\frac{z-a}{b}\right)\right].$$

We can solve

$$z = F_Z^{-1}(z) = a - b \ln\left(-\ln F_Z(z)\right), \tag{11.108}$$

sample $F_Z(z)$, and calculate z.

Bayesian parameter

If the random generator is performing well, the values obtained should be *iid*. If we have a
Bayesian situation and know the distribution of the uncertain parameter we can sample a value
of the parameter first, then make the usual *iid* assumption given the parameter value, and weight
over the distribution of the parameter.

11.5.5 Variance reduction

We saw above that the 'hit-and-miss' and expected value methods resulted in different variances. It is generally the case that the variance will depend on the simulation method. The expected value method can be seen as a numerical approximation of

$$\mu_G = \langle g(X) \rangle = \int f_X(x) g(x) \, dx, \tag{11.109}$$

where $f_X(x)$ is a *pdf*. We take a random simulation

$$x_1, x_2, \ldots, x_i, \ldots, x_n \tag{11.110}$$

in a similar way to the sample (11.2). Then we calculate

$$g(x_1), g(x_2), \ldots, g(x_i), \ldots, g(x_n) \tag{11.111}$$

and use

$$\langle g(X) \rangle \simeq \frac{\sum_{i=1}^{n} g(x_i)}{n} \tag{11.112}$$

to approximate (11.109). The standard deviation associated with this calculation is

$$\frac{\sigma}{\sqrt{n}}, \tag{11.113}$$

where

$$\sigma^2 = \int g^2(x) f_X(x) \, dx - \langle g(X) \rangle^2. \tag{11.114}$$

If this cannot be calculated (as is usual) one can use s^2 computed from the simulation, as in dealing with other data.

Example 11.18 For the simulation of (11.107) above,

$$g(x) = \sqrt{1 - x^2}, \tag{11.115}$$

with $f_X(x) = 1$, $0 \le X \le 1$. From (11.114),

$$\sigma^2 = \int \left(1 - x^2\right) dx - \left(\frac{\pi}{4}\right)^2 = 0.0498. \tag{11.116}$$

(We happen to know the mean and we can complete the integration for the standard deviation, but these can always be simulated.)

Importance sampling

We noted that the variance of a method depends on the way a problem is set up. There are several ways to reduce the variance; see for example Kalos and Whitlock (1986). Importance sampling has become an important part of engineering analysis. The idea is to focus the sampling of the most important part of the distribution. For example, there might be areas where $g(x)$ in

(11.109) is zero and sampling will only produce a string of zeros. The idea of importance sampling is to model using a *pdf* that is different from $f_X(x)$, say $f_1(x)$. Then (11.109) is written

$$\mu_G = \langle g(X) \rangle = \int \frac{f_X(x) g(x)}{f_1(x)} f_1(x) dx. \tag{11.117}$$

The function $f_1(x)$ has to be chosen carefully.

Example 11.19 We use again Equation (11.115):

$$g(x) = \sqrt{1 - x^2} = 1 - \frac{1}{2}x^2 + \cdots. \tag{11.118}$$

we take

$$f_1(x) = \frac{6}{5}\left(1 - \frac{1}{2}x^2\right); \tag{11.119}$$

where the factor 6/5 normalizes the distribution. The variance is reduced from 0.05 calculated above in Equation (11.116) to 0.011. By using

$$f_1(x) = \frac{1 - \alpha x^2}{1 - \frac{1}{3}\alpha} \tag{11.120}$$

with $\alpha = 0.74$, this can be further reduced to 0.0029; see Kalos and Whitlock (1986).

This introduction to importance sampling is intended to give a flavour of the method. It has been used extensively in virtually all areas of simulation. For application in reliability calculations, see for example Melchers (1999).

11.5.6   Conclusion

Monte Carlo simulation presents a powerful and flexible method for solving engineering problems. It is one of the most important members of the engineer's toolkit in the solution of practical problems. There is a need for new packages to assist in Bayesian updating using numerical methods.

11.6   Exercises

11.1 Become familiar with 'stem-and-leaf' and 'box-and-whiskers' plots using one of the data sets below.

11.2 Show that in classical analysis the unbiased estimates of skewness and kurtosis are

$$\bar{g}_1' = \frac{n}{(n-1)(n-2)s^3} \sum_{i=1}^{n}(x_i - \bar{x})^3$$

$$= \frac{n^2}{(n-1)(n-2)}\bar{g}_1, \text{ and} \tag{11.121}$$

$$\bar{g}_2' = \frac{n(n+1)}{(n-1)(n-2)(n-3)s^4} \sum_{i=1}^{n} (x_i - \bar{x})^4 - \frac{3(n-1)^2}{(n-2)(n-3)} + 3$$

$$= \frac{n^2(n+1)}{(n-1)(n-2)(n-3)} \bar{g}_2 - 3\left[\frac{(n-1)^2}{(n-2)(n-3)} - 1\right] \qquad (11.122)$$

respectively. (Reference: Cramér, 1951.)

11.3 For the following data, which represent readings of three-day mortar compressive strength (normalized scale with no units) from a cement plant, follow all the steps in Section 11.2 above, including the fitting of a distribution.

2930	2820	2730	2780	2580
3040	2500	2680	2790	2730
2880	2660	2790	2730	2780
3060	3000	2660	2910	2720
3060	2960	2820	2820	3050
2820	3020	2890	3140	2920
2700	3060	3050	2790	2810
2880	2750	2530	2900	2940
2770	2820	3030	2700	2960
2890	2810	2760	2770	2850

11.4 For the following data, which represent readings of 10 minute average windspeed in the east–west direction at the location of a building project, follow all the steps in Section 11.2 above, including the fitting of a distribution.

−8.93	−2.52	0.54	0.04	5.14
−5.19	1.58	5.68	0.65	0.76
−4.83	−11.05	3.71	−8.61	2.03
2.73	−4.21	−4.72	−9.95	−6.74
−5.37	5.36	2.56	−5.85	1.38
4.90	7.03	8.25	−2.04	−2.78
12.42	0.23	−4.24	−1.66	−4.73
1.93	−3.87	3.67	2.86	−9.22
−7.80	0.19	−3.44	0.81	−4.29
−1.53	6.71	−8.37	−7.49	−2.26
−0.73	2.28	−8.29	−3.52	4.42
−1.83	−11.28	−4.31	−7.00	−0.12

11.5 For the following data, which represent readings of wave hindcast data, follow all the steps in Section 11.2 above, including the fitting of a distribution. Give particular attention to the double exponential (Gumbel) distribution. The data represent estimates of the maximum annual significant wave height in metres, and were supplied by Val Swail and discussed in Swail *et al.* (1998).

9.212	10.484	11.441	10.511	13.122	10.625	10.726
10.499	11.105	12.382	11.428	13.390	10.956	11.644
10.438	13.637	11.356	10.349	9.583	11.500	9.688
11.638	13.427	10.360	9.652	12.854	13.600	10.220
9.823	11.005	10.694	11.138	10.793	12.152	10.190
10.711	7.474	9.046	10.189	10.800	12.652	9.652%

11.6 The following data are from the work of von Bortkiewicz; see Jeffreys (1961). This concerns the number of men kicked to death by horses in a Prussian cavalry unit during a year. Fit a Poisson distribution and carry out a chi-square goodness-of-fit test.

Number of deaths	Number of occurrences (frequency)
0	144
1	91
2	32
3	11
4	2
5 or more	0

11.7 Prove the result that (11.20) or (11.21) tends to a chi-square distribution as $n \to \infty$ using an inductive approach. (See Lindley, 1965, Section 7.4.) Also prove the result that (11.20) or (11.21) tends to a chi-square distribution as $n \to \infty$ using a mechanistic approach. (See Cramér, 1951, Chapter 30.)

11.8 Sketch how 'regression to the mean' in Galton's experiments would be manifested in the best fit line.

11.9 Show that in regression analysis,

$$\hat{a} = \frac{\sum_{i=1}^{n} y_i - \hat{b} \sum_{i=1}^{n} x_i}{n}, \text{ where} \tag{11.123}$$

$$\hat{b} = \frac{\sum_{i=1}^{n} (x_i - \bar{x})(y_i - \bar{y})}{\sum_{i=1}^{n} (x_i - \bar{x})^2}. \tag{11.124}$$

11.10 Carry out a regression analysis of the results in Exercise 8.32 (paired test) and comment on assumptions made there.

11.11 In a study of activation energy of creep movements in a viscoelastic material, the following relationship has been used:

$$\dot{e} = A\exp\left(\frac{-Q}{RT}\right), \tag{11.125}$$

$p = 15$ MPa		$p = 35$ MPa		$p = 55$ MPa		$p = 65$ MPa		$p = 70$ MPa	
$\frac{1}{T}$ (K^{-1})	ln $\dot e$	$\frac{1}{T}$ (K^{-1})	ln $\dot e$	$\frac{1}{T}$ (K^{-1})	ln $\dot e$	$\frac{1}{T}$ (K^{-1})	ln $\dot e$	$\frac{1}{T}$ (K^{-1})	ln $\dot e$
0.003798	−7.30	0.003822	−7.24	0.003797	−6.90	0.003902	−8.10	0.003922	−8.25
0.004060	−9.36	0.003974	−8.38	0.004026	−9.15	0.003956	−8.80	0.004016	−8.45
0.003755	−6.54	0.003881	−8.80	0.003726	−6.10	0.003791	−5.40	0.003953	−8.90
0.003964	−8.50	0.003736	−6.65	0.003900	−8.25	0.003828	−8.15	0.004032	−9.45
0.003812	−7.35	0.004022	−9.60	0.003826	−7.68	0.004036	−9.40	0.003781	−5.35
0.003887	−8.00			0.003962	−8.60	0.003880	−8.00	0.003792	−5.15
0.003738	−5.90			0.003855	−8.35	0.00402	−8.80	0.003861	−7.65
				0.003811	−8.45	0.003925	−9.10		

where $\dot e$ is the strain rate, Q is the activation energy, T is the temperature (degrees Kelvin), R is the universal constant ($R = kN_0$, where k is Boltzmann's constant and N_0 is Avogadro's number) and A is a constant.

(a) Take the logarithm of Equation (11.125) and determine the slope of regression lines, ln $\dot e$ versus $(1/T)$, obtained for each of the five values of p (hydrostatic pressure) below. Determine confidence intervals for each of the slopes. From the slopes, determine the corresponding values of Q with associated variance and confidence intervals.

(b) It is supposed that Q will increase as p increases, due to the effect of pressure melting in grain boundaries. Plot Q versus p, and comment. Pool those results that are similar, including pooled values for the variance of Q, describing how you make these judgements. Pool all the values, including the variance, on the hypothesis that there is no difference in Q for the various p. Comment on the result.

(c) Comment on the role of scientific knowledge in making inferences.

11.12 Complete the proof of Equation (11.68).

11.13 Research the background for the F-test in regression analysis (Table 11.10).

11.14 Use the analysis of uncertainty in the mean response to derive equation (11.75). Let V represent $(1/W)$ or variance. Find confidence intervals for V.

11.15 *Circular regression.* In placing a gravity based offshore structure (GBS) on the seabed, a least-squares fit is required to place the structure on pegs located on the seabed. Formulate an algorithm for least squares around a circle to find the best fit centre location and radius for the GBS, given the location of a number of pegs.

11.16 *Bayesian regression.*

(a) Research methods of conjugate prior analysis for the parameters in regression analysis; see for example Raiffa and Schlaifer (1961) or Box and Tiao (1973); develop an analogy to the normal-gamma processes of Chapter 8.

(b) Research methods for multidimensional analysis using Bayesian methods; use the references in (a) as a starting point.

11.17 Assume that $Y|x$ is $N\left(\alpha+\beta x, \sigma^2\right)$ (in other words, assume that there are sufficient data to estimate $\hat{a}$ and $\hat{b}$ with little uncertainty), and that X is $N\left(\mu_X, \sigma_X^2\right)$. Find the distribution of Y. Note that

$$f_{X,Y}(x, y) = f_{Y|X=x}(x, y) f_X(x).$$

(a) Show that this is a two-dimensional normal distribution of the kind in Equation (A1.140) (Appendix 1) with

$$\mu_Y = \alpha + \beta\mu_X,$$
$$\sigma_Y^2 = \sigma^2 + \beta^2\sigma_X^2 \text{ and}$$
$$\rho = \beta\frac{\sigma_X}{\sigma_Y}.$$

(b) Show that the distribution of $Y|x$ can also be written as

$$N\left[\mu_Y + (x - \mu_X)\frac{\rho\sigma_Y}{\sigma_X}, \sigma_Y^2\left(1 - \rho^2\right)\right]$$

and that of $X|y$ as

$$N\left[\mu_X + (y - \mu_Y)\frac{\rho\sigma_X}{\sigma_Y}, \sigma_X^2\left(1 - \rho^2\right)\right].$$

(c) Show (11.70) using the results of this exercise.

11.18 In Cunnane's 1978 paper, he discusses Hazen's plotting position. Gumbel objected to this because the largest value is plotted at $F = 1 - 1/2n$, which corresponds to a return perion of $2n$. Review Gumbel's and Cunnane's response (in the 1978 paper).

Monte Carlo methods are rich in variety and can be formulated in any area of interest. They can be used to assist in the solution of important engineering problems. The following are introductory, and it is suggested that you compose problems in your area of expertise.

11.19 Use Monte Carlo simulation to generate 100 000 values uniformly distributed in the interval from 0 to 1. Do a chi-square analysis to test that the results conform to a uniform distribution.

11.20 *Buffon needle problem.* A needle of length d is dropped at random onto a horizontal surface ruled with straight lines a distance s apart. Show that the probability that the needle touches or intersects one of the lines is $\left(\frac{2d}{\pi s}\right)$. Formulate a Monte Carlo simulation program to estimate π using this principle. It should be noted that the problem is identical to that of finding the probability that a scouring iceberg or ridge keel interacts with a pipeline on the seabed.

11.21 Simulate Thomas Bayes' billiard table experiment of Section 8.3.3 and compare the results with (8.35).

11.22 Use Monte Carlo simulation to replicate the demon's maximum entropy experiments of Chapter 6.

11.23 Use Monte Carlo simulation to model the Central Limit Theorem.

11.24 *De Méré's problem (Section 8.2).* The problem concerns gambling with dice, which subject interested de Méré considerably. In tossing a die of six sides, the die must be tossed four times in order to achieve a six with a probability greater than one half. In gambling with two dice, it seemed natural to extrapolate from this result; in order to obtain a double-six, multiply the result of four by the ratio of the number of possible results (i.e. by 36/6). The answer, 24, is then supposed to indicate the number of throws required to achieve the double-six. De Méré found, probably by trial in gambling, that this logic did not work. Simulate both the one- and two-dice problems and compare the results to exact solutions. In this case the problem also simulates (for example) the number of years to exceed the capacity of a spillway if the probability of this happening in a year is 1/6 or 1/24.

11.25 *Simulation of extremal distributions*; see Exercises 9.2, 9.6 and 9.7.

11.26 In the design of a dam for flood control, the height of the dam depends on two factors. The first is the current level of water in the dam, and the second is the increase caused by flood waters coming into the reservoir. The first can be modelled using the following discrete distribution, and the second with a uniform distribution in the range 0 to 30 m.
For a height of the dam of 60 m, what is the probability that it will be overtopped?

Water level, m	20	25	30	35	40	45
Probability	0.20	0.30	0.30	0.10	0.06	0.04

11.27 The cumulative distribution function of annual maximum wave height Z in a study aimed at determining the design wave height is given by

$$F_Z(z) = \exp\left[-\exp\left(-\frac{z-\alpha}{\beta}\right)\right]$$

with $\alpha = 10.4$ m (mode) and $\beta = 1.04$ m ($= \text{s.d.} \times \sqrt{6}/\pi$). Simulate this distribution using a uniform distribution on the interval (0, 1) using the integral transform theorem.

(a) Plot a histogram and compare this to the distribution given.
(b) Estimate the 100-year wave height.

Note: These are significant wave heights. The maximum wave height will the largest in a storm with given significant wave height (Chapter 8).

11.28 Use the method in Exercise 7.10 using two uniform distributions to simulate two normal distributions.

11.29 A method for inference that has been receiving a lot of attention is the Markov Chain Monte Carlo (MCMC); see for example Wilks *et al.* (1996). Consider applications in your area of research.

12 Conclusion

Sir, I have found you an argument. I am not obliged to find you an understanding.

J. Boswell, *The Life of Samuel Johnson*

Uncertainty accompanies our lives. Coherent modelling of uncertainties for decision-making is essential in engineering and related disciplines. The important tools have been outlined. It is neither possible nor desirable to attempt to give a recipe that can be used for specific problems. The fun of engineering is to use the tools so as to create a methodology that can be used for a particular problem. Probability is seen as the measure of uncertainty. Decisions are based on an analysis of uncertainties using probability and of desires and aversion using utility. These factors guide our thinking in approaching problems.

In estimating probabilities, the beacon that guides us is the definition of probability as a fair bet. Frequencies can be used only to assist in evaluating probabilities but do not constitute a definition of probability. It is important to address all uncertainties without taking a problem out of reasonable practical proportion. If done in a blind manner without judgement, this can lead to gross overestimation of the uncertainty regarding our quantity of interest. Our estimates of mean values, including variances and other moments, should accord with our judgement, and we should beware of compounding uncertainties that might result in unrealistic engineering judgements.

Flowing from the definition of probability as a fair bet, Bayesian thought should guide our modelling. The inductive world treats any object regarding which we are uncertain as being random.

$$\mathbf{Pr}\,(\text{hypothesis} \mid \text{information}) \propto \frac{\mathbf{Pr}\,(\text{information} \mid \text{hypothesis})}{\mathbf{Pr}\,(\text{information})} \cdot \mathbf{Pr}\,(\text{hypothesis})\,. \qquad (12.1)$$

This can be applied quite generally. Although we have shown often mechanistic models equivalent to inductive ones, inductive reasoning is infinitely richer since it can guide our steps from first hypothesis to solution of a problem. It can assist in guiding our thinking. Bayesian thinking results in a much wider purview than the formal solutions to problems of parameter estimation. At the same time in problems of inference, the Bayesian approach corrects many aspects of classical estimation. It is time to make the transition from the classical methods to a proper Bayesian formulation, especially in education. This presents a challenge but brings forward the methods of decision theory to the engineer in a clear and appropriate manner.

One important tool in practical situations is the analysis of sensitivity. Often when faced with an uncertainty, this can assist considerably in aiding one's judgement. For example, one might have a Monte Carlo simulation of a process and be concerned regarding uncertainty in a parameter or other aspect of the model. A few sensitivity runs might assist enormously in identifying the importance of the feeling of uncertainty as regards the final estimate of a design parameter. It is a question of being a good 'detective', in which Bayesian thought is fundamental.

Appendix 1   Common probability distributions

Introduction

In this appendix we summarize the properties and uses of some common probability distributions. The treatment is brief, concentrating on the main distributions. Generally one-dimensional distributions are given so as to illustrate the character of the form, but the multidimensional normal distribution is added for reference purposes. There are texts such as Johnson *et al.* (1993, 1994) which contain large amounts of detail on distributions, including their history. Feller (1968, 1971) represents a great store of information. Patel *et al.* (1976) summarize properties of many distributions.

The shapes of the distributions are illustrated in Figures A1.1 and A1.2.

A1.1   Discrete distributions

A1.1.1   Uniform distribution

This describes a set of n equally likely outcomes.

pmf $p_X(x) = \dfrac{1}{n}$, for n possible outcomes.   (A1.1)

cdf   Stepped function with jumps of $1/n$ at the possible outcome points.

Mean and variance   For a unit probability mass at $x = a$:

$$\mu = a,$$ (A1.2)
$$\sigma^2 = 0.$$ (A1.3)

The moments for any uniform distribution can be obtained as those for a set of equal point masses at $x_i, i = 1, \ldots, n$:

$$\mu_X = \frac{1}{n} \sum_{i=1}^{n} x_i$$ (A1.4)

$$\sigma_X = \frac{1}{n} \sum_{i=1}^{n} (x_i - \mu_x)^2$$ (A1.5)

Characteristic function   For a unit probability mass at $x = a$:

$$\exp(iua).$$ (A1.6)

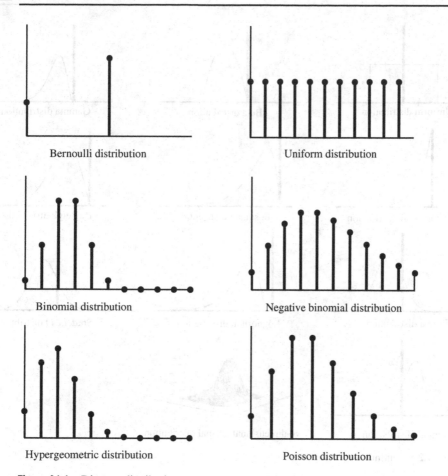

Figure A1.1   Discrete distributions

For a set of n equal point masses at x_i, $i = 1, \ldots, n$, this becomes

$$\phi_X(u) = \frac{1}{n} \sum_{i=1}^{n} \exp(iux_i).$$
(A1.7)

Uses and remarks   We often have a situation where the outcomes are equally likely. Tossing of coins, throwing of dice, and appearance of cards in games are examples. A set of target points where arrivals at each is considered equally likely: asteroids entering the earth's atmosphere, missiles arriving in a region, positions of icebergs within equal areas of a degree-rectangle, and so on. In the Monte Carlo method we choose equally likely points within a certain range.

A1.1.2 Bernoulli distribution

This was introduced in Section 3.2.2. The process consists of a single trial with probability p of success.

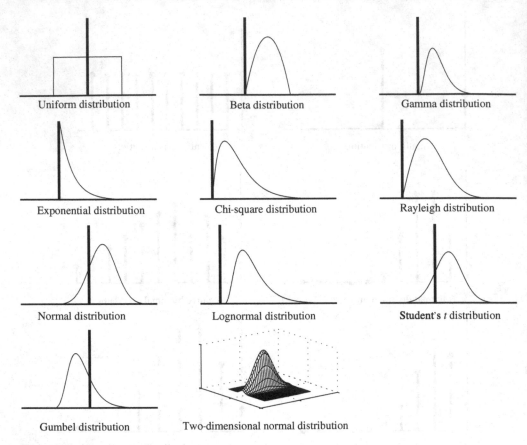

Figure A1.2 continued (Uniform distribution, Beta distribution, Gamma distribution, Exponential distribution, Chi-square distribution, Rayleigh distribution, Normal distribution, Lognormal distribution, Student's t distribution, Gumbel distribution, Two-dimensional normal distribution)

Figure A1.2   Continuous distributions

pmf   For $x = 0, 1$:

$$p_X(x) = 1 - p = q, \quad \text{for} \quad x = 0,$$
$$\qquad\quad = p, \quad \text{for} \quad x = 1. \tag{A1.8}$$

Mean and variance

$$\mu_X = p, \tag{A1.9}$$
$$\sigma_X^2 = p(1 - p) = pq. \tag{A1.10}$$

Characteristic function

$$\phi_X(u) = q + p \exp(iu). \tag{A1.11}$$

Uses and remarks   The distribution is the fundamental building block for various processes, notably the binomial one. It corresponds to a single shot at a target with known probability giving a hit or miss, a toss of a coin, drawing of a ball from an urn with balls of two colours, or any single trial with two outcomes.

A1.1.3 Binomial distribution

The distribution was introduced in Section 3.2.2, Equation (3.3).

pmf $\quad p_X(x) = \binom{n}{x} p^x (1-p)^{n-x}$

$$= \binom{n}{x} p^x q^{n-x}, \tag{A1.12}$$

where $q = 1 - p$.

Mean and variance

$$\mu_X = np, \tag{A1.13}$$
$$\sigma_X^2 = np(1-p) = npq. \tag{A1.14}$$

Characteristic function

$$\phi_X(U) = [q + p\exp(it)]^n. \tag{A1.15}$$

Uses and remarks Sum of n Bernoulli trials; classical model of a series of independent trials, with a probability of success p in each trial. Hits and misses in various processes, defective or not, above or below a certain value, all in independent trials with known probability.

A1.1.4 Negative binomial and geometric distributions

Introduced in Sections 3.3 and 3.4; stated in Equations (3.6) and (3.7).

pmf For the negative binomial distribution

$$p_V(v) = \mathbf{Pr}(V = vk, p) = \binom{v-1}{k-1} p^k q^{v-k}, \tag{A1.16}$$

for $v = k, k+1, k+2, \ldots$

For the geometric distribution

$$p_V(v) = \mathbf{Pr}(V = v|p) = (1-p)^{v-1} p \tag{A1.17}$$
$$= pq^{v-1}$$

for $v = 0, 1, 2, \ldots$

Mean and variance For the negative binomial distribution

$$\mu_V = k\frac{q}{p}, \tag{A1.18}$$

$$\sigma_V^2 = k\frac{q}{p^2}. \tag{A1.19}$$

For the geometric distribution, put $k = 1$.

Characteristic function   For the negative binomial distribution

$$\phi_X(u) = \left(\frac{1 - q \exp it}{p}\right)^{-k}.$$ (A1.20)

For the geometric distribution

$$\phi_X(u) = \left(\frac{1 - q \exp it}{p}\right)^{-1}.$$ (A1.21)

Uses and remarks   This distribution represents trials to the kth success in binomial process, for example the number of defectives sampled until a good sample is found. Useful in design, particularly with regard to the number of trials (years) to the first extreme event; and a basis of return period analysis.

A1.1.5   Beta-binomial distribution

Introduced in Section 8.3.1 and Equation (8.32).

pmf $p_X(x) = \Pr(X = x|n)$

$$= \binom{n}{x} \frac{B(x + r', n + n' - x - r')}{B(r', n' - r')}$$ (A1.22)

$$= \binom{n}{x} \frac{\Gamma(n')}{\Gamma(r')\Gamma(n' - r')} \cdot \frac{\Gamma(x + r')\Gamma(n + n' - x - r')}{\Gamma(n + n')}$$ (A1.23)

Mean and variance

$$\mu_X = n \cdot \frac{r'}{n'},$$ (A1.24)

$$\sigma_X^2 = n \cdot \frac{r'(n' - r')(n' + n)}{(n')^2(n' + 1)}.$$ (A1.25)

Uses and remarks   Used in decisions in binomial process with parameter modelled by the beta distribution (mixtures of the binomial process). See Bernardo and Smith (1994).

A1.1.6   Hypergeometric distribution

Introduced in Section 3.2.5; see Equation (3.8). Note the four forms of this distribution; see Exercise 3.6.

pmf $p_X(x) = \Pr(X = x|b, n, g) = \dfrac{\frac{g!}{x!(g-x)!} \cdot \frac{(b-g)!}{(n-x)!(b-g-n+x)!}}{\frac{b!}{n!(b-n)!}}$

$$= \frac{\binom{g}{x}\binom{b-g}{n-x}}{\binom{b}{n}}.$$ (A1.26)

Mean and variance

$$\mu_X = n\frac{g}{b},$$ (A1.27)

$$\sigma_X^2 = n\frac{g}{b}\left(1 - \frac{g}{b}\right)\frac{b-n}{b-1}.$$ (A1.28)

Characteristic function

$$\phi X(u) = \frac{\binom{b-g}{n}}{\binom{g}{x}} F[-n, -g; b-g-n+1; \exp(iu)],$$ (A1.29)

which

$$F(\alpha, \beta; \gamma; z) = 1 + \frac{\alpha\beta}{\gamma} + \frac{\alpha(\alpha+1)}{\gamma(\gamma+1)}\frac{z^2}{2!} + \frac{\alpha(\alpha+1)(\alpha+2)}{\gamma(\gamma+1)(\gamma+2)}\frac{z^3}{3!} + \cdots$$ (A1.30)

Uses and remarks   Classically, drawing of balls from urns. This also models drawings of defectives from a lot. Also used in discussion of exchangeability; trials without replacement are not independent but are exchangeable.

A1.1.7 Poisson distribution

Introduced in Section 3.2.6 and in Equations (3.13) and (3,14).

pmf $\mathbf{Pr}\,(X = x) = p_X\,(x) = p\,(x, v) = \dfrac{\exp(-v)\,v^x}{x!}$ (A1.31)

$$\text{for } x = 0, 1, 2, 3, \ldots,$$

with

$$v = \lambda t$$ (A1.32)

for a process with constant rate; or

$$v = \int_0^t \lambda(\tau)\,d\tau$$ (A1.33)

without.

Mean and variance

$$\mu_X = v,$$ (A1.34)

$$\sigma_X^2 = v.$$ (A1.35)

Characteristic function   Equation (7.107):

$$\phi_X(u) = \exp\left\{-v\left[1 - \exp(iu)\right]\right\}.$$ (A1.36)

Uses and remarks   A very large number of uses; see Section 3.2.6, such as the distribution of particles or flaws in a fluid or solid; arrivals of telephone calls, computer message, traffic, ships, aircraft, accidents, icebergs at a given position; radioactive decay; errors and misprints; failures; occurrence of storms and floods.

A1.2   Continuous distributions

A1.2.1   Uniform distribution

Introduced in Section 3.3; Equation (3.17) considers the 'equally frangible rod'.

$$pdf \quad f_X(x) = \frac{1}{a}, 0 \le x \le a. \tag{A1.37}$$

$$f_X(x) = 1, 0 \le x \le 1, \tag{A1.38}$$

$$f_X(x) = \frac{1}{b-a}, a \le x \le b, \tag{A1.39}$$

$$f_X(x) = \frac{1}{2}, -1 \le x \le 1. \tag{A1.40}$$

$$cdf \quad F_X(x) = \frac{x}{a}, \tag{A1.41}$$

$$F_X(x) = x, \tag{A1.42}$$

$$F_X(x) = \frac{x-a}{b-a}, \tag{A1.43}$$

$$F_X(x) = \frac{x+1}{2} \text{ for the above four cases, respectively.} \tag{A1.44}$$

Mean and variance   For the four cases, respectively,

$$\mu_X = \frac{a}{2}, \frac{1}{2}, \frac{a+b}{2}, 0, \tag{A1.45}$$

$$\sigma_X^2 = \frac{a^2}{12}, \frac{1}{12}, \frac{(b-a)^2}{12}, \frac{1}{3}. \tag{A1.46}$$

Characteristic function   For the four cases, respectively,

$$\phi_X(u) = \frac{\exp(iua) - 1}{aiu}, \tag{A1.47}$$

$$\phi_X(u) = \frac{\exp(iu) - 1}{iu} \left(= \frac{\exp\left(\frac{iu}{2}\right)\sin\left(\frac{u}{2}\right)}{\frac{u}{2}} \right), \tag{A1.48}$$

$$\phi_X(u) = \frac{\exp(iub) - \exp(iua)}{(b-a)iu}, \tag{A1.49}$$

$$\phi_X(u) = \frac{\exp(iu) - \exp(-iu)}{2iu}. \tag{A1.50}$$

Exercise 7.17 posed the question of showing that the *cf* for a random quantity with a uniform distribution on [0, 1] is given by (7.209), repeated as (A1.48).

Uses and remarks   The measure X could be over distance, area or volume. The collisions of objects on earth, missiles, meteorites from space; many situations require a uniform distribution spatially. In simulation, uniform distributions can be used to generate random values, and subsequently transformed using the integral transform theorem: see Example 7.2 and Chapter 11, Section 11.5.4.

A1.2.2   Triangular distribution

In the shape of a triangle, commencing at a, peaking at c, and diminishing to zero at b.

$$pdf \quad f_X(x) = \frac{2(x-a)}{(c-a)(b-a)} \quad \text{for } a \leq x \leq c,$$

$$= \frac{2(b-x)}{(b-c)(b-a)} \quad \text{for } c \leq x \leq b. \tag{A1.51}$$

Mean and variance

$$\mu_X = \frac{a+b+c}{3}, \tag{A1.52}$$

$$\sigma_X^2 = \frac{a^2 + b^2 + c^2 - ab - ac - bc}{18}. \tag{A1.53}$$

Characteristic function   For a distribution symmetrical about the origin where it peaks, in the range $[-a, a]$;

$$\phi_X(u) = \frac{2[1 - \cos(au)]}{a^2 u^2}. \tag{A1.54}$$

Uses and remarks   This is the sum of two independent uniformly distributed random quantities; see Section 7.5.1. It is a rough-and-ready distribution with a peak at the centre; used in PERT.

A1.2.3   Beta distribution

This distribution usually covers the $[0, 1]$ space but can be transformed to cover $[a, b]$.

$$pdf \quad f_X(x) = \frac{x^{r-1}(1-x)^{n-r-1}}{B(r, n-r)}, \quad 0 \leq x \leq 1, \tag{A1.55}$$

where

$$B(r, n-r) = \frac{\Gamma(r)\Gamma(n-r)}{\Gamma(n)}, \quad \text{where } r, n > 0. \tag{A1.56}$$

Mean and variance   For the shifted version

$$\mu_X = \frac{r}{n}, \tag{A1.57}$$

$$\sigma_X^2 = \frac{r(n-r)}{n^2(n+1)}. \tag{A1.58}$$

Characteristic function

$$\phi_X(u) = M(r; n; iu), \tag{A1.59}$$

where $M(\alpha; \beta; z)$ is the confluent hypergeometric function

$$M(\alpha; \beta; z) = 1 + \frac{\alpha\beta}{\gamma}\frac{z}{1!} + \frac{\alpha(\alpha+1)}{\gamma(\gamma+1)}\frac{x^2}{2!} + \frac{\alpha(\alpha+1)(\alpha+2)}{\gamma(\gamma+1)(\gamma+2)}\frac{z^3}{3!} + \cdots \tag{A1.60}$$

Uses and remarks   Used in several places in Chapter 8; see (8.21) for example. The uniform distribution is a special case, as is the arcsine distribution. The beta distribution serves as the conjugate prior distribution for certainty in the parameter of the binomial distribution. Used in various studies requiring a distribution over $[a, b]$. PERT is one such application.

A1.2.4   Exponential distribution

Introduced in Section 3.3.2; see Equation (3.23); also (7.12).

pdf $f_T(t) = \lambda \exp(-\lambda t), t \geq 0,$ (A1.61)

$\qquad = 0$ for $t < 0.$

Shifted version $f_X(x) = \dfrac{1}{\beta} \exp\left(-\dfrac{x-\alpha}{\beta}\right)$ for $x \geq \alpha.$ (A1.62)

cdf $F_X(x) = 1 - \exp(-\lambda t).$ (A1.63)

Mean and variance

$\mu_X = \alpha + \beta,$ (A1.64)

$\sigma_X^2 = \beta^2.$ (A1.65)

Characteristic function   For

$f_X(x) = \dfrac{1}{\beta} \exp(-x/\beta)$ (A1.66)

the *cf* is as follows:

$\phi_X(u) = (1 - iu\beta)^{-1}.$ (A1.67)

Uses and remarks   There are numerous uses; time between events in Poisson process (special case of Erlang distribution). The exponential distribution forms the basis of extremal analysis; tails of most distributions tend to exponential form, some rather slowly.

A1.2.5   Laplace (double-exponential) distribution

It should be noted that the Gumbel distribution of extremes is sometimes also referred to as the double-exponential distribution, for quite different reasons. In engineering, the Gumbel distribution usually has this designation.

pdf $f_X(x) = \dfrac{1}{2\beta} \exp\left(-\dfrac{|x-\alpha|}{\beta}\right)$ for $-\infty < x < \infty, -\infty < \alpha < \infty, \beta > 0.$ (A1.68)

Mean and variance

$\mu_X = \alpha,$ (A1.69)

$\sigma_X^2 = 2\beta^2.$ (A1.70)

Characteristic function

$$\phi_X(u) = \frac{\exp(iu\alpha)}{1 + \beta^2 u^2}. \tag{A1.71}$$

Uses and remarks Like the exponential distribution, only spread out symmetrically, on both sides about the mean, ranging to $+\infty$ and $-\infty$. See Example 7.7 in which failure probability is discussed, giving a variation of this distribution.

A1.2.6 Gamma distribution

Used in Chapters 7 and 8. See Example 7.20 and Equation (7.123).

pdf $\quad f_X(x) = \dfrac{x^{\alpha-1}\exp(-x/\beta)}{\beta^\alpha\Gamma(\alpha)}, \quad$ for $\quad x \geq 0, \alpha, \beta > 0.$ $\hfill$ (A1.72)

Mean and variance

$$\mu_X = \alpha\beta, \tag{A1.73}$$
$$\sigma_X^2 = \beta^2\alpha. \tag{A1.74}$$

Characteristic function

$$\phi_X(u) = (1 - iu\beta)^{-\alpha}. \tag{A1.75}$$

Uses and remarks This distribution models a random quantity in the space $[0, \infty)$, but can be shifted to start at $x = a$; see Equation (7.22). Conjugate prior for scale parameter; see Sections 8.5.3 and 8.6.2. The sum Y of two gamma random quantities X_1 (parameters: α_1, β) and X_2 (parameters: α_2, β) that are independent is another gamma density with parameters $(\alpha_1 + \alpha_2, \beta)$. The parameter β is the scale parameter, while $(\alpha_1 + \alpha_2)$ is the shape parameter resulting from the original two shape parameters α_1 and α_2. It is also said that the family of gamma distributions is 'closed under convolution'. The Erlang distribution is a special case; see Exercise 3.34.

A1.2.7 Chi-square distribution

The chi-square distribution is a special case of the gamma distribution with shape parameter $\nu/2$ and a scale parameter of 2.

pdf The chi-square distribution with ν degrees of freedom is written as

$$f_Y(y) = \frac{y^{\frac{\nu}{2}-1}\exp\left(-\frac{u}{2}\right)}{2^{\frac{\nu}{2}}\Gamma\left(\frac{\nu}{2}\right)}, \quad \text{for} \quad y \geq 0; \quad \nu \text{ represents the degrees of freedom.} \tag{A1.76}$$

Mean and variance

$$\mu_Y = \nu, \tag{A1.77}$$
$$\sigma_Y^2 = 2\nu. \tag{A1.78}$$

Characteristic function

$$(1 - 2iu)^{\frac{\nu}{2}}.$$ (A1.79)

Uses and remarks   This distribution is used extensively in inference regarding the variance and other scale parameters; see Sections 8.5.3 and 8.6.2. The sums of squares of independent normal random quantities is chi-square. For the sum of three of these squares, the square of the speed of particles can be found, from which the distribution of the speed itself can be deduced, see Feller (1971, p. 48).

A1.2.8   Rayleigh distribution

Introduced in Equation (7.48).

$$pdf \quad f_X(x) = \frac{x}{\sigma^2} \exp\left(-\frac{x^2}{2\sigma^2}\right) \quad \text{for} \quad x \geq 0, \sigma > 0.$$ (A1.80)

$$cdf \quad F_X(x) = 1 - \exp\left(-\frac{x^3}{2\sigma^2}\right).$$ (A1.81)

Mean and variance

$$\mu_X = \sigma\sqrt{\frac{\pi}{2}},$$ (A1.82)

$$\sigma_X^2 = \sigma^2\left(2 - \frac{\pi}{2}\right).$$ (A1.83)

The mode is at σ.   (A1.84)

Uses and remarks   This gives the amplitude in a Gaussian process; also a wave height distribution. It is a special case of the Weibull distribution.

A1.2.9   Inverted gamma distribution

If Y has a gamma distribution, then $(1/Y)$ has an inverted gamma distribution. The symbol α below corresponds to that in the gamma distribution, with θ corresponding to $1/\beta$ there.

$$pdf \quad f_X(x) = \frac{\theta^\alpha}{\Gamma(\alpha)} x^{-(\alpha+1)} \exp\left(-\frac{\theta}{x}\right) \quad \text{for} \quad x > 0.$$ (A1.85)

Mean and variance

$$\mu_X = \frac{\theta}{\alpha - 1}, \alpha > 1,$$ (A1.86)

$$\sigma_X^2 = \frac{\theta^2}{(\alpha - 1)^2(\alpha - 2)}, \alpha > 2.$$ (A1.87)

The mode is at $\dfrac{\theta}{\alpha + 1}$.   (A1.88)

Uses and remarks   The main use is in inference; we saw in Exercise 8.24 that this was needed in the proper Bayesian analysis of uncertainty in variance. See also Bernado and Smith (1994). The inverted chi-square distribution is the inverted gamma distribution with $\alpha = \nu/2$ and $\theta = 1/2$.

A1.2.10 Logistic distribution

Used in Example 7.1, Equation (7.5).

$$\textit{pdf} \quad f_X(x) = \frac{\exp\left(-\frac{x-\alpha}{\beta}\right)}{\beta\left[1+\exp\left(-\frac{x-\alpha}{\beta}\right)\right]^2}, \quad \text{for} \quad -\infty < x < \infty, -\infty < \alpha < \infty, \beta > 0. \tag{A1.89}$$

$$\textit{cdf} \quad F_X(x) = \frac{1}{1+\exp\left(-\frac{x-\alpha}{\beta}\right)} = \frac{1}{2}\left[1+\tanh\left(\frac{x-\alpha}{2\beta}\right)\right]. \tag{A1.90}$$

Mean and variance

$$\mu_X = \alpha, \tag{A1.91}$$

$$\sigma_X^2 = \beta^2 \frac{\pi^2}{3}. \tag{A1.92}$$

Characteristic function

$$\phi_X(u) = \exp(\alpha iu)\pi\beta iu \, \csc(\pi\beta iu). \tag{A1.93}$$

Uses and remarks   This is a symmetrical bell-shaped curve but with more probability in the tails than the normal distribution. It is used for growth process and population theory. But see Feller (1971, Vol. II, pp. 51–52).

A1.2.11 Cauchy distribution

The Witch of Agnesi! Used in Exercise 3.14, Equation (3.136).

$$\textit{pdf} \quad f_X(x) = \frac{a}{\pi(x^2+a^2)}, \text{ for } -\infty < x < \infty, \, \alpha > 0. \tag{A1.94}$$

Shifted version:

$$f_X(x) = \frac{1}{\pi\beta\left[1+\left(\frac{x-\alpha}{\beta}\right)^2\right]}, \quad \text{for } -\infty < x < \infty, \, -\infty < \alpha < \infty, \beta > 0. \tag{A1.95}$$

cdf   For the form (A1.94),

$$F_X(x) = \frac{1}{\pi}\tan^{-1}\left(\frac{x}{a}\right) + \frac{1}{2}. \tag{A1.96}$$

Mean and variance   Moments do not exist. By symmetry, $\mu_X = \alpha$ for form (A1.95). Integration shows $\sigma_X = \infty$ (Exercise 3.14).

The mode and median are α.   (A1.97)

Characteristic function

For form (A1.95), $\phi_X(u) = \exp(iu\alpha - \mid u \mid \beta)$.   (A1.98)

Uses and remarks   This distribution can be derived by transformation of a random quantity. Consider a light ray projected on a wall. If the angle describing the direction of the ray is uniform in $(-\pi/2, \pi/2)$, the position where the ray intersects the wall is given by a Cauchy distribution; see Feller (1971) and Tribus (1969). It is a symmetrical bell-shaped distribution with fat tails, as compared to the normal distribution. It has rather pathological aspects, such as infinite variance. It is closed under convolution; the sum of n iid Cauchy random quantities has the same distribution as any of the members! See Feller (1971, p. 51). Also found by putting $v = 1$ in student's t-distribution.

A1.2.12 Pareto distribution

See also Exercise 3.15 and Equation (3.137).

pdf $f_X(x) = \dfrac{kt^k}{x^{k+1}}$ for $x \geq t; = 0$ for $x < t, k > 0$.   (A1.99)

cdf $F_X(x) = 1 - \left(\dfrac{t}{x}\right)^k$.   (A1.100)

Mean and variance

$$\mu_X = \frac{kt}{k-1}, \ k > 1,$$ (A1.101)

$$\sigma_X^2 = \frac{kt^2}{(k-1)^2(k-2)}, \ k > 2.$$ (A1.102)

The mode is at t.   (A1.103)

The median is $t(2)^{\frac{1}{k}}$.   (A1.104)

Uses and remarks   Has been used for distribution of incomes above a threshhold. It has rather fat tails.

A1.2.13 Normal distribution

Introduced in Section 3.3.3; see Equation (3.25). Also in Chapter 6 as maximum entropy distribution, and in Chapter 7 in the Central Limit Theorem.

pdf $f_X(x) = \dfrac{1}{\sqrt{2\pi}\sigma} \exp -\dfrac{1}{2}\left(\dfrac{x-\mu}{\sigma}\right)^2, \ \sigma > 0$.   (A1.105)

Mean and variance

$$\mu_X = \mu, \tag{A1.106}$$
$$\sigma_X^2 = \sigma^2. \tag{A1.107}$$

Characteristic function   For the standard form

$$f_Z(z) = (1/\sqrt{2\pi})\,\exp(-z^2/2), \tag{A1.108}$$

the *cf* is given by (7.109)

$$\phi_Z(u) = \exp\left(\frac{-u^2}{2}\right). \tag{A1.109}$$

Uses and remarks   This is widely used in many applications. The 'model of sums', in modelling errors of measurements, and other quantities that are the sum of many others.

A1.2.14 Lognormal distribution

If a random quantity $X = \ln Y$ is normally distributed (μ, σ^2), then Y is said to be lognormally distributed (μ, σ^2).

$$pdf \quad f_Y(y) = \frac{1}{\sigma\sqrt{2\pi}}y^{-1}\exp\left[-\frac{(\ln y - \mu)^2}{2\sigma^2}\right], \text{ for } y > 0. \tag{A1.110}$$

Mean and variance   Let $a = \exp\mu$, $b = \exp(\sigma^2)$:

$$\mu = a\sqrt{b} = \exp\left(\mu + \frac{\sigma^2}{2}\right), \tag{A1.111}$$
$$\sigma_Y^2 = a^2 b(b-1) = \exp(2\mu + \sigma^2)[\exp(\sigma^2) - 1]. \tag{A1.112}$$

The mode is at

$$\exp(\mu - \sigma^2). \tag{A1.113}$$

The median is at

$$\exp\mu. \tag{A1.114}$$

Uses and remarks   Model of products, analogously to the normal distribution as the model of sums. Often used for resistance of structural components.

A1.2.15 Student's t-distribution

Used in the expression (8.111) and in similar situations.

$$pdf \quad f_X(x) = \frac{\Gamma\left(\frac{v+1}{2}\right)}{\Gamma\left(\frac{v}{2}\right)\sqrt{v\pi}}\left(1 + \frac{x^2}{v}\right)^{-\frac{1}{2}(v+1)} \tag{A1.115}$$

$$= \frac{1}{\sqrt{v}B\left(\frac{1}{2}, \frac{v}{2}\right)}\left(1 + \frac{x^2}{v}\right)^{-\frac{1}{2}(v+1)}. \tag{A1.116}$$

The parameter v is termed the degress of freedom.

Mean and variance

$$\mu_X = 0 \text{ for } v > 1, \tag{A1.117}$$

$$\sigma_X^2 = \frac{v}{v-2}. \tag{A1.118}$$

Uses and remarks   This is widely used in inference regarding means where the variance is also uncertain.

A1.2.16 *F*-distribution

Used in Section 8.7.6 under 'Common classical inference problems'; also Section 11.3.3 and Exercise 11.13.

$$\textit{pdf} \quad f_X(x) = \frac{\left(\frac{v_1}{v_2}\right)^{\frac{v_1}{2}} x^{(v_1-2)}}{B\left(\frac{v_1}{2}, \frac{v_2}{2}\right)} \left(1 + \frac{v_1}{v_2}x\right)^{(v_1+v_2)/2} \quad \text{for } B > 0, v_1, v_2 = 1, 2, \ldots \tag{A1.119}$$

The parameters v_1 and v_2 are the degress of freedom for the numerator and denominator respectively.

Mean and variance

$$\mu_X = \frac{v_2}{v_2 - 2}, \quad v_2 > 2, \tag{A1.120}$$

$$\sigma_X^2 = \frac{2v_2^2(v_1 + v_2 - 2)}{v_1(v_2 - 2)^2(v_2 - 4)}, \quad v_2 > 4. \tag{A1.121}$$

A1.2.17 Extremal – Gumbel

Also known as the extreme value type I, or the double-exponential distribution. (Note that the Laplace distribution also has this name.) For maxima, see Equation (7.13) and Chapter 9. The following is for maxima.

$$\textit{pdf} \quad f_Z(z) = \frac{1}{b} \exp\left[-\exp\left(-\frac{z-a}{b}\right) - \frac{z-a}{b}\right] \text{ for } -\infty < z < +\infty, b > 0. \tag{A1.122}$$

$$\textit{cdf} \quad F_Z(z) = \exp\left[-\exp\left(-\frac{z-a}{b}\right)\right]. \tag{A1.123}$$

Mean and variance

$$\mu_Z = a + b\gamma, \text{ where } \gamma \text{ is the Euler constant} = 0.5772\ldots, \tag{A1.124}$$

$$\sigma_Z^2 = \frac{\pi^2 b^2}{6}. \tag{A1.125}$$

The mode is at a. $\tag{A1.126}$

The median is at $a - b \ln(\ln 2)$. $\tag{A1.127}$

Characteristic function

$$\exp(iau)\Gamma(1 - ibu). \tag{A1.128}$$

Uses and remarks This is used extensively in extremal analysis, and is the most fundamental of the limiting extremal distributions. See the limit theorem in Chapter 9.

A1.2.18 Extremal – Fréchet

Also known as the extreme value type II distribution. See Equations (9.78) and (9.115). The following is for maxima.

$$pdf \quad f_Z(z) = \frac{\alpha}{b_n}\left(\frac{z - a_n}{b_n}\right)^{-\alpha-1} \exp\left[-\left(\frac{z - a_n}{b_n}\right)^{-\alpha}\right], \text{ for } \alpha > 0 \text{ and } -\infty < z < \infty. \tag{A1.129}$$

$$cdf \quad F_Z(z) = \exp\left[-\left(\frac{z - a_n}{b_n}\right)^{-\alpha}\right], \quad \text{for } \alpha > 0 \quad \text{and} \quad -\infty < z < \infty. \tag{A1.130}$$

Mean and variance

$$\mu_Z = a_n + b_n\Gamma\left(1 - \frac{1}{\alpha}\right) \quad \text{for } \alpha > 1, \tag{A1.131}$$

$$\sigma_Z^2 = b_n^2\left[\Gamma\left(1 - \frac{2}{\alpha}\right) - \Gamma^2\left(1 - \frac{1}{\alpha}\right)\right] \quad \text{for } \alpha > 2. \tag{A1.132}$$

The median is $a_n + b_n(\ln 2)^{-\frac{1}{\alpha}}$. \tag{A1.133}

Uses and remarks This is used in analysis of extremes. It is derived from pathological parents and has fat tails.

A1.2.19 Extremal – Weibull

Introduced in Equation (9.79) and used in Section 9.7 for weakest-link theories. See also (9.119). This is given here for minima; the maximum is given in Equation (9.116).

$$pdf \quad f_W(w) = \frac{\alpha}{b_n}\left(\frac{w - a}{b_n}\right)^{\alpha-1} \exp\left[-\left(\frac{w - a}{b_n}\right)^{\alpha}\right] \text{ for } a \leq w < \infty, \alpha > 0, b_n > 0. \tag{A1.134}$$

Generally a is a physical limit.

$$cdf \quad F_W(w) = 1 - \exp\left[-\left(\frac{w - a}{b_n}\right)^{\alpha}\right], \quad \text{for } a \leq w < \infty. \tag{A1.135}$$

Mean and variance

$$\mu_W = a + b_n\Gamma\left(1 + \frac{1}{\alpha}\right), \tag{A1.136}$$

$$\sigma_W^2 = b_n^2\left[\Gamma\left(1 + \frac{2}{\alpha}\right) - \Gamma^2\left(1 + \frac{1}{\alpha}\right)\right]. \tag{A1.137}$$

Mode $a + b_n \left(\dfrac{\alpha - 1}{\alpha} \right)^{\frac{1}{\alpha}}$ for $\alpha > 1$. $\qquad$ (A1.138)

Median $a_n + b_n (\ln 2)^{\frac{1}{\alpha}}$. $\qquad$ (A1.139)

Uses and remarks   Used extensively in extremal analysis where there is a natural maximum or minimum value, for example a minimum strength.

A1.2.20 Two-dimensional normal distribution

pdf $f_{X,Y}(x, y) = \dfrac{1}{2\pi \sigma_1 \sigma_2 \sqrt{1 - \rho^2}} \exp - \left\{ \dfrac{1}{2(1 - \rho^2)} \right.$

$$\times \left[\frac{(x - \mu_X)^2}{\sigma_X^2} - \frac{2\rho(x - \mu_X)(y - \mu_Y)}{\sigma_X \sigma_Y} + \frac{(y - \mu_Y)^2}{\sigma_Y^2} \right] \right\}. \qquad (A1.140)$$

Uses and remarks   Basic form with two random quantities. See also Example 7.9 and (7.69).

A1.2.21 Multidimensional normal distribution

See Example 7.9, Equations (7.68) and (7.69).

$$f_Y(\mathbf{y}) = \frac{1}{(2\pi)^{(n/2)}} |\Sigma|^{-1/2} \exp \left[-\frac{1}{2} (\mathbf{y} - \boldsymbol{\mu})^T [\Sigma]^{-1} (\mathbf{y} - \boldsymbol{\mu}) \right]. \qquad (A1.141)$$

The matrix $[\Sigma]$ is the covariance matrix

$$[\Sigma] = \begin{bmatrix} \sigma_1^2 & \rho_{12}\sigma_1\sigma_2 & \cdots & \rho_{1j}\sigma_1\sigma_j & \cdots & \rho_{1n}\sigma_1\sigma_n \\ \rho_{21}\sigma_2\sigma_1 & \sigma_2^2 & \cdots & \rho_{2j}\sigma_2\sigma_j & \cdots & \rho_{2n}\sigma_2\sigma_n \\ \vdots & \vdots & & \vdots & & \vdots \\ \rho_{i1}\sigma_i\sigma_1 & \rho_{i2}\sigma_i\sigma_2 & \cdots & \rho_{ij}\sigma_i\sigma_j & \cdots & \rho_{in}\sigma_i\sigma_n \\ \vdots & \vdots & & \vdots & & \vdots \\ \rho_{n1}\sigma_n\sigma_1 & \rho_{n2}\sigma_n\sigma_2 & \cdots & \rho_{nj}\sigma_n\sigma_j & \cdots & \sigma_n^2 \end{bmatrix}, \qquad (A1.142)$$

where the σs are the standard deviations and the ρs are the correlation coefficients.

Appendix 2   Mathematical aspects

These integrals provide a neat way to consider integrations and sums under a single umbrella. We consider generally integrals of the kind $\int_a^b g(x)f_X(x)dx$. We write this as $\int h(x)dx$ for the time being, with $h(x) = g(x)f_X(x)$. The definition of the Riemann integral,

$$\int_a^b h(x)dx, \tag{A2.1}$$

will be given for a bounded function $h(\cdot)$, in the range $a \leq x \leq b$, say. This range is divided into a set of division points

$$a = x_0 < x_1 < x_2 < x_3 < \cdots < x_n = b. \tag{A2.2}$$

Two bounds on the integral are formed. These are named after the French mathematician Gaston Darboux. For every interval $[x_{i-1}, x_i]$ let M_i and m_i denote the maximum and minimum value of $h(\cdot)$ in the interval. Then the following two (upper and lower) Darboux sums constitute an upper and lower bound to the integral:

$$\sum_{i=1}^n M_i(x_i - x_{i-1}), \ \sum_{i=1}^n m_i(x_i - x_{i-1}). \tag{A2.3}$$

The length of the largest subinterval is denoted δ; when this becomes smaller, the number of subintervals becomes larger. The upper Darboux sum will decrease or stay unchanged in value; conversely the lower sum will increase or stay the same. The function $h(x)$ has a Riemann integral if the two sums converge to the same value.

In order to include 'jumps' in the function $h(x) = g(x)f_X(x)$, corresponding in the present case to the 'spikes' in a *pmf*, a weight function in the Darboux sums is included. This is simply the *cdf* $F_X(x)$, so that the function $[F_X(x_i) - F_X(x_{i-1})]$ is considered in each Darboux sum as follows:

$$\sum_{i=1}^n M_i'[F_X(x_i) - F_X(x_{i-1})], \ \sum_{i=1}^n m_i'[F_X(x_i) - F_X(x_{i-1})], \tag{A2.4}$$

where M_i' and m_i' are the maximum and minimum, respectively, of $g(x)$ in the interval $[x_{i-1}, x_i]$.

In the limit when the largest subinterval δ tends to zero, the integral is written as

$$\int_a^b g(x)\mathrm{d}F_X(x). \tag{A2.5}$$

This integral is termed the Stieltjes integral and includes 'jumps' in $F_X(x)$ because of the manner in which the Darboux sums are derived. It therefore combines summations and integrals into one expression. It should be noted that if $F_X(x)$ is differentiable, and since $f_X(x) = \mathrm{d}F_X(x)/\mathrm{d}x$, then Equation (A2.5) becomes

$$\int g(x)f_X(x)\mathrm{d}x.$$

The Riemann integral is therefore a special case of the Stieltjes integral.

An example of the use of Stieltjes integrals, which is very close to the present application, is the calculation of the moments of a group of particles, or 'point masses'. Considering the moment of inertia of n particles,

$$I = \sum_{i=1}^n \mu_i x_i^2; \tag{A2.6}$$

where μ_i is the ith mass and x_i is the distance from axis under consideration. If the particles are small and numerous, we may wish to use expressions of the kind

$$I = \int \rho(x)x^2\mathrm{d}x, \tag{A2.7}$$

where $\rho(x)$ is the mass per unit distance x. Expressions (A2.6) and (A2.7) can be combined in the Stieltjes integral

$$I = \int x^2\mathrm{d}\mu(x), \tag{A2.8}$$

where $\mathrm{d}\mu(x) = \rho(x)\mathrm{d}x$ in the continuous case and corresponds to the jumps μ_i for concentrated masses. Stieltjes integrals have widespread application in engineering and contain a convenient shorthand notation, as illustrated above.

A2.1.1 Dirac delta and Heaviside step functions

These are commonly used in electrical circuit analysis, viscoelastic theory, and in other areas of engineering and physics, and provide an alternative way of considering 'jumps' in functions. The unit step function is defined by the two equations:

$$\Delta(x) = 0 \text{ for } x < 0,$$
$$\Delta(x) = 1 \text{ for } x \geq 0. \tag{A2.9}$$

This function is illustrated in Figure A2.1(a), in which a small interval τ is shown to the left of

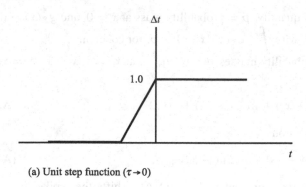

(a) Unit step function ($\tau \to 0$)

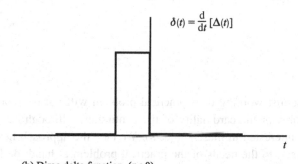

$$\delta(t) = \frac{\mathrm{d}}{\mathrm{d}t}[\Delta(t)]$$

(b) Dirac delta function ($\tau \to 0$)

Figure A2.1   Jumps and spikes

the origin $t = 0$. The definition of Equations (A2.9) can be conceived as a limit of the function of Figure A2.1(a) with $\tau \to 0$.

The derivative of the unit step function is a highly singular function called the Dirac delta function, defined by the following conditions:

$$\delta(t) = 0 \text{ for } t \neq 0$$
$$\delta(t) = \infty \text{ for } t = 0$$
$$\int_{-\infty}^{\infty} \delta(t)\mathrm{d}t = \int_{0^-}^{0^+} \delta(t)\mathrm{d}t = 1. \tag{A2.10}$$

This function is illustrated in Figure A2.1(b); in the limit as $\tau \to 0$ the Dirac function becomes a 'spike' of mass unity at the origin. (As pointed out by Butkov (1968), a more correct form for the definition of $\delta(t)$ should be based on delta sequences which give the same shifting property.)

Probability distributions such as the one illustrated in Figure 3.10 can be written as

$$f_X(x) = (1 - p)\delta(x) + g_X(x) \tag{A2.11}$$

in the case where X = random quantity, p = probability mass at $x = 0$, and $g_X(x)$ = distribution of probability for $x > 0$, with $\int_{x>0} g_X(x)\mathrm{d}x = 1 - p$, for coherence.

We could denote a set of probability masses $p_1, p_2, p_3, \ldots$ at $x = x_1, x = x_2, x = x_3, \ldots$ by means of the function

$$f_X(x) = p_1\delta(x - x_1) + p_2\delta(x - x_2) + p_3\delta(x - x_3) + \cdots, \tag{A2.12}$$

with cumulative distribution function

$$F_X(x) = p_1\Delta(x - x_1) + p_2\Delta(x - x_2) + p_3\Delta(x - x_3) + \cdots \tag{A2.13}$$

The insertion of $(x - x_i)$ as the argument in $\delta(\cdot)$ and $\Delta(\cdot)$ shifts the 'spike' or 'jump' respectively to $x = x_i$. We see that Equations (A2.12) and (A2.13) can be used to include 'spikes' and 'jumps'.

A2.2   Integration

The engineer, scientist or economist working on a practical problem will not be concerned with special sets such as Cantor's, or the cardinality of the continuum (although he might, as in the case of the writer, be interested in these subjects). We take the approach here that mathematics should conform itself to the needs of the practical problem at hand; de Finetti (1972) has made detailed studies of the mathematical implications of various assumptions. His analysis goes much further than our needs here. In short, there is a question of the kind of integral to be used in dealing with certain kinds of situations. For example, if we define a density on the space [0, 1] such that it takes certain values on the rational numbers, a countably infinite set, with other values on the irrationals, which have the cardinality of the continuum, special techniques have to be used for the integration of such a function. Feller (1971, p. 34) shows (as one would expect) that the set of all rationals have a probability of zero, if the probability is distributed over all numbers. The way around the problem of 'different values on the rationals and the irrationals' is the use of Borel sets and Lebesgue integration. Thus the probability associated with a set A is associated with the Lebesgue measure of the set A. This requires the acceptance of complete (or σ-) additivity, i.e. one allows a countably infinite number of intersections and unions of sets, rather than a finite number. This point has caused some controversy amongst mathematicians, with de Finetti opposing σ-additivity as an axiom. For example, he cites the case of an infinite number of possibilities, such as the set of integers. One is chosen 'at random', say X, and therefore $\mathbf{Pr}(X = 1, 2, 3, \ldots$ or any other integer$) = 0$. The sum of these values is also zero, yet the probability of the union, the certain event that one of the numbers occur, is unity. In this case, σ-additivity does not hold. Details of the arguments can be found in de Finetti (1972, and 1974, Vol. II – especially the appendix in the latter, Section 18).

The view is taken here that in studying practical problems requiring integration, one never encounters 'pathological' cases such as the one where a density is defined differently on the rationals and on the irrational numbers. We therefore consider Lebesgue integration to be interesting but by no means necessary for the study of probability theory. Indeed, de Finetti, while excluding σ-additivity as an axiom, states (de Finetti, 1972, p. 105) that one might nevertheless 'be interested in studying those particular distributions which happen to be completely additive on some domain'. This is quite enough for our purposes, and we can therefore study distributions that include an infinite number of possibilities, a simple example being the Poisson distribution.

We have dealt with Stieltjes integrals above as they have the attractive possibility of unifying the treatment of discrete and continuous random quantities.

A2.3   Solution to Lewis Carroll's problem and discussion

The answer to the problem as given by Charles Dodgson goes along the following lines. The rod is divided into $(n + 1)$ parts, where n is odd. These n points are considered to be the only possible division points, that is, the division points are the only points at which the rod will break. If $n = 5$, then there are six intervals, and we are considering the probability that the rod will break at the central one of these. The probability in five trials of breaks at the outer points, other than the centre, is $(4/5)^5$; and at the centre the probability is then $\left[1 - \left(\frac{4}{5}\right)^5\right]$. It is assumed in this that the number of rods tested is equal to n, the number of possible division points. The probability of a central break is

$$p_c = 1 - \left(1 - \frac{1}{n}\right)^n,$$

(A2.14)

which tends to $(1 - 1/e) = 0.6321\ldots$ as n tends to infinity. (We have assumed independence with regard to the probabilities of successive breaks in the outer and central regions.)

The first point of discussion is the linking of the number of rods to the number of division points. Had we tested $2n, 3n, \ldots$ rods each with n division points, we would have obtained

$$\left(1 - \frac{1}{e^2}\right), \left(1 - \frac{1}{e^3}\right), \ldots$$

(A2.15)

If we understand the question to mean that we wish to find the probability of one of the infinity of rods failing at exactly $x = 1/2$, we should first decide on what we mean by 'infinity'. Let us continue to assume that it means a countably infinite number of rods, that is, corresponding to the integers $\{1, 2, 3, \ldots\}$. If the possible locations of a break correspond to all the numbers, rational and irrational, then the probability of a break at $x = 1/2$ reduces to zero. In reality, a break, however clean, will correspond to a certain interval. A proper definition of 'what is possible' should be made at the outset in any study. This would likely lead to a finite set corresponding to certain defined intervals; if there are m possible points, for our 'equally frangible rod' the probability of a break in a particular one of these is $1/m$, similar to Dodgson's assumption.

Appendix 3 Answers and comments on exercises

Chapter 2

2.1 5/6.

2.2 3/7.

2.3 1:2 on.

2.4 Sum of qs is 49/40 – incoherent.

2.5 Sum = 1.50 – incoherent; \$50; 29:1, 14:1, 1:1, 13:2, all against.

2.6 See de Finetti (1937).

2.12 $E_1 \vee E_2 \vee E_3 = E_1 + E_2 + E_3 - E_1 E_2 - E_1 E_3 - E_2 E_3 + E_1 E_2 E_3$.

2.13 (b) $A \cap B$, 1. (c) $B \cap C = C$. (d) $B \cup C = B$. (e) $A \cap C = \varnothing$. (f) $\tilde{A} \cap \tilde{B} \cap \tilde{C} = \{1, 9\}$. Note that '1' is not prime because it does not have exactly two factors.

2.14 (a) 0.53. (b) 0.47. (c) 0.63. (d) 0.37. (e) 0.10.

2.15 (a) 0.9966. (b) 0.0006. (c) 0.0001.

2.16 $\mathbf{Pr}(E_1) + 2\mathbf{Pr}(E_1 E_2) + \mathbf{Pr}(E_3) = 1$.

2.17 0.096.

2.18 (a) 2/5 for θ_1. (b) 1. (c) 32/35.

2.20 Probability of guilt is very high since the evidence comprises an independent series of allegations.

2.21 Yes, odds change to 2:1 on the other closed door (Bayes).

2.22 $\mathbf{Pr}(T|L) = (pt)/(pt + \tilde{q}\tilde{t})$, etc.

2.23 Probability of correctly positioned landing gear given light off is $1/1001.001 \simeq 0.001$. Poor system.

2.24 36/43, 6/43, 1/43.

2.25 (a) +2, +1, 0. (b) 9/50.

2.26 $\mathbf{Pr}(S|I = 0) = 0.989, 0.007, 0.004$; $\mathbf{Pr}(S|I = 1) = 0.784, 0.199, 0.017$, etc.

2.27 (a) 92%. (b) 0.583, 0.942.

2.28 0.9604, 0.9996.

2.29 An offshore rig has failed due to two possible causes: earthquake or hurricane. A seismic station nearby indicates earthquake unlikely. Then the probability of a hurricane increases.

2.30 (a) 63. (b) 70.

2.31 1/2.

2.32 1/4, 5040.

2.33 (a) 0.3182. (b) 0.4273. (c) 0.5091. (e) $a = b$.

2.35 (a) 50 400. (b) 16 627:33 $\simeq$ 504:1.

2.36 See Feller (1968) and Blom *et al.* (1994).

2.38 0.031 03 . . .

Chapter 3

3.1 (a) $k = 1/10$. (c) 0.7.

3.2 $k = 3/2, 2, 2$; probabilities $= 0.3125, 0.0677$ (4 dp), 1/2.

3.3 $f_T(t) = 2(t - 1), 1 \leq t \leq 2$; 0.25.

3.4 (a) 0.40. (b) 0.006 046 . . . (c) 0.0256. (d) 0.000 796 . . . (e) 0.1663. (f) 17.5 s.

3.5 (a) Hypergeometric, can be approximated by binomial. (b) 0.4095 . . . , 0.4042 . . . (exact). (c) 0.008 56. (d) Approximate 0.45, exact 0.4459 (4 dp).

3.6 See de Finetti (1974).

3.7 (a) Exponential. (b) Same exponential, shifted by τ. (c) 0.699, 0.0907 (3 sf). (d) 0.950 (3 sf).

3.8 (a) 1.67, 1.292 (3 dp). (b) 0.188 (3 dp). (c) 3.327 (3 dp).

3.9 7.52 $\times 10^{-7}$ (odds of less than one in a million).

3.10 Model as a Poisson process with rate $10 \exp(-x/2)$, 0.6077 (4 dp).

3.11 (a) 0.0618. (b) 0.6739. (c) 0.2643. (d) Assume independence, use multinomial, 0.00170.

3.12 (a) 241.2. (b) 0.04875. (c) 0.049375 $\simeq 5\%$.

3.13 1, 1, 1; 3/8, 19/320, 0.2437.

3.14 $a/\pi, \frac{1}{\pi} \tan^{-1}\left(\frac{x}{a}\right) + \frac{1}{2}, 2a, 0, \infty$.

3.15 (a) $k > n$. (b) $k > 2$, $\left(kt^2\right) / \left[(k - 1)^2 (k - 2)\right]$.

3.17 $\pi \left(\mu_R^2 + \sigma_R^2\right)$.

3.18 (a) 4/25. (b) 1/5. (c) 1/2. (d) 21/50. (e) 0.

3.19 $1 - (1 - \tau/t)^2$.

3.20 (a) $69.5/24^2 = 0.1207$ (4 dp). (b) 5/9.

3.21 (a) 0.00124 (3 sf). (b) 88%, 558.

3.22 2.

3.23 (a) $p_X(x)$: 8/27, 4/9, 2/9, 1/27, $p_Y(y)$: 1/9, 2/3, 2/9. (b) $p_{X|Y}(x, y)$ for $y = 1$: 2/3, 0, 0, 1/3, $y = 2$: 1/3, 1/3, 1/3, 0, $y = 3$: 0, 1, 0, 0; $p_{Y|X}(x, y)$ for $x = 0$: 1/4, 3/4, 0, $x = 1$: 0, 1/2, 1/2, $x = 2$: 0, 1, 0, $x = 3$: 1, 0, 0; $p_{Y|X=0}(y)$: 1/4, 3/4, 0. (c) 19/9, $\sqrt{2/3}$, $\sqrt{234}/27 \simeq$ 0.567, 19/9. (d) $\sigma_{X,Y} = 0$. (e) No. In addition to $\sigma_{X,Y} = 0$ we need $p_{X,Y}(x, y) = p_X(x) \cdot p_Y(y)$. This is found not to be true.

3.24 By integration, $f_X(x) = 2 \exp(-2x)$; $f_Y(y) = 2 \exp(-y)\left[1 - \exp(-y)\right]$; $f_{X,Y}(x, y) \neq f_X(x) \cdot f_Y(y)$; X, Y are not independent.

3.25 Proof in standard texts; $\rho_{XY} = 0$ implies absence of a linear relationship.

3.26 *EMV* for proposed option is 30.40 – greater than the others and therefore optimal.

3.27 5 units is optimal (1% growth) on assumption that 5 or 10 units are added if necessary; *EMV*s are -4.83, -4.86 and -5.08.

3.28 (a) It is optimal to test, accept if result is positive, reject if negative. $EMV = 8.22$.

3.29 $\Pr(Y|R) = 0.792, 0.151, 0.057$, for $R = 1$; $0.152, 0.810, 0.038$, for $R = 2$; $0.088, 0.118$, 0.794, for $R = 3$. Optimal decision $\{e_0, 75 \text{ m}\}$.

3.30 $\{e_0, \tilde{R}\}$ is optimal; *EMV*s 6.89 vs. 7.45.

3.31 There will be a spike of probability equal to 0.1587 at 75 kmh^{-1}.

3.32 There will only be one value of each expected value, e.g. the $\langle X_h X_i \rangle$ will be the same for each h and i.

3.34 $f_T(t) = 0.004 \, t^2 \exp(-0.2t)$.

3.35 (a) First row: 5/8, 1/8, 1/8; second row: 3/8, 1/2, 1/2; third row: 0, 3/8, 3/8. (b) 1/4, 15/32, 9/32. (c) 1, 0, 0; 5/8, 3/8, 0; 7/16, 27/64, 9/64.

3.36 (a) $\frac{1}{2}\langle A^2 \rangle \cos(\omega t)$. (b) $\sum \sigma_h^2 \cos(\omega_h t)$.

Chapter 4

4.1 Alternative 1, benefits: 31.5×10^6, costs: 28.2×10^6, ratio: 1.12. Alternative 2, benefits: 34.5×10^6, costs: 37.1×10^6, ratio: 0.93. (b) Use of subjective judgement, reduction to equivalent dollar values, ordinal scales plus judgement, interval/ratio scales and utility. (c) Very arbitrary. No risk aversion. Interest rates used can be arbitrary and give a variety of solutions.

4.2 Scenario 1, $NPW = 130$, scenario 2, $NPW = 7$ (in millions), $EMV = 105.4$.

4.3 Maximin, A; maximax, C; Hurwicz, C; Regret, B or C. Utility dominance occurs if all the utilities are greater than or equal to those for another, e.g. B $\geq$ D (strict dominance use > not $\geq$).

4.4 *EMV* of α_1 is optimal ($\$24\,000$).

4.6 If 35 000 m^3 is accepted as certainty equivalent of $\langle 0, 50\,000 \rangle$, one obtains $a = 0.036, b = 0.198, c = 6.05$.

4.7 Decibel scale is risk-prone.

4.8 $\gamma(x) = 1/(x + b)$ might be more appropriate for money than constant aversion in exponential function.

4.9 $\gamma(x) = 2a/(-2ax + b)$; increasing risk aversion.

4.10 See Keeney and Raiffa (1976).

4.12 $\sigma_1 = \{e_0, \alpha_1\}$, $\sigma_2 = \{e_0, \alpha_2\}$, $\sigma_3 = \{e_1, \alpha_1 \text{ if } R, \alpha_1 \text{ if } P\}$, $\sigma_4 = \{e_1, \alpha_1 \text{ if } R, \alpha_2 \text{ if } P\}$, $\sigma_5 = \{e_1, \alpha_2 \text{ if } R, \alpha_1 \text{ if } P\}$, $\sigma_6 = \{e_2, \alpha_1 \text{ if } R, \alpha_2 \text{ if } P\}$; σ_1 optimal if $\Pr(\theta_2)$ is in the range 0 to 0.49, σ_5 if in 0.49 to 0.65, σ_1 if in 0.65 to 1.00.

4.13 $\{e_0, \text{do not build}\}$.

4.14 (a) $\sigma_1 = \{e_0, \alpha_1\}$, $\sigma_2 = \{e_0, \alpha_2\}$, $\sigma_3 = \{e_1, \alpha_1 \text{ if } Y = 0.2, \alpha_1 \text{ if } Y = 0.4, \alpha_1 \text{ if } Y = 0.6\}$, $\sigma_4 = \{e_1, \alpha_2 \text{ if } Y = 0.2, \alpha_1 \text{ if } Y = 0.4, \alpha_1 \text{ if } Y = 0.6\}$, $\sigma_5 = \{e_1, \alpha_1 \cdots, \alpha_2 \cdots, \alpha_1 \cdots\}$, $\sigma_6 = \{e_1, \alpha_1 \cdots, \alpha_1 \cdots, \alpha_2 \cdots\}$, $\sigma_7 = \{e_1, \alpha_2 \cdots, \alpha_2 \cdots, \alpha_1 \cdots\}$, $\sigma_8 = \{e_1, \alpha_2 \cdots, \alpha_1 \cdots, \alpha_2 \cdots\}$, $\sigma_9 = \{e_1, \alpha_1 \cdots, \alpha_2 \cdots, \alpha_2 \cdots\}$,

$\sigma_{10} = \{e_1, \alpha_2 \cdots, \alpha_2 \cdots, \alpha_2 \cdots\}$; σ_2 is optimal. (b) $\sigma_2, \sigma_1, \sigma_7, \sigma_7$. (c) σ_1 optimal if $\mathbf{Pr}(M = 0.2)$ is in the range 0 to 0.33, σ_7 if in 0.33 to 0.70, σ_2 if in 0.70 to 1.00.

4.15 (b) accept, $\langle U \rangle = 9.35$. (c) reject, $\langle U \rangle = -38.6$. (d) $\sigma_1 = \{P\text{–accept}, \tilde{P}\text{–accept}\}$, $\sigma_2 = \{P\text{–accept}, \tilde{P}\text{–reject}\}$, $\sigma_3 = \{P\text{–reject}, \tilde{P}\text{–accept}\}$, $\sigma_4 = \{P\text{–reject}, \tilde{P}\text{–reject}\}$; σ_1 is optimal for $\mathbf{Pr}(S) > 0.94$, σ_2 for 0.54–0.94, σ_4 for < 0.54.

4.16 For example, a plane through three points in 3D can be written in determinantal form, and then expanded in terms of the three variables, say in the first row. The direction cosines are given by the coefficients of these variables divided by the square root of the sums of the coefficients squared.

4.17 The probability that A wins is $p_A / [1 - (1 - p_A)(1 - p_B)(1 - p_C)]$. The fairest way is $p_A : (1 - p_A) \cdot p_B : (1 - p_A) \cdot (1 - p_B) \cdot p_C$. (a) A: \$108, B: \$90, C: \$75 (b) A: \$20, B: \$40, C: \$36.

Chapter 5

5.2 $x_1 = b/3 = x_2$, $x_3 = b/6$, $V = b^3/54$.

5.3 $r = \sqrt[3]{v/2\pi}$, $\ell = 2\sqrt[3]{v/2\pi}$.

5.4 See Chapter 6.

5.5 There are four stationary points: $x_1 = 1.00$, $x_2 = x_3 = 0$, $z = 3.00$ (max); $x_1 = -1.00$, $x_2 = x_3 = 0$, $z = -1.00$ (saddle point); $x_1 = -0.50$, $x_2 = 0.87$, $x_3 = 0$, $z = -1.50$ (min); $x_1 = -0.50$, $x_2 = -0.87$, $x_3 = 0$, $z = -1.50$ (min).

5.6 $x_1 = 4$, $x_2 = 1$, $x_3 = x_4 = 0$, $\hat{x}_5 = 11$, $z^* = -36$.

5.7 $x_1 = 0$, $x_2 = -13$, $x_3 = -8$, $x_4 = 0$, $x_5 = 0$, $x_6 = 19$, $x_7 = 0$, $x_8 = 36$, $(z')^* = 54$, $z^* = -54$.

5.8 Solution is unbounded.

5.9 $x_1 = 4$, $x_2 = 10/3$, $x_3 = 2$, $z^* = -34/3$.

5.11 (a) $F_1 = 60$, $F_2 = 200$, $F_3 = 90$, $(z')^* = 350$. (b) $A = 720$ N, $B = 300$ N, $C = 0$ N; another optimal solution: $A = 620$, $B = 250$, $C = 150$.

5.14 (a) $x_1 = 2$, $x_2 = 1$, $x_3 = x_4 = 0$, $x_5 = 2$, $(z')^* = 4$. (b) Min $(5y_1 + 3y_2 + 4y_3)$ s.t. $2y_1 + y_2 \geq 1.5$; $y_1 + y_2 + 2y_3 \geq 1.0$. (c) $y_3 = 0$, $y_1 = 0.5$, $y_2 = 0.5$, $w^* = 4$.

5.15 (a) Min cost: Min $(100x_1 - 100x_2)$; Min mass: Min $(x_1 + 0.20x_2)$ s.t. $x_1 + x_2 = 200$, $x_1 \leq 160$, $x_2 \leq 120$; compatible objectives at $x_1 = 80$, $x_2 = 120$, $z_1^* = -\$4000$, $z_2^* = 56$ tonnes.

5.16 Saddle point at $\{\sigma_3, \tau_2\}$; value $= 6$.

5.17 No saddle point. Strategy for A: $\{\sigma_1, 1/2; \sigma_2, 1/2\}$; for B: $\{\tau_1, 2/3; \tau_2, 1/3\}$; value $= 0$.

5.18 No saddle point. Strategy for A: $\{\sigma_1, 2/3; \sigma_2, 1/3\}$; for B: $\{\tau_1, 1/3; \tau_2, 2/3; \tau_3, 0; \tau_4, 0; \tau_5, 0\}$; value $= 5/3$.

5.19 No saddle point; solution $\{1/3, 1/3, 1/3\}$ for each player.

5.20 (a) $\mathbf{Pr}(\sigma_1)$: 1 to 5/7, τ_3 is optimal; 5/7 to 1/3, τ_2; 1/3 to 0, τ_4; τ_1 never optimal. (c) $\{5/7, 2/7\}$; value $= 17/7$.

Chapter 6

6.2 $p = 1/2$. Max entropy solution corresponds to Equation (6.85) which has a very sharp peak for large N.

6.3 Combine Equations (6.31), (6.35) and (6.36) to obtain a quadratic in $\exp(-\lambda_1)$. Solve for various μ and plot.

6.4 See Jaynes (2003) and Kapur and Kesavan (1992). Positive definiteness implies convexity.

6.5 $p_i = \exp \lambda_0$; $\lambda_0 = \ln n$, where $n = $ number of possibilities.

6.6 Fix $\langle X \rangle$, $\langle \ln X \rangle$ for gamma. Fix $\langle \ln X \rangle$, $\langle \ln(1 - X) \rangle$ for beta. See Tribus (1969) and Kapur and Kesavan (1992).

6.7 See Kapur and Kesavan (1992).

6.10 See Jaynes (1963a).

6.12, 6.13 Follow procedure in Section 6.9.

6.15 Derive the Maxwell–Boltzmann (classical) distribution of energies. Then calculate the probability of surmounting an energy barrier of magnitude ϵ_A.

Chapter 7

7.2 $y = g(x) = \exp(-y)$; use to generate realizations of the exponential distribution.

7.3 $f_Y(y) = ky^{k-1} \exp(-y^k)$.

7.4 $f_Y(y) = \frac{2y}{\mu} \exp\left(-\frac{y^2}{\mu}\right)$.

7.5 $y = \frac{x-a}{b-a}$.

7.7 $f_Z(z) = 3\left[1 - \frac{1}{t}\ln\left(\frac{z}{c}\right)\right]^2$ for $c < z < c \exp t$.

7.8 $f_Y(y) = \frac{1}{3}$ on transformed parallelogram. Affine because parallelism is preserved.

7.9 $f_Y(y) = y \exp(-y)$, $y > 0$.

7.11 $f_R(r) = r \exp\left(-\frac{r^2}{2}\right)$, $f_\Theta(\theta) = \frac{1}{2\pi}$.

7.12 See Feller (1971).

7.14 $f_P(p) = \left[p^{\alpha_1-1}(1-p)^{\alpha_2-1}\right]/B(\alpha_1, \alpha_2)$.

7.15 (a) $[pt + (1-p)]^n$. (b) $[rt + (1-p)]^{n_1+n_2}$.

7.18 See Feller (1971).

7.19 $1/(1+u^2)$.

7.20 $\exp(-a|u|)$.

7.21 $0.8413\ldots, 0.4214\ldots$

7.25 See Johnson *et al.* (1994); Feller (1971).

7.27 (a) 0.1715, binomial 0.1789. (b) 0.9722, binomial 0.9753.

7.28 $\exp\left\{-\frac{1}{2}\mathbf{u}^T[\Sigma]\mathbf{u} + i\mathbf{u}^T\mu\right\}$.

7.29 (a) Total time in telephone calls or electronic messages – number and length of each is random. (c) $\sum_{n=0}^{\infty} \frac{\exp(-\mu)\mu^n}{n!} \cdot \exp\left[\nu n(\exp t - 1)\right]$; $\langle Y \rangle = \mu\nu$; $\sigma_Y^2 = \mu\nu(1 + \nu)$.

7.30 0.0015 (4 dp).
7.31 See de Finetti (1974, vol. 2), Feller (1971).
7.32 See de Finetti (1974, vol. 1), Papoulis (1965).

Chapter 8

8.1 Then result is the beta distribution; see Section 8.3.
8.4 (a) $\mu_f = 5.68, \sigma_f = 0.0316\ldots$ (b) $5.60 < A < 5.76$ with 99% probability (2 dp). (c) $5.69 < A < 5.91$ with 99% probability (2 dp). (d) 96.
8.6 We have to consider both σ and σ_f.
8.10 (a) Take the first moment. (b) Use the parallel axis theorem.
8.11 $2^{-\left(\frac{\nu}{2}+\frac{1}{2}\right)} \sqrt{\frac{\pi}{\pi}} \left(\nu\sigma^2\right)^{\frac{\nu}{2}} \frac{1}{\Gamma\left(\frac{\nu}{2}\right)}$.
8.15 (a) Posterior mean $= 0.518$, variance $= 0.0239$, s.d. $= 0.155$.
8.17 $f_\Lambda(\lambda) \propto 1/\lambda$; one observation in specified period; maximum entropy subject to mean $\bar{\lambda}$ known.
8.18 (a) $c_2/\left[(c_1 - c_3)n\right]$. (d) If the utility is linear with money over the range considered.
8.20 0.030, 0.064 (2 sf).
8.24 See Lindley (1965, vol. 2), Box and Tiao (1973), Bernardo and Smith (1994).
8.28 (a) Yes. $z = 3.99\ldots, z_{0.05} = 1.645$. (b) $t = 3.99, t_{0.05}(4 \text{ sf}) = 2.132$.
8.29 3209.
8.30 Yes. Comparison of means with pooled sample variance. $t = 1.986, t_{0.05,9} = 1.833$.
8.31 Test for equality of variances – OK. Compare means with pooled variance. $z = -1.603$, p-value (two-sided) $= 0.110$. Accept equal population means.
8.32 Paired two-sample t-test. $t = -6.92, t_{0.05,6} = 1.943$. Accept difference.
8.33 No. Test for difference in proportions.
8.34 $\mu_f = 5.50, \sigma_f = 0.05345\ldots$ Distribution of Y is lognormal.

Chapter 9

9.8 7.596 (3 dp).
9.9 (a) 7.60. (b) $\check{z} = a_n + b_n \left[\ln\left(\frac{p_{ne}}{n}\right)\right]^{\frac{1}{\alpha}}$.
9.10 $\hat{x} = \frac{\sigma}{\sqrt{2}}$.
9.16 $\mu(x) = kx^{k-1}$.
9.18 1, 0, 0; 0, 1/4, 0; 0, 0, 1/3.
9.19 See Freudenthal (1968).
9.20 Torsion of cylindrical specimen: $\frac{v_*}{v} = \left[\frac{2\alpha+1}{\pi(\alpha+2)}\right]^{\alpha(\alpha+1)}$; rectangular beam under point load:
$\frac{v_*}{v} = \frac{1}{2(\alpha+1)^2}$.
9.21 See Madsen *et al.* (1986).

Chapter 10

10.2 The distribution of capacity should be updated.

10.4 Let f_T and f_S be uniform in $[a, b], [c, d]$. Then $p_f = [(b-c)(b+c-a)] / [2(b-a)(d-c)]$.

10.5 0.027.

10.6 The ratio of two chi-square distributions (divided by their d.f.) is an F-distribution. Use the third of (10.2) to obtain the result.

10.10 See Exercise 2.12.

10.11 For n identical units, $(n\lambda)^{-1}$ (series), and $\sum (n\lambda)^{-1}$ (parallel).

10.12 0.9802 (4 dp).

10.13 Use Poisson processes.

Chapter 11

11.3 $\bar{x} = 2841$, $s = 145$.

11.4 $\bar{x} = -1.515$, $s = 5.24$.

11.5 $\bar{x} = 11.00$, $s = 1.34$, mode = 10.40.

11.6 $y = 3.751 + 0.8136x$.

11.8 Slope will be less than 1 : 1.

11.10 $y = 1.204x - 4.419$.

11.11 For $p = 15$ Mpa, $y = -9623.1x + 29.574$; for $p = 35$ Mpa, $y = -9432.9x + 28.531$; for $p = 55$ Mpa, $y = -8839.5x + 26.209$; for $p = 65$ Mpa, $y = -11600x + 37.223$, for $p = 70$ Mpa, $y = -15648x + 53.554$.

11.15 Given a number of 'pegs' as data points (x_i, y_c), the best-fit centre (x_c, y_c) and radius r can be found for the GBS. The error for each data point can be expressed as $e_i = \sqrt{(x_i - x_c)^2 + (y_i - y_c)^2}$. Use partial derivatives and minimize $\frac{\partial \delta_e}{\partial x_c}$, $\frac{\partial \delta_e}{\partial y_c}$, and $\frac{\partial \delta_e}{\partial r}$, where δ_e is the sum of the squares of errors, $\sum_{i=1}^{n} e_i^2$.

11.26 0.057 (2 sf).

References

Aston, J. G. and Fritz, J. J. 1959. *Thermodynamics and Statistical Thermodynamics*. Wiley.

Baker, M. J. 1977. *Rationalisation of Safety and Serviceability Factors in Structural Codes*. CIRIA (Construction Industry Research and Information Association), Report 63, London, England.

Barlow, R. E., Fussell, J. B. and Singpurwalla, N. D. (eds.) 1975. *Reliability and Fault Tree Analysis*. Society for Industrial and Applied Mathematics.

Barnett, A. and Wang, A. 2000. Passenger-mortality risk estimates provide perspectives about airline safety. *Flight Safety Digest*, **19**, No. 4, April.

Bayes, T. 1763. An essay towards solving a problem in the doctrine of chances. *Phil. Trans. Royal Society London. Series A53*. Reproduced with biographical note by G. Barnard in *Biometrika*, **45** (1958).

Bea, R. 1998. Human and organization factors in the safety of offshore structures. In *Risk and Reliability in Marine Technology*, ed. C. Guedes Soares. Balkema, pp. 71–91.

Benjamin, J. R. and Cornell, A. 1970. *Probability, Statistics and Decision for Civil Engineers*. McGraw-Hill.

Beer, F. P. and Johnston, E. R. 1981. *Vector Mechanics for Engineers: Statics and Dynamics*. McGraw-Hill Ryerson.

Bernardo, J. M., and Smith, A. F. M. 1994. *Bayesian Theory*. Wiley.

Bernstein, P. L. 1996. *Against the Gods: The Remarkable Story of Risk*. Wiley.

Beveridge, G. S. and Schechter, R. S. 1970. *Optimization: Theory and Practice*. McGraw-Hill.

Billinton, R. and Allan, R. N. 1992. *Reliability Evaluation of Engineering Systems*, second edn. Plenum Press.

Blom, Gunnar 1958. *Statistical Estimates and Transformed Beta-Variables*. John Wiley and Almqvist and Wiksell.

Blom, Gunnar, Holst, Lars, and Sandell, Dennis 1994. *Problems and Snapshots from the World of Probability*. Springer-Verlag.

Boeing Commercial Airplanes Group 2000. *Statistical Summary of Commercial Jet Airplane Accidents, 1959–1999*. www.boeing.com/news/techissues

Bolotin, V. V. 1969. *Statistical Methods in Structural Mechanics*. Holden-Day. (Translated from the Russian by Samuel Aroni.)

Box, G. E. P. and Tiao, G. C. 1973. *Bayesian Inference in Statistical Analysis*. Addison-Wesley.

Bridgman, P. W. 1941. *The Nature of Thermodynamics*. Harvard University Press.

— 1958. *The Logic of Modern Physics*. (First published 1927) Macmillan.

Brillouin, L. 1962. *Science and Information Theory*. Academic Press.

Butkov, E. 1968. *Mathematical Physics*. Addison-Wesley.

Canadian Standards Association 1992. Reissued in revised form 2004. *General Requirements, Design Criteria, the Environment, and Loads*. Standard CAN/CSA-S471-92. Part of Code for Offshore Structures.

Carter, J. E., Frederking, R. M. W., Jordaan, I. J., Milne, W. J., Nessim, M. A. and Brown, P. W. 1992. *Review and Verification of Proposals for the Arctic Shipping Pollution Prevention Regulations*.

Memorial University of Newfoundland, Ocean Engineering Research Centre. Report prepared for Canadian Coast Guard, Arctic Ship Safety, Transport Canada Publication TP 11366E.

Cartwright, D. E. and Longuet-Higgins, M. S. 1956. The statistical distribution of the maxima of a random function, *Proc. Royal Society*, A, 237.

Castillo, E. 1988. *Extreme Value Theory in Engineering*. Academic Press.

Churchill, R. V. 1972. *Operational Mathematics*. McGraw-Hill.

Chvátal, V. 1980. *Linear Programming*. W. H. Freeman.

Cohen, M. H. 1961. The relative distribution of households and places of work – a discussion of the paper by F. G. Wardrop. *Proceedings of the Symposium on the Theory of Traffic Flow*, ed. R. Herman. Elsevier, pp. 79–84.

Cohn, M. Z. and Maier, G. 1979 (eds.). *Engineering Plasticity by Mathematical Programming*. Proceedings of the NATO Advanced Study Institute, University of Waterloo, Canada, 1977. Pergamon.

Cornell, C. A. 1969a. Bayesian statistical decision theory and reliability-based structural design. *Proceedings, International Conference on Structural Safety and Reliability*, Washington D.C.

— 1969b. A probability-based structural code. *Journal of the American Concrete Institute*. **66**(12): 974–985.

Courant, R. and Robbins, H. 1941. *What is Mathematics?* (tenth printing, 1960) Oxford University Press.

Cramér, H. 1951. *Mathematical Methods of Statistics*. Princeton University Press.

Cullen, Hon. Lord 1990. *The Public Inquiry into the Piper Alpha Disaster*, The Department of Energy (UK).

Cunnane, C. 1978. Unbiased plotting positions – a review. *Journal of Hydrology*, **37**, 205–222.

D'Agostino, R. B. and Stephens, M. A. (eds.) 1986. *Goodness-of-fit Techniques*. Marcel Dekker.

Dalal, S. R., Fowlkes, E. B. and Hoadley, B. 1989. Risk analysis of the space shuttle: pre-Challenger prediction of failure. *Journal of the American Statistical Association*, **84**(408): 945–957.

Davenport, A. G. 1964. Note on the distribution of the largest value of a random function with application to gust loading. *Proceedings of the Institution of Civil Engineers*, **28**, 187–196.

David, H. A. 1981. *Order Statistics*, second edition. Wiley.

Dawkins, R. 1989. *The Selfish Gene*. Oxford.

de Finetti, B. 1937. La prévision: ses lois logiques, ses sources subjectives. *Annales de l'Institut Henri Poincaré*, **7**, 1–68. Translated in Kyberg, H. E. and Smokler, H. E. (eds.) 1964. *Studies in Subjective Probability*. Wiley.

— 1972. *Probability, Induction and Statistics*. Wiley.

— 1974. *Theory of Probability*. Vols. 1 and 2, translated from Italian by Machi, A., and Smith, A. Wiley.

— 1977. Probabilities of probabilities: a real problem or a misunderstanding? *New Developments in the Applications of Bayesian Methods*, eds. A. Aykac and C. Brumat. North-Holland, pp. 1–10.

DeGroot, M. H. 1970. *Optimal Statistical Decisions*. McGraw-Hill.

DeGroot, M. H. and Schervish, M. J. 2002. *Probability and Statistics*, third edn. Addison Wesley.

de Neufville, R. 1990. *Applied Systems Analysis*. McGraw-Hill.

de Neufville, R. and Keeney, R. L. 1972. Use of decision analysis in airport development for Mexico City. *Analysis of Public Systems*, eds. A. W. Drake, R. L. Keeney and P. M. Morse. MIT Press.

Eatwell, J., Milgate, M. and Newman, P. 1990. *The New Palgrave: Utility and Probability*. MacMillan Press, pp. 227–239.

Eisberg, R. and Resnick, R. 1974. *Quantum Physics of Atoms, Molecules, Solids, Nuclei, and Particles*. Wiley.

Evans, I. G. and Nigm, A. M. 1980. Bayesian prediction for two-parameter Weibull lifetime models. *Communications in Statistics – Theory and Methods*, **A9**(6), 649–658.

Feller, W. 1968. *An Introduction to Probability Theory and its Applications*. I, third edn. Wiley.

— 1971. *An Introduction to Probability Theory and its Applications*. II, second edn. Wiley.

Feynman, R. P. 1986. Personal observations on reliability of shuttle. *Report to the President by the Presidential Commission on the Space Shuttle Challenger Accident,* **2**, Appendix F.

Finnie, B. W. and Johnston, I. H. A. 1990. The acceptance of software quality as a contributor to safety. In *Safety and Reliability in the 90s.* Proceedings of the Safety and Reliability Symposium, Elsevier, pp. 279–289.

Fisher, R. A. and Tippett, L. H. C. 1928. Limiting forms of the frequency distributions of the largest or smallest member of a sample. *Proc. Cambridge Philos. Soc.* **24**, 180–190.

Flint, A. R. and Baker, M. J. 1976. Risk analysis for offshore structures – the aims and methods. *Design and Construction of Offshore Structures, Proceedings of the Conference.* The Institution of Civil Engineers, London, 1977.

Frankenstein, G. E. 1959. *Strength data on lake ice.* CRREL Technical Report 59.

— 1961. *Strength data on lake ice.* CRREL Technical Report 80.

Fraser, D. A. S. 1976. *Probability and Statistics: Theory and Applications.* Duxbury Press (Wadsworth Publishing).

Fréchet, M. 1927. Sur la loi de probabilité de l'écart maximum. *Ann. de la Societé Polonaise de Mathématique.* Cracow, 6:93.

Freudenthal, A. M. 1968. Statistical approach to brittle fracture. In *Fracture,* ed. H. Liebowitz, vol. II, pp. 591–619.

Freudenthal, A. M., Garrelts, J. M. and Shinozuka, M. 1966. The analysis of structural safety. *Journal of the Structural Division, Proceedings of the American Society of Civil Engineers,* **92**(ST1): 267–325.

Galambos, J. 1978. *The Asymptotic Theory of Extreme Order Statistics.* Wiley.

German, J. G. and Sukselainen, J. 1985. *M.V. Arctic Test Results and Analysis.* Transport Canada Report No. TP 6727E. Prepared for Coast Guard Northern, by German and Milne Inc., Ottawa, and Technical Research Center of Finland, Helsinki.

Gnedenko, B. V. 1941. Limit theorems of the maximal term of a variational series. *Comptes Rendus de l'Académie des Sciences de l'URSS,* **32**, 7–9.

— 1943 Sur la distribution limite du terme maximum d'une série aléatorie. *Ann. Math.* **44**, 423–453.

Goldsmith, James 1994. *The Trap.* Carroll and Graf.

Good, I. J. 1950. *Probability and the Weighing of Evidence.* Hafner.

— 1965. *The Estimation of Probabilities.* Research Monograph No. 30, MIT Press.

Gopal, E. S. R. 1974. *Statistical Mechanics and Properties of Matter.* Ellis Horwood.

Gow, A. J., Udea, H. T., Govoni, J. W. and Kalafut, J., 1988. *Temperature and structure dependence of the flexural strength and modulus of freshwater model ice.* CRREL Report 88–6.

Grayson, C. J. 1960. *Decisions Under Uncertainty: Drilling Decisions by Oil and Gas Operators.* Harvard.

Greub, W. 1975. *Linear Algebra.* Springer-Verlag.

Gribble, J. N. and Preston, S. H. (eds.) 1993. *The Epidemiological Transition: Policy and Planning Implications for the Developing Countries.* National Academy Press.

Griffith, A. A. 1920. The phenomenon of rupture and flow in solids. *Philosophical Transactions of the Royal Society,* A221, pp. 163–168.

Gringorten, I. I. 1963. A plotting rule for extreme probability paper. *Journal of Geophysical Research,* **68**, 3, 813–814.

Gumbel, E. J. 1958. *Statistics of Extremes.* Columbia University Press.

Hall, A. D. 1962. *A Methodology for Systems Engineering.* Van Nostrand.

Hambly, E. C. and Hambly, E. A. 1994. Risk evaluation and realism. *Proceedings, Institution of Civil Engineers, Civil Engineering,* **102**, May: 64–71.

Hammersley, J. M. and Handscomb, D. C. 1964. *Monte Carlo Methods.* Methuen.

Heath, D. and Sudderth, W. 1976. De Finetti's theorem on exchangeable variables. *American Statistician,* **30**, No. 4, 188–189.

Hong, Han Ping. 1994a. A note on extremal analysis. *Structural Safety.* **13**: 227–233.

— 1994b. Estimate of extremal wind and wave loading and safety level of offshore structures. *Risk Analysis, Proceedings of a Symposium*, ed. A. S. Nowak. University of Michigan, Ann Arbor, Michigan.

Hunt, R. A. and McCartney, L. N. 1979. A new approach to Weibull's statistical theory of brittle fracture. *International Journal of Fracture,* **15**, 365–375.

Jaynes, E. T. 1957. Information theory and statistical mechanics. *Physics Review.* **106**, 620–630 and **108**, 171–190.

— 1963a. New engineering applications of information theory. *Proceedings of the First Symposium on Engineering Applications of Random Function Theory and Probability.* Wiley.

— 1963b. Information theory and statistical mechanics. In *Statistical Physics,* ed. K. W. Ford, W. A. Benjamin. pp. 182–218.

— 1976. Confidence intervals vs. Bayesian intervals. In *Foundations of Probability Theory, Statistical Inference, and Statistical Theories of Science*, eds. W. L. Harper and C. A. Hooker. D. Reidel Publishing Company. Also in Rosenkrantz (1983).

— 2003. *Probability Theory: the Logic of Science,* ed. G. Larry Bretthorst. Cambridge University Press.

Jeffreys, H. 1961. *Theory of Probability*, third edn. Clarendon Press.

Jensen, F. V. 1996. *An Introduction to Bayesian Networks.* UCL Press (also distributed by Springer).

Jessop, A. 1990. *Decision and Forecasting Models with Transport Applications.* Ellis Horwood.

Johnson, N. L., Kotz, S. and Kemp, A. W. 1993. *Univariate Discrete Distributions.* Wiley.

Johnson, N. L., Kotz, S. and Balakrishnan, N. 1994. *Continuous Univariate Distributions.* Wiley.

Jones, A. J. 1980. *Game Theory: Mathematical Models of Conflict.* Ellis Horwood.

Jordaan, I. J. 1988. Safety levels and target probabilities implied in codes of practice. *Civil Engineering Systems,* **5**, No. 1, March 1988, 3–7.

— 2000. Mechanics of ice-structure interaction. *Engineering Fracture Mechanics,* 2001; **68**: 1923–1960.

Jordaan, I. J. and Maes, M. A. 1984. Probability, exchangeability and extremes: with discussion of iceberg loading. *Civil Engineering Systems,* **1**, 234–241.

— 1991. Rationale for load specifications and load factors in the new CSA code for fixed offshore structures, *Canadian Journal of Civil Engineering,* **18**, No. 3, 404–464.

Jordaan, I. J. and Pond, J. 2001. Scale effects and randomness in the estimation of compressive ice loads. *Proceedings IUTAM Symposium on Scaling Laws in Ice Mechanics and Dynamics*, eds. Dempsey, J. P. and Shen, H. H. Kluwer, pp. 43–54.

Jordaan, I. J., Nessim, M. A., Ghoneim, G. A. and Murray, A. M. 1987. A rational approach to the development of probabilistic design criteria for Arctic shipping. *Proceedings, Sixth OMAE (Offshore Mechanics and Arctic Engineering) Symposium*, Houston, **IV**, pp. 401–406.

Jowitt, P. W. 1979. The extreme value type 1 distribution and the principle of maximum entropy. *Journal of Hydrology,* **42**: 23–38. Discussion, **47**: 385–390.

— 1999. A conceptual systems model of rainfall-runoff on the Haast River. *Journal of Hydrology (NZ),* **38**(1): 121–144.

Jowitt, P. W. and Munro, J. 1975. The influence of void distribution and entropy on the engineering properties of granular media. *Proceedings of 2nd International Conference on Applications of Statistics and Probability in Soil and Structural Engineering.* Aachen.

Kalos, M. H. and Whitlock, P. A. 1986. *Monte Carlo Methods.* Wiley.

Kapur, J. N. and Kesavan, H. K. 1992. *Entropy Optimization Principles with Applications.* Academic Press.

Karlin, S. 1959. *Mathematical Methods and Theory in Games, Programming and Economics.* Vol. 1, Matrix Games, Programming and Mathematical Economics. (Vol. II deals with infinite games.) Addison-Wesley.

Kaufman, G. M. 1963. *Statistical Decision and Related Techniques in Oil and Gas Exploration.* Prentice-Hall.

Keeney, R. L. and Raiffa, H. 1976. *Decisions with Multiple Objectives*. John Wiley and Sons. Reprinted Cambridge University Press, 1993.

Khinchin, A. I. 1957. *Mathematical Foundations of Information Theory*. Dover.

Lancaster, J. 1996. *Engineering Catastrophes*. Abington Publishing.

Laplace, P. S. de 1812. *Théorie Analytique des Probabilités*. (first edn. 1812 and subsequent edns.) Courcier.

— 1914. *Essai Philosophique sur les Probabilités*. (first edn., 1814 and subsequent edns.) Courcier. English translation – A philosophical essay on probabilities. Wiley 1917, reprinted by Dover 1952.

Lathi, B. P. 1998. *Modern Digital and Analog Communication Systems*. Oxford University Press.

Leadbetter, M. R., Lindgren, G. and Rootzén, H. 1983. *Extremes and Related Properties of Random Sequences*. Springer-Verlag.

Lee, P. M. 1997. *Bayesian Statistics*. Arnold.

Lind, N. C. 1972. *Theory of Codified Structural Design*. Notes pertaining to Invited Lectures to Academy of Sciences of Poland.

— 1983. Models of human error in structural reliability. *Structural Safety*, **1**: 167–175.

Lind, N. C., Nathwani, J. S. and Siddall, E. 1991. *Managing Risks in the Public Interest*. Institute for Risk Research, University of Waterloo, Waterloo, Ontario, Canada.

Lindley, D. V. 1965. *Introduction to Probability and Statistics from a Bayesian Viewpoint*. Part 1, Probability, Part 2, Inference. Cambridge University Press.

— 1987. The probability approach to the treatment of uncertainty in artificial intelligence and expert systems. *Statistical Science* **2**, No. 1, 17–24.

Lipson, Charles and Sheth, Narendra J. 1973. *Statistical Design and Analysis of Engineering Experiments*. McGraw-Hill.

Lloyd, Emlyn 1980. *Handbook of Applicable Mathematics*. Vol. II, Probability. Wiley.

Luce, D. and Raiffa, H. 1957. *Games and Decisions*. John Wiley and Sons.

Maddock, W. P. and Jordaan, I. J. 1982. Decision analysis applied to structural code formulation. *Canadian Journal of Civil Engineering*, **9**, No. 4, December: 663–673.

Madsen, H. O., Krenk, S. and Lind, N. C. 1986. *Methods of Structural Safety*. Prentice-Hall.

Maes, M. A. 1985. *Extremal Analysis of Environmental Loads on Offshore Structures*. Ph.D. Thesis. University of Calgary, Canada.

— 1992. Probabilistic behavior of a Poisson field of flaws in ice subjected to indentation. *Procs. 11th IAHR Int. Symposium on Ice*, Vol. 2, pp. 871–882.

Margenau, H. and Murphy, G. M. 1956. *The Mathematics of Physics and Chemistry*, second edn. Van Nostrand, p. 604.

Markham, W. E. 1980. *Sea Atlas – Eastern Canadian Seaboard*. Environment Canada.

Martin, N. F. G. and England, J. W. 1981. *Mathematical Theory of Entropy*. Vol. 12 of *Encyclopedia of Mathematics and its Applications*, ed. Gian-Carlo Rota. Addison-Wesley.

Matthews, R. A. J. 1998. Fact *versus* factions: the use and abuse of subjectivity in scientific research. *ESEF (The European Science and Environment Forum)*. Working paper.

McCormick, N. J. 1981. *Reliability and Risk Analysis*. Academic Press.

Melchers, R. E. 1993. Society, tolerable risk and the ALARP principle. In *Probabilistic Risk and Hazard Assessment*, eds. Melchers, R. E. and Stewart, M. G. Balkema, pp. 243–252.

— 1999. *Structural Reliability Analysis and Prediction*, second edn. Wiley.

Monahan, G. E. 2000. *Management Decision Making*. Cambridge University Press.

Moan, T. 2004. Safety of floating offshore structures. *Proceedings of the PRADS Conference*, Sept 12–18, Lübeck-Travemünde, Germany.

National Research Council of Canada. 1989. *Hydrology of Floods in Canada: A Guide to Planning and Design*. Associate Committee on Hydrology.

Nessim, M. A. 1983. *Decision-making and Analysis of Errors in Structural Reliability*. Ph.D. thesis, University of Calgary, Calgary, Alberta, Canada.

Nessim, M. A. and Jordaan, I. J. 1985. Models for human error in structural reliability. *Journal of Structural Engineering*, ASCE, **111**(6): 1358–1376.

Nessim, M. A. and Stephens, M. J. 1995. Risk-based optimization of pipeline integrity maintenance. *Proceedings, OMAE*.

Nessim, M. A., Hong, H. P. and Jordaan, I. J. 1995. Environmental load uncertainties for offshore structures. *Journal of Offshore Mechanics and Arctic Engineering*, **117**: 245–251.

Ochi, M. K. 1990. *Applied Probability and Stochastic Processes*. Wiley.

O'Hagan, Anthony 1994. *Kendall's Advanced Theory of Statistics*. Vol. 2B, Bayesian Inference. Edward Arnold.

Owen, G. 1982. *Game Theory*, second edn.

Palmer, A. 1996. The limits of reliability theory and the reliability of limit state theory applied to pipelines. *Proceedings, Offshore Technology Conference*, pp. 619–626.

Papoulis, A. 1965. *Probability, Random Variables, and Stochastic Processes*. McGraw-Hill.

Parnell, G. S., Frimpon, M., Barnes, J., Kloeber, J. M., Deckro, R. F. and Jackson, J. A. 2001. Safety risk analysis of an innovative environmental technology. *Risk Analysis*, **21**(1): 143–155.

Patel, J. K., Kapadia, C. H. and Owen, D. B. 1976. *Handbook of Statistical Distributions*. Vol. 20 in Statistics series, ed. D. B. Owen. Marcel Dekker.

Paulos, John Allen 1990. *Innumeracy: Mathematical Illiteracy and its Consequences*. Vintage.

Perrow, C. 1999. *Normal Accidents*. Princeton University Press.

Press, W. H., Teukolsky, S. A., Vetterling, W. T. and Flannery, B. P. 1992. *Numerical Recipes in C: the Art of Scientific Computing*, second edn. Cambridge University Press.

Pugsley, A. G. 1951. Concepts of safety in structural engineering. *Proceedings, Institution of Civil Engineers*, London, pp. 5–51 (including discussion).

— 1966. *The Safety of Structures*. Edward Arnold.

— 1973. The prediction of proneness to structural accidents. *The Structural Engineer*, **51**, No. 6, 195–196.

Putcha, C. S. and Patev, R. C. 2000. Risk assessment methodology for flood control and reservoir dam gates. *Proceedings, Third Annual Conference, Canadian Dam Association*, Regina, Saskatchewan, pp. 18–25.

Raiffa, H. 1968. *Decision Analysis: Introductory Lectures on Choices Under Uncertainty*. Addison-Wesley (reprinted 1970).

Raiffa, H. and Schlaifer, R. 1961. *Applied Statistical Decision Theory*. Harvard University Press. Reprinted by M.I.T. Press, 1968.

Rosenblueth, E. 1987. What should we do with structural reliabilities? *Reliability and Risk Analysis in Civil Engineering*, Proceedings of ICASP 5 (International Conference on Applications of Statistics and Probability in Soil and Structural Engineering), Vancouver, Vol. 1, pp. 24–34.

Rosenkrantz, R. D. (ed.) 1983. *E. T. Jaynes: Papers on Probability, Statistics and Statistical Physics*. D. Reidel Publishing Company.

Rowe, W. D. 1977. *An Anatomy of Risk*. Wiley.

Rubinstein, M. F. 1975. *Patterns of Problem Solving*. Prentice-Hall.

Rubinstein, R. Y. 1981. *Simulation and the Monte Carlo Method*. Wiley.

Sanderson, T. J. O. 1988. *Ice Mechanics: Risks to Offshore Structures*. Graham and Trotman.

Sarpkaya, T. and Isaacson, M. 1981. *Mechanics of Wave Forces on Offshore Structures*. Van Nostrand Reinhold.

Savage, L. J. 1954. *The Foundations of Statistics*. Wiley. Reprinted by Dover, 1972.

— 1962. Bayesian statistics. In *Recent Developments in Decision and Information Processes*, eds. Machol, R. E. and Gray, P. MacMillan Company, pp. 161–194.

Schlaifer, R. 1959. *Probability and Statistics for Business Decisions*. McGraw-Hill Book Company.

— 1961. *Introduction to Statistics for Business Decisions*. McGraw-Hill Book Company.

Schrödinger, Erwin 1952. *Statistical Thermodynamics*, second edn. Cambridge University Press, reprinted Dover Publications, 1989.

Segré, E. 1980. *From X-Rays to Quarks*. Freeman.
— 1984. *From Falling Bodies to Radio Waves*. Freeman.
Shannon, C. E. 1948. The mathematical theory of communications. *Bell Systems Technical Journal*, July and October.
Snyder, D. L. 1975. *Random Point Processes*. Wiley.
Starr, C. 1969. Social benefit versus technological risk. *Science*, **165**: 1232–1283.
Stedinger, J. R., Vogel, R. M. and Foufoula-Georgiou, E. 1993. Frequency analysis of extreme events. In *Handbook of Hydrology*, ed. D. R. Maidment. McGraw-Hill.
Stewart, M. G. and Melchers, R. E. 1997. *Probabilistic Risk Assessment of Engineering Systems*. Chapman & Hall.
Stigler, S. M. 1982. Thomas Bayes' Bayesian inference. *Journal of the Royal Statistical Association*. **145**, Part 2, 250–258.
— 1986. *The History of Statistics*. Harvard University Press.
Swail, V. R., Cardone, V. J. and Cox, A. T. 1998. A long term North Atlantic wave hindcast. In *Proceedings of the 5th International Workshop on Wave Hindcasting and Forecasting*, Melbourne, FL, January 26–30, 1998.
Tolman, R. C. 1979. *The Principles of Statistical Mechanics*. Dover, reprinted from original version published by Oxford University Press in 1938.
Tribus, M. 1961. *Thermostatics and Thermodynamics*. Van Nostrand.
— 1969. *Rational Descriptions, Decisions and Designs*. Pergamon.
Turkstra, C. J. 1970. *Theory of Structural Design Decisions*. Study No. 2, Solid Mechanics Division, University of Waterloo.
Venn, J. 1888. *The Logic of Chance*. Macmillan.
Vinnem, J. E. 1998. Introduction to risk analysis. In *Risk and Reliability in Marine Technology*, ed. C. Guedes Soares. Balkema, pp. 3–17.
von Mises, R. 1957. *Probability, Statistics and Truth*, second revised English edn. Macmillan. (Original German edn., 1928.)
von Neumann, J. and Morgenstern, O. 1947. *Theory of Games and Economic Behaviour*, second edn. Princeton University Press (first published 1944).
Webber, S. J. 1996. Rockburst risk assessment on South African gold mines: an expert system approach. *Proceedings Eurock 96. Prediction and Performance in Rock Mechanics and Rock Engineering*, ed. G. Barla. Balkema.
Weibull, W. 1939. A statistical theory of the strength of materials. *Proceedings of the Ingeniorsvetenskapsakademin,* Handlingar Nr 151, Stockholm.
— 1951. A statistical distribution function of wide applicability. *Journal of Applied Mechanics*, **18**.
Wells, Geoff. 1996. *Hazard Identification and Risk Assessment*. Institution of Chemical Engineers, Rugby.
Wen, Yi-Kwei 1990. *Structural Load Modeling and Combination for Performance and Safety Evaluation*. Elsevier.
Wilks, W. R., Richardson, S. and Spiegelhalter, D. J. (eds.) 1996. *Markov Chain Monte Carlo in Practice*. Chapman and Hall.
Williamson, G. F. 1997. Software safety and reliability. *IEEE Potentials,* October/November, pp. 32–36.
Wilson, A. G. 1970. *Entropy in Urban and Regional Modelling*. Pion.
Wilson, R. 1979. Analyzing the daily risks of life. *Technological Review*, **81**(4): 41–46.
Wirashinghe, S. C. and Jordaan, I. J. 1985. Common risk and risk associated with air travel in Canada. *Transportation Forum*, **1–4**, 52–56.
WOAD Statistical Report 1996. *Worldwide Offshore Accident Databank*. Veritas Offshore Technology & Services, 1996, Høvik, Norway.

Index

Numbers in *italics* indicate tables or figures without text on that page
Number followed by 'ap' are in the appendices